FIELD GUIDE
TO THE
FOREST TREES
OF
GHANA

This field guide is dedicated to the late John Hall and to Mr Albert Enti in respectful appreciation of the extent to which they have advanced our knowledge of Ghana's plants, particularly through their extensive herbarium collections. I would also like to express special gratitude to Bill Howard, for encouraging and promoting the preparation of the guide from the earliest stage. Without their contribution, and without the support of Ghanaian foresters and others acknowledged on page 7, this guide would not have been written.

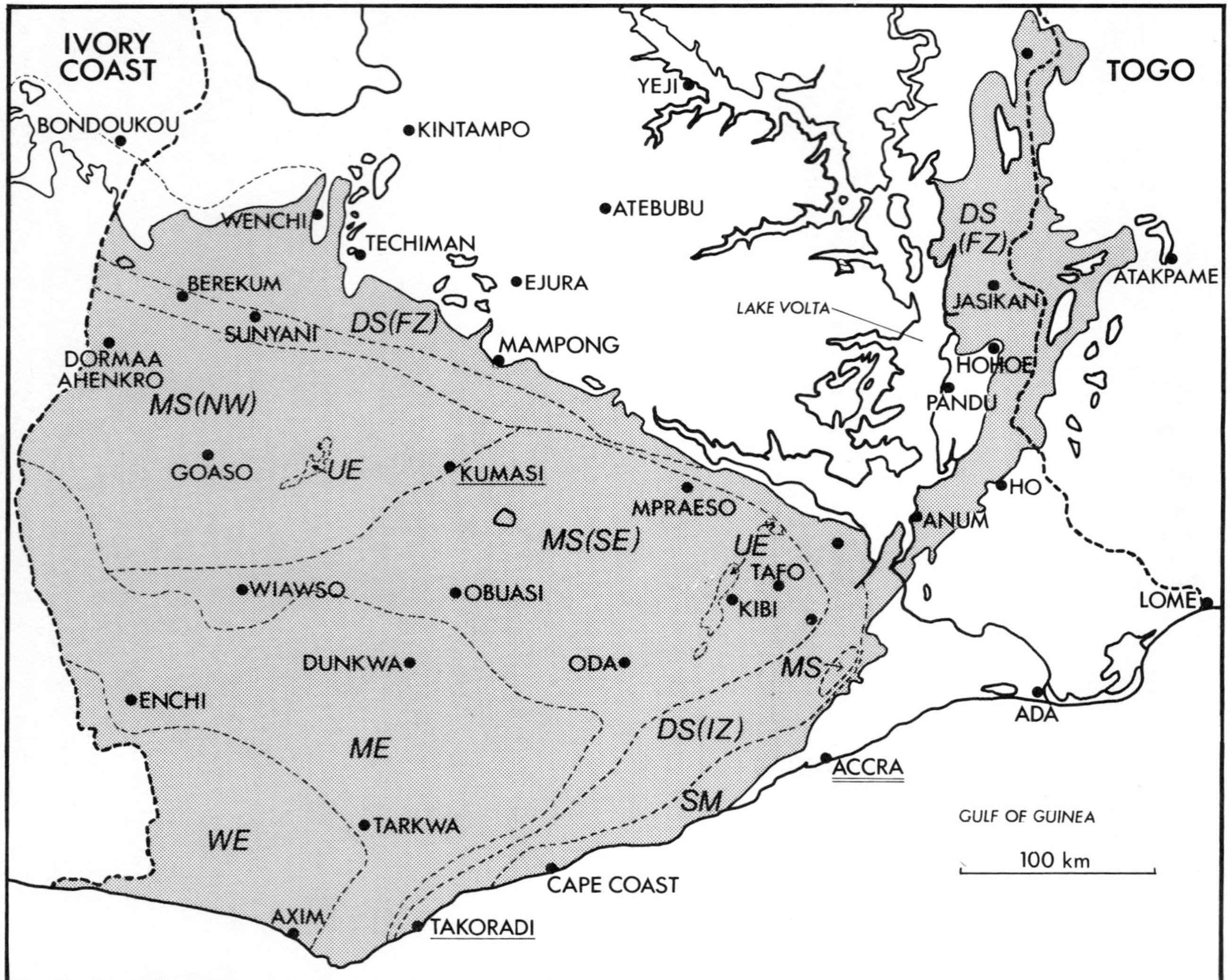

MAP OF FOREST ZONE IN GHANA

(STIPPLED)

(after Hall + Swaine 1981)

Forest zones demarcated by thin broken line, simplified in text as follows:

WE	(Wet evergreen)	EVERGREEN FOREST	Moist Forest
ME	(Moist evergreen)		
UE	(Upland evergreen)		
MS(NW)	Moist Semideciduous (North West Subtype)		
MS(SE)	Moist Semideciduous (South East Subtype)		
DS(Z)	Dry Semidecid. (Inner zone)		Dry Forest
DS(F2)	Dry Semidecid. (Fire zone)		
SM	Southern marginal		
(SO South East Outlier)—not mapped			

FIELD GUIDE TO THE FOREST TREES OF GHANA

WILLIAM HAWTHORNE
Overseas Development Administration

Illustrated by Rosemary Wise

Published by the Natural Resources Institute for the Overseas Development Administration. The Natural Resources Institute (NRI) is the scientific arm of Britain's Overseas Development Administration. NRI's principal aim is to increase the productivity of renewable natural resources in developing countries through the application of science and technology. Its areas of expertise are resource assessment and farming systems, integrated pest management, and food science and crop utilization.

No charge is made for single copies of this publication sent to governmental and educational establishments, research institutions and non-profit making organisations working in countries eligible for British Government Aid. Free copies cannot normally be addressed to individuals by name but only under their official titles. For further details contact:

Publications Distribution Office
Natural Resources Institute
Central Avenue
Chatham Maritime
Kent ME4 4TB
United Kingdom

Printed by Hobbs the Printers of Southampton
(3249/89)

ISBN 0 902500 34 1

CONTENTS

SECTION 1

INTRODUCTION

SECTION 1: INTRODUCTION

This guide has been produced to help foresters identify trees in Ghanaian rain forest. The range and definition of forest types covered are the same as those described by Hall and Swaine (1981). Although the guide is designed primarily for use by Technical Officers in the Forestry Department, it is hoped that other interested parties will find it useful as well: technical jargon has been kept to a minimum and the leaves of most species are illustrated. It is designed to be used in conjunction with other works on the trees of Ghana and neighbouring countries (see bibliography below), but nevertheless the aim is for it to be used primarily in the field, and to preclude the need to take a sackful of books into the forest.

The range of species covered is based on a list of 674 species of woody plants attaining 5 cm dbh (diameter at breast height), compiled by Hall, Abbiw and Enti for the Forestry Department (unpublished), with a few additions (e.g. climbing palms). Local names are mostly taken from this same list, although again a few alterations and additions have been made, as will be discussed below.

Standard botanical works are of limited use to anyone interested in identifying trees in the field, because they rely heavily on details of the flowers and fruits, which are usually not available. 'Sterile material' is generally greeted disdainfully by most herbarium taxonomists. It is not acceptable to name species new to science without flowers and fruits, and it is not possible to make with leaves alone an identification which has, so to speak, the full weight of botanical law behind it. Nevertheless, most people would accept without argument that they had a mango tree in their garden if the tree had the overall habit of the mango and mango leaves, even if the tree was without flower and fruit. This guide therefore furnishes the reader with details of leaves and ecology adequate to name most Ghanaian forest trees most of the time.

Bark and other 'whole tree' characters cannot be defined and described with enough precision to specify all Ghana's forest trees, at least with the current state of botanical knowledge. Bark structure, branching patterns and similar features do, of course, correlate well with the flowers and fruits of the taxonomist's herbarium. It is on this basis that most tree-spotters name trees, often with little idea of the nature of the tree's flowers or leaves, sometimes 35 metres above their head. However, such features are very variable in single species, or sometimes constant among several species, and it is for these reasons that most tree-spotters often make mistakes.

The solution attempted in this guide is to provide one (subsidiary), main species key based on gross features of the branching pattern, crown, bole, bark, etc., for approximately 200 of the larger trees, and another, 'main' key for all trees, which requires that leaves and significant details of the rest of the tree are considered. The '200 Main Species Key' is crude and more liable to error, but does not require close examination of the leaves. The Main Key comes closer to providing definitive answers, and should be used where possible to confirm the results of the subsidiary key.

THE LAYOUT AND USE OF THE GUIDE

The guide is in four sections. First is this introductory section, in which the use of a key and the layout of the book are explained. Abbreviations and other textual conventions are explained at the end of Section 1, after the acknowledgments and literature references.

The second section is the 'Main Key', which is arranged in Groups which correspond closely to standard botanical families. Flowers and fruits of most species are described, but knowledge of only the leaves, twigs and certain other vegetative characters are needed for the main key to be useful. This section is arranged so that a separate section summarizing each plant is not needed. However, to make most use of this arrangement, it is important that readers are familiar with the use of a key, both normally and 'backwards'. Both modes are explained below.

The third section includes the 200 Main Species Key together with some explanatory comments on the identification of 'whole trees'. Finally, the fourth section includes a glossary and the index to scientific and local names.

Absolute beginners will have to start identifying a tree by using the 'Key to Major Groups' at the beginning of Section 2, if they have a twig with leaves. Otherwise, if for instance they are standing at the base of an unknown forest giant, they must turn to the beginning of Section 3 and, having worked through the key there, look up in Section 2 the details of the leaves of those species to which they have been directed. Hopefully, they will be able to find evidence supporting the tentative identification made when using Section 3, for example fallen leaves, or bark features they had not previously noticed.

Illustrations of most species can be found opposite the part of Section 2 where they are mentioned. Every tree species has been given a unique number which is listed on the illustration, in the text and in the index for ease of cross-reference. For those species which are mentioned in more than one Group in Section 2, the index lists first the Group where the species is illustrated. The illustrations show 'average' leaves of mature trees, or two or more leaves overlapped for species with very variable leaves. For a few of the more important species, sapling leaves are shown if they differ greatly from those of the adult. Details, which will only be seen clearly with a hand lens, are drawn in small boxes.

Once the Main Key has directed users to a satisfactory answer, further confirmation of the true identity can be made by waiting (up to several years) for the tree to flower. An intermediate solution is to take a specimen of leaves to a herbarium to compare with specimens there. There is always more indefinable information in the overall appearance, or 'facies' of a leaf than can be recorded on paper. Although there should be enough information listed in this book to avoid the need for such confirmation, the Main Key should be helpful in the herbarium as well.

Tree identification may seem time-consuming at first, but budding tree-spotters who are familiar with a few species should find themselves reaching answers substantially more quickly. It has been a major consideration in the design of the guide that recognition of certain families or family groups should be encouraged. With a little experience these families/Groups should be easily recognized, and the user should then be able to turn straight to the relevant Group in Section 2, and examine the options on one or a few pages. In this context, it is hoped that the key will have some value as an *aide-mémoire* even to experienced tree identifiers.

NOMENCLATURE

In the main text, scientific names are written in lower case italics, except for the first letter of the genus name (e.g. *Khaya ivorensis*). Family names are written in capital non-italic letters (e.g. MELIACEAE) in the botanical keys. In some places family names have been abbreviated, as explained in the index (e.g. MELI). Local names are written in italics in large letters, except for the TWI syllables ɔ (written as 'o') and ε (written as 'e') (e.g. *KANe*).

Scientific names

Scientific names are forever changing, but these changes are more often a nuisance than a boon for practical ecologists, foresters and other people with an interest in Ghana's trees. Faith in the necessity of name changes has not been encouraged by the instability of the generic name of one of Ghana's most important trees, known to most as *Chlorophora* or *ODUM*. Having spent some time in *Maclura*, this species has again been transferred to *Milicia*. Meanwhile, most timber merchants continue to use the older names.

I have not sought in every case to provide the most up-to-date name available. In virtually all cases, nomenclature follows Hall and Swaine (1981), but some changes have been made to maintain compatibility with other, more recent reference works (particularly the revised edition of *Nigerian trees*). In these, and many other cases, both new and old names have been kept in the index for easier cross-reference with earlier books.

Local names

Initially, lists of local names were taken from a Forestry Department list (see above). This list was clearly derived from many sources and will almost certainly be met with controversy, as would any attempt at standardizing local name/scientific name equivalences. The list is biased towards Akan dialects, but many names have been invented and added to the list for rarer trees during the course of inventory work. Many names are used differently in different areas, and other names show dialectical changes from place to place. For instance, *Khaya grandifoliola* [*KRUBA*] and *Khaya anthotheca* [*KRUMBEN*] are often not distinguished and the different local names are quite probably derived from dialectical differences for a tree perceived to be equivalent.

Nevertheless, local names are more convenient and easier for non-botanists to learn, and it is hoped that this guide will provide a stimulus for many names to be made more standard. Names modified by gender and other suffixes (like *-BERE* (female), *-NiNi* (male), *-AKOA* (slave), *KoKoo* (red), *-FUFUo* (white),

-*KOBIRI* (black)) are in many cases not distinguished thus in the course of normal village life, and many of the invented names have suffixes of this type.

Forest types

Names of forest types are abbreviated versions of those of Hall and Swaine (1981) (see map for summary). In the keys, the subtypes of moist **semideciduous forest** are rarely differentiated, '**dry forest**' includes **dry semideciduous, southern marginal** and **southern outlier** types, '**moist forest**' covers all the forest that is not dry and '**evergreen forest**' includes **upland, moist** and **wet evergreen** types, unless specified more accurately.

EQUIPMENT NEEDED and WHAT TO DO WITH IT

Four botanical 'tools' are almost indispensable if this guide is to be used successfully: a x10 hand lens or magnifying glass for looking at the details of leaves and the slash; a cutlass; a catapult (readily available in many Ghanaian markets), for shooting down leaves and (less important) binoculars, for studying the arrangement of leaves in the crown.

Where possible, technical terms have been avoided, but some terms of shape and structure are unavoidable. The meanings of most will be made apparent by the illustrations accompanying the key. Some short definitions are given in the Glossary. A few concepts which may be unfamiliar to some readers are rather crucial to successful identification, however, and these will be discussed in more detail. In particular, the user of this book must be able to make accurate observations on the slash of the tree, and on the arrangement of leaves. **It is important that the slash and the leaf arrangement are examined "on the spot"**, whereas more detailed study of leaf specimens can be examined with the key later. Some notes on leaf arrangement will be found below the main Key to Groups in Section 2. Here, it is necessary to describe what is required in observations of the slash.

Observations on slash details

A 'slash' is made by cutting into the bark obliquely downwards with a sharp cutlass. Usually, such a cut will penetrate all the various layers of the bark, will often reveal a whitish layer of cambial tissue, but will generally make little or no progress into the sapwood. The oblique cut effectively 'spreads out' the layers of the bark like a deck of cards, for easy examination. It is essential that the slash details are always noted, even on small trees with little bark. *For the Main Key, however, one only needs to be aware of whether a tree produces latex, and, in a few cases whether the slash is sweetly scented, red or with a black layer, but for the 200 Main Species Key more information on the slash is required.*

The most important component of the slash is the nature of any exudate that appears in the first 30 seconds or so. In particular look for **latex**, which is a thick, opaque and milky liquid often sticky or rubbery when rubbed between the fingers. Latex can be any colour from pure white to bright yellow. Often it will turn brown after a minute or so of exposure, but the latex of some species turns brown almost immediately. Latex sometimes appears only very slowly, in small spots. If a species produces latex in cut petioles and other fresh parts, then it is safe to assume that the species is also one listed as having latex in the slash, even if none has been seen in the slash (as sometimes happens during the peak of the dry season). Presence or absence of latex is important also in the Main (leaf) Key.

Yellow and red exudates which are not opaque (i.e. not milky) are not counted as latex. Exudates described as red in Section 3 range from red to red-brown.

The smell of the slash can be important. Sweetly perfumed slashes and 'vegetable' smelling slashes are distinctive when once experienced, but can cause problems for beginners. For this reason they are listed in the 200 Main Species Key only as a last resort. As a crude rule of thumb, perfumed slashes are sharp and pleasant, like perfume itself, whereas musky vegetable smells are not sharp, and are likely to be more closely associated with cooking or the rubbish bin. Ginger is close to the borderline but, being to most people's noses, both pleasant, slightly hot and ethereal, it would belong to the perfumed group. Tobacco, on the other hand, however sweet (i.e. even pipe tobacco), is a prime example of a 'musky vegetable' smell. Other specific examples are included in the key.

Similarly, slash texture can be very hard to ascertain, as so many are part fibrous and part granular. Slash texture is again only used in the key as a last resort. The perfumed and fibrous nature of most

Annonaceae, however, is so distinctive that it has been used in the Main Key to segregate this family from others which would otherwise be confused with it. Very often, the annonaceous slash smells of various combinations of strong eau-de-Cologne, ginger and black pepper.

The colour of the slash is important. Take particular note if it is red to red-brown, or has strong vertical bands of this colour, as opposed to yellow to orange to orange-brown. Some shades of brown are ambiguous, but if there is doubt it will usually be a colour classified here as on the orange side of brown. Presence or absence of grit-like streaks and other features will be used and explained in the 200 Main Species Key. During the research for this key, slash colours have been compared to the standard 'Royal Horticultural Society' colour chart. This has revealed the reliability of a red slash or non-red slash as a field character, and it is the definition of red (particularly in 'greyed reds') in that chart which is adopted here.

HOW TO USE A KEY

A key is a method which saves reading too many irrelevant descriptions. Imagine you live in a country with only four animals, and you wish to write a guide to enable visitors to identify them. The animals are the four listed in the mock key below. You could write out a description of each animal, which would be convenient for the imaginary country, but useless for a subject like 'Ghana's trees', because it would take several hours to sift through the descriptions in such a book (try with Irvine, 1961). The idea behind a key will be shown for the four animals of the imaginary country.

Example key for imaginary country with four animals

Face of animal with trunk, and often with tusks; usually more than 1 metre long; skin thick and not furry; ***Elephant***

Face of animal without trunk and tusks.

  Animal with fur and four legs.

    Body >50cm long; >2 metres tall, with long neck and hump on back; tail <1/2 as long as body; most common in drier areas; rare in towns. ***Camel***

    Body <50cm long; two long, straight teeth at front of mouth, tail almost as long as body; common in houses, towns and villages. ***Rat***

  Animal with wings and feathers; able to fly, often coloured with grey and red; hooked, sharp beak; lives in trees; lays eggs. ***Parrot***

The keys in Sections 2 and 3 are the same as this in principle, except that they are divided up into small keys, or Groups (most of which fit onto one page). In order to work out which small key to use the user works through a 'Key to Groups'.

Read through the example key for a while, and you will probably be able to see how it works. It will be seen that the key is a series of statements. These statements are interpreted with the following facts and rules in mind:

1) Each statement has an opposing statement somewhere in the same key.
- The *opposing statement* is one which begins the **same distance from the left hand margin**;
- The pair of statements are **closer to the left hand margin than any intervening statements**
- The pair of statements **begin with the same or similar words.**

(With these rules, the three pairs of statements are easily spotted in the example.)

2) Start the key with the pair of statements which are closest to the left-hand margin. One of these statements will be at the top of the key.

3) Decide which statement in the pair is more correct.

4) If the correct statement ends
a) with a species name, then this name is the solution.
b) with a Group name, then turn to this Group and start the key there.
c) with neither of these, then move onto the next pair of statements.

5) The next statement always begins immediately below the one chosen as correct, and is always indented slightly relative to that one. Find the opposing statement in the pair and continue from 3) above.

In a few places in Section 3 (200 Main Species Key), more than two contrasting statements are presented at the same level. In these cases, the aim is still to choose a single correct statement, and to carry on from there as described in 4) above.

This is the straightforward way of using the key. People with little experience of Ghanaian trees, or at least who have little idea of the affinities of the tree to be identified, will probably want to use the key in this way. However, it will also be apparent that the experienced reader can use the key 'backwards', to answer questions like, for example, 'how does *Parinari excelsa* differ from *P. congensis*?'.

To find how to distinguish between two species

In many cases, a study of the illustrations of the two species will reveal important differences between them. Even if this appears to be the case, it is useful to check for the differences suggested by the text as follows:

1) Locate the two species names in the key; the Groups in which species are found are listed in the Index to Scientific Names in Section 4. (e.g. *Parinari* spp. are in Gp 14B).

2) Find the single pair of statements which separates the species as follows:

a) If the two species are in the **same Group** find the single statement, if any, between them which starts most closely to the left-hand margin, and find its opposing statement in the manner described above. If there is no intervening statement and they start at equal levels, then the crucial statements are those terminating in the species names.

In our example, the intervening statement is 'Base of leaf not cordate', which applies to *P. excelsa* (amongst other species). The choice is between this statement and 'Base of leaf cordate', which applies to *P. congensis.*

Further characteristics of the species, which will not necessarily differentiate them, can be found by reading the relevant statements between the crucial ones and the species names. So, in our example we learn that leaves of both *Parinari* species are often discolorous due to dense hairs, but that whereas *P. congensis* is found by rivers, *P. excelsa* does not have such an ecological preference.

b) If the two species are in **different Groups**, turn to the relevant 'Key to Groups' (or to the Key to Main Groups at the beginning of Section 2) and continue as above.

In some cases a single species is mentioned in two keys. In these cases, the above procedure can be followed for each reference, but the reader must remember that the species occurs in two forms (e.g. sun and shade leaves).

ACKNOWLEDGEMENTS

The information in this book comes from a variety of sources. It has been written as part of the Ghana National Inventory Project, based in the Forest Department in Kumasi and supported by the UK Overseas Development Administration. I would particularly like to thank Albert Enti, who introduced me at the start of the project to the trees which are for him intimate friends gained over many years of experience in the Forest Department. Subsequently, I have learnt a great deal from Forest Guards and Technical Officers and other personnel of the Inventory Project, either through their knowledge of the trees, or through their assistance in the collection of information and specimens, or in providing constructive criticisms of earlier drafts. Thanks to Ntim Gyakari, Kwasi Orgle, Paul Oppong, Kusi Asare, De Graft Owusu-Ansah, J.C. Ameyaw, Kofi Donkor (both of you), Martin Anane and Messrs. Sarbeng, Nkrumah, Ganoh and Osei-Tutu. Thanks also to Drs Swaine, Keay, Hepper, Verdcourt, Bridson, Abbiw, Howard and Styles for looking over various parts of the manuscript.

BIBLIOGRAPHY – SOURCES OF FURTHER INFORMATION ON WEST AFRICAN PLANTS

Information has also been taken from published sources (listed below), although where possible this information has been checked in the field. Published information on field characters is sparse, and this guide presents only a tentative and incomplete summary of this type of information available for Ghana. Further information on such matters, in the form of detailed notes accompanying pressed and dried specimens of leaves, will be gratefully received by the Forest Herbarium in Kumasi.

Aubréville, A. (1959) *La flore forestière de la Côte d'Ivoire.* 2nd edn. Nogent-sur-Marne: C.T.F.T.

Aubréville A. *et al.* (1961-) *Flore du Gabon.* Paris: Mus. Nat de l'Hist. Nat.

Brunnel, J. F. *et al.* (1984) *Flore analytique du Togo. Phanérogames.* Englera 4.

Burkill, H.M. (1985) *Useful plants of West Tropical Africa.* Kew: Royal Botanic Gardens.

Hall, J.B. and Swaine, M.D. (1981) Distribution and ecology of vascular plants in a tropical rain forest vegetation in Ghana. *Geobotany* 1. The Hague: Junk.

Hallé, F., Oldeman, R.A.A. and Tomlinson, P.B. (1978) *Tropical trees and forests.* London: Springer-Verlag.

Hepper, F.N. ed. (1954-72) *Flora of West Tropical Africa.* 2nd edn. London: HMSO.

Houssain, M. and Hall, J.B. (1969) *A field guide to the trees of Mole Game Reserve, Damongo, Ghana.* Legon: University of Ghana.

Irvine, F.R. (1961) *Woody plants of Ghana.* Oxford: O.U.P.

Keay, R.W.J., Onochie, C.F.A. and Stanfield, D.P. (1960-4) *Nigerian trees.* Ibadan: Dept. Forest Research.

Kunkel, G. (1965) *The trees of Liberia.* Report no. 3 of the German Forestry Mission to Liberia, BLV Bayerischer Landwirstschaftsverlag.

Letouzey *et al.* ed. (1963-) *Flore du Cameroun.* Paris; Mus. Nat. de l'Hist. Nat.

Savill, P.S. and Fox, J.E.D. (1967) *Trees of Sierra Leone.* Unpubl. Manuscript. Oxford: Forestry Library.

Steentoft, M. (1988) *Flowering plants in West Africa.* Cambridge: C.U.P.

Taylor, C.J. (1960) *Synecology and silviculture in Ghana.* Edinburgh: Nelson.

Vivien, J. and Faure, J.J. (1985) *Arbres des forêts denses d'Afrique Centrale.* Paris: Agence de Coop. Culturelle et Technique.

Voorhoeve, A.G. (1965) *Liberian high forest trees.* Wageningen: Centre for Agric. Publ.

ABBREVIATIONS

/	Per, e.g. stipules 2/node
>	Greater than
<	Less than
->	Becoming
±	Plus or minus; with or without; more or less
br.	Branch
f.	Female (flower)
flwr	Flower
for.	Forest
f.r.	Forest Reserve
ft(s)	Fruit(s) (Fruit dimensions are maximum diameter, unless specified.)
Gp	Group
infl.	Inflorescence
lf	Leaf
lvs	Leaves
lft	Leaflet
ls	Lower surface
m.	Male (flower)
nr	Near
pr	Pair
sl.	Slightly
us	Upper Surface
yng	Young

SECTION 2

MAIN KEY

VARIOUS TYPES OF LEAF ARRANGEMENT

SIMPLE

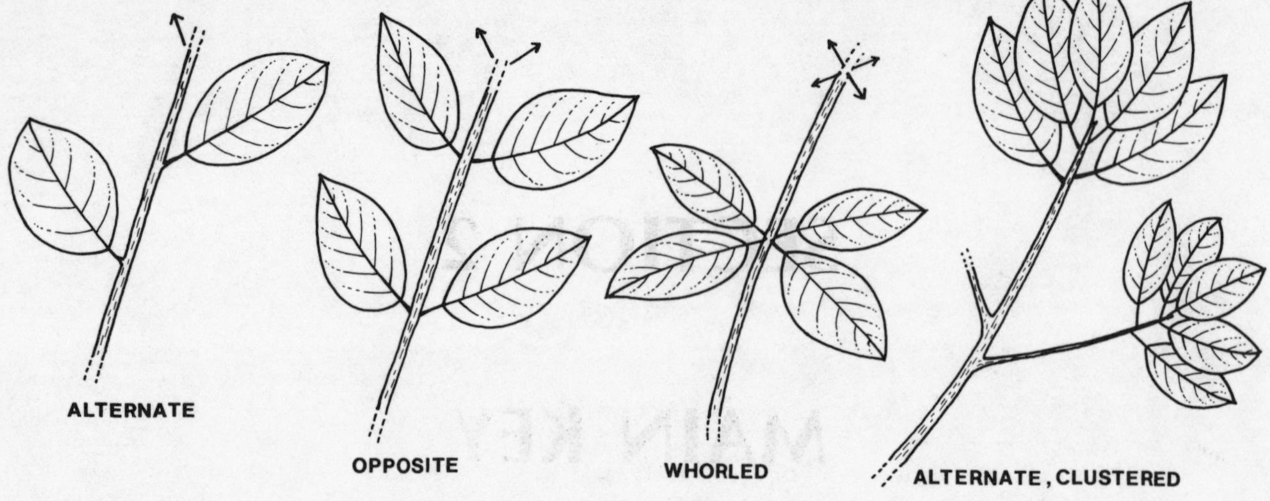

COMPOUND

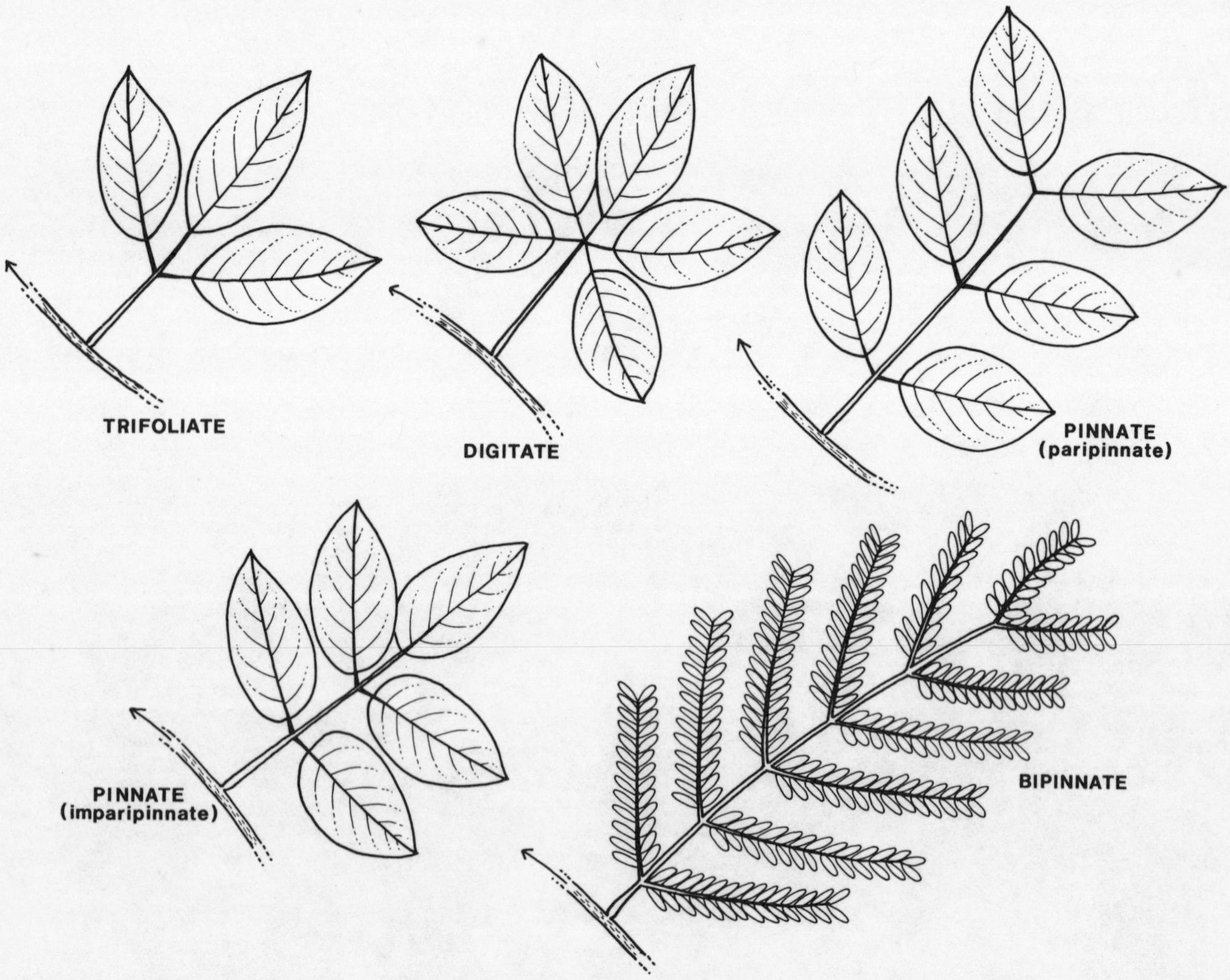

SECTION 2: MAIN KEY

STARTING KEY USING LEAF CHARACTERISTICS

NORMAL (i.e. dicotyledonous) TREES: NOT PALM TREES, BAMBOOS nor OTHER MONOCOTYLEDONS	
LEAVES ALL SIMPLE	
LEAVES OPPOSITE or IN WHORLS	TURN TO **KEY A**
LEAVES ALTERNATE.	
Petiole short (<2 cm long)	TURN TO **KEY B**
Petiole long (>2 cm long)	TURN TO **KEY C**
LEAVES COMPOUND	TURN TO **KEY D**

MONOCOTYLEDONOUS TREES like palms or bamboos with many main nerves running from the bottom of the lamina to the top. See Group 39.

Notes for beginners: the arrangement of leaves

It is easy to be deceived by the leaf arrangement of a tree. The diagrams opposite, in the various Groups, and in the glossary should make clear most of the terms, but the simple/compound question requires more explanation.

Compound leaves are similar to two or more 'ordinary', simple leaves on specialized twigs of limited length. The parts like simple leaves on compound leaves are called **leaflets**. The twig-like structure which forms the backbone of the compound leaf is the **rachis**. Whole leaves, whether simple or compound, grow from nodes on twigs, whereas leaflets grow on the rachis which lack nodes of the type found on twigs. To decide whether you have a simple or compound leaf, therefore, you have to decide whether the axis is a twig, with nodes and growing points, or a rachis.

A rachis is of fairly fixed length, and is not able to grow after it has initially unfolded, either at its apex (where a leaflet sometimes terminates convincingly the compound leaf) or at the base of the leaflets. The unfolding compound leaf generally has all the part-formed leaflets, which then grow together. The whole compound leaf will usually fall off the tree together. A rachis (i.e. a compound leaf) never bears flowers or fruits, and does not have resting buds or stipules (nor, therefore, stipule scars). All compound leaves from the same part of a tree are of a similar length and overall appearance. (However, as with simple leaves, there are usually differences between the leaves on young and old or exposed and shaded trees.) Very often, the rachis is swollen at the base, where it joins the nodes of the twigs, and narrows to a pointed tip. Often, the many similar rachises can be found in the leaf litter below a compound-leaved tree.

A rachis is normally unbranched, but the **bipinnate leaves** of Group 38 have one order of branching (primary and secondary rachises).

Contrast this with a twig bearing simple leaves. The twig has a growing point at one end, or in the axils of the leaves. In these places there are often signs of new leaves unfolding, of new branches appearing, of budscales or stipules, or of flowers or fruits. Therefore, a leaf never terminates a twig, nor is it normal for twigs to fall off from greater branches when laden with leaves. If this does happen (e.g. in a storm) each leaf-bearing twig will tend to be of a different size.

Nevertheless, a few cases are not so clear-cut, particularly when there is no leaf terminating the axis of a compound leaf. In *Rothmannia* a simple leaf appears to terminate a twig, but we can see the leaf is simple from the branches that appear immediately below this leaf. Certain species of simple-leaved plants mimic compound leaves, producing branches of similar length, with leaves dispersed along them. This is the case, for example, in *Psydrax*, *Diospyros*, certain Annonaceae, and *Panda*. At the other extreme, certain compound leaves can resemble simple ones. *Balanites* produces compound leaves with two leaflets which resemble strongly opposite, simple leaves (see Gp 31). Certain legumes with alternate leaflets (e.g. see *Crudia* spp. in 37G) can also be deceptive, particularly if they produce 'stipels' which resemble thread-like stipules. The long compound leaf of *Pierreodendron* seems to maintain a growing point at its tip, but this species is unusual and the tree is in any case rare.

Most cases are easily ascertained, though. It may help to mention a few traits shown by simple leaves which are rarely shown by leaflets and vice versa.

Traits much commoner in (trees with) simple leaves	*Traits much commoner in pinnate leaves*
Stalks (i.e. petioles) >2 cm or reflexed	
Long lamina (>15 cm), particularly if symmetrical	Lamina shorter than 5 cm
Pit domatia	
Rounded, ovate or obovate shape	V. asymmetric base + oblong lamina (but c.f. Gp 17)
Serrations; basal nerves and basal glands (for exceptions see key D)	Axis (rachis) with grooves on one surface
Discolorous crown; dense hairs below leaf (but see *Anthonotha*)	Axis with glands, basal swelling, constrictions
Stipules at base of lamina (but see 'stipels' of certain legumes)	

KEY A: (GROUPS WITH OPPOSITE OR WHORLED, SIMPLE LEAVES)
(GROUPS 1 to 9)

- Trees without latex[1,2]
  - Twigs with (interpetiolar) stipules or lvs serrated; lvs never whorled
    - Margin entire; tree sometimes spiny — Group 1 — RUBIACEAE
    - Margin serrated. — Groups 2, 3 — RHIZOPHORACEAE, RHAMNACEAE
  - Twigs without (interpetiolar) stipules OR lvs whorled; lvs entire
    - Mangrove trees, in swamps by the sea (often with stilt roots) — See Group 3
    - Trees in ordinary forest, or at least not mangroves
      - Leaves or petioles v. long, tufted at branch ends; little-branched, 'cabbage-trees' often with short spines and stilt roots — Group 4 — *Anthocleista* (LOGANIACEAE)
      - Not 'cabbage trees'; plants unarmed and lvs and petioles not v. long
        - Lvs strongly trinerved OR with venation obscure; not whorled — Group 6 — MELASTOMATACEAE
        - Lvs not trinerved, and finer venation visible; OR lvs whorled
          - Lvs in whorls — See Group 3
          - Lvs opposite
            - Leaves with many fine, parallel laterals and often with a sub-marginal nerve; usually glabrous; often gland-dotted — Group 7 — MYRTACEAE, GUTTIFERAE
            - Leaves without finely parallel venation; without sub-marginal nerve; sometimes with domatia — Group 5 — OLEACEAE, etc.

- Tree with latex
  - Latex orange; lvs glandular and hairy — Group 8A — GUTTIFERAE
  - Latex yellow; lvs glabrous — Group 8B, C — GUTTIFERAE
  - Latex white; (or lvs with tuft domatia) — Group 9 — APOCYNACEAE

NOTES: 1) It is important to distinguish clustered (alternate) and whorled leaves. Clustered, alternate leaves are never quite opposite at the nodes, and will usually be clearly alternate away from the twig ends. Long petioles are not typical of whorled or opposite-leaved trees, but are typical of trees with clustered, alternate leaves. Some trees, particularly in Group 4, have clustered opposite leaves.

2) Latex will often only arise as specks in the slash of understorey trees; for these, it is often worth cutting a fresh petiole to examine for latex there as well.

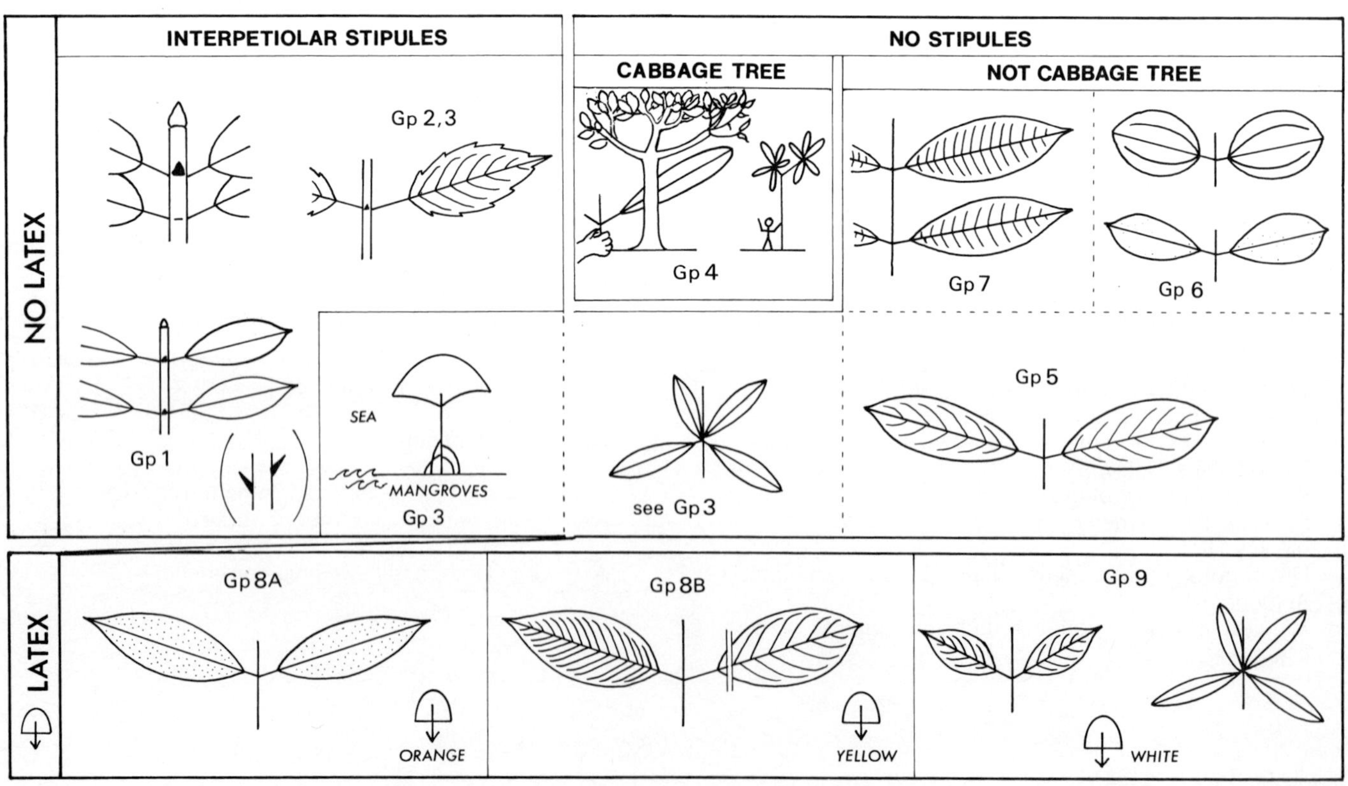

KEY B: (GROUPS WITH SIMPLE, ALTERNATE (SOMETIMES CLUSTERED) LEAVES WITH SHORT PETIOLES)
(GROUPS 10 TO 21)

NB: Make sure you have found the largest length of petiole on your specimen. Some species, typically those with very clustered leaves, have short-petioled leaves at the twig ends, and long-petioled leaves further down the twigs: these are included in the long-petiole key (KEY C).

- Tree with spines or prickles — Group 15 — miscellaneous
- Tree unarmed
  - **Leaves without basal nerves**
    - Margin entire
      - Tree with latex (lvs usually clustered) — Group 10 — SAPOTACEAE
      - Tree without latex (lvs rarely clustered)
        - **Slash not sweet-scented OR plant with stipules**
          - Lvs with basal glands AND slash reddish[1]; stipules present and often persistent — Group 14 — CHRYSOBALANACEAE
          - Lvs without basal glands OR slash far from red
            - Lateral nerves and veins finely transverse — Group 16 — OCHNACEAE
            - Lateral nerves and veins not like *KAKU*, etc.
              - –Bark with a black layer; brs whorled; twigs without stipules — Group 11 — *Diospyros* (EBENACEAE)
              - –Not *Diospyros*; (or if in doubt) species sometimes with stipules or red exudate — Group 13 — miscellaneous
        - **Slash characteristically sweet-scented, bitter or hot; no stipules;** flower parts in 3s. — Group 12 — ANNONACEAE
    - Margin serrated
      - Tree without latex — Group 17 — EUPHORBIACEAE-FLACOURTIACEAE
      - Tree with latex — Group 19 — MORACEAE
  - **Leaves with basal nerves**
    - Tree with latex — Group 19 — MORACEAE
    - Tree without latex
      - Leaves without stellate hairs, scales and glands; slash often v. brittle, sometimes yellow and brown layered — Group 18 — ULMACEAE, etc.
      - Leaves with stellate hairs or dotted with glands or scales below; slash usually v. fibrous
        - Margin serrated — Group 20 — TILIACEAE
        - Margin entire — Group 21 — miscellaneous

NOTE: 1) If leaf has basal glands, but the tree is without stipules or without a red slash, see notes at top of Group 14.

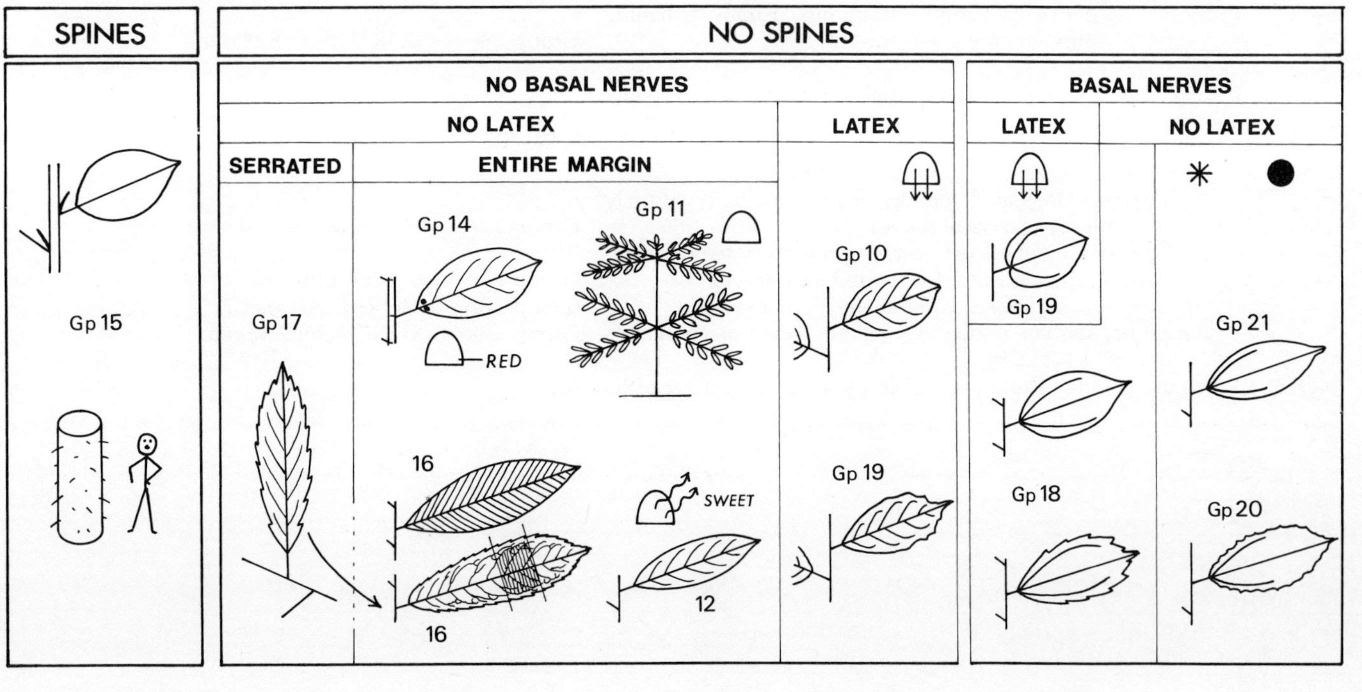

KEY C: (GROUPS WITH SIMPLE, ALTERNATE (USUALLY CLUSTERED) LEAVES WITH LONG PETIOLES)
(MAINLY GROUPS 19 TO 27)

Tree with spines	Group 23	*Macaranga* (EUPHORBIACEAE)
Tree without spines		
Petiole only 2-3cms long[1] (tree with latex or lvs serrated)		
Trees with latex		
Leaves entire, elliptic and symmetrical; twigs without stipules OR lvs very discoloured below	Group 10	SAPOTACEAE
Leaves serrated OR ovate OR asymmetric; with (ring) scars at nodes, where stipules have fallen	Group 19	MORACEAE
Trees without latex; lvs serrated, with stellate hairs	Group 20	TILIACEAE
Petiole, at least on some leaves, >3 cm long (OR tree without latex AND leaves entire)		
Tree with latex	Group 19	MORACEAE
Tree without latex		
Leaves without basal nerves: (not usually so broad relative to length)		
Lower surface of lf discolorous with dense silvery or golden scales	Group 21	
Lower surface without obvious scales		
Leaf serrated	Group 22B	EUPHORBIACEAE
Leaf not serrated (sometimes lobed or undulate)		
Leaf margin wavy or irregular; nerve axils without domatia		
Leaf hairy; weedy tree without stilt roots	Group 26	BORAGINACEAE-SOLANACEAE-COMPOSITAE
Leaf glabrous; (mature) tree with stilt roots	Group 24	*Uapaca*, etc. (EUPHORBIACEAE)
Leaf margin entire, or nerve axils with domatia		
Petiole swollen at tip OR leaf glabrous and >15 cm long OR with tuft domatia		
Swamp-loving trees normally with stilt roots; lvs v. rarely acuminate, but usually rather rounded towards apex	Group 24	*Uapaca*, etc. (EUPHORBIACEAE)
Trees without stilt roots. Leaves either with hairy domatia, (in timber trees) or lvs markedly acuminate	Group 27	STERCULIACEAE
Petiole not swollen (sometimes with glandular flange); (pioneering trees)		
Lf base symmetric; lf sometimes with pit domatia or glands on petiole; petiole not otherwise flattened; lvs very clustered at twig ends, and branches in whorls; slash yellowish fibrous	Group 25	COMBRETACEAE, etc.
Lf ±asymmetric at base; petiole rather flattened; margin ±irregular, with dense hairs or completely glabrous; slash fibrous, usually darkening through grey-green shades	Group 26	BORAGINACEAE, etc.
Leaves with 2 or more strong basal nerves: (usually broadly ovate-elliptic)		
Margin serrated		
Leaves not lobed (stellate hairs often present)		
Leaves with (basal) glands, never very hairy; usually with only two basal nerves	Group 22A	EUPHORBIACEAE
Leaves without glands, with dense hairs or teeth thread-like; typically with 4 or more basal nerves	Group 27C	STERCULIACEAE
Leaves strongly lobed (without stellate hairs)	Group 28A	MORACEAE (part 2)
Margin not serrated (though sometimes lobed or undulate)	Group 27	STERCULIACEAE, etc.

NOTE: 1) See the warning about petiole length at the beginning of Key B.

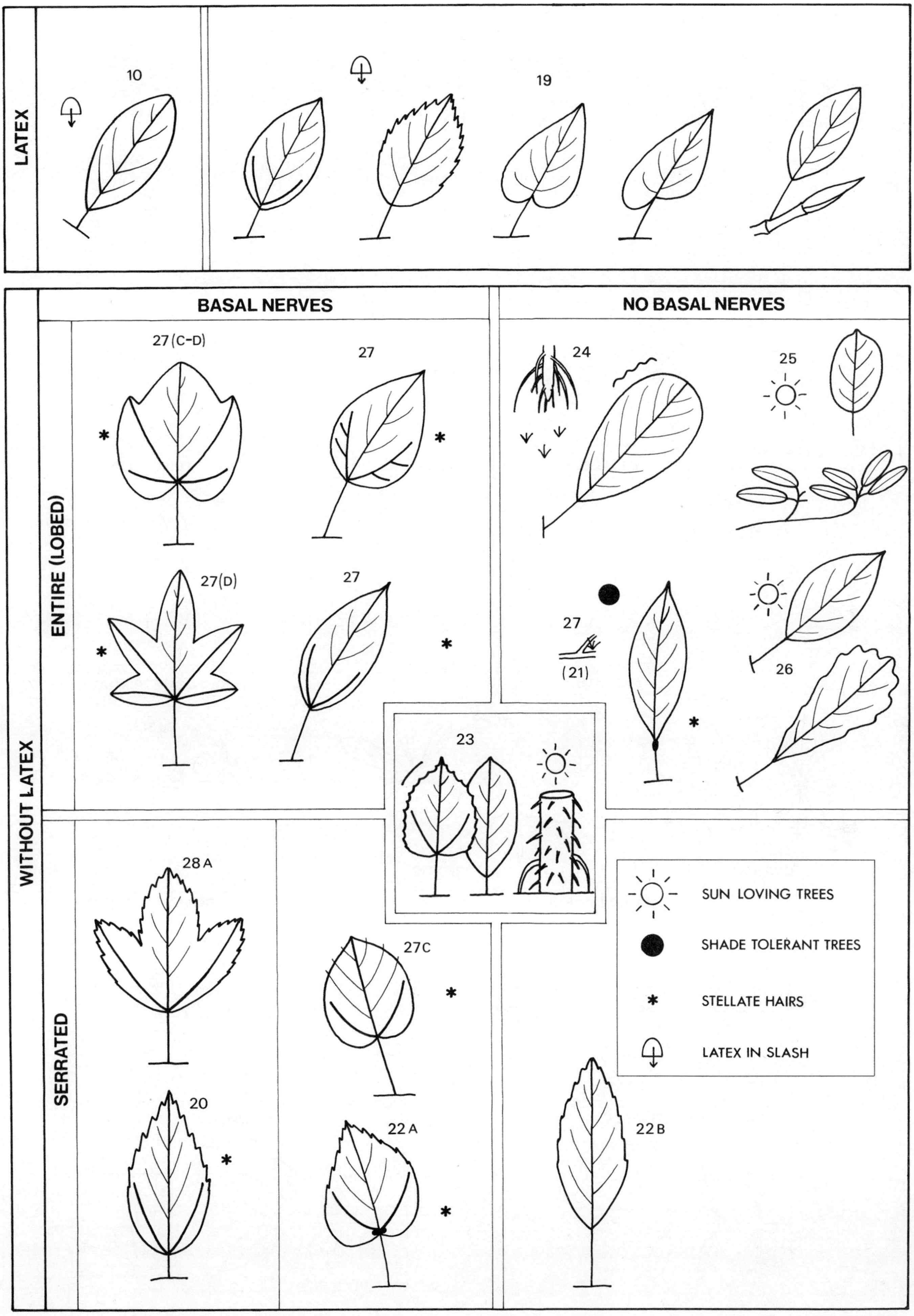

LATEX
10
19
WITHOUT LATEX
BASAL NERVES
NO BASAL NERVES
ENTIRE (LOBED)
27(C-D)
27
27(D)
27
24
25
27
(21)
26
23
SERRATED
28 A
27C
20
22 A
22 B
SUN LOVING TREES
SHADE TOLERANT TREES
STELLATE HAIRS
LATEX IN SLASH

BIPINNATE

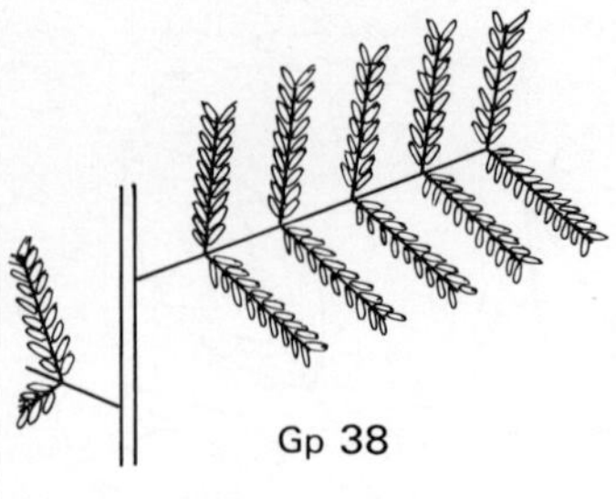

SPINES OR PRICKLES

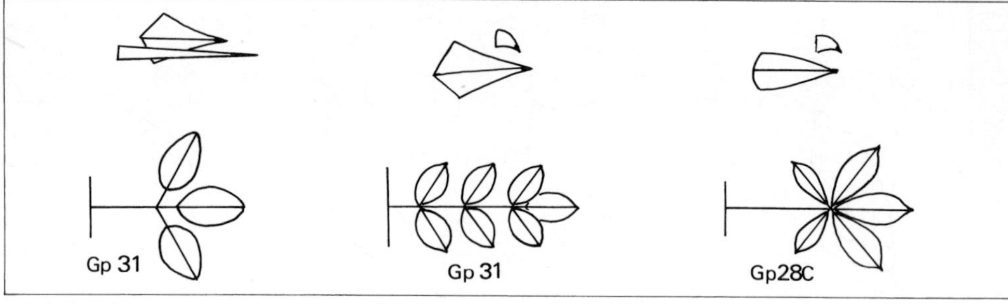

LATEX

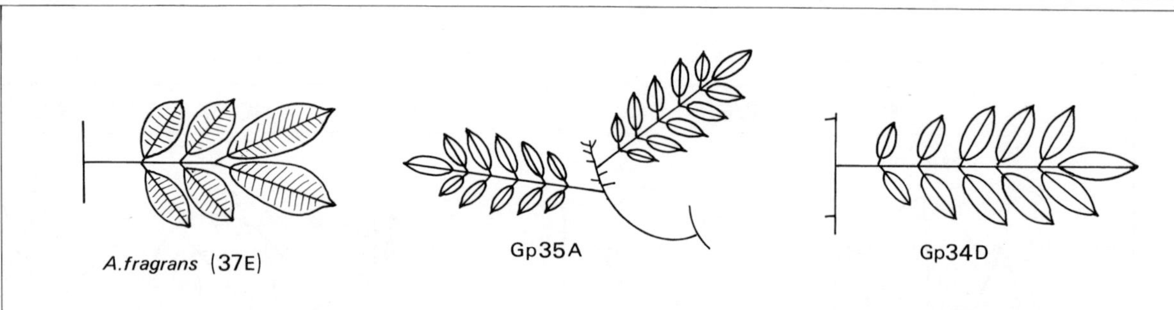

2 LEAFLETS

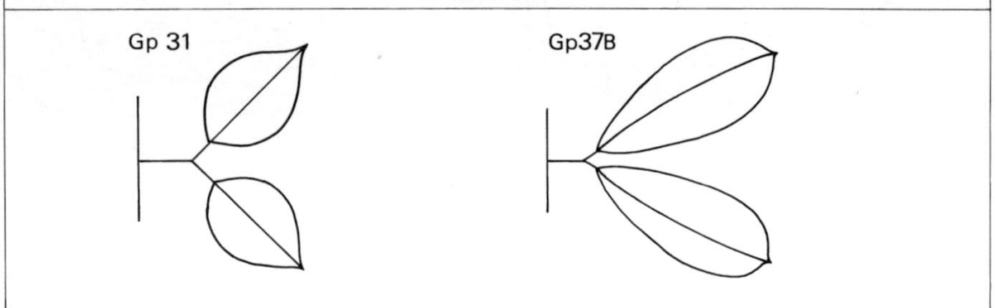

TRIFOLIATE

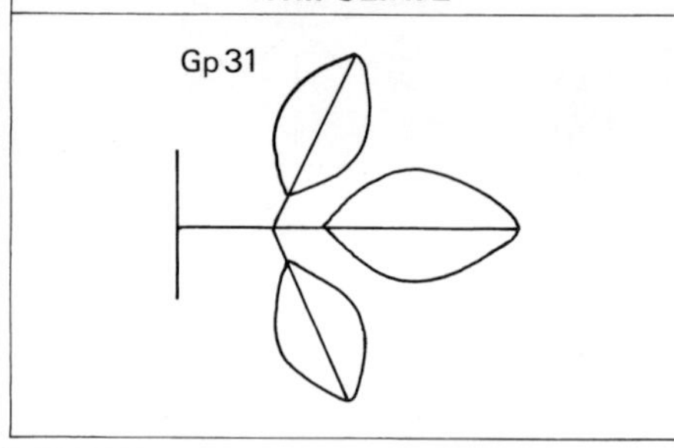

DIGITATE

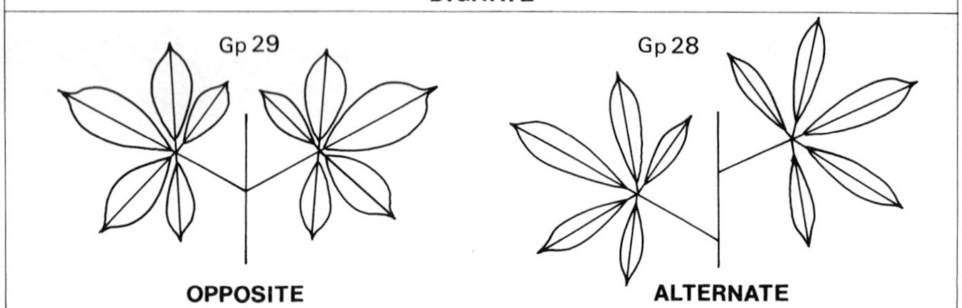

OPPOSITE, PINNATE

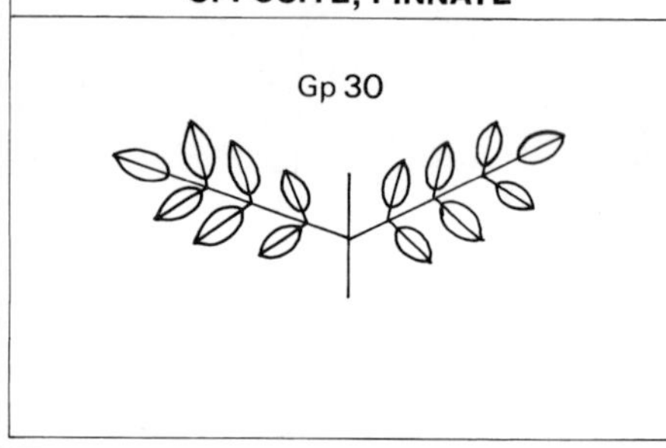

WINGED RACHIS

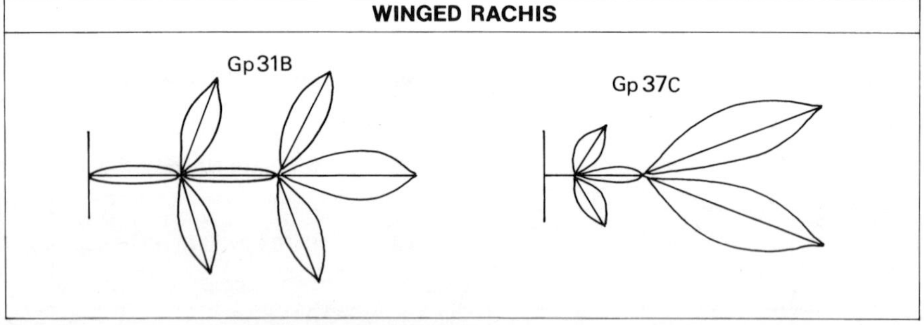

SMALL LEAFLETS

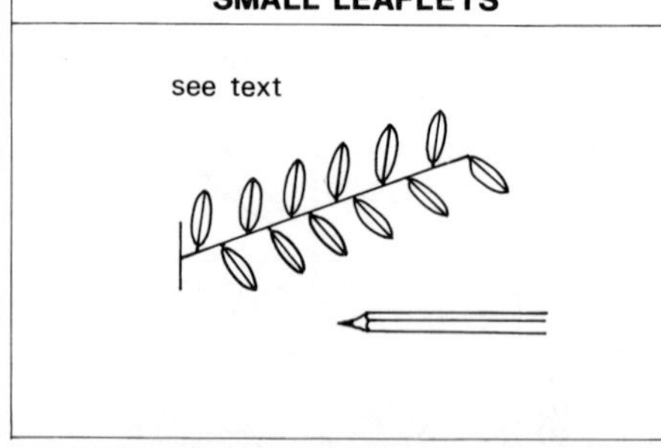

SWOLLEN PETIOLULE

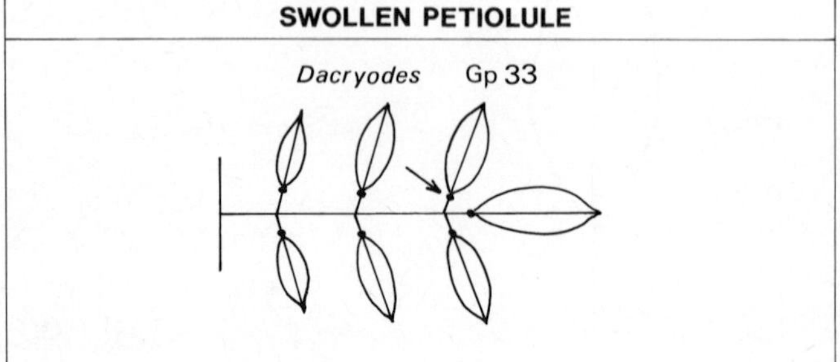

STIPELS

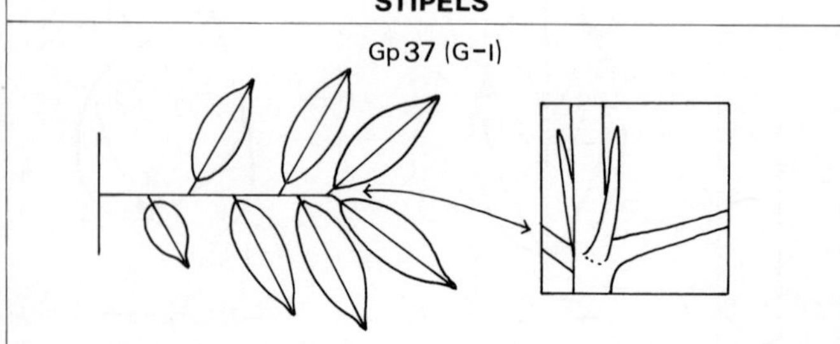

KEY D: GROUPS WITH COMPOUND LEAVES

The following key is more complicated than the key for the preceding groups because compound leaves are more complicated than simple leaves, and there is less chance of retrieving a whole compound leaf (e.g. with a catapult) than there is for simple leaves. Whilst the practised eye can normally differentiate, for instance, a legume leaf from those of other groups, the facets being recognized are too complex and numerous to explain briefly. A glance through the illustrations may help pinpoint the correct group. Otherwise, the following key picks first on obvious characters of the whole compound leaf (either visible with binoculars or good eyesight, or on smaller trees). Then features of the leaflets are listed, to be ruled out in turn. Finally, for species without distinctive features on their leaflets, the key returns to questions about the whole compound leaves. As a double-check, it is worth running through the keys to the indicated subgroup, if any, at the top of the Group in question.

If your tree has pods (look on ground below) turn straight to Group 37.

D1. Initial key, based on leaf and whole tree

Lvs bipinnate	38 (MIMOSACEAE)
Lvs not bipinnate	
Tree with spines or prickles	
Lvs digitate (with 4 or more leaflets joined to the top of the petiole)	Group 28C
Lvs not digitate (pinnate or trifoliate or only two leaflets)	Group 31
Tree unarmed	
Slash with white or brownish, milky latex	
Crown deeply red-discoloured; lvs not clustered, usually with 3 pairs of lflets	37E (*Anthonotha fragrans*)
Not *Anthonotha fragrans*	
Slash red, contoured, v. fibrous; trees in **evergreen forest**; lvs v. clustered	35A (*Trichoscypha* spp.)
Slash brownish, soft, darkening rapidly, often slightly scented; common outside evergreen forest	34D (*Trichilia* spp.)
Slash without latex	
a) Leaves with only 2 lflets	
Leaflets v. falcate or with folded apex; tree unarmed	37B (*Guibourtia* etc.)
Leaflets almost rounded or broadly elliptic; tree usually with spines	31 (*Balanites*)
b) Leaves trifoliate: with 3 leaflets	31 (RUTACEAE, etc.)
c) Leaves with 4 or more leaflets	
Leaflets all arranged around top of petiole (**digitate**)	
Leaves alternate	28
Leaves opposite; slash pale, darkening through greenish shades	29 (Vitex spp.)
Leaflets not digitate: pinnate	
Leaves (i.e. as a whole: NOT NECESSARILY LEAFLETS) opposite; slash pale, ±fibrous, often darkening through greenish shades	30 (BIGNONIACEAE)
Leaves alternate	
Rachis with broad, green, leaf-like wings	
Leaflets in 2 pairs; the largest at end of rachis	37C (*Hymenostegia afzelii*)
Leaflets >2 pairs (e.g. 1 terminal lflet) or serrated	31B (*Bersama*, etc.)
Rachis without wings of this type.....	

check now whether the leaflets have any of the following characteristics, and select group (N.B. it is important to rule out questions in the order they are given)

D2. Key based on distinctive leaflets of compound leaves

Lflets all v. **narrow**: <15 mm wide	
Lflets <5 mm wide OR lflets almost rhombic (lvs bipinnate)	38 (MIMOSACEAE)
Lflets >5 mm wide AND not rhombic (lvs once pinnate)	
Lflets glossy, emarginate, with twisted petiolule	37D (*Copaifera*)
Lflets not emarginate	
Lflets all lanceolate; tree normally with latex	34D (*Trichilia martineaui*)
Lflets not lanceolate	
Lflets in 5-7 pairs, gregarious in **dry forest**	37C (*Talbotiella*)
Not *Talbotiella*	37I (*Millettia*, etc.)
Most leaflets >15 mm wide...	
Petiolule swollen (and often bent) at apex, c. 1 cm or more long	33 (*Dacryodes*)
Petiolule not swollen at apex...	
Leaves with stipels	37 (G-I)
Leaves without stipels (or whole leaf not available...)	

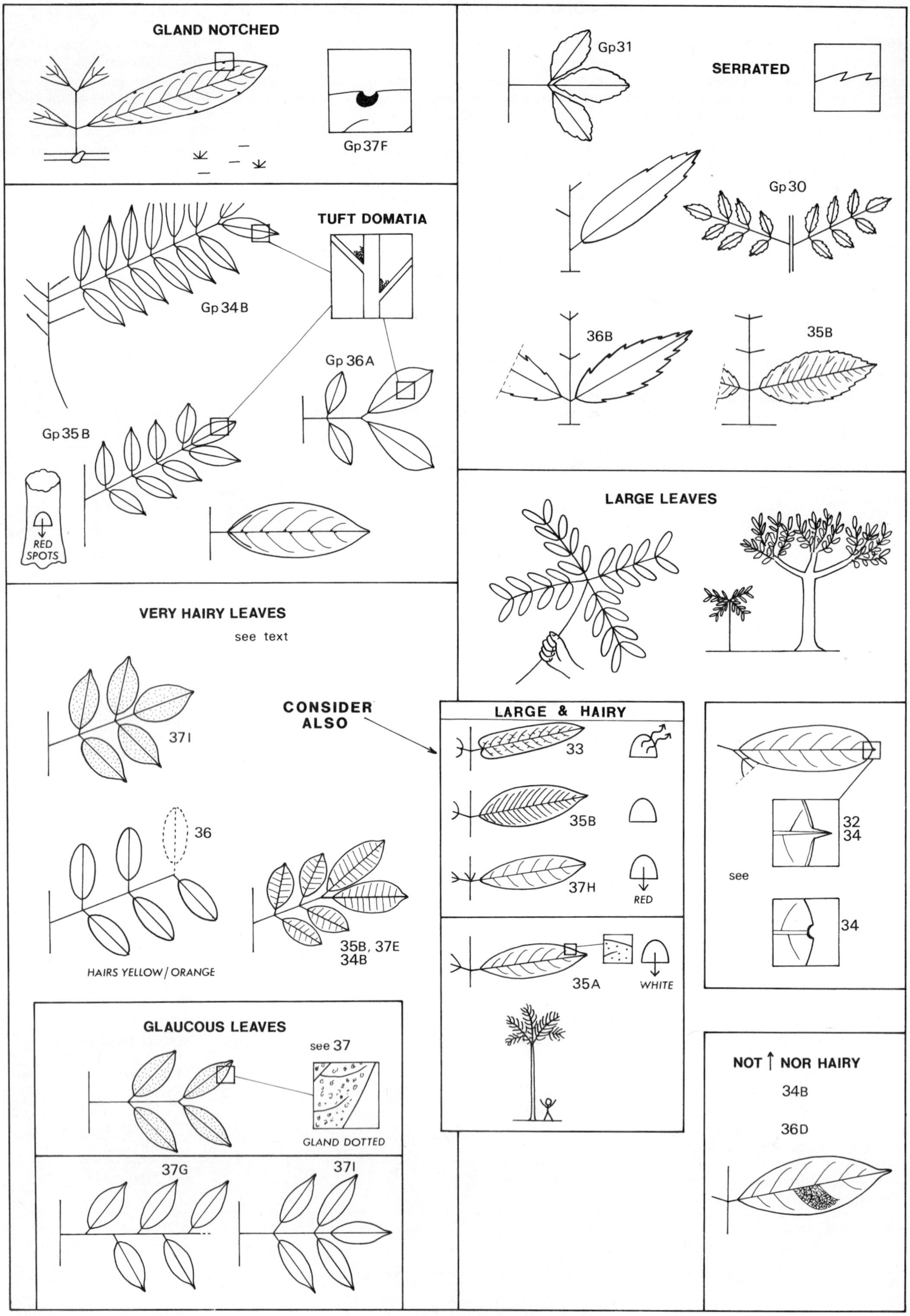
GLAND NOTCHED
Gp 37F
SERRATED
Gp31
Gp30
36B
35B
TUFT DOMATIA
Gp 34B
Gp 36A
Gp 35B
RED SPOTS
LARGE LEAVES
VERY HAIRY LEAVES
see text
371
CONSIDER ALSO
LARGE & HAIRY
33
35B
37H
RED
35A
WHITE
32
34
see
34
36
HAIRS YELLOW / ORANGE
35B, 37E
34B
GLAUCOUS LEAVES
see 37
GLAND DOTTED
37G
37I
NOT ↑ NOR HAIRY
34B
36D

Lflets serrated or gland-notched	
Lflets v. large, or with large stipules, gland-notched; trees in **wet places**	37F (*Gilbertiodendron* spp.)
Lflets <15 cm long, or with serrations as opposed to gland-notches along margin	
Venation ± scalariform OR teeth well-defined	
Lflets with tuft domatia (lvs trifoliate)	31 (*Allophylus*)
Lflets without tuft domatia	
Lvs (and most lflets) opposite (or lflet + basal glands)	30 (BIGNONIACEAE)
Lvs OR all lflets alternate	36B (*Lychnodiscus* etc.)
Venation fine-transverse, or with a nerve closely parallel to margin	35B (ANACARDIACEAE)
Lflets entire...	

Lflets with tuft domatia in nerve or vein axils	
Slash without red exudate: lflets without two almost basal nerves	
Large, v. clustered leaves (>4 prs lflets) on large trees	34B (*Entandrophragma*)
Small tree with unclustered lvs; <5 prs symmetrical lflets	36A (*Blighia unijugata*)
Slash with spots of red exudate; lflet base v. unequal sided, often with 2 nerves almost basal; small spreading tree of **swamps**	35B (*Pseudospondias*)
Lflets without tuft domatia... (Note: only simple-leaved spp. have pit domatia in Ghana)	

Lvs v. large, >50 cm long, clustered at the extremities of little-branched trees	
Lflets, and often twigs, coarsely or softly hairy, especially when young	
1) slash v. sweetly scented; lflets ± cordate-acuminate	33 (sapling *Canarium*)
2) lflets with many parallel laterals	35B (sapling *Antrocaryon*)
3) Lvs with large, leafy stipules at base (and stipels)	37H (sapling *Amphimas*)
4) Not sapling *Canarium*, etc. (mature treelets)...	
Evergreen (or coastal) forest trees; lvs pustulate or with veins obscure; (slash normally red, with latex); (check first if in doubt)	35A (*Trichoscypha*)
Trees outside these forest types, **OR** slash orange-granular	36C (*Chytranthus, Deinbollia*)
Lflets not coarsely or softly hairy	
Tip of leaflet with thickened projection or folded drip tip	
Lflets with some hairs	34B (saplings), 32
Lflets glabrous	34B (saplings), 34C, 32
Tip of leaflet without thickened projection	
Venation prominent above, minutely, regularly reticulate	36D (*Placodiscus*)
Venation not like this; lflets (broadly) elliptic	34B (saplings)
Lvs smaller...	

Leaflets soft or rough with **long, dense hairs OR glaucous below**	
Lflets gland-dotted, narrow, very fragrant, v. small weedy tree with narrow lflets	31 (*Clausena*)
Lflets not gland-dotted or tree in shade or >10 cm dbh or with broad lflets	
Leaflets mostly <1.5 cm wide	37I
Leaflets larger	
Leaflets cordate or obtuse at base	
Lflets hairy, but without woolly stellate hairs; bark even of young plants, with strong, sweet scent	33 (*Canarium*)
Leaflets woolly, with stellate hairs - rare plant	(28C - see *Polyscias fulva*)
Not this combination	
Hairs very fine or apparently absent; lower lf surface glaucous	
–Lflets all paired	
Lflets gland-spotted; twigs with apex wrapped up in leafy stipules, falling to leave scars at nodes	37D (sapling *Daniellia*)
Lflets not gland-spotted, or twigs without such apical buds	37E
–Lflets alternate	37G (*Pericopsis*, etc.)
–Lflets paired, except for terminal one	37I
Hairs coarse, rough or soft to touch	
Lateral nerves closely parallel, at steep angle to midrib; lflets narrow	35B (*Antrocaryon*) (check 34)
Lateral nerves not dense, or lflets broad	
Hairs yellowish to pale orange, not making crown very discoloured	36 (Sapindaceae) (34D)
Hairs reddish brown; slash usually with vegetable scent; lflets rather rounded towards apex, in <5 pairs	
Tree with brownish crown; slash often with spots of latex; lflets all in pairs	37E, (34D) (*Anthonotha fragrans, Trichilia*)
Tree outside evergreen forest OR lvs with one odd terminal lflet	37I (*Lonchocarpus*) (34D)
Hairs not so dense and leaf not glaucous...	

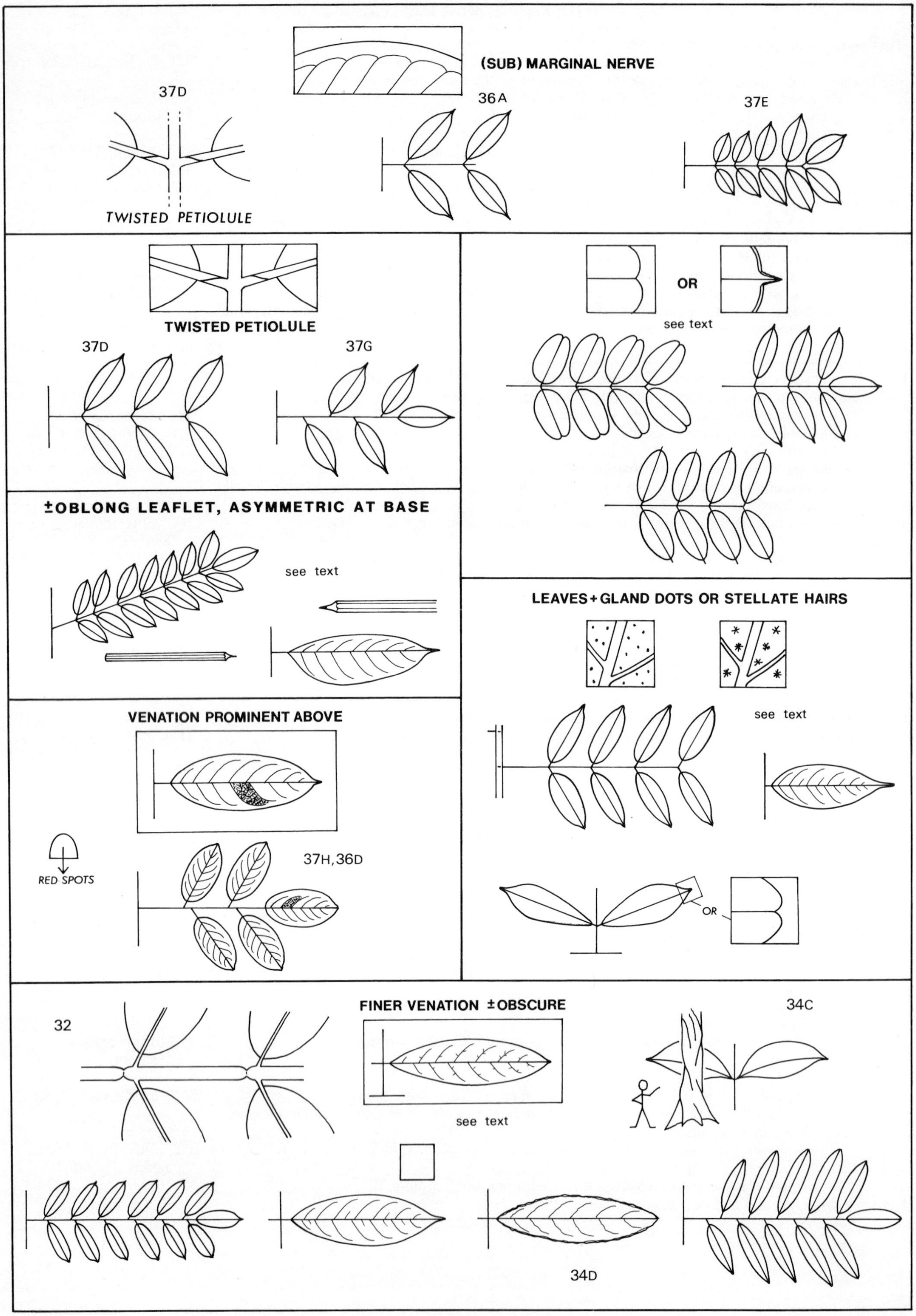

(SUB) MARGINAL NERVE
37D
TWISTED PETIOLULE
36A
37E
TWISTED PETIOLULE
37D
37G
OR
see text
±OBLONG LEAFLET, ASYMMETRIC AT BASE
see text
LEAVES+GLAND DOTS OR STELLATE HAIRS
see text
VENATION PROMINENT ABOVE
RED SPOTS
37H,36D
OR
FINER VENATION ±OBSCURE
32
see text
34C
34D

D2. Key: distinctive leaflets (contd.)

Leaflets with strong nerve closely parallel to margin or clearly running round margin	
Petiolules twisted, or leaflet emarginate or with translucent spots	37D
None of these	
Lvs imparipinnate - with terminal lflets	35B (*Spondias*)
Lvs paripinnate	
Leaflets narrowly elliptic, in <3 pairs	36A (*Pancovia*)
Leaflets broadly elliptic, in >3 pairs	37E (*Chidlowia*)
Leaflets without marginal etc. nerve...	

Leaf apex emarginate or with thickened point, or folded drip tip	
Rachis <25 cm long; lflets often hairy; without folded drip tip	
Lflets with marginal nerve or gland-dotted or lflets ovate	37D (*Detarium,* etc.)
Lflets without marginal nerve; broadest above middle; slash soft, pale fibrous	32 (SIMAROUBACEAE)
Lflets tip acuminate + apical notch like a fork	37C (*Cynonetre*)
Rachis >25 cm long OR lflets with folded drip tip	
Lflets broad, with soft hairs alongside midrib	34B (*Entandrophragma angolense*)
Lflets narrow-oblong, glabrous	34C (*Turraeanthus,* etc.)
Leaf apex not unusual; sometimes acuminate, but never thickened or emarginate...	

Petiolules twisted	
Margin thickened (marginal nerve); lflets opposite	37D (*Copaifera,* etc.)
Margin not thickened; lflets alternate	37G (*Crudia*)
Petiolules not twisted...	

Lflets <3 cm wide, ±lanceolate-oblong, acute-acuminate AND with v. asymmetric base (obtuse and cuneate)	
Lflets gland-dotted	See next box
Not *Daniellia*	
Venation not v. visible	
Lflets not all strongly lanceolate; slash camphor-scented	See 34C (*Majidea, Trichilia*)
Lflets strongly lanceolate; (slash normally with latex)	34D
Venation clearly visible; slash not sweetly scented	32 (*Nothospondias*)
Lflets not like this...	

Lflets with gland dots at end of veins or plant with stellate hairs	
Lflets rhombic or emarginate; with few or no hairs	37C
Lflets asymmetric at base, oblong; dots small and many; cylindrical bole	37D (*Daniellia*)
Lflets ± symmetric, with hairs; gland dots few, towards lf apex	37E (*Isomacrolobium*)
Lflets or twigs with stellate hairs	
Lflets not gland-dotted (or, at least, not these species)...	See Gp 33 and notes

Lflets with v. regular, reticulate venation, prominent ABOVE as a neat mesh of small squares, diamonds, etc.	
Tree with reddish exudate; lvs (mostly) imparipinnate, sometimes with stipels or with rounded or notched apex; twigs rounded, lenticellate	37H (*Dialium etc.*)
Tree without reddish exudate; lvs paripinnate, often with small point at end of rachis; apex acute to acuminate; lvs never with stipels or stipules (but cf. 'pseudostipules'); twigs often thick, triangular or winged; few-no lenticels	36D (*Placodiscus*)
Lflets with finest veins either not prominent ABOVE or not arranged as a regular, equal-celled mesh...	

Lflets with finer venation ± obscure, except for some of the larger interlateral veins	
Venation ± impressed; lflets in precise pairs, entirely glabrous, acuminate, asymmetric – see next box	
Venation not *Khaya*-like	
Rachis v. brittle or hollow, often compressed at nodes	32 (*Hannoa*)
Rachis corky; lvs strongly clustered at twig tips; lflets usually pustulate	35A (*Trichoscypha*)
Rachis neither v. corky, nor brittle, sometimes hairy	
Margin irregularly undulate, usually recurved	34D
Margin entire	
Lflets v. asymmetric; twisted, fluted tree	34C (*Trichilia prieuriana*)
Lflets ± symmetric	
lvs large and clustered; usually >30 cm long	34C (*Turraeanthus*)
lvs much smaller, not clustered	37I (, 37G)
Venation not so obscure...	

Venation not v. visible, usually with larger veins impressed, thread-like on surface, drying to a distinctive appearance, a little like leather	
Lflets broadly elliptic; lvs normally with one, odd terminal lflet; young parts with stellate hairs	36B (*Lannea*)
Lflets more slender, always in pairs, and markedly acuminate: completely glabrous	34A (*Khaya*, spp.)
Venation not like this...	

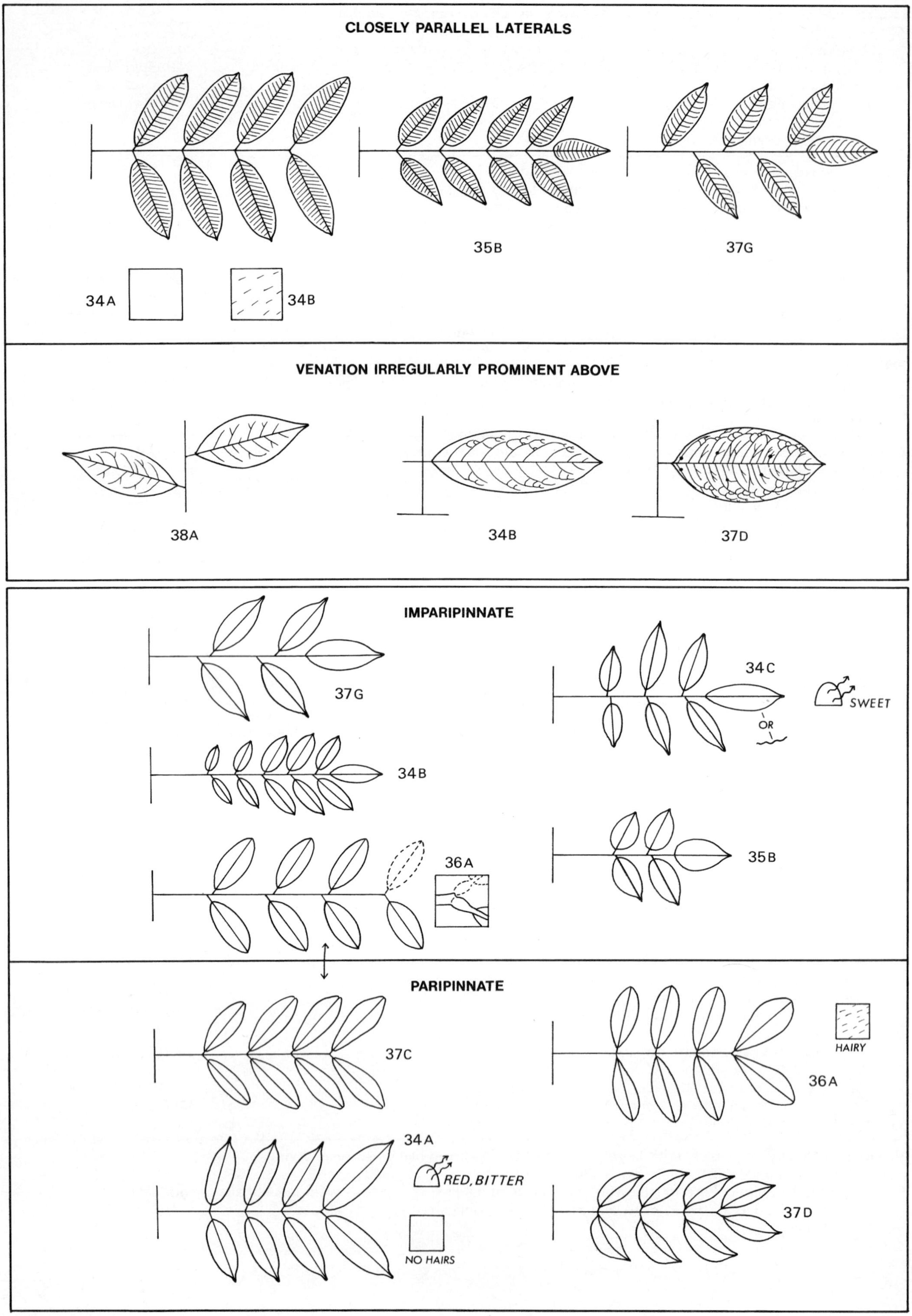
CLOSELY PARALLEL LATERALS
35B
37G
34A
34B
VENATION IRREGULARLY PROMINENT ABOVE
38A
34B
37D
IMPARIPINNATE
37G
34C
OR
SWEET
34B
36A
35B
PARIPINNATE
37C
HAIRY
36A
34A
RED, BITTER
NO HAIRS
37D

Lateral nerves closely parallel	
Lflets strongly paired; petiole slightly winged	
Lflets glossy, glabrous and elliptic; rachis irregularly winged; thick	34A (*Lovoa*)
Lflets with hairs; lvs v. clustered; nerves and midrib impressed above	34B (*Entandrophragma candollei*)
Lflets imparipinnate or otherwise not paired; petiole not v. winged	
Lflets paired + terminal one; nerves not v. impressed; lvs clustered; lflts thin; finer veins obscure	35B (*Antrocaryon*)
Lflets alternate; base often obtuse, ±glaucous; lvs not clustered.	37G (*Distemonanthus*, etc.)
Nerves not closely parallel...	

Lflets with irregular, v. prominent venation above	
Lflets ovate with acute apex (bipinnate, often armed)	38A (*Cylicodiscus*)
Lflets with thickened margin or glands	37D (*Afzelia*, etc.)
Lvs v. clustered; none of above two	34B (*Entandrophragma* spp.)
Venation not unusually prominent above...	

D3. Key based on other characteristics of compound leaves
If your leaf(-let) has none of the above features, continue with the following key...

Lvs imparipinnate – usually with leaflets (at least one of them) not in pairs	
Lflets symmetrical, with regular margin (or v. strongly alternate)	
Lflets markedly alternate, <10 cm long; usually broadly elliptic or ovate	37G
Lflets sub-opposite or longer or not this shape	
Usually oblanceolate; lflets precisely paired except for a longer terminal one; rachis without apical point; slash NOT gritty; (normally with latex) soft-fibrous, darkening rapidly	34D (*Trichilia* spp.)
Usually with projection at tip of rachis where one lflet falls; lflets often broadly elliptic, with yellow-orange hairs; slash granular and gritty	36A (*Lecaniodiscus,* etc.)
Lflets with irregular margin, or highly asymmetrical	
Lflets much narrower than long, acuminate, v. asymmetric; lflets papery or tree small and crooked; slash cedar-scented	34C (*Guarea/Trichilia prieuriana*)
Tree with 'pock-marks'; lflets rounded, asymmetric with leather-like venation; 'corned beef' slash	35B (*Lannea*)
Lvs paripinnate – lflets in pairs, even if sometimes not perfectly opposite.	
Lflets in 2-4 pairs, sometimes glandular, and <4 cm wide; sl. rhombic, glossy, with one almost straight edge tending to lie parallel to rachis	37C (*Hymenostegia*, etc.)
Lflets larger, or evenly elliptic or >4 pairs	
Lflts asymmetric, utterly glabrous, leathery with sharp, often drip-tipped apex; slash pink to red, bitter, sometimes streaked or cedar-scented; venation distinctive often poorly-defined	34A (*Khaya*)
Not *Khaya*; lvs with a few (inconspicuous) hairs, or lvs broadest well above middle	
Lflts in 2-5 pairs; nerves impressed above or petiolule and rachis hairy; lamina >3 cm wide; often with point at end of rachis	36A
Not *AKYE* and relatives	
Lvs often broadest towards tip: oblanceolate and slightly curved to one side; young lvs red	37D *Berlinia*
Lvs rather rounded or elliptic, and thin, papery, usually glaucous	37E

FLOWERS AND FRUIT OF RUBIACEAE (part 1)

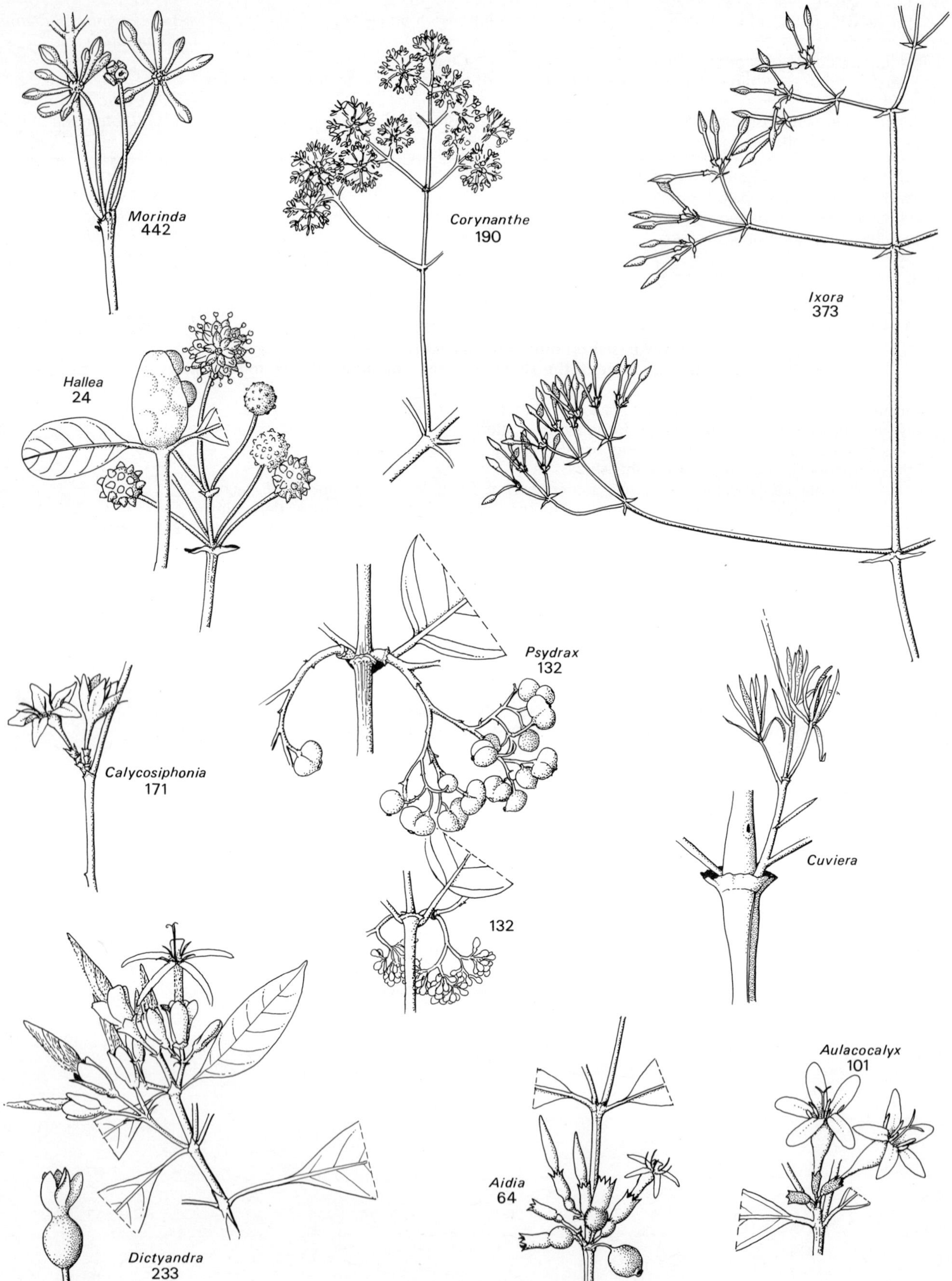

GROUP 1: RUBIACEAE

Rubiaceae is the 'coffee family' and most species are small understorey trees of similar stature to coffee plants. A few species (in Groups 1A, 1B), notably *Nauclea* and *Hallea* (=*Mitragyna*) species, become timber-sized trees with straight, cylindrical boles, narrow crowns and very small (if any) buttresses.

Several species have distinctive branching patterns. Certain *Psydrax* (=*Canthium* – see Group 1A) species branch in a regular way, producing a crown of what appear like very large compound leaves. One leaf regularly falls away from the pair at each node in *Morinda* (1A). Similarly, in *Rothmannia, Schumanniophyton, Gardenia* (sometimes) and *Aulacocalyx*, branches terminate with one leaf of the normal pair missing, or reduced, very close to a pair of leaves at the next node; this can look like three leaves in a whorl. In most species, however, the opposite arrangement of leaves with the characteristic stipules between the (younger) leaves, is obvious. The slash of the trees varies considerably, from those with fibrous bark (1B), to brittle barks, often with orange gritty flecks. In many species (e.g. *Aidia, Aulacocalyx, Morinda, Rothmannia*) there is a yellow exudate with a hot, distinctive smell. The slash of the common tree *Corynanthe* is characteristic: leathery, peelable, and rapidly darkening with a thick brown exudate appearing over the sapwood. Red exudates are recorded sometimes in *Psydrax* and *Tarenna*. Small ants are often found occupying the hollowed twigs of *Psydrax* spp. (1A), *Vangueriopsis* spp. and *Cuviera* spp. (1G).

The flowers have a corolla joined to the axis above the ovary (i.e. the ovary is inferior, except in *Gaertnera*), with stamens attached to the inside of the corolla tube. In *Gardenia* and *Rothmannia* the tubes can be more than 10 cm long, like small trumpets, but in most cases the tube is not so obvious. Flowers are mainly white, except in a few species like *Ixora* (species of which are common in gardens, with their scarlet flowers). The fruits are very often crowned by the remains of the calyx, superficially like a guava fruit, but there is a wide range of fruit sizes.

For Key to subgroups see next page.

Genus	Group	Notes on flowers and fruits	
A) Small flowers and fruits in globose heads			
Morinda	1A	Calyx tubes not all fused together	Ripe fruits rather dry; infls. opposite single leaves
Hallea	1B	Calyx tubes not all fused together	Ripe fruits splitting to release winged seeds
Nauclea	1B	Calyx tubes all fused into a single mass	Ripe fruits fleshy (eaten by elephants)
B) Long, narrow, terminal inflorescence			
Bertiera	1C	Long, thick, terminal, drooping infl. with short branches	Fts <1 cm wide, ovoid
C) Small flowers (<2 cm long) in much branched terminal inflorescences			
Corynanthe	1D	Flwrs + small (1 mm) pockets on lobes	Fts 1 cm long, thin capsules splitting between internal walls
Pausinystalia	1D	Flwrs + 1 cm long thread-like lobes	Fts 1 cm long, thin capsules splitting along internal walls
Ixora	1D	Infls + 5 mm linear bracts; corolla tube 1-2 cm, red	Fts ½ cm red, fleshy
Gaertnera	1E	Infls + 5 mm linear bracts; corolla tube <1 cm long	Fts ½ cm red, fleshy
Psychotria	(1E)	Corolla tube < 5 mm	Fts ½ cm red, fleshy
D) Small flowers in v. short-branched axillary inflorescences, or in fascicles, 2 per node; fruits globose, <1.5 cm			
Belonophora	1D	Corolla tube *c.*1 cm long, enclosing the stamens	
Calycosiphonia	1E	Corolla tube *c.*1 cm long, wide-mouthed, glabrous, with stamens protruding (i.e. 'exserted') at end	
Coffea	1E	Corolla tube *c.*1 cm long, wide-mouthed, glabrous, with stamens protruding (i.e. 'exserted') at end	
Tricalysia	1D,1E	Corolla tube with hairs inside, <1 cm long, with exserted stamens	
Craterispermum	1E	Short peduncle strap-like, stamens exserted (corolla lobes not overlapping in bud)	
E) Small-medium flowers (<4 cm long) in well-branched axillary inflorescences (often with curvy branches), 2/node; fts <2 cm			
Corolla tube and corolla lobes<1 cm long			
Psydrax	1A	Inconspicuous bracts	Fts usually 2-lobed, 1 cm
Vangueriopsis	1G	Inconspicuous bracts; calyx deeply lobed	Fts globose
Cuviera	1G	Long bracts; 5 mm calyx tube	Fts globose
Corolla tube or lobes >1 cm long, hairy			
Robynsia	1G	Infls + conspicuous linear bracts	Fts globose, 5 mm
Dictyandra	1G	Corolla lobes *c.*1 cm long; calx + large, rounded, overlapping lobes	Fts 1 cm, globose, + lobed calyx

PTO for continuation of flower and fruit notes...

FRUITS AND FLOWERS OF RUBIACEAE (part 2)

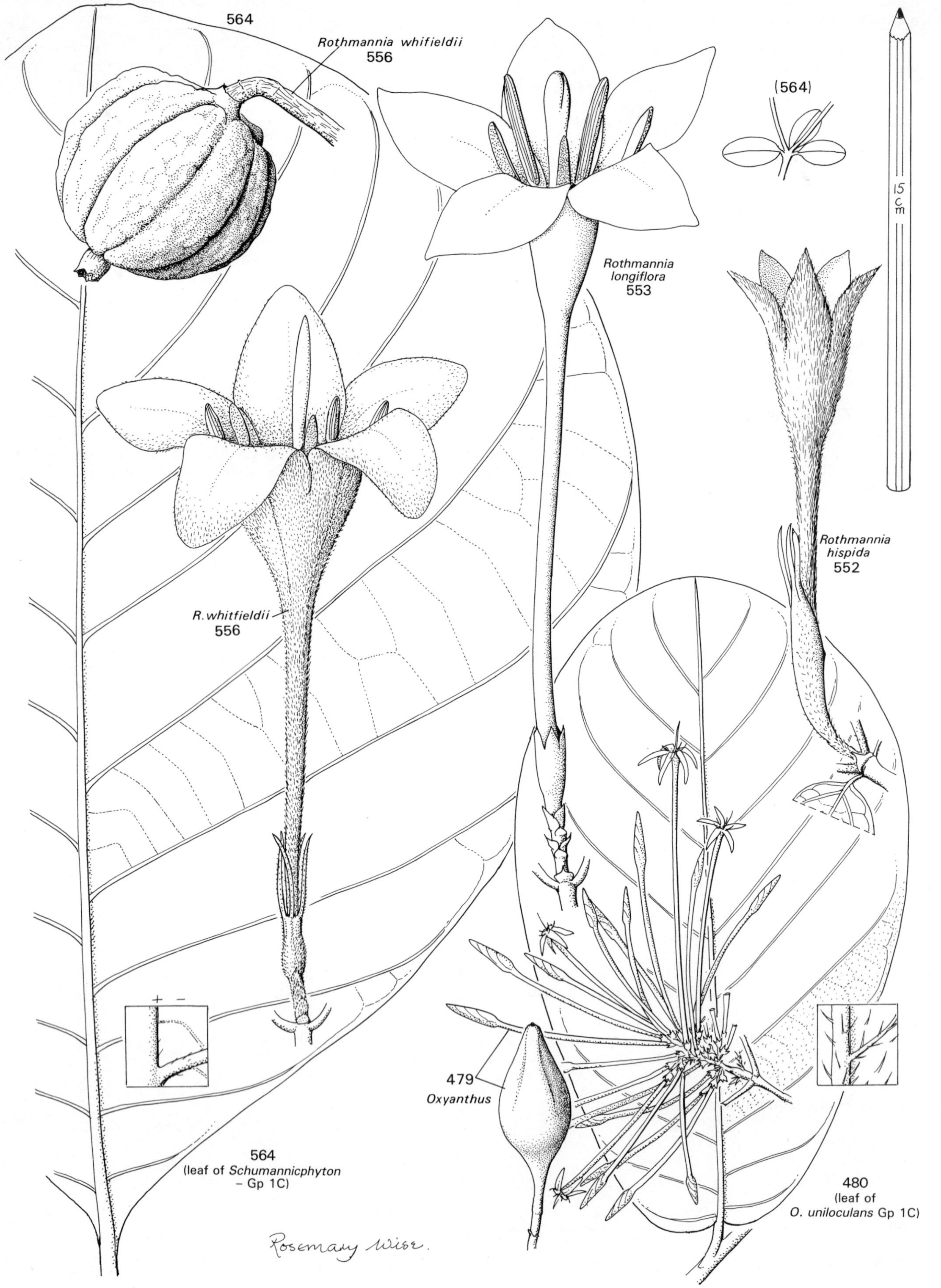

Genus	Group	Notes on flowers and fruits (contd.)	
F) Small-medium flowers (1 cm+ long), inflorescences 1 per node			
Aidia	1E	Infls congested; flwrs hairy in throat +1 cm lobes	Fts globose, 1.5 cm
Aulacocalyx	1E	Infls lax; flwrs v. hairy outside, 2 cm tube + 2 cm lobes	Fts globose, 1.5 cm
Massularia	1F	Few infl branches; flwr bud 1 cm, + v. tubular calyx, pointed	Fts >2 cm wide, ridged
G) Long flowers (>5 cm long); fts woody or leathery, often ridged, several cm long			
Flowers 1(-2) in lf axils			
Gardenia	1C	Calyx tube *c.*1 cm or longer.	
Rothmannia	1C,1F	Calyx tube v. hairy inside; corolla often purple-dappled (fts yielding dye)	
Euclinia	1F	Calyx tube v. short	
Flowers in dense inflorescences			
Oxyanthus	1C,1F	Corolla lobes <7, many flowered	
Schumanniophyton	1C	Corolla lobes 7-12; infls congested, opposed to lvs + large bract	

Key to subgroups of Rubiaceae
(Leaves simple, opposite, entire, with interpetiolar stipules)

Plant not spiny (spines include regular, $\pm$sharp, branch bases found on lower bole)

- Many twigs infested by **ants; crown regularly branched**, in layers like large compound lvs. See *Psydrax* spp. — (Group 1A)
- Twigs not normally ant-infested
  - Lf base cordate AND lf usually about 15 cm long, acuminate — Group 1F[1]
  - Lf base not cordate, OR leaf much more than 15 cm long OR apex not acuminate
    - *Long leaves* (>15 cm long)
      - Leaves with v. few hairs, generally narrowly elliptic, acute to acuminate, with cuneate base. Petiole often >2 cm long; venation often scalariform; **shade-tolerant understorey trees** — Group 1D
      - Leaves rounded at tip OR cordate at base OR v. hairy OR v. asymmetric OR v. large, and not slender OR >20 pairs laterals, OR with domatia away from midrib; **sun-loving (sometimes large) trees typical of secondary or swamp forest**
        - Small trees or shrubs, with >20 laterals OR acute apex; or trees flowering when <10 m tall; or lvs coarsely hairy or irregular, or arranged almost in groups of three — Group 1C
        - Timber trees with straight boles; lf apex rather rounded; gregarious in wet places or in forest canopy (try here first if in doubt) — Group 1B
    - *Short leaves* (<15 cm)
      - Lvs appearing to be arranged in 3s at some nodes (where branches diverge), OR in clusters (pseudo-whorls) of more than 3 at end of brs — Group 1F
      - Lvs not commonly appearing to be arranged in 3s nor in pseudowhorls
        - *Lvs with* domatia in the axils at forks of laterals as well as in nerve axils; sun-loving trees — Group 1A
        - *Lvs without* domatia in forks of lateral nerves; typically understorey trees (but some sun-lovers in this Gp as well) — Group 1E

Plant with spines, either on twigs or main bole (stems often inhabited by **ants**) — Group 1G

NOTE: 1) The uncommon shrub *Mallotus subulatus* (Gp 22A) has opposite, stipulate, cordate lvs, but they are **ovate** and bear stellate hairs.

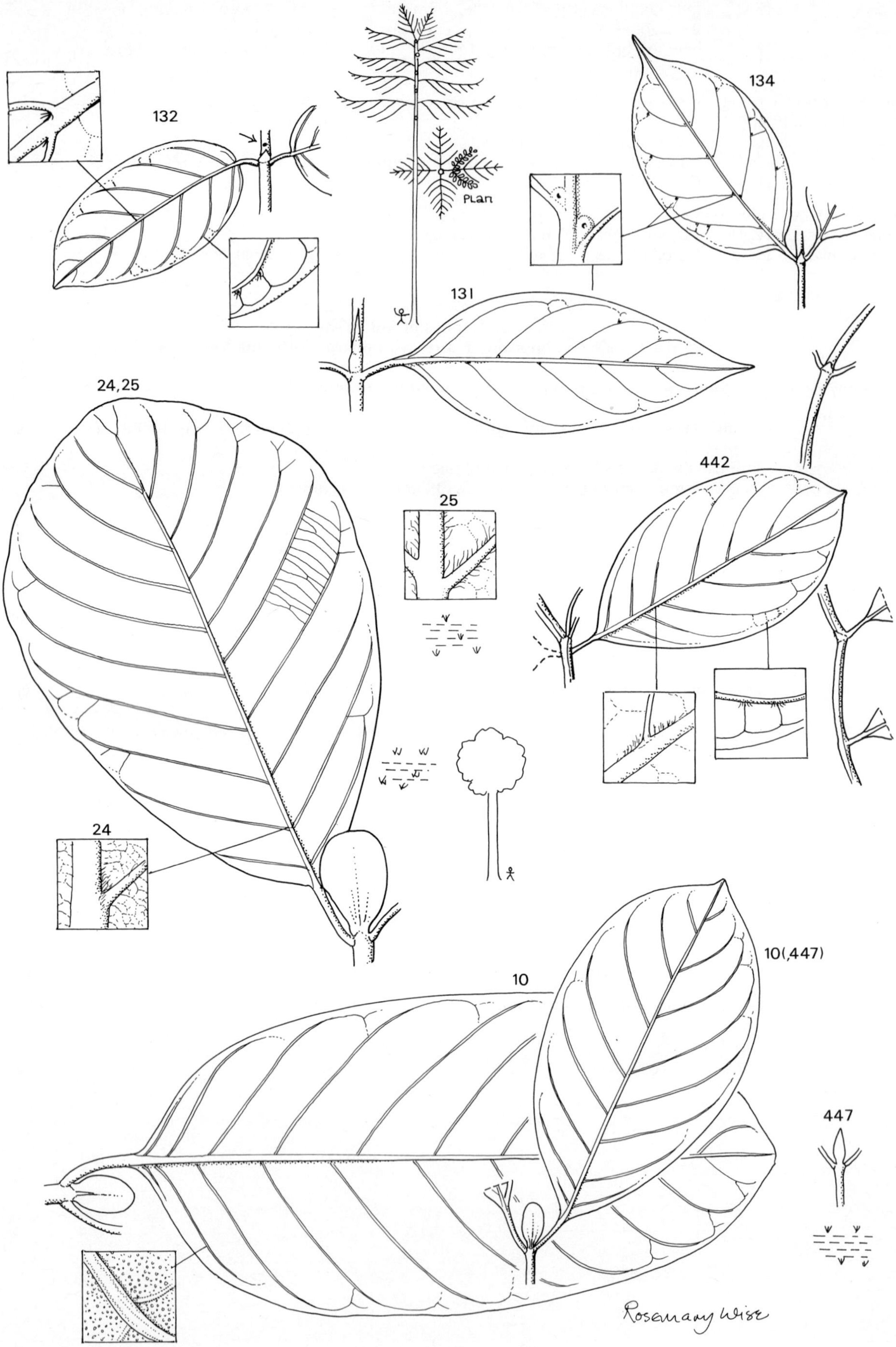
132
134
Plan
131
24,25
25
442
24
10(,447)
10
447
Rosemary Wise

Group 1A: *Psydrax-Morinda*
(Rubiaceae with small lvs with some domatia away from midrib)

The species of this Group are strongly sun-loving trees of **secondary forest** with either regular, or at least domed crowns, and brittle slashes which often have a distinctive, almost paint-like smell.

Domatia merely pits, without hairs; slash brown and white, brittle, slightly scented	***Psydrax*** spp. (see Notes), previously called ***Canthium*** spp.	
Lf base obtuse: widest very close to base, ovate and asymmetric; sometimes slightly decurrent at base, but not along whole petiole; common weedy tree with ± regular crown, but crown not v. 'palm-like'; slash brown with grey spots, darkening through greenish shades, sometimes with reddish exudate	***Psydrax parviflora***[3] [*OGYAPAM-NINI*]	134
Lf base drawn out – cuneate: blade lanceolate and decurrent as fine wing along whole petiole; nerves and midrib often curvy; domatia of nerve axils close to apex of lf v. conspicuous on top surface; younger twigs often flat or slightly winged; uncommon slender tree of **evergreen forest**, sometimes reaching upper canopy	***Psydrax arnoldiana***[2]	131
Domatia with tufts of hair		
Hairs of domatia covering pits; tree with angled stems and symmetric horizontal branches; hollow stems regularly infested with black ants; young internodes with small entrance holes; bark smooth with reddish lenticels; slash creamy with orange gritty streaks, darkening rapidly, brittle, sometimes with red watery exudate; common as small tree in exposed sites; slender bole with raised ring-ridges, and v. regular branches clustered at the top, *like a flat-topped palm, with upturned branch tips*	***Psydrax subcordata***[1] [*oGYAPAM, TETIADUPON*]	132
Hairs arranged as fringes adjacent to and over axils; young stems v. loopy, several-sided with swollen nodes; one lf at each node often falling before the other; medium-sized tree in **secondary forest** or farms etc. with rough fissured and scaly, grey bark; slash dark yellow or brown-orange, darkening and hot-tasting, with a moist, yellowish sapwood, with conspicuous yellow exudate	***Morinda lucida*** [*KONKROMA*]	442

NOTES: 1) *P. arnoldiana* can be distinguished when in flower by its much smaller inflorescences, never reaching more than 3 cm from the twigs. *Psydrax manensis*, a slender tree with horizontal branches, similar to *P. subcordata* has been found in the Atewa range forests; it lacks domatia in the outer axils, but has ovate lvs almost cordate or rounded at base, with stiff, dense hairs, such that it is discolorous below.
2) The seedling of *P. subcordata* has cuneate lf bases.
3) *P. parviflora* was previously *Canthium vulgare*. More information on *Psydrax* can be found in Bridson's paper (*Kew Bulletin*, vol. 40, pp. 687-725 (1985))

Group 1B: *Hallea (Mitragyna) – Nauclea*
(Large[1], broad rather rounded lvs: canopy or swamp trees)

A group of often tall, straight and cylindrical trees with compact, not v. spreading crown of dark lvs; with rough bark and thick slash. *Hallea* spp. and *S. pobeguinii* are found only in **swamps**, whereas *N. diderrichii* is common in **secondary forest**. The seedlings/saplings of *N. diderrichii*, which have much larger leaves than the adults, are often very common along old forest roads, where the soil has been disturbed; the seeds can be dispersed by elephants.

Leaf lower surface without dots; with hairs in vein axils or many long hairs over nerves and veins; margin sometimes slightly undulate; young lvs and stipules reddish, not v. glossy nor fleshy; slash thick, fibro-granular, pink-yellow turning brown; **swamps** etc. (with knee roots); **lvs not glossy;** young lvs red		
Long hairs on nerves + veins, not especially in axils; swamps, esp. of **dry forest** zone; flowers with glabrous calyx	***Hallea stipulosa***[2] [*SUBAHA-AKOA*]	25
Long hairs concentrated in nerve + some vein axils; along rivers and in swamps esp. of **evergreen forest** zone; flowers with calyx hairy around margin	***Hallea ledermannii***[2] [*SUBAHA*]	24
Leaf lower surface covered in v. fine dots, never with long hairs on nerves, nor in axils, but sometimes with short hairs on veins; lvs slightly fleshy and **glossy above**; slash pale yellow-orange, sticky, fibrous		
Small tree of **swamps**, riversides etc.; stipules narrowing from base; young lvs reddish	***Sarcocephalus pobeguinii***[2] [*SUKUSIA*]	447
Large tree, stipules broadest above 1/2 way; young leaves NOT red.	***Nauclea diderrichii*** [*KUSIA*]	10

NOTES: 1) Cultivated Teak – *Tectona grandis* – has opposite, large lvs, but without stipules (see Gp 29).
2) These *Hallea* species were previously in *Mitragyna* (*H. ledermannii* = *M. ciliata*) and *Sarcocephalus pobeguinii* was in *Nauclea*.

Group 1C
(Rubiaceae: small trees with long and broad, cordate or hairy lvs)
(Tree of exposed places, e.g. swamps or roadsides)

Lvs v. large (often >20 cm long), and broad (>15 cm wide) with <20 prs laterals; usually broadest above the middle (obovate or broadly oblanceolate) with a ±rounded apex; base cordate or obtuse; little-branched treelets

Hairs not coarse and bristly all over lower surface; lvs often in 3s

Lvs ±glabrous, or with a few hairs on midrib etc.; very large with wavy margin and obtuse base; **evergreen forest**; habit striking, with 2 or 3 vertical files of horizontal branches; **3 lvs per node on young vertical shoots and "3"** lvs at twig ends like *Rothmannia* spp.; slash brittle yellow with yellow gritty streaks, very bitter — ***Schumanniophyton problematicum***[3] [*KWAETAWA-KESE*] 564

Lvs with hairs in nerve and vein axils; base cordate; finer veins obscure; **evergreen forest** — ***Rothmannia megalostigma*** 554

Hairs coarse and bristly all over lower surface; base cordate often v. asymmetric; slash darkening, yellowish, with watery fruity-acid exudate — ***Oxyanthus unilocularis***[3] [*KWAETAWA*] 480

Lvs long, but not exceptionally broad, usually elliptic and <15 cm wide or with >20 prs laterals; apex acute

Lvs with <20 prs laterals. Veins obscure; long brown hairs on nerves and (4-sided) stem; stipules large and persistent; lf base cuneate to cordate; bark with papery flakes — ***Bertiera racemosa***[2] [*KAKADUA*] 114

Lvs with >20 prs laterals; **evergreen forest**

Base of lf without lobes; midrib at base high and narrow; young stem flattened; stipules large; short hairs flat alongside nerves; tree of disturbed or swampy **evergreen forest** — ***Pauridiantha hirtella*** 491

Base of lf with two small lobes, or pockets[1], or minutely cordate; surface below rough with stiff hairs on laterals; lvs clustered at twig ends and young buds v. sticky; wet places in **evergreen forest** — ***Gardenia imperialis***[1] 337

NOTES:
1) The basal pockets of *Gardenia imperialis* are said to be places where ants maintain farms of scale insects (Steentoft, 1988).
2) ***Ixora brachypoda*** from **fringing forest in dry areas** has similar lvs, but with veins visible above and below; the lvs have few hairs.
3) *S. problematicum* and *O. unilocularis* are illustrated opposite the key to subgroups (page 27).

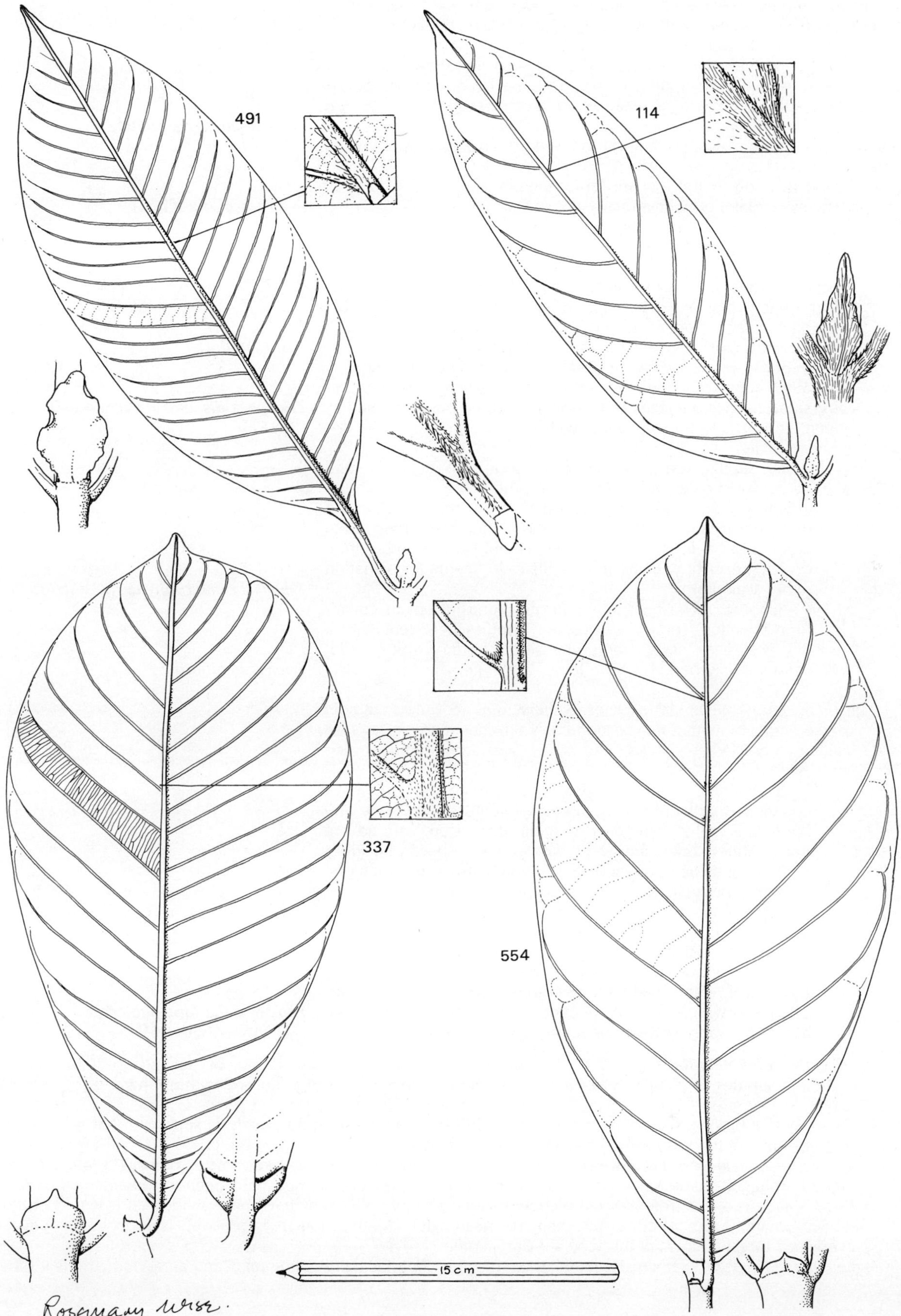
491
114
337
554
15 cm
Rosemary Wise

Group 1D: *Tricalysia-Corynanthe*
(Understorey Rubiaceae with medium-sized to large, slender, acuminate, not v. hairy lvs)

Finer venation (virtually) obscure below

  Tufts of hairs in domatia; petiole long (sometimes >2 cm); lf base usually asymmetric and decurrent along the petiole; lvs drying black; lower surface covered with tiny raised, short dots and dashes (lens); stipules v. broad-based, even attached to base of petiole, with a sharply pointed mid-vein — ***Dictyandra arborescens*** [1] [*see Group 1G*]

  No tuft domatia; petiole shorter; stipules ± 2 cm[5] long linear threads falling soon to leave $\pm$triangular stipules $\pm$glabrous; twigs becoming very corky; dry forest treelet; slash with a green outer layer, orange-brown → red brown — ***Belonophora hypoglauca*** 107

Fine veins distinct, usually almost scalariform (but see first two species below)

  No domatia

    Lf with short hairs on midrib and very short petiole; twigs not v. corky; lf base usually cordate, but occasionally not so — See ***Massularia*** (Gp 1F)

    Not *Massularia*; lvs usually with tiny spots below (lens)

      Venation obviously reticulate, although not particularly prominent or conspicuous; petioles typically *c.*5 mm long, but not well-defined because base of lamina decurrent into petiole wings on top; young twigs often corky or scaly, contrasting with base of petiole and making the latter appear jointed at base; undersurface of lf with hairs or spots (barely visible with lens); **shrub or small tree** — ***Ixora laxiflora*** 373

      Venation not v. visible OR scalariform; lf narrowly elliptic; petiole typically 1 cm or longer; undersurface with metallic glint or discolorous; **small to medium-sized tree**; slash contoured dark grey and green, gritty and crumbly-chunky, with watery exudate — ***Tricalysia discolor***[2] [*KWAE-KOFI BERE*] 608

  Domatia in nerve axils

    Tuft domatia; stipules long, pointed; slash fibrous-spongy, brittle and soft, yellow with paler sapwood, scented and bitter

      Lvs (younger ones) with many short hairs on midrib (x10) and often on lamina also; lf base obtuse, then decurrent; often around 15 cm long; lf apex merely acuminate; lvs rather thin, drying greenish; with *1 cm* thread-like tip to stipules; venation rather scalariform — ***Tricalysia macrophylla***[2] [*KWAE-KOFI*] 610

      Lvs $\pm$glabrous; lf base (evenly) cuneate; 'rungs' (steps in scalariform venation) sometimes obscure; stipules persistent with a *3-5 mm sharp point*; abrupt transition from young to old, yellow, corky bark; lvs often long acuminate — ***Tricalysia elliotii*** 609

    Pit or pocket domatia; slash soft or leathery, pink to yellow, rapidly darkening to brown[3]; finer venation finely prominent above; stipules falling early leaving a ring around nodes; slash taste quinine-bitter; new lvs red

      Midrib channel v. fine, raised with a fine central groove or channelled with thick edge; *petiole often long (>2 cm)* and finely channelled; bole often irregular or twisted, with adventitious shoots often arising nr base; flwrs with corolla lobes with broad apex; fts as capsules splitting half-way between one wall and the next, so that each opened section has two parts; very common and widespread tree — ***Corynanthe pachyceras*** [*PAMPRANA*] 190

      Midrib channel broad and shallow (>2 mm wide nr petiole); petiole not reaching 3 cm (on mature lvs at least); slash soft, pink-yellow turning brown, with smell of apples; bole cylindrical, with pitted bark; **evergreen forest**; flwrs with corolla lobes with long, thin appendages; capsules splitting open along the walls between the loculi (compartments) — ***Pausinystalia lane-poolei***[4] [*PAMPRANA-NUA*] 493

NOTES:
1) ***Dictyandra involucrata*** is similar to *D. arborescens* but lacks the distinct bumps of that species. It has a long petiole and large stipules (+ similar 'involucre' at the base of the infl.), and is known within Ghana only from the Volta region.
2) *KWAE-KOFI* ('bush coffee') is normally applied indiscriminantly to these *Tricalysia* spp., and that is by a good tree-spotter. Many of the Rubiaceae are often given this name.
3) The lvs of ***C. canephora*** (cultivated coffee), with pocket/pit domatia will key to this point (190/493); it sometimes escapes from cultivation. It has tiny shiny, raised bumps on the discolorous, pale green lower surface, like *T. discolor.*
4) *Pausinystalia* is a rare tree from the **wettest forests**. So few collections have been made that it would be unwise to rely on anything but a fertile specimen for meaningful identification. It is probably safe to assume that outside **evergreen forest** all trees of this type are *Corynanthe.*
5) *Sericanthe toupetou* (See Gp 1E) may key to this point but has hairy young parts and an asymmetric leaf base.

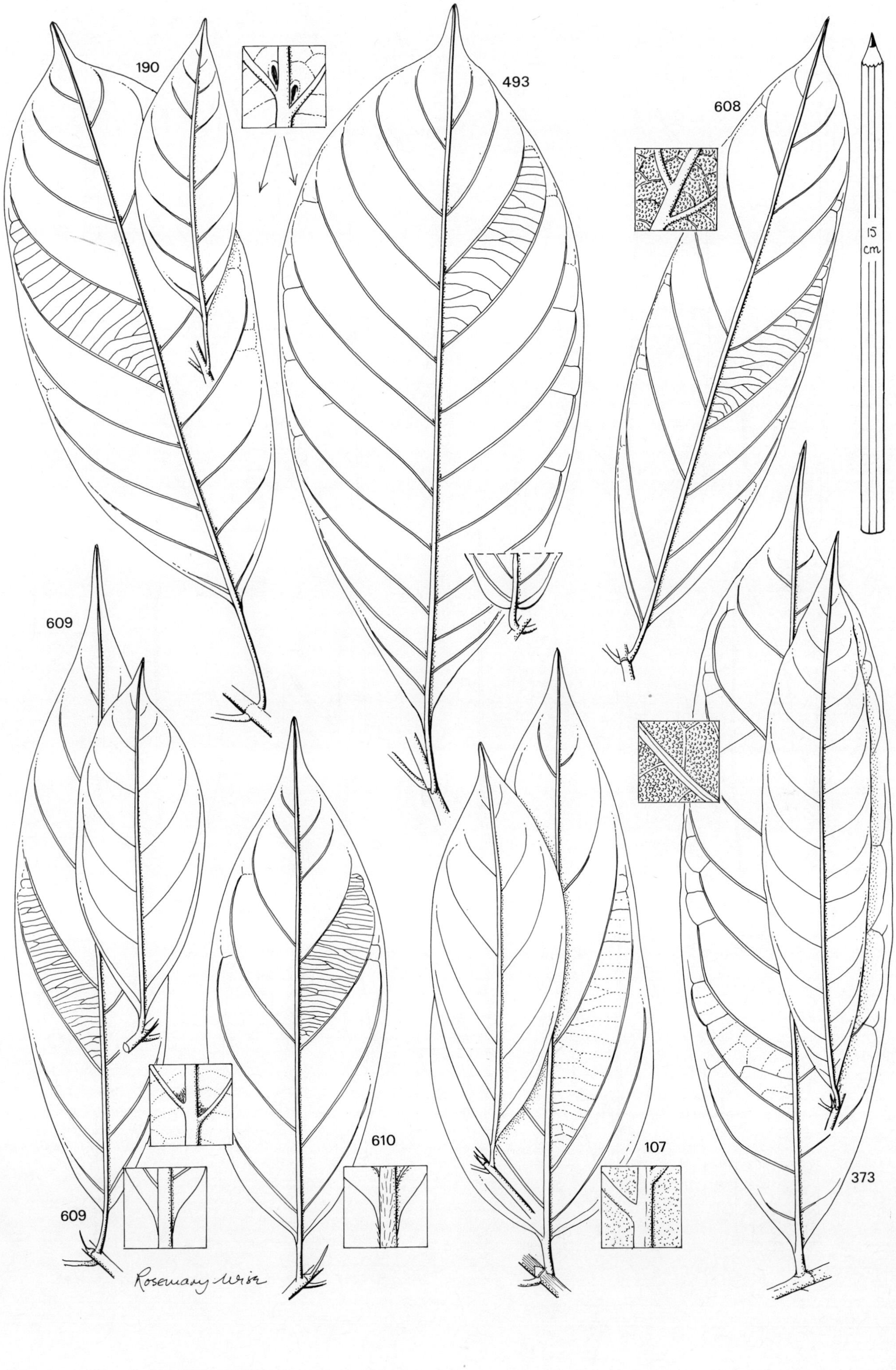
190
493
608
15 cm
609
609
610
107
373
Rosemary Wise

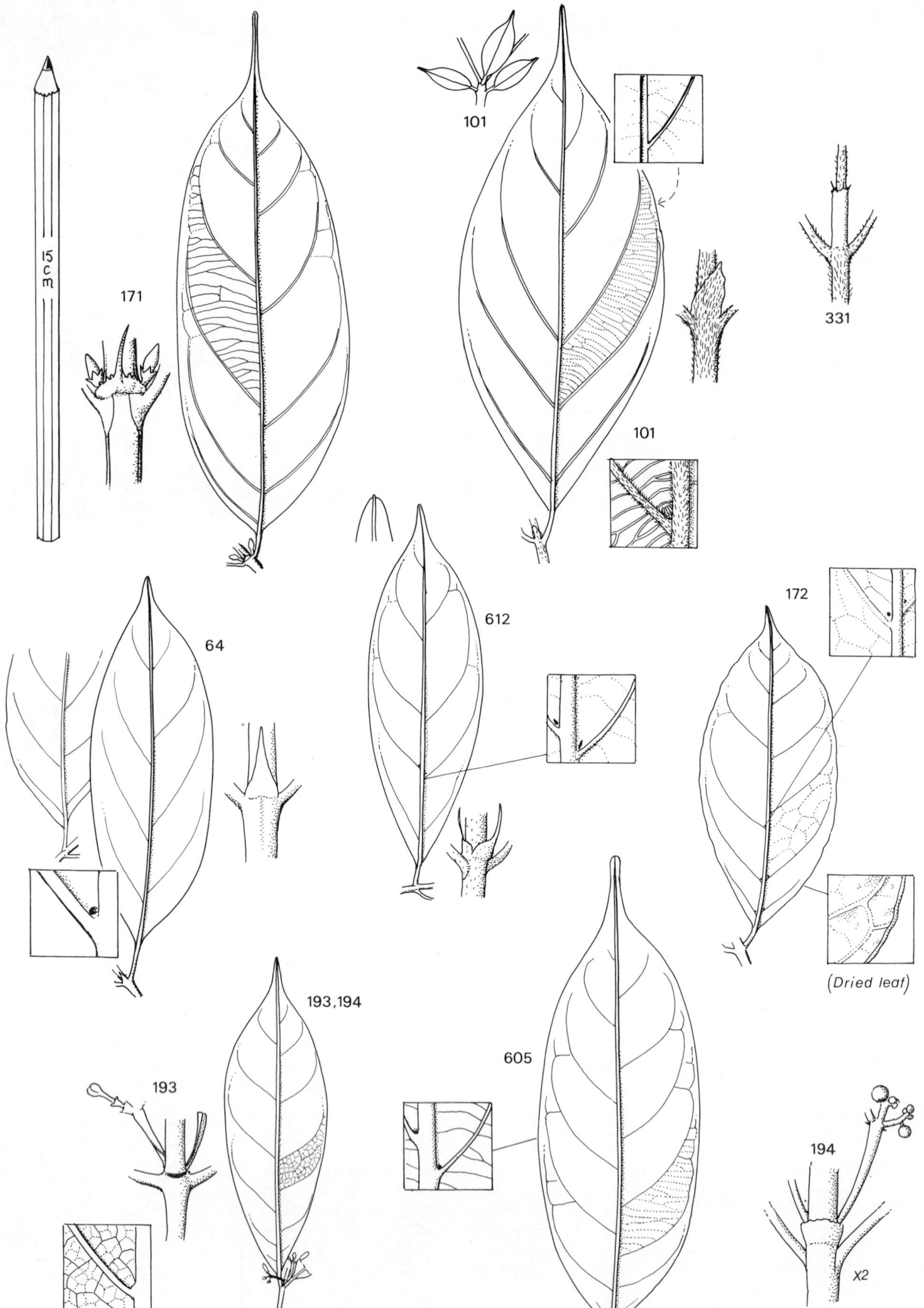
15 cm
171
101
331
101
64
612
172
(Dried leaf)
193,194
605
193
194
X2
Rosemary Wise

Group 1E
(Rubiaceae with small lanceolate-elliptic leaves)

This Group is made up of shrubs and small trees. Many other, rarer or smaller species will key to this point, but of these only some, particularly riverine species are indicated in the footnotes.

Lvs broadly elliptic and acuminate, with markedly ascending nerves and with the finer veins[1] parallel and transverse; laterals often reaching from *c.*½ way to near apex of leaf; lvs with yellow hairs		
Leaf with v. marked drip-tip; often with pit domatia; nerves prominent above; lvs drying black; twigs reddish, slightly winged; stipules broad based with a sharp apex; small tree common in **dry forest**	***Calycosiphonia spathicalyx***[2]	171
Leaf without pronounced drip-tip.		
Stipules arranged as a tube 1 cm or more long, with teeth around edge; yng lvs emerging from this tube, with white hairs on nerves	***Gaertnera paniculata***	331
Stipules not arranged as a tube; laterals and midrib impressed above, young stems + silky dense orange-brown hairs; tree with *Rothmannia*-like branching (lvs often in 3s, see 1F); slash brittle, gritty, orange-cream with yellow gritty streaks; bitter tasting and with distinctive bitter-sweet scent	***Aulacocalyx jasminiflora*** [*NTWESON*]	101
Lvs without this distinctive type of venation		
Pit domatia present (inspect several leaves)		
Finer venation mostly or entirely obscure, and not prominent above		
Pits conspicuous in most axils of lateral nerves; lf thin, plastic-like ±asymmetric or irregular; slash contoured yellow or orange with pink or brown, granular and gritty, with yellow sapwood darkening with exudate, and with distinctive bitter-sweet taste and smell, like that of *Aulacocalyx*	***Aidia genipiflora*** [*OTWENSONO*]	64
Not *Aidia*, e.g. because pits in only a few axils		
Apex of acuminate leaf appearing rounded at arm's length, but with a fine apical point when viewed more closely; twigs pale yellow, rather corky, with dense fine hairs when young; stipule points thread-like; lvs elliptic-oblanceolate, drying brown, and paler below	***Tricalysia pallens***[3] [*TUROMDUA*]	612
Apex sharply acuminate, with a tiny apical point (climber or rare treelet)	***Pauridiantha sylvicola***	492
Finer venation visible and slightly prominent above		
Lvs without obvious black spots when held up to light		
Veins mainly transverse; lvs oblong-elliptic; domatia in most axils of the 5-6 laterals, papery, with drip-tip slightly rounded at apex; bark peelable in strips	***Tricalysia biafrana***[4] [*DAKONUA*]	605
Venation not transverse		
Petiole < 1 cm long, margin ±undulate; dried lamina creased near midrib; domatia in lower axils; twigs + fine hairs; dry forest shrub	***Coffea togoensis***	172
Petiole (often) >1 cm long; lf narrowly elliptic, acuminate; tree to 10 m tall with pale bark and darkening slash (reddish exudate)	***Tarenna laurentii***[5]	
Lvs with obvious black spots when held up to light, acuminate	See ***Pavatta (corymbosa)***	Note 3, Gp 1F
Pit domatia always absent		
Tuft domatia always absent; lf base cuneate		
Veins visible; lvs ±glabrous lvs symmetric and narrowly elliptic; twigs smooth ±orange; margin sharply recurved; fine reticulate venation prominent above and below (but often not from a distance); lvs dry yellowish or light green; bark flaky; slash yellow to orange, contoured, ±scented, thick		
Stipules not persistent; lvs drying grey-green; inflorescence with peduncle <1 cm long	***Craterispermum caudatum*** [*DUADE*]	193
Stipules slightly persistent, like a collar with small lobes; lvs drying yellow-green; basal peduncle >1 cm long	***Craterispermum cerinanthum***[6] [*AFRA-NI-AFEI*]	194
Veins ±obscure ±hairy[7]; lf base slightly asymmetric; lf elliptic; stipules, petioles (broad-based), midrib etc evenly covered in flat, brownish hairs; medium-sized tree with flaky or corky outer bark, with thin, pale orange, brittle slash	***Sericanthe toupetou***	567
Tufts of hair in domatia, or lvs clearly asymmetric or cordate at base; SEE GROUP 1F		

NOTES:
1) **Gardenia nitida** is a small shrub along **streams in drier areas**, with pale twigs forking often in threes. *G. vogelii* is similar, but neither really qualify as trees. Their lvs are sometimes whorled.
2) *C. spathicalyx* was previously known as *Coffea spathicalyx.*
3) ***Psychotria guineensis*** sometimes attains 5 cm dbh, unlike most other species in this genus. It has elliptic-obovate lvs with **±rounded apices**, pustules below and ±tuft domatia. It is rare, apparently typical of rivers and has a crumbly slash darkening with a scented yellow exudate.
4) Other *Tricalysia* spp. are distinctly rare or small. *T. coriacea* is a leathery-leaved swamp/riverine species. *T. reticulata* is a small **evergreen forest** shrub with papery lvs like *T. macrophylla*, but with densely reticulate raised venation.
5) Three ±shrubby *Tarenna* spp. are recorded in Ghana (see index). Their lvs and the spine of the stipules dry from dark green to black, and the lvs are narrowly elliptic. At least one becomes a medium-sized tree, but collections are too few to make the key more precise at this point.
6) ***C. laurinum*** is a third species of this genus very similar to *C. cerinanthum*, with veins more densely reticulate on more leathery leaves, with a stouter peduncle; calyx with no teeth and fruits ±sessile (where they are clearly stalked in *C. cerinanthum*). It is a species of **riverine forest in drier areas**.
7) *S. toupetou* lvs often dry yellow and vary in size. ***Cremaspora triflora*** is a variable climbing shrub or treelet which will key here. The lf is usually <15 cm long; lf base is obtuse, or cordate at base of the side branches. The basal parts of the side branches are characteristically reflexed and thickened to assist in climbing.

Group 1F: *Rothmannia,* etc.
(Three lvs around nodes or lvs cordate, asymmetric or clustered)

- Leaves in strong tufts or rosettes at the end of twigs, with the rosettes arranged in layers like a *Terminalia*; lvs thin, papery, with veins not particularly clear; the clusters of lvs surrounded by a ring of chaffy stipules; the lvs often with tuft domatia; drying black; petioles and lamina with raised spots (lens); small tree — ***Euclinia longiflora***[1] [*oGYANEYA*] 289
- Not *Euclinia* (lvs not strongly in rosettes, but sometimes (next few spp.) in flat layers
  - 1) Lvs arranged in 3s at many nodes; lf base cuneate and symmetrical; often flat-topped treelets
    - Nerves ascending, with obviously transverse veins (see *Aulacocalyx* (1E))
    - Nerves and finer venation not like this
      - Hairs long and conspicuous, on papery lvs; lvs drying blackish
        - Hairs all over lamina and twigs; slash pale yellow, very gritty, darkening with red to brown exudate — ***Rothmannia hispida*** [*TUKOBO*] 552
        - Hairs conspicuously aggregated in nerve axils, with some on midrib and on venation as well; hairs on yng twigs dense; **dry forest** tree with rough, flaky bark; slash brittle, scented, brown to yellow-contoured, darkening slightly with yellow to brown exudate — ***Rothmannia urcelliformis*** [*TUKOBO-BERE*] 555
      - Hairs short or absent; lvs thicker than paper
        - Lvs with tuft domatia (see last species)
        - Lvs without tuft domatia; lower surface with minute spots barely visible with lens
          - Lf with pit domatia; 4-6 prs laterals; slash soft, yellow-brown with gritty lines, darkening with brownish exudate — ***Rothmannia longiflora*** [*SAMANKUBE*] 553
          - Lf without domatia; 7-10 prs laterals; yng lvs sometimes with very dense short hairs beside midrib, disappearing rapidly with age; slash cream, brittle, with brown gritty lines — ***Rothmannia whitfieldii*** [*SABOBE*] 556
  - 2) Lvs asymmetric or cordate at base; not grouped into 3s at nodes
    - **Lf base usually cordate**, but symmetrical and lf acuminate
      - Lvs glossy, not v. papery, with a few hairs on nerves, etc.; oblanceolate; petiole v. short; slash yellow turning brown, gritty, with darker gritty spots; sapwood hard and creamy; small treelet — ***Massularia acuminata***[4] [*POBE*] 420
      - Lvs with obvious long hairs, rather papery and not glossy.[2]
        - Lvs with coarse, stiff hairs, even on the **prominent** nerves, midrib, etc. of the upper surface; stipules at end of twigs 1.5–2 cm long; medium-sized tree of **hilly or dry areas**; slash with a red and green speckled outer layer, orange-brown, contoured, darkening — (See Gp 1G) ***Robynsia glabrata*** [*oGYAPAM-NWI*] 551
        - Lvs with rather soft hairs in **midrib channel** and elsewhere above, and on midrib below; lvs with conspicuous black spots (bacterial[3] nodules; hold up to light and use lens); small tree (<10 cm dbh) — ***Sericanthe chevalieri***
    - **Lf base either cuneate** or, if cordate on one side, then normally asymmetric[3]
      - Veins obscure; domatia often absent; common shrubs
        - Shrub <4m tall; one side of lf obtuse to cordate, and the other side cuneate; stipules >1cm long — ***Oxyanthus formosus*** [*KORANTEMA-NINI*]
        - Small tree often >4m; lvs not so asymmetric, ±cuneate, sometimes with domatia; stipules <1 cm long; lvs silvery-green below; bud glutinous; apex ±apiculate — ***Oxyanthus pallidus***
      - Veins clearly visible; base of some lvs obtuse-asymmetric; medium-sized tree, but occasionally fertile as shrub; tuft domatia usually present; stipules large and persistent; slash brittle, contoured, yellowish turning brown with slightly pungent smell; often fluted or twisted — ***Oxyanthus speciosus***[4] [*KORANTEMA*] 479
  - 3) Leaves not as above (flowers and fruits small) see next Group

NOTES:
1) Small shrubs of the genus *Pavetta* also, typically, have v. clustered lvs.
2) Small, v. thin, ovate-acuminate lvs with obscure venation and tuft domatia – see **Rytigynia umbellata**
3) ***Pavetta corymbosa*** is a small, pale-stemmed tree or bush which has lvs with a slightly asymmetric base (albeit variable) and tuft domatia. There are distinct dark spots (bacterial nodules) in the leaves. The species is common in thickets in savanna, and found also in secondary forest. There are five other species of *Pavetta* in Ghana, which are either climbing shrubs, or at least very rarely encountered as forest trees.
4) ***Massularia*** has very variable leaves, but is usually a small enough tree or shrub for the typical leaves to be obtained. The presence of tuft domatia on younger leaves of *O. speciosus* normally segragates this from *Massularia*.

289
552
556
555
l.s.
555
553
552
15 cm
420
479

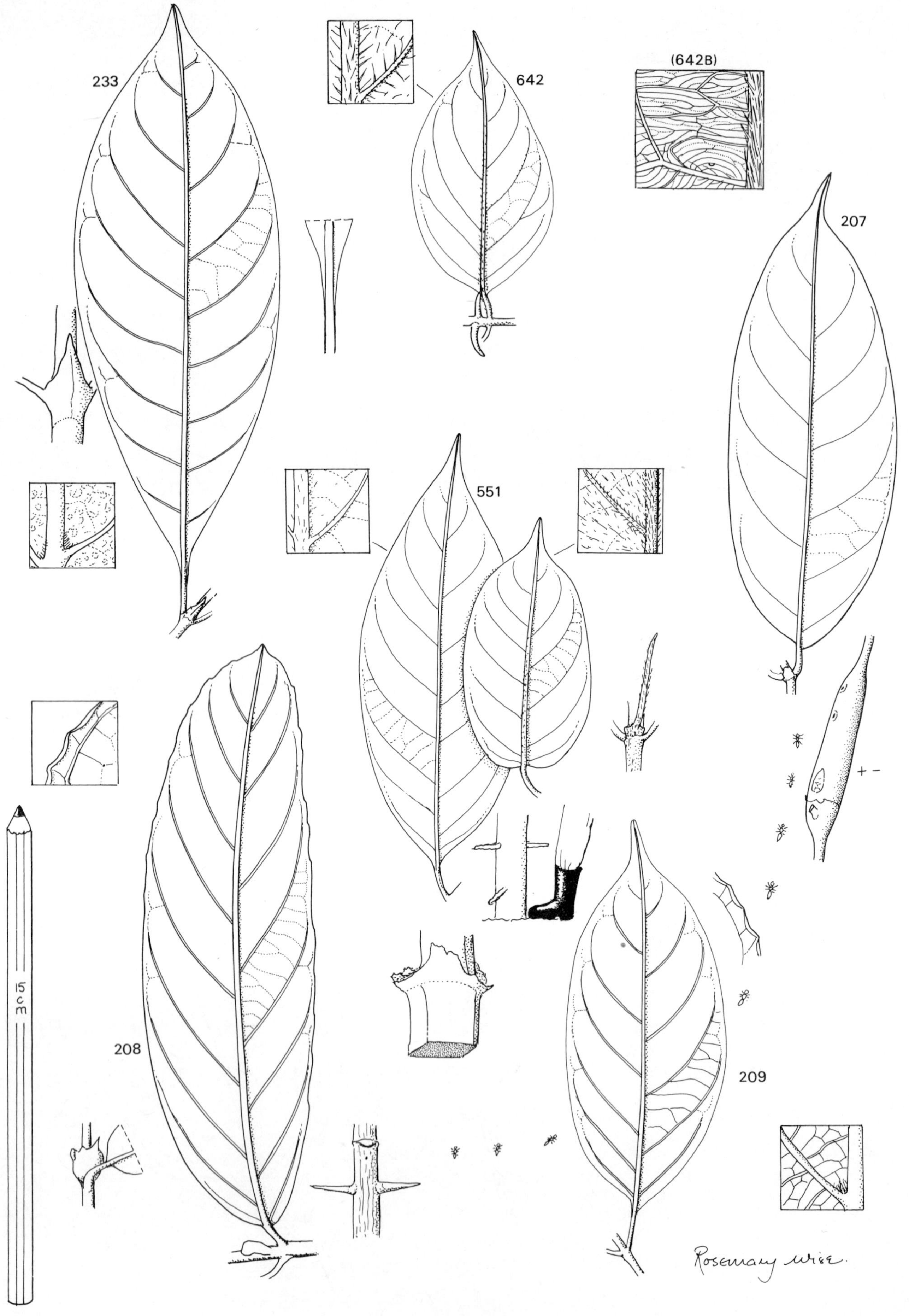
233
642
(642B)
207
551
208
209
15 cm
Rosemary Wise.

Group 1G *Cuviera*, etc.
(Spiny Rubiaceae)

In *Dictyandra* and *Robynsia* particularly, the spines are often missed as they are not usually on the twigs and other smaller branches, but are regular, stubby, short, usually sharp branch-like outgrowths of the bole below leafy branches.

Stipules not inflated nor distorted. Lvs hairy, at least in the axils of the nerves, or venation obscure		
Leaf ±papery, with long brownish hairs on both surfaces		
Lf base ±cordate: spines on older branches or trunk; stipules >1 cm long; young lvs with v. dense long hairs, disappearing with age; twigs with widened, 4-sided nodes ±remains of stipules; medium-sized tree with gritty slash, darkening with exudate	***Robynsia glabrata***[3] [*oGYAPAM-NWI*]	551
Lf base not cordate		
Spines recurved, above lvs even on twigs; bark on twigs becoming creamy and corky; **v. dry forest** or **savanna**	***Vangueriopsis spinosa***	642
Spines not recurved on young twigs; twigs with long hairs; lvs with v. distinctive finer venation (lens), with many fine fingerprint-like raised veinlets like contours of a hilly area	***Vangueriopsis vanguerioides***[1]	(642B)
Leaf without long hairs on both surfaces, but sometimes hairy in axils of nerves and veins; lvs often black when dry		
Veins ±obscure, and at least not prominent		
Leaf long, acuminate, with raised dots and dashes on lower surface (lens); stipules large and persistent, like bishop's hat; slash with corky brown outer bark over pale orange-brown, fibro-granular and gritty, with cream sapwood; older bole with spines and often with small adventitious shoots	***Dictyandra arborescens*** [*KWAKUO-ASENABA*]	233
Leaf oblong-elliptic, often >15 cm long, ±leathery; without raised dots and dashes; with laterals and midrib recessed above; veins not very visible; sometimes with swollen node occupied by ants; lf base slightly asymmetric; prefers **riverbanks**	***Cuviera acutiflora***[3] [*KWAKUO-ASRA*]	207
Veins ±visible; lvs thin papery, <15 cm long and acuminate; + tuft domatia; oblong elliptic; lf base asymmetric; laterals and midrib recessed above; with tiny tooth-like indentations near apex due to margin being minutely folded up; young twigs drying black but older twigs yellowish and corky; slash cream, with orange grit	***Cuviera nigrescens*** [*KOTO-BOWERE*]	209
Stipules persistent, becoming curiously distorted like ears, sometimes with developing inflorescence beneath, and regularly infested with ants; venation reticulate, visible; spines on mature stems ½ way between stipules; margin irregularly recurved, appearing sinuous; fine 'false teeth' nr apex (see last species); lamina thick; oblong	***Cuviera macroura***[3]	208

NOTES: 1) *V. vanguerioides* is normally a scrambler, but records of 10 m trees have been made in Ghana, although the identification of those records needs checking. The distinctive venation is found also in *V. nigerica*, typical of fringing forest in savanna.

2) *Psydrax (=Canthium) orthacanthum* is a spiny shrub in **evergreen forest** (especially **upland evergreen** – e.g. Atewa and Tano Ofin) with a ciliate margin and very recurved spines at nodes.

3) A fourth species of *Cuviera, Cuviera subuliflora* occurs in, but has been little-collected in Ghana. In Nigeria it has very large leaves with few hairs, ant holes in the twigs, and deeply cordate ±subsessile leaves.

See also Anthocleista species (Gp 2), which have v. long leaves, and no interpetiolar stipules.

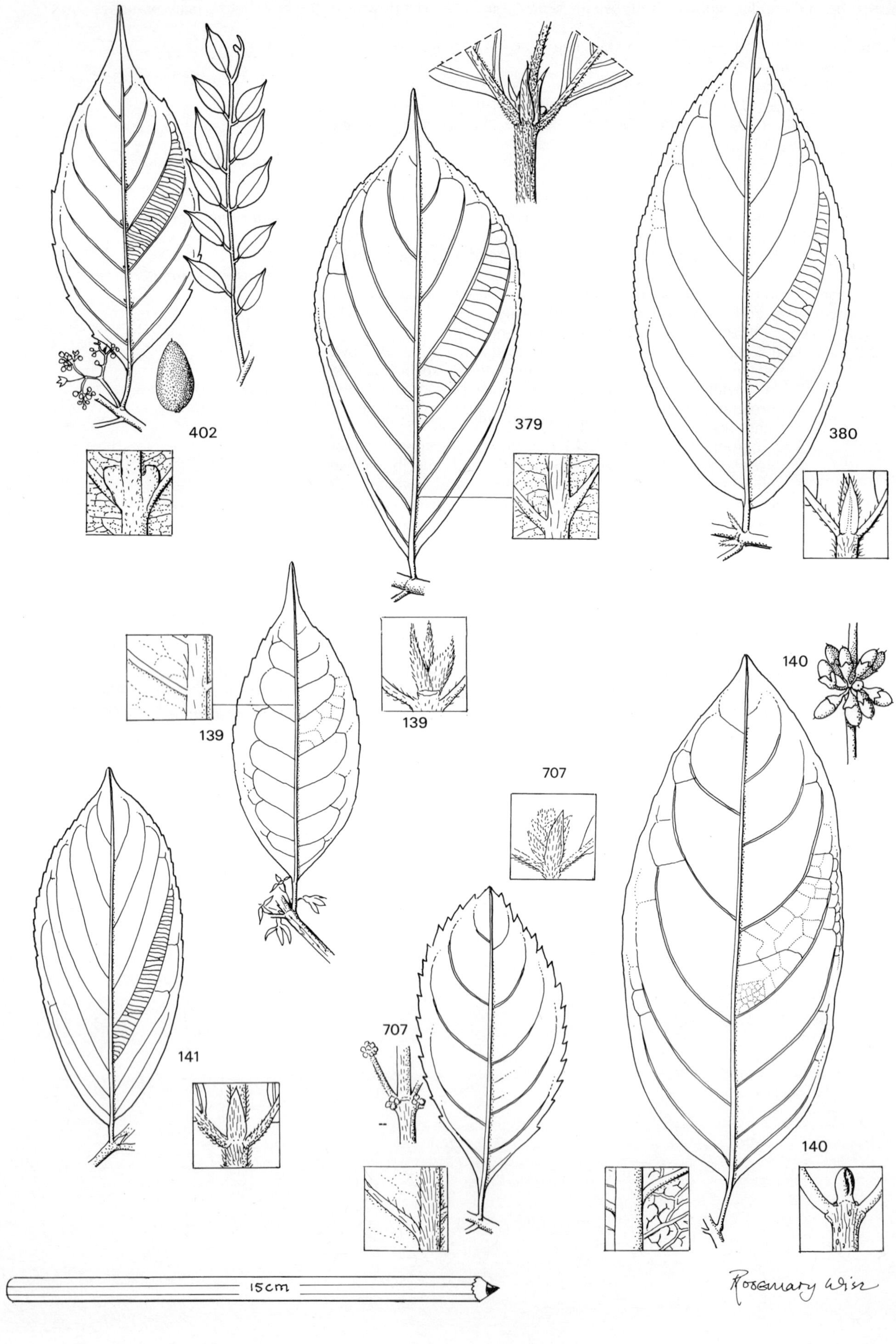
402
379
380
139
139
140
707
707
141
140
15cm
Rosemary Wise

GROUP 2: RHAMNACEAE
(Leaves opposite, serrated; ±interpetiolar stipules; slash reddish)

Rhamnaceae trees have rather reddish, fibrous, scented slashes. The flowers are clustered or on branched infls in the lf axils. *Maesopsis* fruits are 2 cm long, black, fleshy with a hard stone, whereas *Lasiodiscus* fts are ±3-lobed capsules.

Petiole >3 cm long; lvs ovate		
Lf on one side of node usually larger than opposite lf; lvs with scales.	See ***Mallotus*** (Gp 22)	
Lvs in whorls, all ±the same size, without scales	See ***Premna angolensis*** (Gp5)	
Petiole short or lvs not even nearly ovate		
Venation obscure; margin irregular; tree with reddish exudate	***Tetrorchidium*** (Group 13)	
Venation visible; not *Tetrorchidium*		
Venation reticulate; hairs, when present, yellow-orange; slash yellowish and gritty, or mangrove trees	See Group 3	
Venation ±scalariform (or at least with many prominent parallel nerves running between laterals); (hairs, when present, rusty red-brown in colour and slash red except last species)		
Domatia in nerve axils (tufts or glands); lvs sometimes alternate, and sometimes with spiny stipules; straight, cylindrical, unbuttressed tree with light crown with some horizontal boughs; slash thick, fibrous, red with white lines with yellow inner bark, sweet-savoury smell; **disturbed forest**; young branches with paired sharp spines	***Maesopsis eminii*** [*oNMWAMDUA*]	402
No domatia		
Leaf with base cordate or obtuse, at least on one side, ±asymmetric; serrations pronounced; venation sometimes not v. scalariform; stipules like grain husks; lf and stem with long brown hairs; small, twisted tree often in **riverine forest**; slash red, brittle-fibrous, contoured, slight sweet scent, not bitter	***Lasiodiscus mannii*** [*ADAFA*]	380
Leaf with cuneate base		
Stipules not v. conspicuous; twigs, midrib etc. with a few short, velvety brown hairs, or glabrous; small ±slanting tree of drier forests, (especially on hills?) slash reddish, scented distinctively like camphor 'VIC' or 'ROB'.	***Lasiodiscus fasciculiflorus***	379
Stipules *c.*1 cm or longer, ±persistent; **evergreen forest** tree slash gritty, not red	***Cassipourea hiotou*** (see next Gp.)	141

GROUP 3: RHIZOPHORACEAE (ETC.)
(Lvs opposite, stipulate, serrated or whorled, or mangrove trees)

The Rhizophoraceae include *Rhizophora* in mangrove swamps, and *Cassipourea* spp. and *Anopyxis* in normal forests. The forest species at least have a granular slash with large, darker, gritty spots, the whole darkening on exposure. *Cassipourea* flowers are in clusters in the leaf axils, resembling superficially certain Rubiaceae, whereas infls of *Anopyxis* are more branched, on stalks a few cm long from the axils (individual *Anopyxis* flwrs resemble those of Meliaceae). *Anopyxis* fruits are 5-valved ±woody capsules 4 cm long with the calyx persistent at the base. *Cassipourea* fruits are fleshy capsules, with arillate seeds.

Trees in forest, or at least not in mangrove swamps		
Venation scalariform; stipules *c.*1 cm or longer, ±persistent; midrib, twigs, etc. with conspicuous, dense and soft brown hairs; **evergreen forest**	***Cassipourea hiotou***	141
Venation reticulate or lvs trinerved		
Lvs never in whorls: always opposite and ±serrated		
Apical bud not sticky; petiole <1 cm long; twigs with persistent stipules[2], flat and hairy; venation mostly with closed ends, not exceptionally conspicuous		
Moist forest species with dense long hairs on lvs and stems, and esp. on terminal bud	***Cassipourea lescotiana***[1]	707
Dry forest zone streamside species; glabrous or with a few hairs on young shoots	***Cassipourea congoensis***	139
Apical bud sticky; petioles 1-1.5 cm; twigs hairless, v. black + white lenticels; stipules falling rapidly leaving rings at nodes; lvs often >5 cm wide, teeth ±blunt; very glossy, with conspicuous venation, with many 'loose ends'; **evergreen forest**	***C. gummiflua*** [*KoKoTE-NUA*]	140
Lvs in whorls, margin entire and usually recurved		
Lvs trinerved	See Group 5	
Lvs without basal nerves		
Lvs 2 or 4 per node; margin recurved; tall tree with dark green, symmetric crown; bark smooth or scaly; slash thick, granular, pale orange-pink turning rusty brown (esp. on thin edges) + rust-coloured gritty streaks; lvs drying brownish	***Anopyxis klaineana*** [*KoKoTE*] (See page 14)	34
Lvs without stipules; usually 3 per node; margin v. slightly irregular; often apiculate; small fluted tree of **dry forest** understorey; lvs drying pale green; twigs orangeish (mainly in Volta region)	***Nuxia congesta*** (LOGANIACEAE)	456

Trees in mangrove swamps – fringing estuaries, lagoons, etc. nr the sea (See next page)

NOTES: 1) ***Cassipourea afzelii*** is a small, glabrous shrub similar to the other species of its genus; it grows close to the sea.
2) ***Elaeodendron buchananii*** (CELASTRACEAE) will key to this point, but has no stipules: it is a small tree with 1 cm petioles, glabrous lvs and v. lenticellate twigs, lvs with raised venation above, acuminate and drying bluish.

83
83-87
84
903
86
901
83
15 cm
86
87
84
902
15cm
R. Wise

GROUP 3 (contd.)
Trees in mangrove swamps – fringing estuaries, lagoons, etc. nr the sea

Rhizophora fts germinate on the tree, are up to 25 cm long and leathery. *Avicennia* fts are *c.*3 cm long, ovoid and beaked; *Languncularia* fts are *c.*2 cm long, ribbed, obovoid and hairy.

Leaves strongly clustered at the tips of loopy twigs; tree with large prop roots; lf base decurrent; veins ±obscure, with no obvious sub-marginal nerve.	***Rhizophora racemosa*** (Red mangrove) [*AMUTSI*]	903
Leaves not strongly clustered at twig ends; submarginal nerve usually visible (lens)		
Submarginal nerve without glands; lvs slightly white below with v. short felt of hairs (individually indiscernible even with lens); narrowly elliptic; twigs with pithy centre; laterals raised below; tree surrounded by abundant pneumatophores (peg-like breathing roots) ±small prop roots	***Avicennia africana*** (VERBENACEAE) (Olive mangrove) [*ESUKURU*]	901
Submarginal nerve (+ elsewhere) with small, but conspicuous raised glands; base of leaf obtuse or minutely cordate; lf ovate, glabrous; tree without prop roots but with small pneumatophores	***Languncularia racemosa*** (COMBRETACEAE) (White buttonwood) [*ABIN*]	902

GROUP 4: *Anthocleista* spp. (LOGANIACEAE)
(Long, very clustered leaves)

Anthocleista spp. are very characteristic, little-branched trees of **disturbed forest** or **swamp**, particularly in wetter areas. The branching pattern is frangipani-like (Aubreville's model of Halle *et al.* 1978), with clustered leaves which look like narrow, long-leaved (sometimes 1 m or more long) cabbages on stalks, especially when young. Leaves on older branches become much shorter and rounder. Taller trees are often fluted, or have stilt roots or spines. The slash is gritty and brittle, becoming more fibrous towards the sapwood, orange-brown, darkening rapidly – similar to some Rubiaceae (Group 1) (*Gaertnera*, in that family, is sometimes placed in Loganiaceae). The habit of some monocotyledons (Gp 39) is similar, but their venation is very different.

The flowers of *Anthocleista* are borne in terminal panicles, fleshy and tubular but, unlike those of Rubiaceae, the ovary is superior (i.e. with the fruit developing normally above the attachment of the calyx and corolla). The fruits are fleshy and several-seeded, two or more cm in diameter. In these and other ways there is a strong resemblance to the Apocynaceae which, however, produce latex. *Nuxia* (see Gp 3), with whorled lvs, is also in the Loganiaceae.

All the species have very varied leaves with ±obscure finer venation and midribs often forked before reaching the apex. The petiole base is normally considerably widened where it joins the stout twigs.

Plant with spines (2-pronged, at right angles to leaves above them; ±persistent)[1]		
Leaves with leaf blade continued (almost) to stem, i.e. with v. short petiole; often with leafy lobes at the base of the petiole; paired spines below lvs sharp, but usually less so on the stems and bole; prefers wet places	***Anthocleista vogelii*** [*AWORABoNToDEe*]	87
Leaves with distinct petiole		
Spines fused for most of length, with a short apical notch (or spines apparently unpaired); lf apex ±rounded; larger leaves with cordate base; tree of drier areas, particularly in **riverine forest**	***Anthocleista djalonensis*** [*BoNToDEeBERE*]	83
Spines clearly 2-pronged and well-divided; lf apex ±acute to acuminate lf long and slender; large leaves with cuneate or decurrent base; **evergreen forest**, roadsides, etc.	***Anthocleista nobilis*** [*BoNToDEe*]	86
Plant without spines; leaves *narrowly* oblanceolate; lf base decurrent even on large leaves; tree in swamps; (flower buds round)	***Anthocleista liebrechtsiana***	84

NOTES: 1) ***Strychnos spinosa*** (also LOGANIACEAE) has opposite, trinerved, emarginate leaves and sharp spines. It grows in **savanna** and the **driest forest** types. Other trinerved *Strychnos* spp. are lianes in high forest, sometimes found as understorey shrubs.

2) In addition to these there is *A. microphylla* – a straggly tree or epiphyte in upland **evergreen forest**.

388
563
±
521
520
456
34
15 cm
Rosemary Wise

GROUP 5: OLEACEAE, etc.
(Leaves entire; ±domatia, without stipules)

The two genera of Ghanaian trees in the Oleaceae are mentioned here. *Chionanthus* leaves, with their tuft domatia, could easily be mistaken for Rubiaceae, but the lack of interpetiolar stipules precludes that family. *Linociera* flowers (< 1 cm long) are in short inflorescences or clusters in the leaf axils, white, 4-lobed and very fragrant (like jasmine). They are (just) tubular and white (with only 2 stamens). *Schrebera* flowers are larger (up to 2 cm long), more markedly tubular and browner, in few-flowered panicles, but again they have only two stamens. *Linociera* fruits are olive-like, black or bluish drupes 2 cm long. *Schrebera* fruits are distinctive, woody capsules with winged seeds, superficially like some Meliaceae, but slitting into two halves.

The other families are also mentioned elsewhere (Combretaceae – Gp 25; Verbenaceae – Gp 29).

Twigs broadly winged – see Group 7		
Twigs not winged (but see *P. hispida*)		
Leaves with domatia – without dense orange hairs		
Lvs without glands; with pits and tufts of hairs in domatia; petiole corky, yellow, rough; lvs oblong-oblanceolate; laterals meeting; twigs very pale and corky with conspicuous raised lenticels; young twigs hairy; midrib with broad channel above; slash thick and soft, mottled cream and brown, darkening, with slight scent of iodine; *small understorey tree*, usually near rocks or rivers	***Chionanthus africanus***[1] (OLEACEAE)	388
Lvs with 2 or more glands (in upper part); ±tufts but no pit domatia; petiole slender; lvs not perfectly opposite; lvs hairy at first but hairs disappearing rapidly; lvs pale green; venation vague, fine-reticulate or slightly scalariform; slash pinkish brown, granular, turning purple-brown; tree slightly fluted, but with *spreading, graceful crown* like *Terminalia ivorensis*	***Pteleopsis hylodendron*** (COMBRET.) (below)	
Leaves without domatia		
Lvs not trinerved (see above, sometimes with glands).		
Leaves small (< 5 cm long, petiole *c.*1 mm long), typically 2-3 cm long, ovate with a blunt apex; midrib with dense, fine rusty hairs; venation minutely reticulate; lvs often alternate; bark, even on young twigs, flaking off; small twisted tree on margin of **driest forest** types, esp. rocky hills; fts 3-winged	***Pteleopsis habeensis*** (Gp 25) (COMBRETACEAE)	
Leaves larger, or tree in moist forest		
Hairs only dense on young lvs; medium-large trees		
Lvs *c.*5 cm long, narrowly elliptic; with glands; tree in moister forests; fts 2-winged;	***Pteleopsis hylodendron*** (Gp 25) (COMBRETACEAE) [*KWAE-KANE*]	
Lvs > 5 cm long and ±leathery, hairy at first, ±ovate; petiole 2 cm+ long, flattened on top; bole fluted at base; bark white, flaking to leave, on mature trees, pale patches; slash mottled cream and orange, stringy-fibrous, darkening; fts 2-valved, egg-shaped capsules with winged seeds; **dry forest or savannah**	***Schrebera arborea*** (OLEACEAE)	563
Dense, orange hairs even on mature lvs, especially thick and soft on midrib above; stem ±4-angled and grooved; petiole > 1cm long; small, weedy tree with scented lvs darkening when dry; exudate darkening	***Premna hispida*** (VERBENACEAE)	521
Lvs actually or almost trinerved; small-medium tree with whorls of ovate to elliptic lvs; margin sometimes serrated	***Premna angolensis*** (VERBENACEAE)	520
	Nuxia congesta (see Gp3)	456
	Anopyxis klaineana (see Gp 3)	34

NOTES: 1) *Chionanthus* spp. were previously *Linociera* spp. Other species are very similar to *C. africana*:
C. nilotica is a **riverside savanna** species;
C. mannii is a rare tree v. similar to ***L. africana*** but with smaller (< 10 cm) more slender leaves with drip tips and slender twigs (< 1 mm wide).
C. mannii var. ***congestus*** a glabrous shrub in **evergreen forest**;

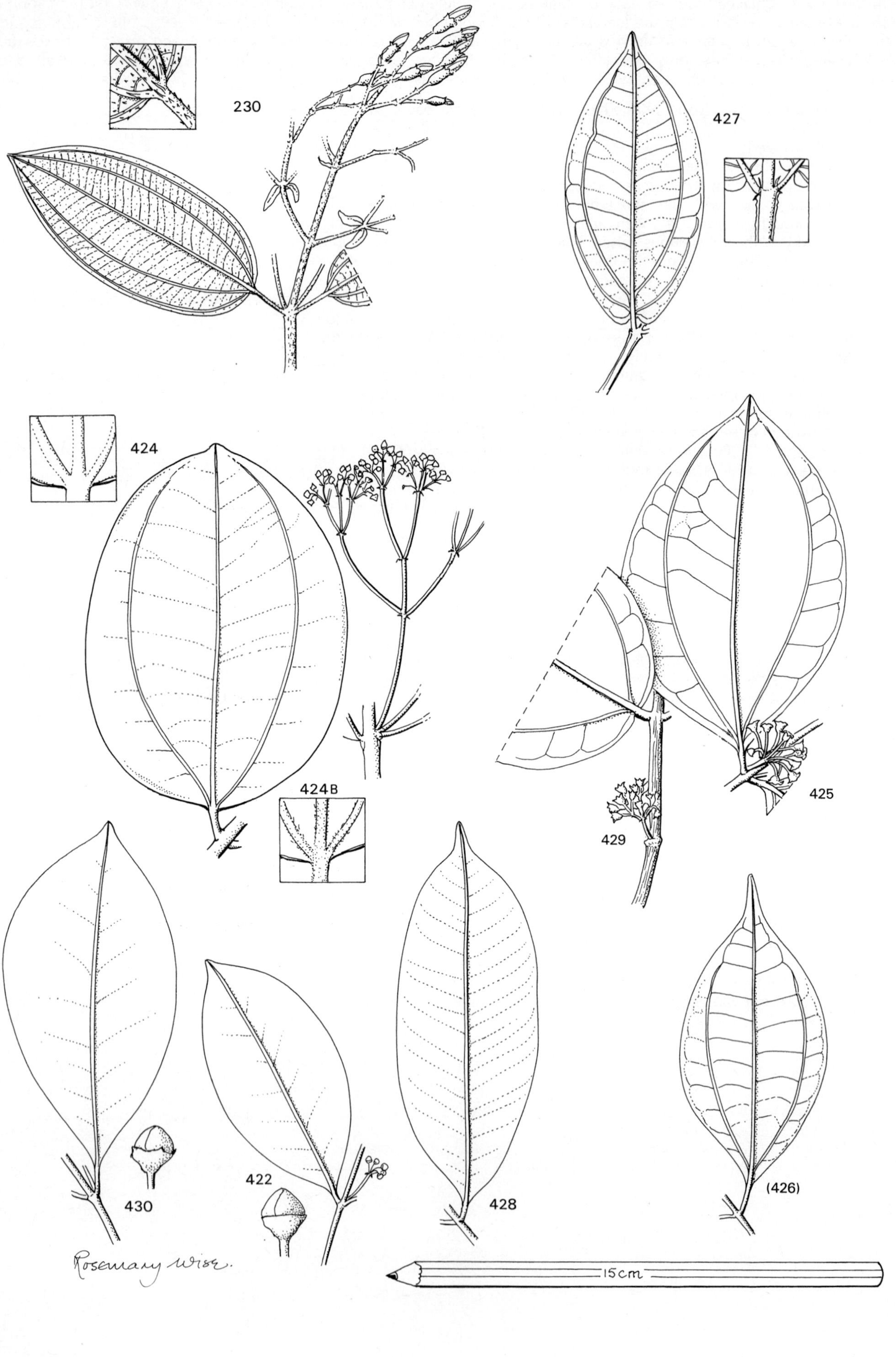
230
427
424
424B
425
429
430
422
428
(426)
Rosemary Wise.
15cm

GROUP 6: MELASTOMATACEAE
(Leaves strongly trinerved, or veins obscure)

This family includes small to medium-sized trees, with very thin bark and hard wood. The bark usually has a reddish brown slash. Often, the apical growing point of the twigs is hidden from view between the two terminal leaves. The flowers are pink or purple with an inferior ovary. In *Dichaetanthera* they are pink in conspicuous panicles, whereas the flowers of the other genera are typically more blue and inconspicuous, in small axillary infls except in *Spathandra*. The fruits are small, often purple berries, with the remains of the calyx at the top (as in Rubiaceae).

The species of *Spathandra* and *Warneckea* were, until recently, included in *Memecylon*. Whilst the genera are easily recognized their species are difficult to separate when sterile, and the following key is liable to be misleading in some circumstances. oTWESE is a local name that can therefore be interpreted as '*Memecylon* spp.', and *oTWESE-NINI* as *Warneckea* (or *Spathandra*) spp.

LEAVES trinerved

- Tree spiny – see *Strychnos spinosa* (see Gp 4)
- Tree without spines[1]
  - Lvs in whorls with petioles >2 cm long – see Group 5
  - Lvs merely opposite, without long petioles
    - Medium-sized tree; lvs (and twigs) rough on both surfaces with bristles, especially on veins (x10 lens); 5-nerved, the outer pair close to margin; twigs with ring scars at nodes; leaves turning reddish when old; **evergreen for.** especially in swamps; slash fibrous-peelable — ***Dichaetanthera africana*** 230
    - Small to medium trees with smooth leaves
      - Lf base minutely cordate; twigs strongly 4-winged with the wings flared out towards the nodes, resembling, therefore, small stipules; lower lf surface with many spots (lens); flowers ±sessile at the nodes — ***Warneckea guineense*** 427
      - Lf base obtuse to cuneate, OR twigs not strongly winged
        - Leaves with finer venation invisible between a few main (basal and lateral) nerves; leathery; medium-sized trees of **riversides or swamps**; twigs rounded; petiole slightly flattened; basal nerves joining midrib at base; slash with red outer layer; mainly found in Western region; infls with well-developed branches
          - Leaves completely glabrous — ***Spathandra blakeoides***[2,3] 424
          - Leaves with some hairs (lens) — ***S. blakeoides*** var. ***fleuryi*** 424B
        - Leaves with finer, reticulate venation visible, although venation sometimes very laxly reticulate; basal nerves often joining midrib well above base
          - Base of lf with a fine, wavy nerve just visible at arm's length, outside the main basal nerves, and running close to the margin; drying usually golden yellow below and reddish above — [*oTWESE-NINI*][3]
            - Lf base obtuse-cordate; **swamp** species; infls. on twigs below lvs — ***Warneckea memecyloides***[2,3] 429
            - Lf base obtuse-cuneate; **understorey** spp.; infls. amongst lvs — ***Warneckea cinnamonoides*** 425
          - Base of lf without even vague nerves running outside the main basal nerves at the base BUT sometimes with submarginal nerve v. close to margin around the centre of the leaf; common understorey treelet — ***Warneckea membranifolium*** (426)

LEAVES not trinerved; with finer veins ±obscure; common understorey treelets

- Laterals usually just visible or prominent below, and 1 or 2 per cm, rather ascending
  - Youngest twigs not v. slender (1mm+ wide), 4-sided and remaining so for several nodes; often almost oblong or even slightly asymmetric or irregular; bark v. corky; (lf normally drying reddish to olive brown); infls normally below the leaf-bearing nodes; mature fts spherical; calyx with short, sharp lobes — ***Memecylon normandii*** 430
  - Youngest twigs v. slender (<1 mm wide), not v. corky and 4-sided, but sometimes slightly so later; lvs regular, ±ovate-elliptic, drying yellowish or olive green; infls amongst lvs; calyx only covering ½ of bud, without 'teeth'; mature fruits ellipsoid. — ***Memecylon afzelii*** [*oTWESE*][3] 422
- Laterals barely discernible, but if so then dense (2-5 per cm), almost perpendicular to midrib; youngest twigs 1 mm or more wide; lvs usually drying reddish brown, ±oblong elliptic; fts rounded; calyx covering >½ of bud. — ***Memecylon lateriflorum*** 428

NOTES: 1) *Gmelina arborea* is an introduced tree with opposite, trinerved leaves (see notes for Gp 29).
2) *W. memecyloides* and *W. cinnamonoides* can be effectively indistinguishable in the field. Similarly, *Spathandra barteri* is a rarer species that may key to *S. blakeoides*, but the leaves are more similar in size to *W. membranifolium*.
3) See notes at top about *Warneckea* and *Spathandra* spp.

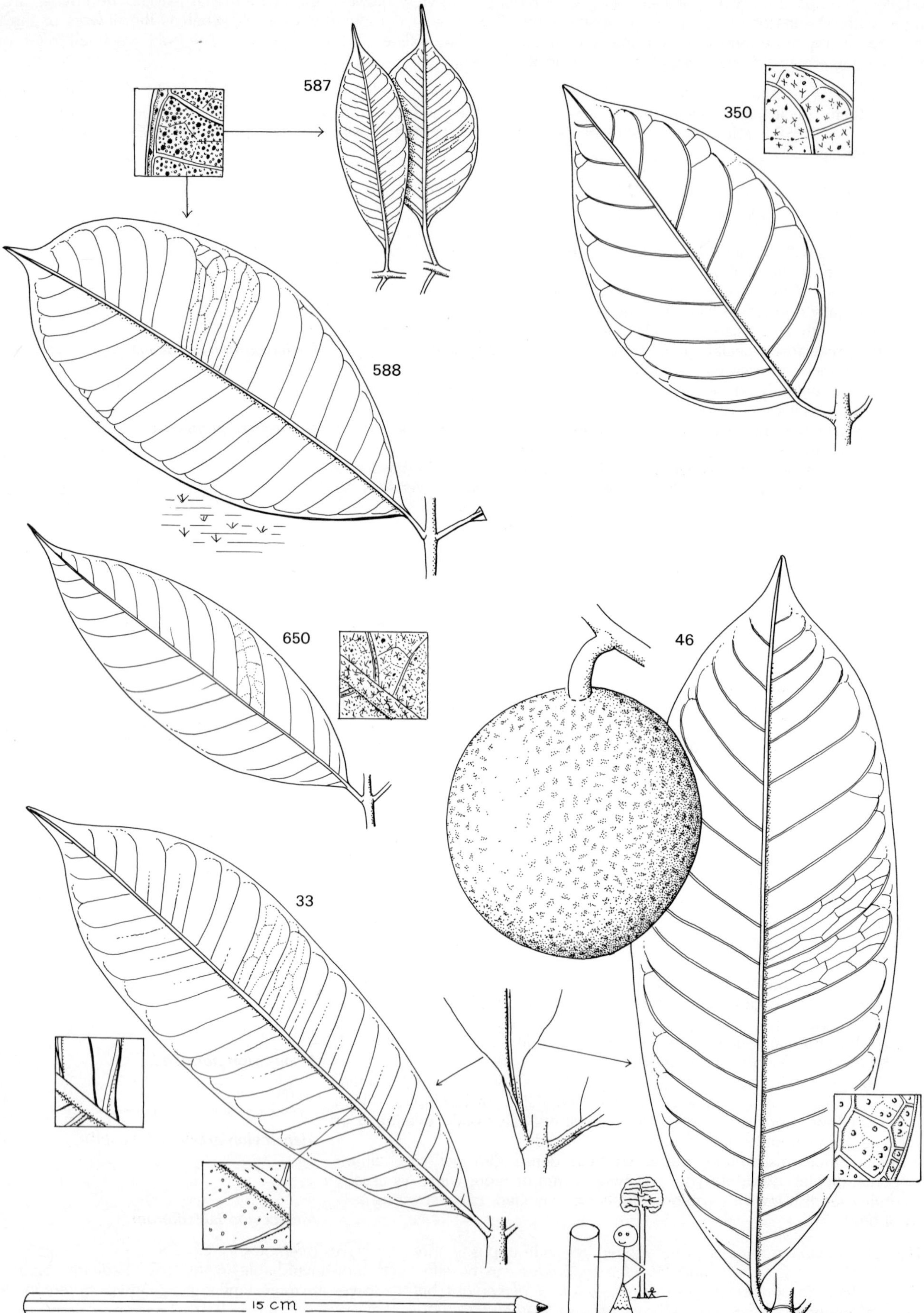
587
350
588
650
46
33
15 cm
R. Wise

GROUP 7: MYRTACEAE; GUTTIFERAE (part)
(Laterals many and parallel, or with many ±transverse intermediate veins and meeting near margin)

Leaves with many fine translucent spots (lens) and usually scented when crushed; widely spreading trees, sometimes with twisted bole with adventitious shoots; young twigs often square or 4-winged; slash red		
Lvs long acuminate, and often slightly falcate, thin and papery	***Syzygium guineense***[1,2,3]	587
Lvs rather thick, not normally acuminate nor falcate	***Syzygium rowlandii*** (MYRTACEAE) [*ASIBENYANYA*]	588
Leaves without translucent spots (or >15 cm long and leathery)		
Understorey, low trees, or tall slender trees with symmetrical clustered boughs		
Lvs long and leathery (>15 cm); slash fruity-acid and sweet-tasting, with fine pink lines; medium-sized tree; lvs with poorly-defined submarginal nerve	see ***Allanblackia*** (Gp 8)	
Lvs small (<15 cm) not v. thick; venation v. finely transverse	see ***Garcinia*** spp. (Gp 8)	

NOTES: 1) *Syzygium guineense* has many variants, from **savanna** to **rainforest**. The key here leads to *Syzygium guineense* spp. *occidentale* from **evergreen forest**. Its slash is thick and dark.

2) There are five *Eugenia* (also Myrtaceae) shrub species in Ghana. These also have close gland-dotted, fragrant lvs. *E. calophylloides* (**evergreen forest**; rusty hairy twigs, *c.*10 prs laterals); *E. coronata* (hairless twigs; Accra plains); *E. kalbreyeri* (loopy-veined, not marginal-nerved, lanceolate); *E. leonensis* (v. small lvs, on rocky hills), *E. obanensis* (pink hairs on young twigs). *Eugenia jambos*, with long narrowly lanceolate lvs is planted in gardens.

3) The introduced guava tree (*Psidium guajava*) with winged twigs and scaly bark on a sinewy bole is also in Myrtaceae.

GROUP 8: GUTTIFERAE
(Opposite (or whorled) lvs, without stipules; yellow or orange latex)

Many books include the two species of Group 8A in the HYPERICACEAE family, and indeed they are very different from Groups 8B and 8C in field characteristics. The species of Groups 8B and 8C include several with straight boles with narrow crowns composed of whorled, almost horizontal branches of very similar length. Although yellow (not translucent, but thick and milky) latex normally immediately distinguishes this family from others of similar habit (e.g. *Diospyros*, Annonaceae), the exudate of some of the species is translucent, and even colourless in *Allanblackia*, so these have been mentioned in Group 7 as well. Hairs are only found in members of Groups B and C on the flowers, and then only in *Garcinia kola*. The flowers are usually large, sometimes unisexual, with many stamens. The fruits are often edible. *Garcinia* spp. are used as chewsticks.

Genus	Flowers	Fruits
Harungana	V. small, in much branched infl.	2-4 seeded, *c.*3 mm
Vismia	Small yellow + red, in branched infl.	Many-seeded small berries
Allanblackia	Pink-red, 5 cm clustered at br. ends	50 cm firm, many-seeded, sausage-shaped
Garcinia	Yellow, various	Many-seeded, 1-5 cm
Mammea	White, clustered (m + f) or solitary(m)	8 cm, leathery, with 2-4 stones containing edible 'aril'
Pentadesma	White, thick, rancid, 5 cm	Pointed. 15 cm, several seeds containing oil
Symphonia	Red, 6 mm, clustered on short twigs	Rough, red, 3 cm, 1-2 seeds + 'arils'

Key to subgroups

Leaves not v. leathery, often hairy; covered in minute dark or translucent spots and stellate hairs; trees of disturbed or dry forest. Latex normally orange	Group 8A
Leaves usually thick, glossy, leathery and hairless; often with translucent vein-like resin channels; latex normally yellowish	
Venation finely transverse (>20 lateral nerves on a lf 10-15 cm long)	Group 8B
Venation not finely transverse. Usually around 10 pairs of laterals	Group 8C

Group 8A
(Guttiferae with latex orange; lvs not leathery)

Mature lvs cordate or ovate (when torn exuding orange latex) with conspicuous, small, black dots, and pale stellate hairs below (lens); small tree of disturbed forest with golden-green crown; bark rough with vertical fissures; slash thin, wet, peelable in long strips, rapidly discoloured by latex	***Harungana madagascariensis*** [*KOSOWA*]	350
Mature lvs elliptic; midrib and nerves below covered in vivid red-brown stellate hairs and scattered spots, especially along margin; medium-sized tree with straight, unbuttressed bole; slash orange-brown, leathery to brittle, with fruity acid, hot taste and scented; crown pale green	***Vismia guineensis*** [*KOSOWA-NINI*]	650

Group 8B
(Guttiferae with venation finely transverse)

Lvs with small raised dots at the ends of many of the finer veins, without resin channels, and with conspicuous fine reticulations between the many parallel laterals; twigs crossed with a fine line (like stipule scar) above each node; base of lf folded over into petiole/midrib channel; lvs drying orange-brown; flwrs yellowish; bark developing concentric swirly ridges below scales with age; slash with **yellow latex** in reddish inner bark; outer slash red, fibrous-brittle and gritty	***Mammea africana*** [*BOMPAGYA*]	46
Lvs without spots of this type, OR with resin channels		
Leaves with resin channels OR tree **not** in swamps or by rivers		
–Medium-sized to large trees with stout twigs and large (>13 cm long) or leathery lvs.		
Laterals not densely and clearly finely transverse or sub-marginal nerve v. wavy; tree sometimes on dry land outside evergreen forest, lacking stilt roots, sometimes without yellow latex		
Lvs not v. leathery, ±oblong-oblanceolate, with 1-2 laterals/cm contributing to marginal nerve, often with fine superficial scales on younger lvs; base of lamina running into top edges of petiole; bark with small rectangular or circular scales over small red pits; **exudate sweet, fruity, clear at first**; slash contoured or with fire-lines, chunky, red, pink and orange; fruits v. large, pendulous, yellowish, sausage-shaped	***Allanbackia floribunda*** [*SONKYI*]	33
Lvs leathery, variable, but usually with venation almost obscure except for rather few laterals arching, interspersed with many transverse resin channels which sometimes resemble laterals; narrowly to broadly elliptic, but rarely oblanceolate or oblong; **yellow latex soon visible**	see ***Garcinia kola*** (below)	
Laterals closely parallel (often >4/cm) and sub-marginal nerve in a smooth curve; resin channels normally parallel to laterals; ±translucent spots; latex thick, bright yellow; new lvs red; twigs ±black; outer bark becoming dark grey or black; often fissured into small rectangles; slash red-brown, turning dark brown, brittle; **tree in evergreen forest** (rarely swamps elsewhere), often stilt-rooted	***Pentadesma butyracea*** [*ABOTOASEBIE*]	502
–Small (-medium-sized) trees; lvs <13 cm long, with many resin channels ascending much more steeply than the very closely parallel laterals; young twigs slender, 4-sided; flowers yellow; exudate often **very slow or not yellow**; bark easily removed		
Evergreen forest species with long acumen (drip-tip); bark thick, gritty, yellow-orange; occasionally with small stilt roots when in swamps	***Garcinia epunctata*** [*NSoKoNUA*]	333
Drier forest species with short acumen; slash creamy, brittle or crumbly	***Garcinia afzelii*** [*NSoKo*]	332
Leaves without resin channels, usually <15 cm long, with minute scales (barely visible with lens); **tree in swamps or by rivers**; midrib shallowly channelled; yng twigs thick, very striate, with short conspicuous perpendicular side-branchlets; slender tree with unusually small crown and often stilt-rooted; bark grey; slash thin, brittle, with yellow latex	***Symphonia globulifera*** [*EHUREKE*]	584

NOTE: *Calophyllum inophyllum*, with extremely conspicuous finely transverse venation, is widely cultivated in gardens. The glossy, large leathery leaves are clustered rather like a *Ficus*, but are opposite and lack stipules.

Group 8C:
(Guttiferae with leathery lvs with venation not finely transverse)

In the following species, the petioles of the terminal lvs often clasp the terminal growing point.

Young stem sharply and conspicuously 4-winged; sometimes with **4 lvs in a whorl**; base of midrib entering into pocket at top of petiole channel; apex drawn out: distinctly acuminate; resin lines not conspicuous; **evergreen forest**; fts ellipsoid	***Garcinia gnetoides*** [*TWEAPIAKOA*]	334
Young stems sometimes wrinkled or v. slightly winged, but without four sharp wings		
Sub-marginal nerve <3 mm from margin (laterals often apparently reaching the margin); **evergreen forest** species with reticulate venation clearly visible; young stem v. wrinkled but not sharply winged; apex merely acute; lvs v. discolorous, with recurved margin; fts ?<2 cm diam.	***Garcinia smeathmannii*** [*BOHWE*]	336
Sub-marginal nerve >3 mm from margin; finer venation v. fine when lf young, along with v. clear resin channels, venation therefore sometimes finely transverse; base of lamina joining fine wing on petiole; bark smooth with fine flakes; slash red-brown gritty (outer) merging with pale brown, hard, granular or contoured, with bright yellow exudate; fts shape and size like orange	***Garcinia kola*** [*TWEAPIA*]	335

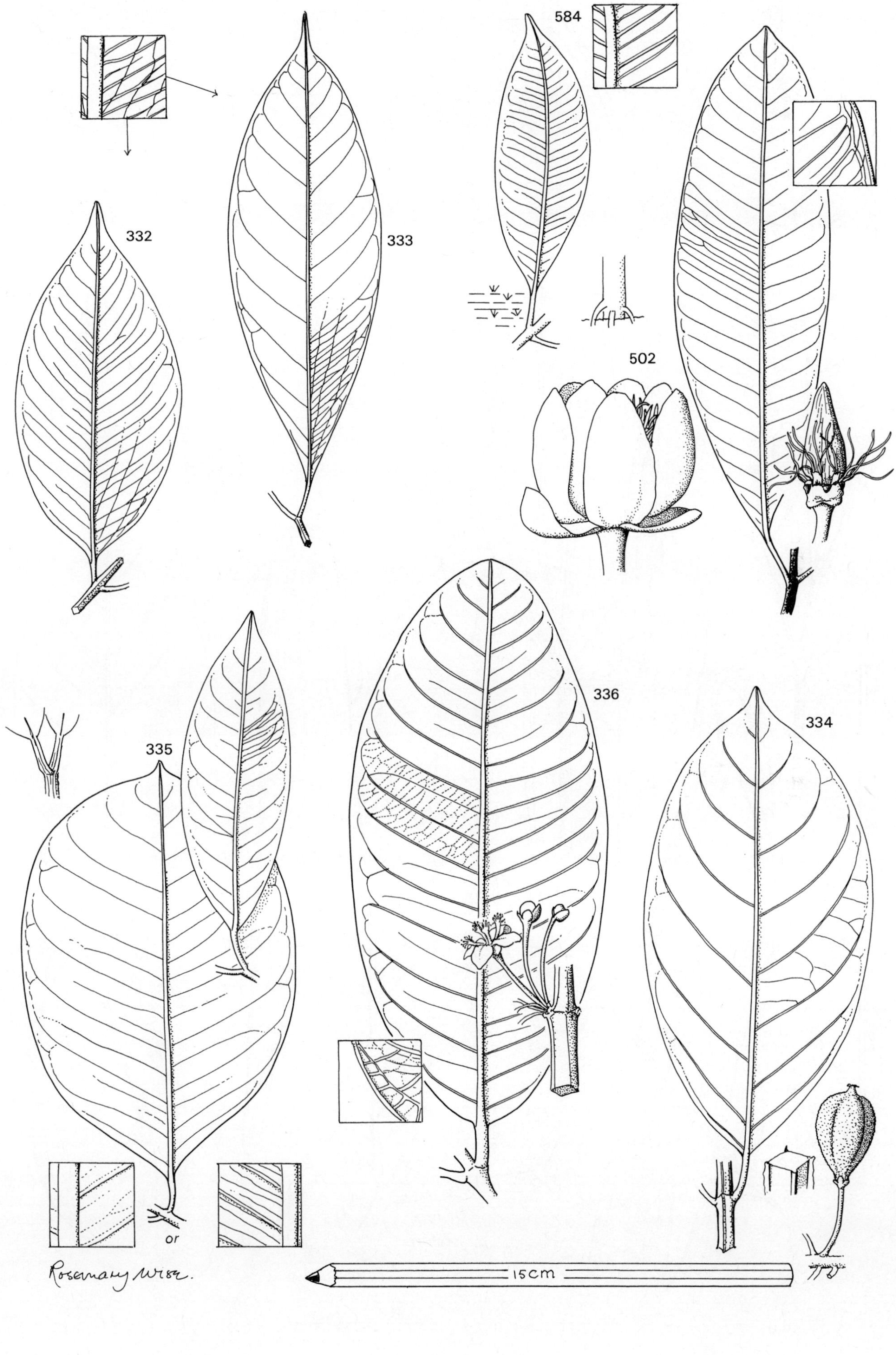
332
333
584
502
335
336
334
or
15cm

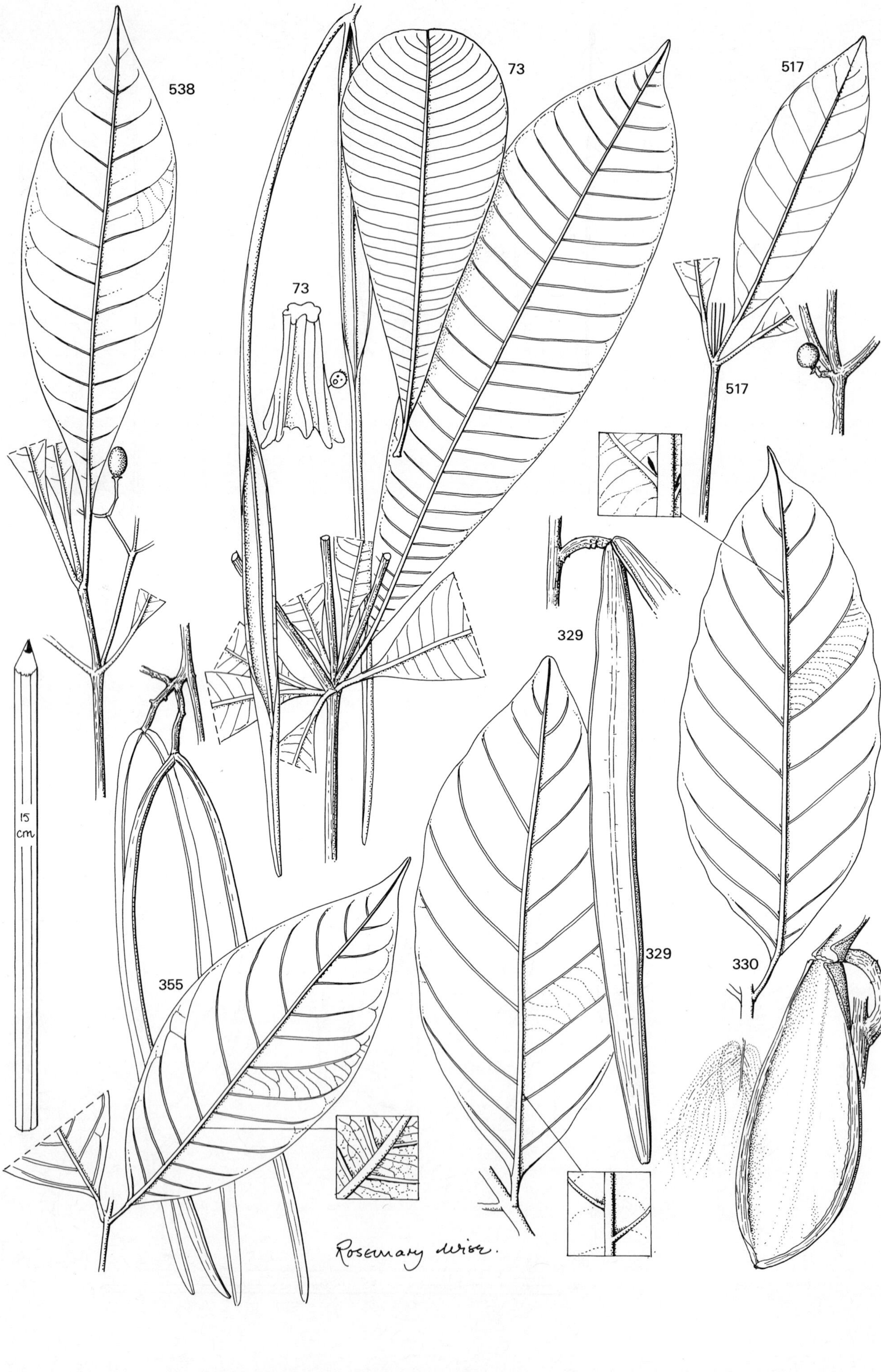
538
73
73
517
517
329
329
330
355
15 cm
Rosemary Wise.

GROUP 9: APOCYNACEAE
(White latex; lvs opposite or whorled)

Key to subgroups

Leaves in whorls of three or more; fts various	Group 9A
Leaves opposite (arranged in pairs at nodes)	
Leaves with domatia; usually with a copious outpouring of latex; canopy trees with plumed seeds	Group 9B
Leaves without domatia; small understorey trees sometimes producing only sparse, almost powdery spots of latex; fts fleshy	Group 9C

APOCYNACEAE

Most of the smaller trees of this family (the garden frangipani is a familiar example) have a distinctive, apparently dichotomous (or trichotomous) branching pattern, with no single main stem but with branches diverging at each node ('Leeuwenberg's model' – like *Anthocleista*, Gp4). In some of the larger trees (*Alstonia, Funtumia*), one of the branches at each node takes over as 'leader', resulting in a different architecture ('Prévost's model') with a single, straight bole. The species are either cylindrical to the ground, or have very contorted buttresses or fluting. Latex is often copious. That of *Funtumia elastica* can be used to make rubber.

The flowers of the Apocynaceae are tubular, with the lobes at the end of the corolla tube overlapping either to the right or the left, a bit like overcrowded fan blades (The frangipani tree (*Plumeria rubra*) again provides a well-known example). *Alstonia* and *Rauvolfia* flowers are small in terminal panicles. *Tabernaemontana* and *Voacanga* flowers are larger, in less branched, stouter inflorescences, and the remaining species produce little-branched clusters of flowers in the leaf axils.

There are two carpels at the base of the flower. In most cases these develop into the characteristically paired fruits, but in *Rauvolfia* often only one from each flower develops; in *Voacanga* and other genera this happens occasionally. Three main types of fruit can be distinguished. The taller, gap-pioneering trees have slender follicles which split open to release plumed, wind-dispersed seeds. The second fruit type, produced by species often found in the understorey shade (except *Voacanga*), have fleshy fruits with many seeds presumably attractive to a wide range of animal dispersers. Thirdly, *Rauvolfia* is a small weedy tree of roadsides producing small red fruits well-suited to bird dispersal.

Genus	Fruit
Alstonia	50 cm slender follicles releasing seeds with tuft of white silky hairs at end
Holarrhena	50 cm slender follicles releasing seeds with tuft of white silky hairs at end
Funtumia	Broad, canoe-like, dark (13 cm) follicles with tufted seeds
Picralima	Large mango-shaped, *c.*15 cm long follicles with seeds in pulp
Hunteria	Spherical, 5 cm diam. with *c.*10 disc-shaped seeds
Voacanga	Spherical, 5 cm diam. with groove; elongated seeds
Tabernaemontana	Spherical, 10 cm diam with groove; elongated seeds
Pleiocarpa	1 cm, single seeded, in tight clusters
Rauvolfia	Small red, berry-like, spherical (<1cm) drupes (1-seeded)

Group 9A
(Apocynaceae with whorled leaves)

(Note: the **evergreen forest** tree *Neolemonniera* (Gp 10D) has v. clustered, alternate lvs which resemble whorled lvs.)

Laterals joining in sub-marginal nerve; lvs glossy and not papery; oblong, oblanceolate or, at least, not sharply pointed (except in saplings)		
Leaf oblong with blunt acumen; venation 'finely transverse' but laterals interspersed with visible finer veins and cross-connected; uncommon tree found in **dry forest** (Mampong-Aburi); branching typically dichotomous (2 brs per node)	***Pleiocarpa pycnantha***	517
Leaf oblanceolate, with tip normally rounded,[1] with laterals almost concentric around a point at the end of the midrib; laterals closely parallel, with finer veins ±obscure: ±glaucous between laterals below; v. common tree of **disturbed forest**, often with highly curved buttresses (older trees), with strongly whorled branches; bark v. lenticellate, slash yellow brown, very gritty; latex copious; branches often 3 or more per node	***Alstonia boonei*** [*SINURO*]	73
Laterals not meeting in a sub-marginal nerve; leaves thinly papery; elliptic with acute apex; twigs with many raised pale lenticels; small sun-loving tree of disturbed forest	***Rauvolfia vomitoria*** [*KAKAPENPEN*]	538

NOTE: 1) *Alstonia* juveniles have lvs with more pointed tips, opposite when v. young, but the nerves are still distinctive.

Group 9B
(Apocynacae with domatia)

The following three trees have cylindrical boles without buttresses, and produce plumed seeds.

Leaves with domatia (tufts of hair or pits) in the axils of nerves		
Vein axils with pit domatia; margin often very wavy or irregular; latex rapidly forming rubbery balls when rolled between fingers; bark often pale brown; common in all forest zones	***Funtumia elastica*** [*FRUNTUM*]	330
Vein axils with tufts of hairs; latex sticking to fingers when rubbed		
Lvs oblong-elliptic; finer venation not v. visible; laterals usually impressed above; latex copious; medium tree; bark very dark, often black, without many pale lenticels	***Funtumia africana*** [*OKAE*]	329
Lvs papery, ovate; venation v. visible; laterals usually finely prominent or recessed; bark with conspicuous, raised pale lenticels; **dry forest**, often small (-medium-sized) tree	***Holarrhena floribunda*** [*SESE*]	355

Group 9C
(Leaves not (normally) whorled; without domatia)

The following trees are small, spreading, understorey or swamp species.

Leaves large OR papery OR with rounded apex, *without conspicuously fine-transverse venation*; twigs thick, often almost hollow; base of petiole normally with very conspicuous pockets (which clasp growing point of twigs on youngest lvs); flowers in branched inflorescences, rather stout or fleshy		
–Lower surface with finer venation obscure, smooth BUT evenly peppered with dark spots; most lvs on mature tree >15 cm long (often >20 cm long), thick and glossy	***Tabernaemontana*** spp.[1] [*OBONAWA*]	589–591
–Lower surface lacking spots, often with corky twigs with conspicuous scars		
Leaf apex rounded; lf thick and glossy; nerves and veins appearing (50 cm away) as dashed lines due to regular undulations on lamina (esp. dried lvs); top surface with many minute pits (lens); small tree gregarious in **wet places**	***Voacanga thouarsii*** [*FOBA*]	656
Leaf apex acute or acuminate; lvs thin + papery, very variable with age, etc.; laterals not meeting; petiole 'clasps' leaving scars on corky twigs; venation ±obscure; sun-loving small tree	***Voacanga africana*** [*OFURUMA*]	655
Leaves with *many parallel laterals* and visible veins; understorey trees with medium-sized leaves; twigs not normally hollow, nor particularly thick; 'clasps' at base of petiole not so conspicuous, nor with obvious pockets; (or flowers small in little-branched clusters)		
Midrib above grooved at base, but with several wrinkle lines submerged there: the same lines emerging and prominent about half-way up leaf (and flattening out towards tip). RARELY 3 lvs in a whorl; **shrub** or treelet; flowers in small axillary fascicles.	***Pleiocarpa mutica*** [*ONWEMNA*]	516
Midrib with simple groove above nr. petiole, sometimes with single line running along base of groove; slash yellow-brown, brittle-granular, **bitter**, ±contoured, ±gritty, with scanty latex		
Midrib channel generally broad: laterals fusing with its thickened edge; laterals usually interspersed with veins which do not reach sub-marginal nerve and are often not quite parallel to laterals; flowers in branched terminal infls		
Lf tip blunt (rounded-acute): midrib clearly branching before end of tip; some laterals clearly channelled above; dots barely visible above	***Hunteria umbellata*** [*KANWENE-NINI*][2]	362
Lf tip sharp, acuminate; dots often visible above with lens		
Lvs slender (see diag.); dots v. conspicuous but venation less so on top surface; **dry forest** (esp. SM + SO)	***Hunteria ghanensis***	363
Lvs broader; dots not v. obvious: and finer veins v. prominent, on top surface; **evergreen forest**	***Hunteria eburnea*** [*KANWENE-AKOA*]	361
Midrib channel v. fine; laterals disappearing into channel when viewed from above; lvs often with raised spots below; pockets at base of petiole clearly visible with lens; corolla 2 cm or more long; calyx lobes 5 mm long	***Picralima nitida*** [*KANWENE*][2]	506

NOTES: 1) There seems to be no basis on which the three species of *Tabernaemontana* can be separated when sterile; when fertile, they can be distinguished as follows:

–Calyx up to 2 cm long; corolla up to 9 cm long.	***T. africana (=T. chippii)***	589
–Calyx OR corolla smaller		
–Calyx <1 cm long; slender corolla broadening to tip, <5 mm wide in middle; glabrous inside.	***T. crassa***	590
–Corolla *c.* 2 cm long and v. hairy inside.	***T. pachysiphon***	591

2) Not surprisingly, given the extreme difficulty with which the species are distinguished when sterile, 'KANWENE' seems to be traditionally applied to all *Hunteria* spp. as well as to *Picralima*.

589 (-591)
591
590
655
656
589
362
506
516
363
361
u.s.
u.s.

GROUP 10: SAPOTACEAE

(Simple, alternate, entire, not trinerved, etc., lvs often clustered at twig ends)
(Slash with white latex, darkening, often with broad vertical bands of dilatation tissue)

The Sapotaceae include several commercially important timber species but, whilst the family is easily recognized, the species are differentiated by tree spotters with great difficulty, if at all.

Many species are fluted trees (e.g. *Chrysophyllum* spp.), but others (e.g. *Tieghemella* and *Aningeria*) can be impressively cylindrical or low-buttressed. All species show some tendency to produce leaves clustered at the ends of twigs, but in a few (e.g. *Tieghemella*) this reaches the extreme of *Terminalia* species, even to the point of the crown being arranged in flat layers, with horizontal, whorled boughs. The outer bark can be rough and deeply fissured (e.g. older *Tieghemella*) or smooth, often with fine fissures. In several species (some 'ASAMFENA' or 'AKASA') it is 'quilted' – i.e. smooth (and often v. pale), but with regular shallow-edged fissures. The slash, which produces darkening latex, is v. pale in some species (*'DUATADWE'*), and yellow to bright red in others, usually with great variation between the ridges and dips of the flutes or buttresses, and typically with broad, darker vertical bands of 'dilatation tissue'. The slash normally has a slightly sweet, characteristic, almost milk-like scent.

The combination of vegetative characters can lead to confusion with Moraceae. However, Ghanaian Sapotaceae never produce trinerved or serrated leaves. The leaves are only markedly asymmetric at the base in *Chrysophyllum beguei*, and only cordate in species with metallic or unusually discolorous leaves, whereas moraceous leaves are often serrated, asymmetric or cordate but not discolorous. The latex of Moraceae is usually more watery and yellowish, and the slash of that family is rarely coloured with broad vertical bands. The flowers and fruits are entirely different, and the stipules of the Sapotaceae, if present, are thread-like, whereas Moraceae typically have broader stipules ensheathing the terminal bud in a pointed, often falcate cone, and falling to leave ring scars at nodes. **Only *Pachystela, Vincentella* and *Neolemonniera* have obvious stipules in the Sapotaceae.**

The flowers are never very large, are often produced on leafless stems, and only in branched inflorescences in *Chrysophyllum giganteum*. Apart from *Aubregrinia* and *Omphalocarpum* the flowers are hermaphrodite. Most fruits have one or more seeds embedded in a fleshy, sometimes edible pulp but in *Gluema* and *Neolemonniera* the single, very glossy seeds are ejected explosively from dehiscing, canoe-like follicles. The seeds have a hard, glossy, often brown coat with a scar on the edge or base. The seeds of *Tieghemella heckelii* and of the sapotaceous savanna tree *Vitellaria paradoxa* (=*Butyrospermum pankii*) provide oil for soap and cooking.

Genus	Notes on fruits
Gluema	In leaf axils; 6 cm; roughly ovoid, pointed, 1-seeded follicle; explosively dehiscent
Neolomonniera	In leaf axils; 3 cm; roughly ovoid, pointed, 1-seeded follicle; explosively dehiscent
Breviea	In leaf axils; 6 cm, 8-seeded
Aubregrinia	In leaf axils; 8 cm, 8-lobed + 8 brown seeds
Chrysophyllum	Various (nr lvs); typically 5(-11) cm; 5-seeded, sometimes lobed, ±ovoid, often delicious
Synsepalum	In leaf axils; ellipsoid; 2 cm, 1-seeded
Tieghemella	In leaf axils; 8 cm, yellow, ovoid; ±1-2 hard, large ellipsoid seeds + 1 flat side; elephant-dispersed
Manilkara	In leaf axils; 1–2 cm; calyx at base; 1-2 seeded
Malacantha	In leaf axils; 2 cm ellipsoid, apiculate, 1 seeded
Aningeria	On twigs amongst + below lvs; 2 cm, spherical + persistent calyx; 1 seed
Vincentella	On twigs amongst + below lvs; 2 cm, ovoid; usually 1-seeded
Pachystela	On twigs below lvs; 2.5 cm, hairy, 1-seeded
Afrosersalisia	On twigs below lvs; 2.5 cm, red, 1-seeded
Bequaertiodendron	On older branches + twigs; 2 cm + persistent calyx; 1-seeded
Ituridendron	On older branches; 5 cm, pointed, globose
Omphalocarpum	On trunks; 10-20 cm, many-seeded spheres with apical 'belly button'

NOTE: *Ituridendron* is treated by many authorities as a synonym of *Omphalocarpum*. Here it is kept separate from that genus because of the (perhaps superficial) differences in habit and overall appearance.

Key to subgroups and similar species

Leaves not at all clustered at twig ends, but spread out evenly in layers slightly like compound lvs; latex copious and remaining white; lvs oblong acuminate with laterals rather closely parallel; tree of swamps	(***Anthostema aubryanum***) (EUPH) [*KYRIKUSA*]	See Group 22
Leaves clustered OR latex darkening rapidly OR tree not in a wet place		
Leaves with finely transverse venation (many closely parallel lateral nerves (>4/cm), OR fewer laterals (but still >2/cm) but with the intermediate finer veins tending in the same direction as the laterals)		
'Strangling figs' or twigs with ring scars at nodes or lvs strongly pustulate	***Ficus*** spp	See Group 19
Not *Ficus* spp.		Group 10A
Leaves without finely transverse venation		
Laterals reaching margin and fusing with **marginal nerve**		Group 10B
No marginal nerve: laterals not merging at margin		
Leaves with lower surface discoloured with hairs, OR at least with silvery or golden sheen		Group 10C
Leaves not discoloured nor v. hairy below (but sometimes with hairs on midrib, etc.), and without a metallic glint		
Leaves emarginate v. thick and succulent; latex v. copious and pure white	***Elaeophorbia grandifolia*** (EUPH) [*AKANI*]	See Group 22
Not *Elaeophorbia*: leaves acuminate or rounded at apex		Group 10D

3 SAPOTACEAE FRUITS

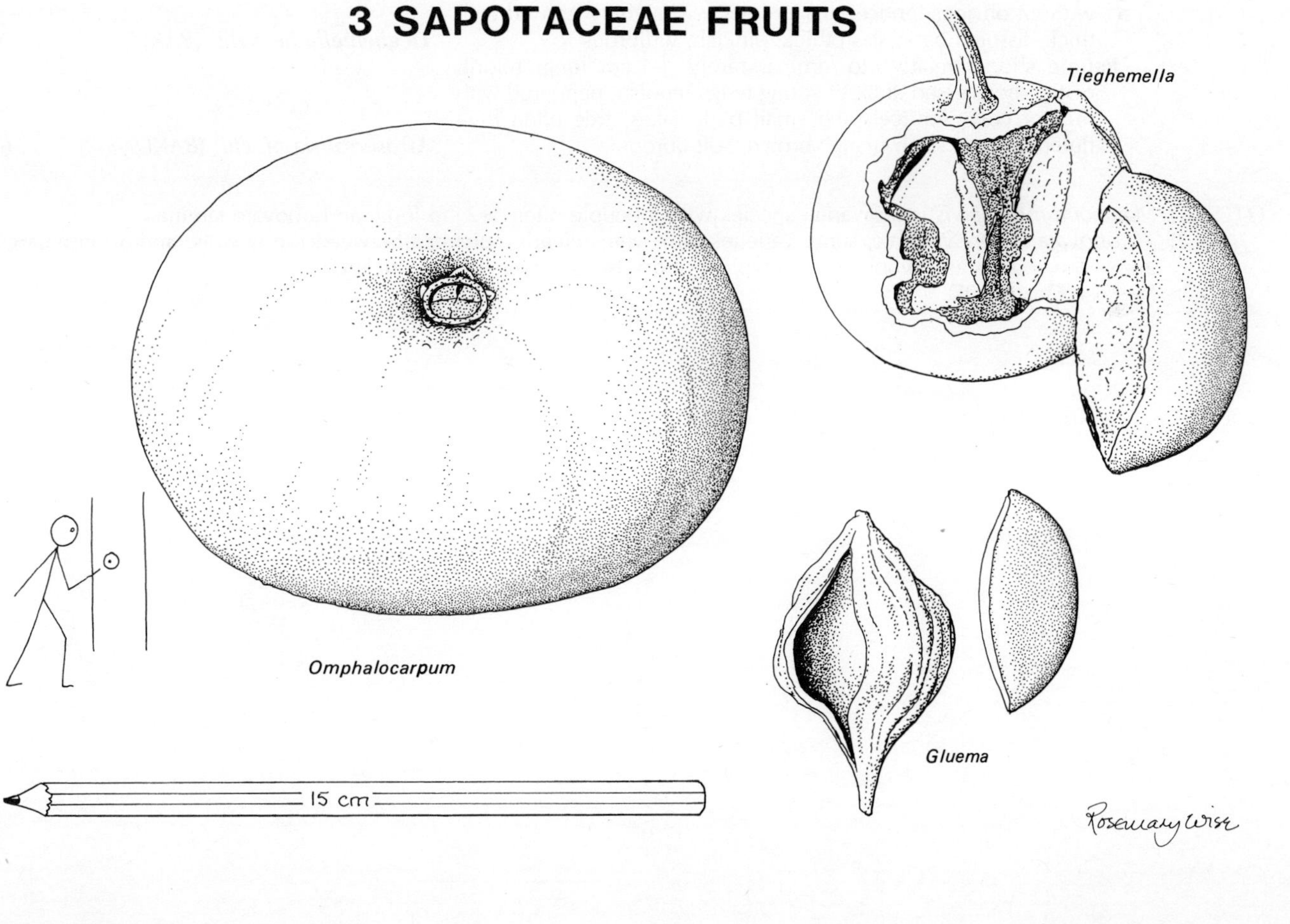

Group 10A
(Sapotaceae with 'finely-transverse' venation)

Key	Species	Page
Leaves with **golden or silvery** shine below; lamina >12 cm long; petiole often 2-3 cm long		
Lvs silvery-white below; (trees mainly in **dry forest**)		
Long taper on leaf base (i.e. lf oblanceolate); bole often fluted, with peeling, papery bark	***Bequaertiodendron oblanceolatum***	109
Short taper on lf base (i.e. lf ±obovate); finer veins obscure; fluted slightly at base or not at all; bark dark with deep, sharp fissures; slash red to pink, fibrous-stringy	***Manilkara obovata***[1] [*BEREKANKUM*]	409
Lvs golden below; trees usually in swampy places		
>30 pairs laterals; lvs >3 times longer than broad; bark not deeply fissured	***Bequaertiodendron magalismontanum***	108
<30 pairs laterals; lvs <3 times longer than broad; apex acuminate; slash red (see above)	***Manilkara obovata*** var.[2]	
Leaves without metallic glint or lamina <12 cm long		
Finely-tranverse venation v. distinct: >8 laterals/cm; Lvs distinctly acuminate; base asymmetric; drying slightly red; bole often v. fluted; slash creamy, rather soft and brittle, slightly sweet-scented; lvs clustered, but not extremely so		
Lvs. with v. abrupt drip tip; usually >10 cm long	***Chrysophyllum pentagonocarpum*** [*DUATADWE-NUA*]	155
Lvs. with steady, shallow curve leading to tip; young twigs etc. with dense, rusty hairs; usually <10 cm long	***Chrysophyllum pruniforme*** [*DUATADWE*]	157
Laterals less densely-packed, rather irregular and finer venation not v. distinct; large trees with straight, cylindrical bole; slash orange to red, with darker bands (but check ***Aubregrinia***)		
(Top surface only with finely wavy, prominent, rippled transverse veins—see *Aubregrinia*, Gp 10D)	***Aubregrinia taiensis*** [*DUATAONE-KESE*] See Gp 10D	98
Not ***Aubregrinia***		
Petiole 1-2 cm or more long, up to *c.*½ as long as lamina; crown (esp. young trees) like *Terminalia* with tiers of tight clusters of lvs; young twigs smooth (except for lf scars), drying black, without obvious lenticels; older trees v. straight to ground with thick, fissured bark; slash thick, pinkish, v. fibrous	***Tieghemella heckelii***[3] [*BAKU*]	9
Petiole shorter relative to lamina, rarely >1 cm long; midrib distinctly recessed at base; young twigs reddish, peppered with minute corky lenticels and small bark flakes; tree often low-fluted at base; slash orange-brown, soft-fibrous.	***Afrosersalisia afzelii***[3] [*BAKUNINI*]	61

NOTES: 1) *Manilkara multinervis* is a **savanna** species with a petiole often >2 cm long, and obovate lamina.
2) *M. obovata* is very variable; some varieties have veins clearly reticulate between the laterals, and/or an emarginate apex. The species as a whole is most typical of **dry forest, riversides or thickets**.
3) See also Group 10D.

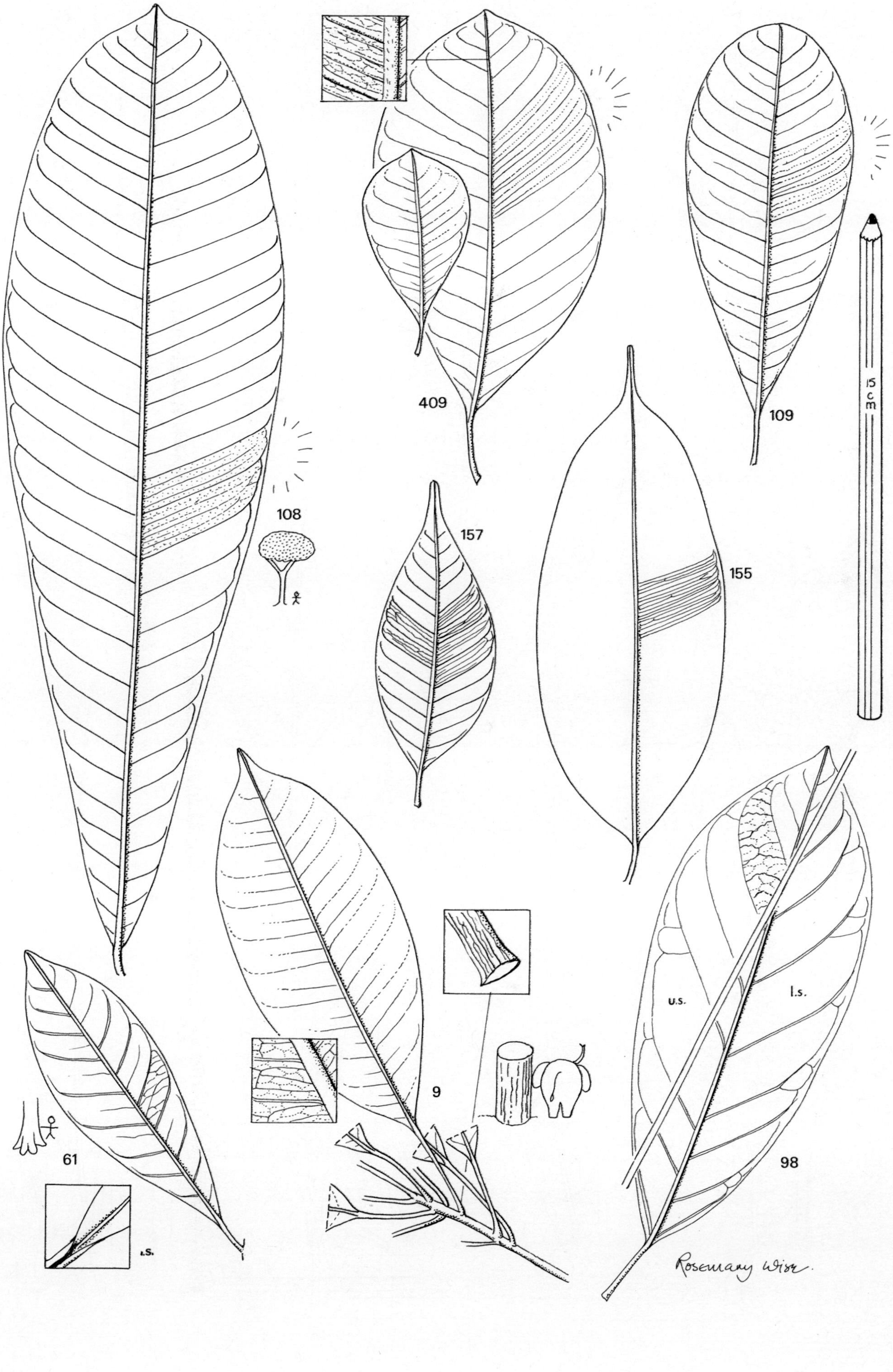
108
409
109
15 cm
157
155
9
61
98
u.s.
l.s.
l.s.
Rosemary Wise.

78
406
152
79
120
585
483
482
15cm

Group 10B
(Sapotoceae: lvs with marginal nerve)

Note: In many species, e.g. *Chrysophyllum* spp., the laterals curve very close to the margin without actually joining a thickened marginal nerve. As well as the marginal nerve, these species often have **translucent spots**.

Leaves with conspicuous, long, orange-brown hairs over lower surface: apex rounded or obtuse, occasionally with a rounded 'hump' or apiculate; fine venation prominent below; lvs often with rough spots on top		
Lvs large (often >9 cm long) usually with coarse or tufted hairs on top surface, nerves and lamina; hairs below often longer than laterals are broad; small, twisted, often fluted understorey tree of **driest forest types**	***Malacantha alnifolia*** [*ASAMFENA-AKOA*]	406
Lvs smaller (usually <9 cm) with no, or v. few, hairs on top apart from midrib; lower surface usually **orange-discolorous with soft hairs** (but hairs usually shorter than the laterals are broad); tall, straight, unfluted, common tree, often with large, often triangular buttresses; bark pale, smooth, but fissured (quilted); slash pale yellow to pink with darker bands, fibrous but slightly brittle	***Aningeria robusta*** [*ASAMFENA-NINI*]	79
Leaves with scattered, inconspicuous and short (not soft or coarse) hairs, especially on the nerves, etc.; looking pale, even glaucous, below with veins not v. prominent; lf apex drawn out to an acute, sometimes almost acuminate point; tree found only in **drier** (e.g. fire zone) **forests** similar in habit to *A. robusta*	***Aningeria altissima***[1] [*ASAMFENA-BERE*]	78

NOTE: 1) The name 'ASAMFENA' is often used by the less experienced for virtually any sapotaceous species. *Aningeria* spp. are often not differentiated by the timber cutters, although the distinction in usually easy. 'Aningueri rouge' (= 'Red Asamfena?') is *Chrysophyllum perpulchrum* (Aubréville, 1953).

NOTES FOR GROUP 10C:

1) Species of *Chrysophyllum*, particularly these last 5, are easily confused; the common name 'AKASA', used widely in the timber trade, means more realistically '*Chrysophyllum* spp.', especially *C. giganteum* or *C. subnudum*; Hall and Swaine (1981) doubt whether *C. albidum*, which is sometimes cultivated, is native to Ghana. It is conceivable that it is a cultivar variety evolved under human selection pressure. The fruits of these *Chrysophyllum* spp. distinguish them as follows:

    Fruits ovoid – pointed away from the stem (4-5 cm long)
        With a short stalk = *C. delevoyi*
        Without a stalk = *C. albidum* (cultivated for fruit)
    Fruits globose, sometimes with a very slight apical bump or hollow
        Mature fruits hairy, with no stalk = *C. perpulchrum*
        Mature fruits glabrous, with a stalk (peduncle) almost as long as fruit = *C. giganteum*.

    Fruits slightly 5-angled, rounded with a blunt apical point = *C. subnudum*

    Both *C. subnudum* and *C. giganteum* fruits are variable, delicious and sold in village markets. Therefore, in practice it would be wise to obtain leaves to confirm the identity of these species.

2) *C. cainito* ('Star apple'), cultivated for its fruits, has ovate leaves with very dense red or golden hairs, many parallel laterals, and ovate-elliptic lvs smaller than these species.

3) *C. subnudum* has rather thin, slender, elliptic oblanceolate leaves with acuminate apex. If the leaf being keyed out is rather thick, glossy above, broadly oblanceolate to obovate with a rather abruptly acuminate or acute apex, and the finer venation obscure, consider the very variable ***Manilkara obovata*** (Gp 10A).

GROUP 10C
(Sapotaceae: conspicuous hairs, metallic glint or discolouration)

Leaf base cordate to obtuse but not asymmetrical
- Small thin lf. (usually <15 cm long) with long drip tip; lower surface v. variable, with dense long brown hairs or silvery shiny; lvs not clustered; fluted tree with silvery grey, fissured and flaky bark; slash pinkish with darker bands, soft, darkening — ***Breviea leptosperma*** [*KANKABIM*] 120
- Larger lf with drip tip blunt or absent on small tree in understorey of **evergreen forest**; lvs clustered — ***Synsepalum aubrevillei*** [*ASAA-NINI*] 585

Leaf base cuneate, or v. asymmetrical
- Leaf base very asymmetrical with one side often cordate; lvs broadly elliptic-oblong, acuminate with dense, long (branched) soft reddish-brown hairs below; medium-sized tree with straight bole ±sl. fluted base; bark thick-scaly; slash pale yellowish (±short orange lines) — ***Chrysophyllum beguei*** [*DUATADWE-NINI*] 152
- Leaf base ±symmetrical
  - Small trees with leaves v. strongly clustered at twig ends, with conspicuous slender stipules at their base; lvs broadest well above the middle, therefore with a long, steady taper to the petiole; metallic, but not softly hairy below; slash red to brown, fibrous-peelable, often gritty
    - Lvs with >15 laterals; slightly folded up at base; golden or silvery below; stipules *c.* 1 mm wide at base; not strongly favouring wet places — ***Pachystela msolo*** [*ASABA*] 483
    - Lvs with <15 laterals; not folded up at base; lower surface minutely bumpy and silvery; veins ±invisible; stipules <1 mm wide; spreading, often fluted, **usually on banks of rivers or streams** — ***Pachystela brevipes*** [*AFRAMSUA*] 482
  - Large trees with lvs not strongly clustered; lvs usually elliptic, or only slightly broadest above the middle, and without stipules (but undeveloped lvs look like stipules); twigs v. grooved; petioles up to 3 cm long, also grooved; lvs sometimes with dense, soft red, white or golden hairs – otherwise metallic; slash thick fibrous reddish to yellow with strong sweet smell like paint mixed with milk; **usually slightly fluted**
    - Lvs normally with 7-10 pairs of laterals (with a few more on (sapling) lvs longer than 15 cm); veins between nerves vaguely scalariform, wavy and transverse between wide-spaced laterals; hairs below rather thick, matted, gold on young lvs, but white and shorter on older ones; midrib channel above usually sharp; widespread, common tree (esp. in elephant areas) sometimes with thick, small buttresses — ***Chrysophyllum giganteum***[1,2] [*KUMFANA*] 154
    - Lvs with more than 10 pairs of laterals on those lvs *c.*15 cm long
      - **Lvs below mostly with 10–20 prs laterals**; mature lvs white to silver below, without soft thick hairs; venation scalariform, reticulate or obscure
        - Venation ±scalariform, transverse (like KUMFANA – above); leaves below NOT golden, but usually silvery due to v. fine hairs which are barely visible even with lens; young twigs with v. sparse hairs also; small tree absent from drier forests — ***Chrysophyllum subnudum***[1,3] [*ADASEMA*] 158
        - Venation obscure or reticulate; trees possibly only natural(-ized) in the **driest forests**, but planted for fruit in villages elsewhere; venation above finely raised-reticulate; midrib channel not normally v. sharp; hairs disappearing with age, but usually whitish, and not brown even on young lvs, but sometimes brown on petioles — ***Chrysophyllum albidum***[1] [*AKASOA*] 150
      - **Lvs below often with >20 laterals**, OR with very reddish, soft hairs; finer venation closely scalariform, or obscure; hairs often reddish, soft and dense, at least on young lvs; often with broad fluting or small buttresses
        - –Tree along rivers or in forest of **evergreen forest zone**; hairs below rather silky; fine venation easily visible, fine ±scalariform; on top surface raised, not appearing pitted between; often v. acuminate — ***Chrysophyllum delevoyi*** 153
        - –Tree widespread, common in **dry forest**, with v. red crown, due to v.obvious red-brown, not particularly silky hairs below; top surface usually appearing finely pitted (lens) in the gaps between the raised veins; acute, or slightly acuminate on shaded trees — ***Chrysophyllum perpulchrum*** [*ATABENE*] 156

For NOTES see page 61.

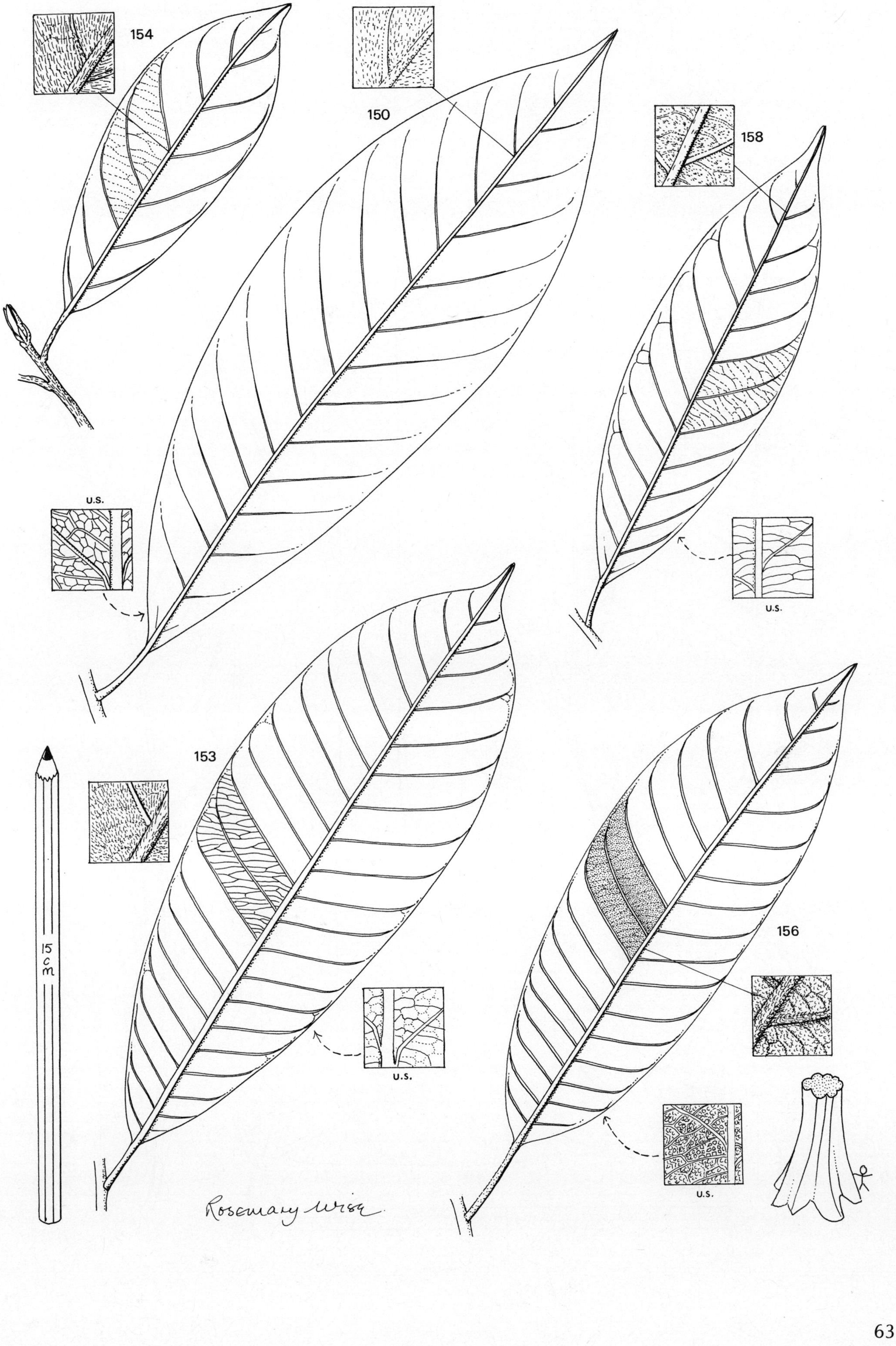
154
150
158
U.S.
U.S.
153
15 cm
156
U.S.
U.S.
Rosemary Wise

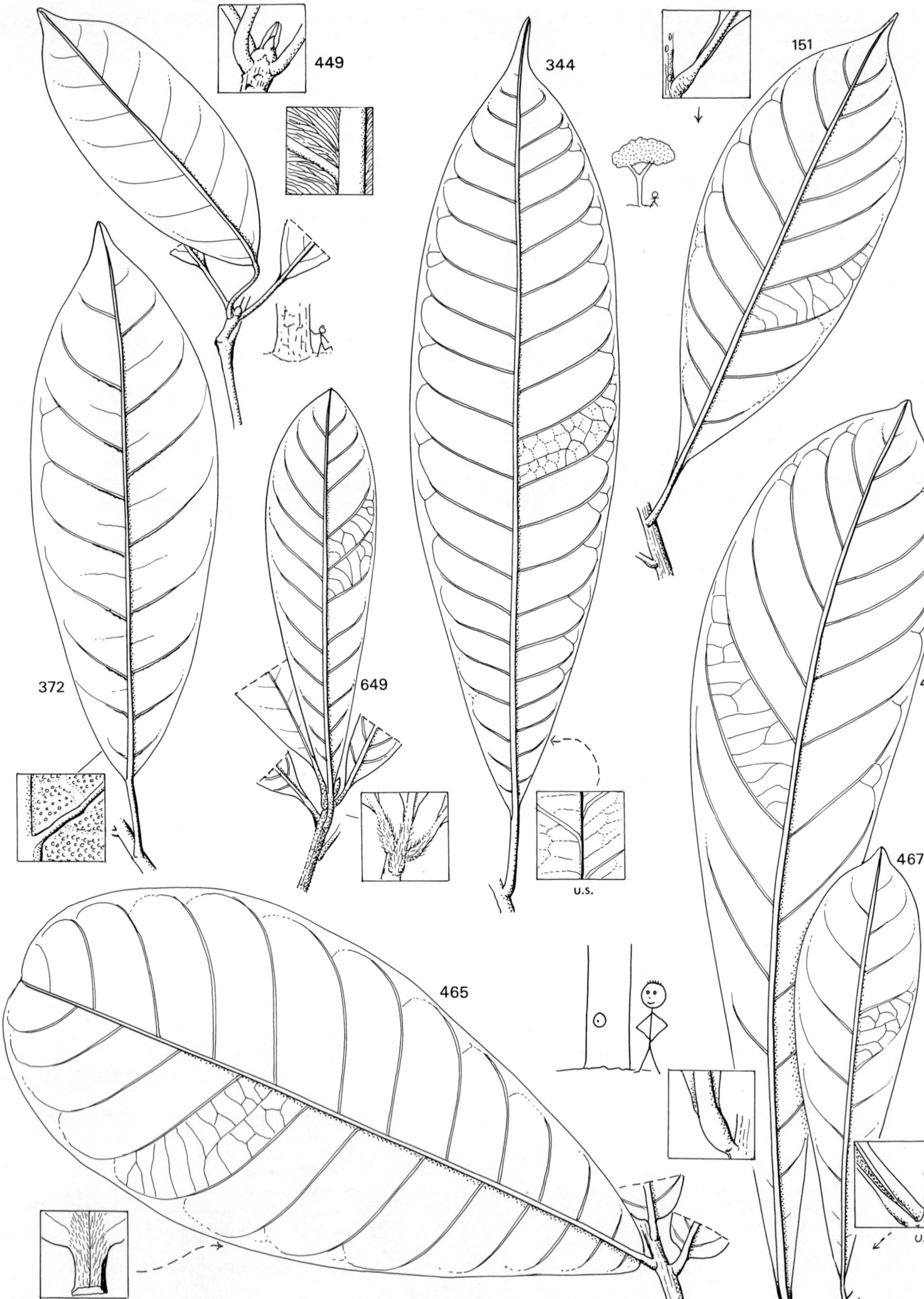
449
344
151
372
649
466
467
465
U.S.
U.S.
U.S
15 cm
R. Wise

Group 10D
(Sapotaceae: other species: venation more normal; lvs not discolorous)

Nodes markedly swollen; finer venation rather hard to see but, on v. close inspection (lens), **venation is unusually dense and 'swirly', like fingerprints**; many fine stipules clustered at twig tips; lf shape variable, but usually acuminate; **evergreen forest**; large tree with scaly, not fissured bark; slash red, fibrous	***Neolemonniera clitandrifolia***	449
Nodes not markedly swollen and venation not so		
Midrib distinctly channelled above; venation prominent both surfaces; laterals forming sub-marginal nerve; lf with sharp drip-tip; **evergreen forest**; (lvs often slightly resembling mango lvs) bole slightly fluted, with flaky bark; slash reddish, fibrous-spongy latex often copious + rubbery	***Gluema ivorensis*** [*NSUDUA*]	344
Midrib not channelled above, although sometimes recessed or guttered OR tree in drier forests		
Lvs with venation on top surface prominent, wavy and slightly transverse, like ripples on water	See *Aubregrinia* (below)	
Lvs without prominent, wavy vein rippling on top surface		
Small, understorey or riverine trees with sharp-tipped lvs usually >15 cm long		
Young twigs and midrib (but not whole leaf) densely, softly hairy; margin wavy; stipules short, falling soon; **riversides**, esp. in **dry forest** and **savanna**	***Vincentella passargei***	649
Young twigs and midrib not densely hairy		
Secondary and finer veins (apart from a few) below obscure or sunk between raised bumps (lens) on the lower surface		
Lvs v. clustered towards branch ends, with thread-like stipules *c.* 1 cm or more long; midrib recessed above; rounded or grooved below		
Lf apex rather blunt; margin wavy; **evergreen forest** tree	***Vincentella revoluta***	
Lf apex usually pointed, often sharply so	***Pachystela brevipes*** (see Gp 10C)	
Lvs not so clustered: twigs without aggregations of thread-like stipules at ends; **laterals bordered by fine dark lines** due to gutters; lower surface with raised spots (lens) except over the (paler) venation; bark pitted or fissured; slash cream turning pale brown	***Ituridendron bequaertii*** (=*Omphalocarpum pachysteloides*)	372
Secondary veins slightly prominent and distinct; passing from one lateral to the next with few forks; lamina decurrent into narrow petiole; **evergreen forest**	***Chrysophyllum azaguieanum***	151
Large trees with leaves small (<15 cm) OR apex rather rounded (at least not sharp-tipped) trees typically v. cylindrical (except sometimes *Afrosersalisia*)		
Lvs with blunt or rounded tips: oblanceolate with short petioles; midrib flat or raised above; fts shaped like rubber balls with the ends pushed in, produced on older wood		
Lvs. very large (often >15cm wide): midrib >5 mm wide at base, flat but striate; slash chunky-fibrous yellow-white with orange-brown bands; medium-sized tree usually in **slightly swampy areas**	***Omphalocarpum ahia*** [*DUAPOMPO*]	465
Lvs. smaller; drying black to leather brown; midrib raised; **slash red**		
Lvs often with >10 prs. laterals and >14 cm long; typically 2 laterals in 3 or 4 cm; petiole >2 mm wide at base, + marked channel; *flowers with 1 cm stalk*	***Omphalocarpum elatum*** [*ESONODOKONO*]	466
Lvs usually with <10 prs. laterals, <14 cm long and; lf base often v. markedly decurrent; petiole <2 mm wide at base with v. slight channel; *flowers sessile*	***Omphalocarpum procerum***[2] [*OGYATAFONKONWA*]	467
Lvs with slightly drawn-out tips or, if rounded, then petiole not v. short (>1 cm); trees sometimes becoming very large; fts produced nr. lvs		
Venation not, or only slightly prominent above. Common		
Petiole 1-2 cm: up to *c.* ½ as long as lamina	***Tieghemella heckelii*** [*BAKU*] (See Gp 10A)	
Petiole shorter relative to lamina	***Afrosersalisia afzelii*** [*BAKUNINI*] (See Gp 10A) (illustrations on page 59)	
Venation very conspicuous and prominent above: orientated mainly away from the midrib, appearing like 'ripples'; lvs sometimes >15 cm long, narrowly obovate; rare, straight, low-buttressed tree of **moist, hilly areas**; bark rough and thick; slash pale orange with whiter stripes	***Aubregrinia taiensis***[1] [*'DUATADWE-KESE'*]	98

NOTES: 1) ***Mimusops kummel*** from fringing forest in **savanna** has similar lvs to the last sp., but has v. hairy yng lvs.

2) ***Synsepalum dulcificum***, a small **dry forest** tree, has lvs <10 cm long like this sp., but fls among lvs. The small fruits have a very unusual sweetening influence on sour tastes experienced subsequently.

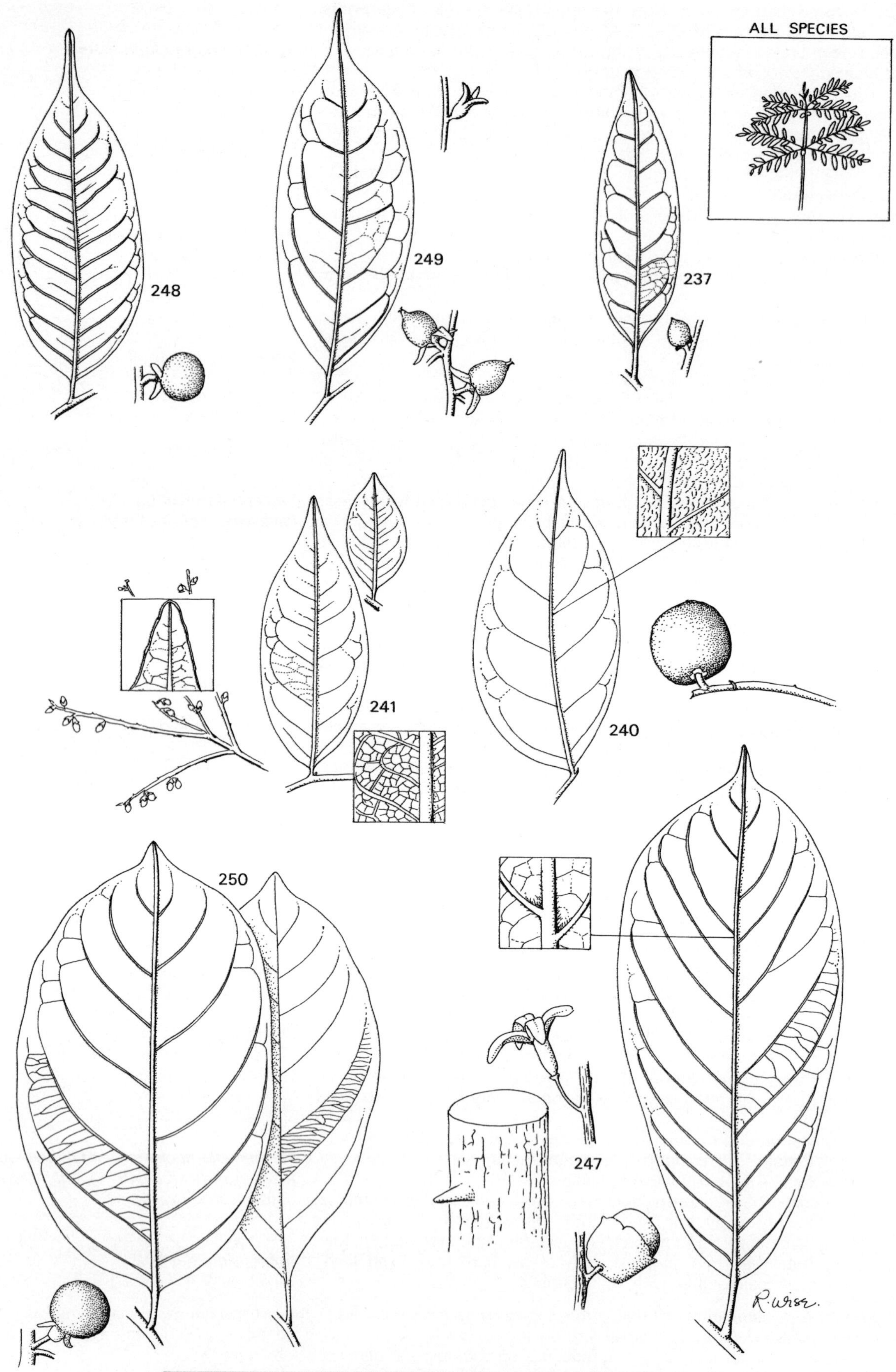
ALL SPECIES
248
249
237
241
240
250
247
15 cm
R. Wise.

GROUP 11: *Diospyros* spp. (EBENACEAE)
(Lvs simple, alternate, entire, short-petioled; bark with a black outer layer, unscented)
(Branches whorled)

Diospyros spp. are very common in Ghana as understorey trees, although only a few reach the canopy. The outer bark is often very black, although in *D. monbuttensis* (which is atypical because it also has spines) it is papery and shiny brown, and in *D. viridicans* it is often v. pale. The black outer bark has an unusual form in the 'flint-bark' species – particularly *D. sanza-minika* and *D. gabunensis*[1]. It is like hard, brittle coal or glass, with regular fissures, and can quite often be seen standing as an empty, black cylinder in evergreen forest long after the wood has rotted away from inside it. The same bark can also be seen on exposed roots. Even where the bole does not appear black, a black line will normally be seen in the outer slash.

The inner bark is pinkish or yellow, and darkens rapidly, often with a yellowish exudate. The yellow inner bark is more typical; pink slashes have been recorded sometimes in *D. canaliculata*, but seem typical only of *D. cooperi*, *D. ferrea*, *D. heudelotii*, *D. kamerunensis*, and *D. mespiliformis*. This is therefore a useful diagnostic character.[2]

Diospyros spp. have a characteristic branching pattern (Massart's model of Hallé *et al.*), with branches in whorls and with leaves (of slightly varied size or shape) spread evenly along the horizontal branches (especially in the papery-leaved species, like *D. vignei*). In this characteristic, *Napoleonaea* (group 17D) resembles *Diospyros* spp., but that species often has serrated leaves, and the slash is not *Diospyros*-like. *Diospyros* spp. can be distinguished from Annonaceae (Gp 12), which also often have a black layer in the outer bark and are also very widespread as lower storey trees, by the characteristic sweet scent in the bark of Annonaceae. Annonaceae trees also often produce horizontal branches with evenly dispersed leaves, but the branches are in a (tight) spiral, not arising (*Terminalia*-fashion) in discrete, whorled layers.

Diospyros flowers are unisexual, and often dioecious. The fruits of *Diospyros* are berries, with seeds often shaped like orange segments around the centre. Fruits are borne usually on older twigs or boughs, in un- (or little-)branched inflorescences, and have at the base a 3 or more lobed calyx which continues to develop around the fruits after the petals have fallen.

In addition to the (often quite small) trees in the Main Key there are two species of shrub, which are restricted to **evergreen forest**. These have glossy and often leathery leaves, and can be distinguished as follows.

Shrubs in evergreen forest with long brown hairs	
Lvs subopposite with glands* in lamina nr. base; yng twigs + dense brown, long hairs	***D. chevalieri***
Lvs alternate, often slightly cordate; glaucous below + long brown hairs on veins	***D. barteri***

* The only other Ghanaian *Diospyros* with similar glands is *D. gabunensis*. Other *Diospyros* species, however, often have 'glands' scattered as spots in the venation in other parts of the leaf. These are said to be 'tannin pockets' (*Flore du Gabon*) and their presence and position is less predictable.

Key to *Diospyros* trees

Plant with spines; nerve axils with tufts of hairs; bark reddish, very flaky, sometimes in papery scrolls; spines unbranched, rather blunt, on older branches or bole; slash yellowish, darkening with exudate	***Diospyros monbuttensis*** [*ATWERE-NANTIN*])	247
Plant without spines; nerve axils without domatia; leaf underside not glaucous .... (continued on next page)		

NOTES: 1) 'KUSIBIRI' is a name widely used for any of the 'flint-bark' *Diospyros*: i.e. *D. sanza-minika*, (*D. canaliculata*) and *D. gabunensis*, although in this key it is used for the latter species.

2) Of the two *Diospyros* species common in **savanna**, *D. elliotii* (**riversides**) can be distinguished from *D. mespiliformis* on the basis of the inner bark colour: *D. elliotii* is the yellow-slash species. *D. mespiliformis* (found also on the periphery of forest, e.g. on rocky outcrops) has v. wavy branched laterals, and young lvs with v. dense orange hairs.

Group 11 (contd.)

Leaf underside not glaucous nor strongly discolorous with hairs (or, if slightly so, then lvs <10 cm long and NOT papery); (lvs in many cases drying black)

- –Veins strikingly **parallel, transverse and prominent above** between *c.* 7 prs laterals; margin slightly undulate; lvs ⩾3 cm wide; broadly elliptic to obovate; midrib impressed; outer bark sometimes pale, soft and corky; slash yellowish; common — ***Diospyros viridicans*** [*ATWEA*] 250
- –Finer venation not finely transverse and prominent OR lvs smaller OR slash pink OR outer bark v. black OR lvs drying black
  - **Leaves <5 cm wide, + <15 cm long, OR lvs thin and papery**; lvs often drying black; flwrs or fts sometimes with 3 calyx lobes
    - LVS slightly pustulate below, thin papery OR with thickened margin; often asymmetric, with yellow hairs on young twigs; **slash reddish**
      - Lvs ovate-acute and thinly papery, with veins ±obscure below, and (3-6) nerves impressed above; clearly pustulate below; young stems sl. hairy; (Bia-Krokosua forests); drying blackish — ***Diospyros cooperi*** [*'French ATWEA-BERE'*] 240
      - Lvs not thinly papery; margin thickened and finely undulate, thus **appearing minutely serrate**; lvs ⩽10 cm long; midrib sharply channelled above; fresh lvs v. pale below, with obvious venation; drying red-brown — ***Diospyros ferrea*** [*'oMENEWA-NINI'*] 241
    - LVS not pustulate below; **slash yellowish**, darkening
      - Leaves often with two nerves at base more ascending than rest, extending 1 cm or so before dissipating into reticulations; venation raised below; apex blunt-acute; lf base v. slightly decurrent into petiole; canopy narrow and dark with shiny lvs; new lvs red; most common in **drier forests and thickets**; slash fibrous and gritty — ***Diospyros abyssinica*** [*GBLITSO*] 237
      - Leaves without two almost-basal nerves, etc., oblong acuminate, drying v. black; not obviously hairy, except for orange hairs on underdeveloped lvs
        - Shrub or small tree; lf only acuminate; <10 prs of main laterals; often fertile; fts red berries in lf axils + 3 calyx lobes — ***Diospyros soubreana*** [*oTWETO*] 249
        - Medium tree with 10 or more prs of main laterals (joining submarginal nerve); **drip-tip** normally very conspicuous; often fluted or hollow — ***Diospyros piscatoria*** [*'oTWETO-KESE'*] 248
  - **Leaves often >5 cm wide or >15 cm long, usually slightly leathery**; midrib shallowly-channelled, strap-like above; outer bark v. hard, brittle, black and glass-like; calyx lobes >3
    - Leaf hairless, often but not always <15 cm long; no glands; venation prominent above; leaves ±oblong or elliptic; lenticels on twigs raised and rounded; petiole slender and channelled; cauliflorous, widespread; slash often pinkish, outer bark **never deeply grooved**; (lvs sometimes staining drying-paper red) — ***Diospyros canaliculata*** [*ATWEA-BERE*] 239
    - Leaf with short brown hairs on YOUNG midrib, etc., ±**small glandular areas at base;** usually >15 cm long, ±broadly oblanceolate, with only a short tip; venation conspicuous, almost scalariform; lenticels on twigs elongated; **evergreen forest only**; slash yellow, darkening; bark like deeply furrowed black glass — ***Diospyros gabunensis*** [*KUSIBIRI*] 242

Leaf underside glaucous (blue or pinkish) or discolorous with golden hairs

- Tree (even twigs 1 cm wide) with v. **brittle bark like black glass**, regularly furrowed (see above); hairs on twigs and midrib not conspicuous; lf lanceolate to elliptic; venation fine and reticulate, prominent or obscure below; lower surface sometimes looks like fine, pinkish (wall) plaster; lamina edge at base folded over into petiole channel; slash (difficult) revealing pinkish inner bark; **evergreen forest** — ***Diospyros sanza-minika*** [*SANZA-MULIKA*] 41
- Tree + twigs without brittle, black-glass bark; smooth or flaky; lvs hairy or papery
  - Leaves with long (½ mm) yellowish hairs on veins and midrib, etc.; often ±glaucous as well (lvs often >15 cm long)
    - –Slash yellowish + slow, thick yellow exudate; finer venation ±visible below and scalariform; hairs below on midrib and veins, etc., but not dense elsewhere on mature lvs; twigs hairy; lamina joining side of midrib; lower surface ±pinkish; outer bark not v. black, flaky; **moister forests** — ***D. mannii*** [*ATWEA-FUFU*] 245
    - –Slash pinkish, soft; finer venation often hard to see, but in any case not close scalariform; yellow hairs v. dense and flat all over lower surface, esp. on young lvs; **crown golden-brownish** (hairy lvs); widespread, even in drier forests — ***D. kamerunensis*** [*oMENEWA*] 244
  - Leaves with only v. short hairs (<¼ mm), thin-papery, glaucous below
    - Tall tree; lvs (even on twig ends) tending to be ovate; hairs on twigs v. short; fine hairs on veins and upper nerves of young lvs; bark sometimes blackish, ±furrowed, but in any case + obvious black layer; inner bark becoming thick, pink to red, brittle fibro-granular — ***D. heudelotii*** [*oMENEWA-HOA*] 243
    - Shrub or small tree of or near **evergreen forest**; leaves tending to be elliptic or lanceolate, especially at twig ends; veins ±obscure; yellowish hairs of discernible length on twigs, midrib, etc., giving the latter a slight soft-furry appearance when closely examined with naked eye — ***D. vignei*** [*oMENEWA-HOAKOA*] 243B

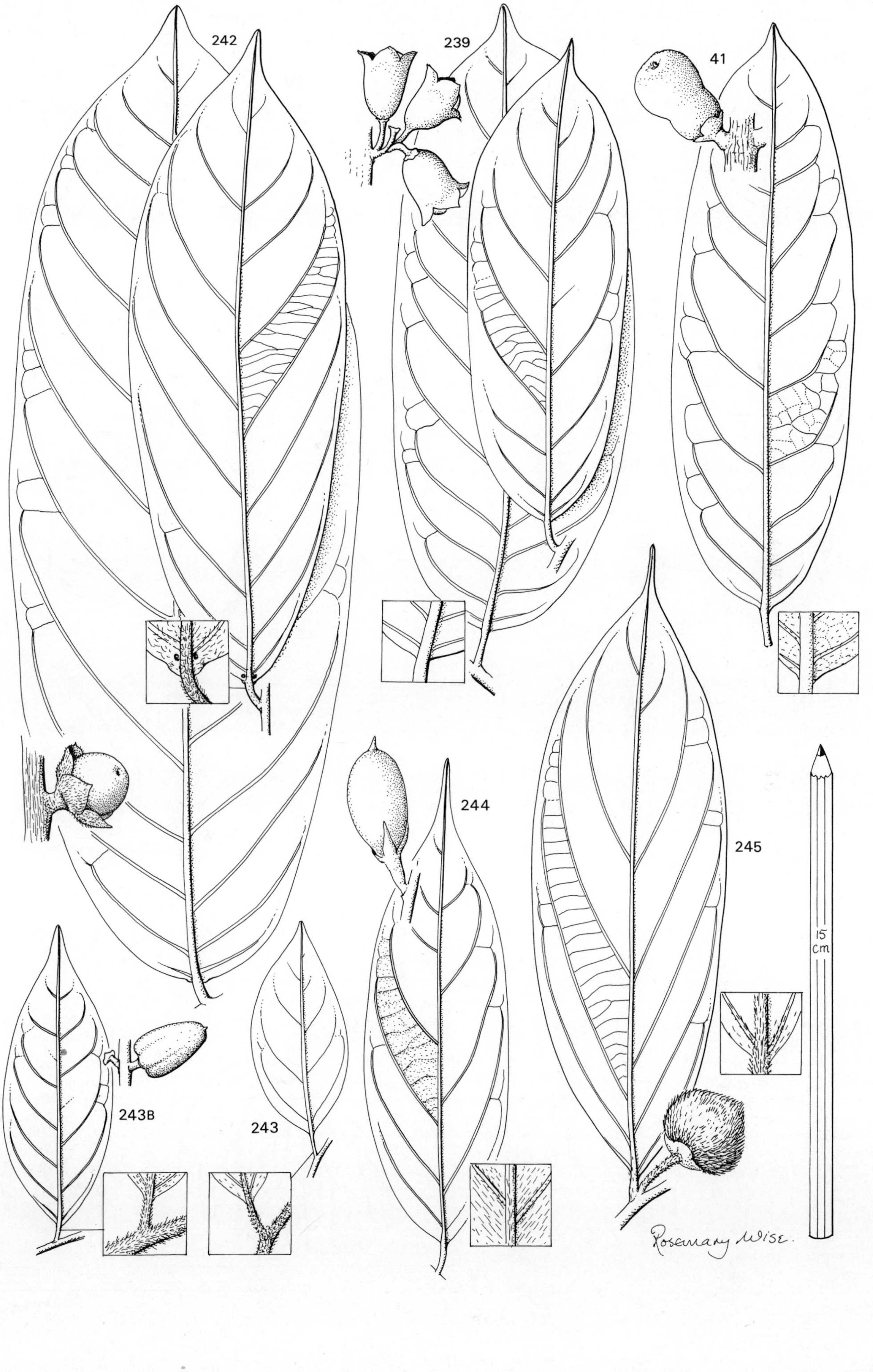
242
239
41
244
245
243B
243
15 cm
Rosemary Wise

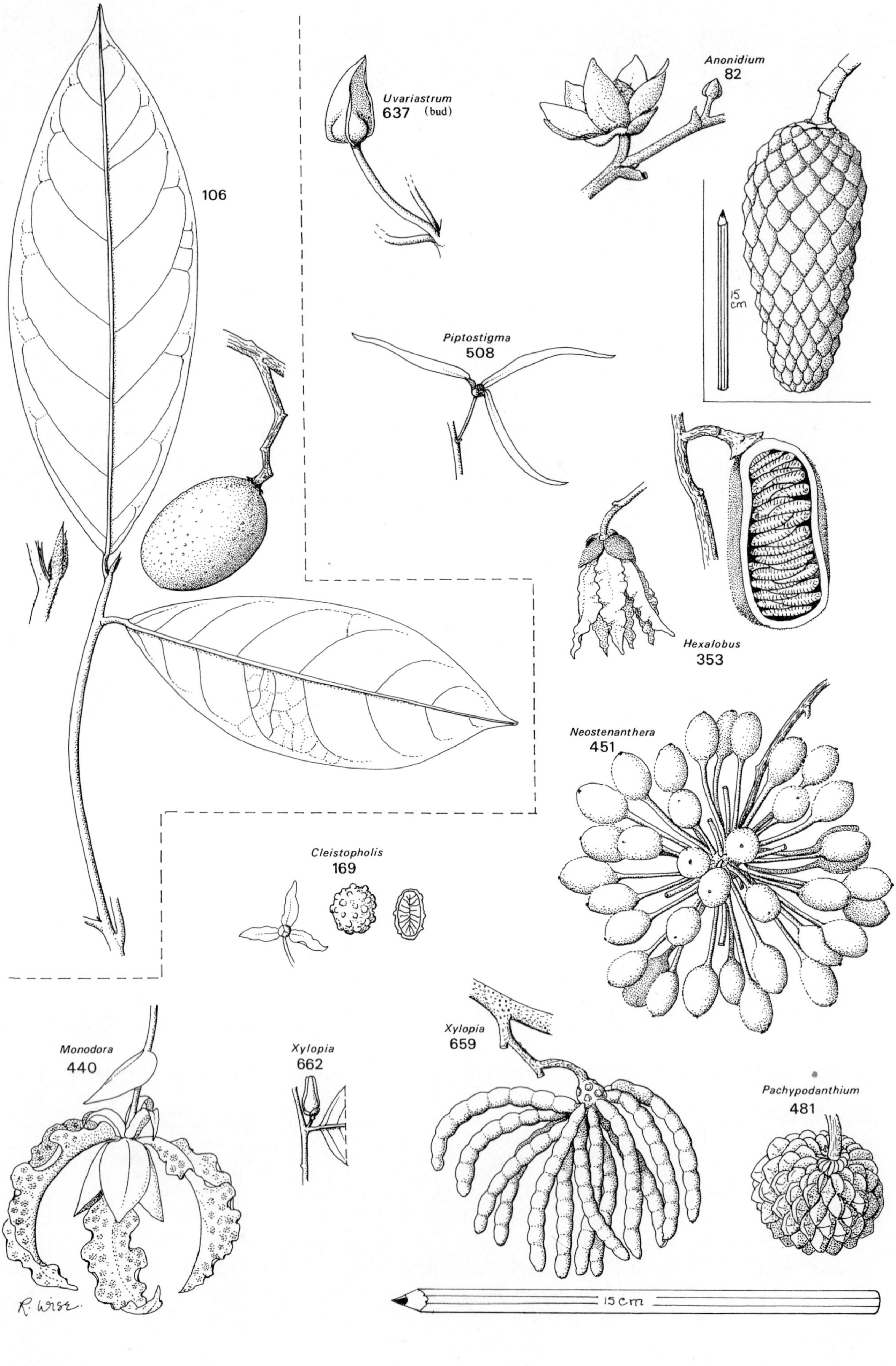
106
Uvariastrum
637 (bud)
Anonidium
82
15 cm
Piptostigma
508
Hexalobus
353
Neostenanthera
451
Cleistopholis
169
Monodora
440
Xylopia
662
Xylopia
659
Pachypodanthium
481
15cm
R. Wise

GROUP 12: ANNONACEAE
(Lvs simple, alternate, margin entire, not trinerved, etc.)
(Slash fibrous, sweetly aromatic)

This is a family of small to medium-sized trees common in the lower storeys of forest. Apart from 12F, all Groups include trees with slender boles, and often short regular branches at the top of the stem. The outer bark of some species is black, like *Diospyros* (see notes under Group 11) but the Annonacaeae can be distinguished by their fibrous, stringy to spongy, yellow to orange darkening slash with strong, characteristic sweet, slightly fruity scent and a hot, peppery taste (although in some species (e.g. *Xylopia villosa*) the slash also has a marked granular component). The branches are not whorled, but often in a spiral of horizontal layers like a spiral staircase (See Note 2, Gp 12A).

Many species produce flowers and fruits on the older wood. The flowers are very characteristic, with the parts arranged in 3's or 6's (again, like a few *Diospyros* spp.). *Uvariopsis* is exceptional with 4 petals and 2 sepals. The flower in most species consists of many carpels which develop into one to many-seeded, often sausage-shaped fruits. The seeds are also characteristic, as they have a 'ruminate endosperm' meaning that, when cut into sections, there are many lines directed inwards from the edge of the seed (also found in Myristicaceae and Scytopetaloceae – Gp 13A).

Beilschmiedia (below) is the only species in the related family, Lauraceae, in Ghana.

Key to subgroups

Lvs clustered close to branch ends; slash reddish, rather granular, strongly perfumed; lvs ±falcate-asymmetric; **twigs with small, tight, sharp, ±cone-shaped bud at tip**; fresh lvs rather glutinous when squashed; usually in wet places	***Beilschmiedia mannii*** (LAURACEAE) [*TWEANKA*] 106
Lvs not clustered (Not TWEANKA)	
Slash yellow to red, not v. fibrous, ±gritty, with hot, 'mustard-like' scent; lvs thick, glossy, **asymmetric at base; outer bark not black**	See *Drypetes* spp. (Gp 17)
Not a *Drypetes*: slash usually fibrous, rarely reddish, with definite sweet perfume component (as opposed to more paint-like smell of *Drypetes*), but usually with peppery taste	
–Leaves with stellate hairs sometimes v. small and spot-like (lens)	
Veins ±obscure; slash bright golden yellow	***Enantia polycarpa*** Gp 12D
Veins visible; lvs (except saplings) long and narrow; yng buds softly hairy, pointed to one side at twig ends; slash dirty yellow, with large pores	***Pachypodanthium staudtii*** Gp 12A
–Leaves without stellate hairs	
Lvs often (look at several) **broadest around or below middle** (elliptic, lanceolate or oblong) OR with a **rounded apex** OR lvs asymmetric OR lvs clearly obovate (check illustrations)	
Leaves of varied shapes and sizes (but <**15 cm long**) along same twigs (NOT resembling pinnate lvs) OR lvs drying black; outer bark often black and smooth; small trees.	Gp 12F
Leaves spread evenly along twigs rather similar in shape and size, often resembling pinnate leaves from a distance; medium-sized, slender trees often with horizontal boughs at top of stem	
Lf normally >15 cm long, narrowly oblong, ±asymmetric, and often with wavy or recurved margin	Gp 12A
Lf normally <15 cm long or broadly elliptic or ovate; trees often with slender bole and narrow regular canopy of short horizontal branches clustered at the top; twigs with raised diagonal lines or tree with pronounced **stilt roots**	***Xylopia*** spp. Gp12C
Lvs normally **broadest well above middle** (oblanceolate) AND acuminate (less commonly acute); symmetrical but not egg-shaped on broader (i.e. NOT obovate)	
Lvs glaucous (e.g. pale blue below)	Gp 12B
Lvs not glaucous	
Base of leaf gradually tapered to petiole: cuneate; lvs often <15 cm long	Gp 12D
Base of leaf with an abrupt 'turn', approaching petiole in a steep angle (obtuse or cordate); (petiole often seeming longer below); lvs often >15 cm long	Gp 12E

Genus	Group	Notes on flowers (green to yellow in colour unless stated)
Hexalobus	12A	In lf axils; 6 v. wavy narrow petals joined at base; flr bud 'beaked'
Monocyclanthus	12D	On older wood; 6 petals in a single ring
Isolona	12F	In lf axils; 6 narrow lobes (±purple inside) on a cupular base
Uvariopsis	12E	On older twigs or bole; 4 petals + 2 sepals, globose
Uvariodendron	12D,E	Globose, on older wood, petals 3 outer + 3 inner not meeting at base, red inside
Anonidium	12E	Several globose fls on stout common stalks from older wood; 3 sepals; 3+3 velvety petals
Uvariastrum	12D	Flower buds strongly 'beaked' where 3 sepals join; 3 thin outer petals like the 3 inner petals
Mischogyne	12D	White, on older wood; ovoid, not beaked; 3 thin outer like 3 inner
Polyceratocarpus	12B	In lf axils; 3 outer larger than 3 inner
Cleistopholis	12A	On slender stalks in lf axils; 3 outer larger than 3 inner
Pacyphodanthium	12A	In small clusters nr lf axils; 3 inner petals smaller
Enantia	12D	3 fleshy ovoid petals (surrounded by 3 thin 'sepals')
Monodora	12E,F	Very showy; outer 3 very wavy, narrow, spreading, dappled red; inner 3 smaller
Neostenanthera	12B	On long hairy pedicels on twigs; outer 3 petals linear, to 7cms; inner 3 much smaller
Piptostigma	12B	3 very slender petals (+3 outer v. inconspicuous) up to 8 cm long
Xylopia	12C	Small and narrow, in lf axils, 3 outer petals linear; 3 inner same or different
Greenwayodendron	12F	Small and narrow, in lf axils, 3 outer linear petals similar to 3 inner petals

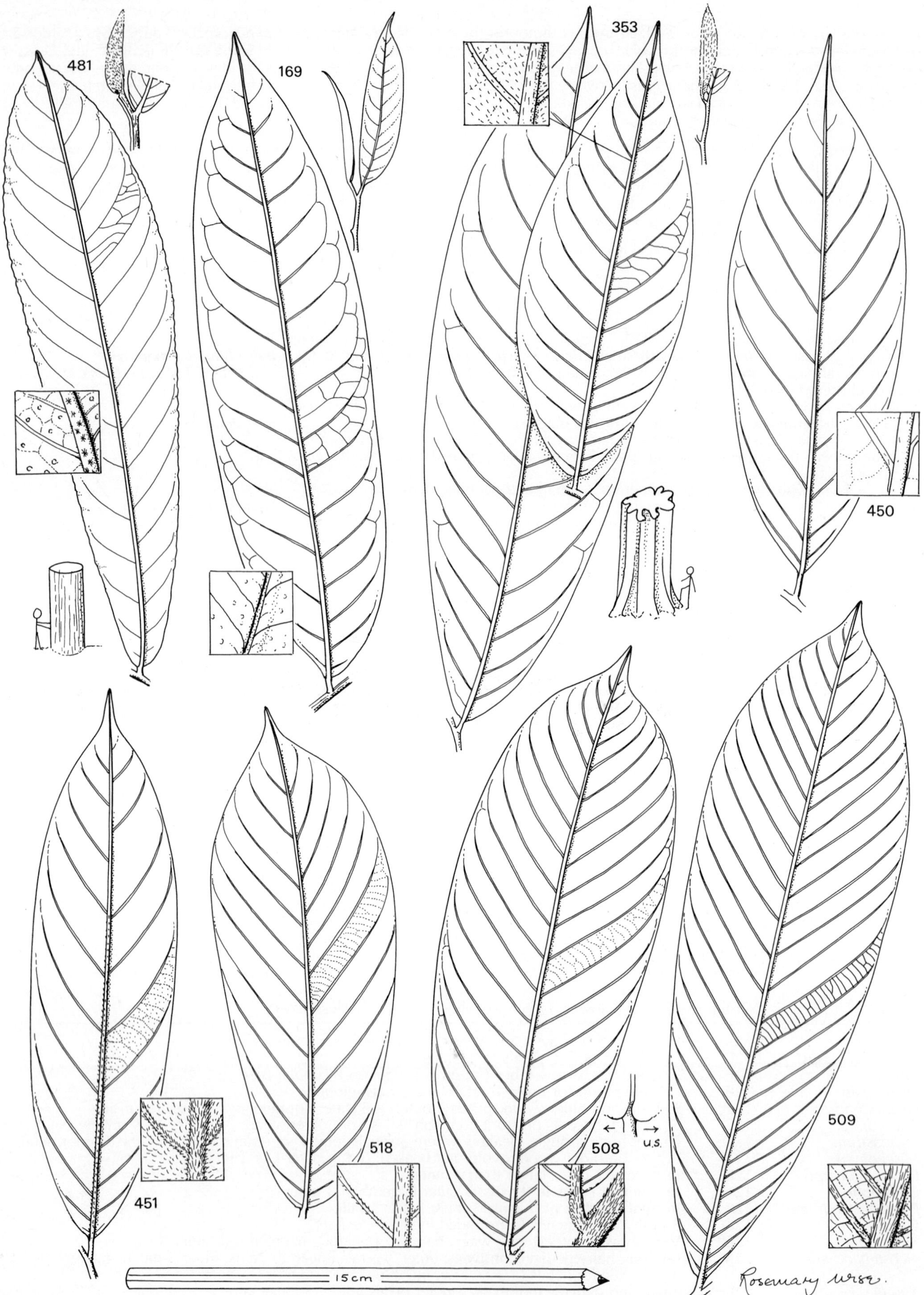
481
169
353
450
451
518
508
U.S.
509
15cm
Rosemary Wise.

Group 12A
(Annonaceae: larger trees; lvs >15 cm long, narrow and usually asymmetric)

Medium-sized trees with branches often appearing like long pinnate leaves (except *Hexalobus*) because of regular arrangement of equal-sized lvs; bole fairly smooth, 'quilted' with shallow fissures, often with vertical rows of lenticels; slash v. strongly sweet-scented and hot tasting. *Hexalobus* is very variable; older trees are v. fluted, with a peelable ribbon-like slash. The other two species are normally found with very cylindrical boles.

Lower surface v. clearly glaucous – Group 12B
Lower surface not v. glaucous
- Margin of lf wavy (up and down); petiole v. short; lf with fine translucent dots and yellow stellate hairs below; slash yellow to orange-brown, extremely fibrous-stringy with thick granular to gritty stripes, and large pores; young unfolding lvs conspicuous at end of twigs, hairy and turned to one side (see Gp 12D); fts like a bundle of woody crystals 5 cm across, many seeded — ***Pachypodanthium staudtii***[1,2] [*KUMDWIE*] 481
- Margin of leaf not wavy: leaf without stellate hairs
  - Leaf glabrous below; young unfolding lf not v. hairy; margin of lf v. recurved; top of lf v. glossy; finely peppered or bumpy (lens); in **swamps or secondary forest**, with leading shoot arching over from vertical to horizontal, and regularly arranged glossy lvs; slash white to pale orange-brown (with thin black layer below outer surface), soft-fibrous, becoming thick (and 'chunky'), gritty and streaked, normally with pleasant scent like eau-de-Cologne, less peppery than the above; fts 1-2 seeded, globose, with very rough, warty seeds — ***Cleistopholis patens***[2] [*NGONONKYENE*] 169
  - Leaf with (straight) hairs lying flat on lower surface; unfolding leaf yellow + hairy; single twigs often with leaves of a variety of shapes and sizes; laterals evenly curved up to nr. margin and sl. impressed above; branches slender and horizontal at first, but tree becoming large and irregular and **deeply fluted;** outer bark with narrow, oblong flakes; slash brown outer merging to yellow inner (with dark outer line over), all darkening, peeling in v. long strips; slash sometimes with slightly salty taste (as well as normal Annonaceae smell and taste); lvs drying orange-ish; fts oblong, rounded ends, 8 cm x 4 cm, woody + hairy — ***Hexalobus crispiflorus***[2] [*DUABAHA*] 353

NOTES: 1) Young *KUMDWIE* trees (in shade) have rather different leaves: oblanceolate, acuminate, and with the base slightly asymmetric where it is twisted around the very short petiole. The stellate hairs and rather thick, juicy young stems and buds are then useful diagnostic characters.

2) *Cleitopholis* and *Monodora* spp. have the leading shoot arched to the horizontal (Troll's model – obvious in saplings) – a habit adopted by many legumes (Gps 37,38) as well. The other trees of the family though, usually have a straight, vertical, leading shoot and horizontal branches (Roux's model – see Hallé *et al*., 1978).

Group 12B
(Small Annonaceae trees: leaves glaucous; oblanceolate)

Note: the last few *Xylopia* species (Gp 12C) also produce glaucous leaves, as do *Friesodielsia enghiana* and *Monanthotaxis* spp. which are climbers occasionally found as erect treelets. **If some lvs are obovate, ovate or lanceolate, check *Xylopia* first.**

Leaves with cuneate ±equal-sided base and distinct, narrow petiole; veins sometimes obscure; fts ellipsoid, 1-seeded, on long stalks (clustered in heads derived from a single flower)
- Lvs with a few fine hairs below; petiole v. slender and finely channelled; lf often <15 cm long; finer veins obscure, margin slightly undulate — ***Neostenanthera gabonensis*** 450
- Lvs underneath, and young stem with dense, long, yellowish hairs; petiole not particularly slender; **evergreen forest** — ***Neostenanthera hamata*** 451

Leaves slightly cordate or rounded or unequal at base with short, thick petiole; veins ±scalariform; fts not long-stalked; ellipsoid but several-seeded
- Lf with <20 laterals
  - Finer veins clearly visible and close scalariform; base asymmetric; fine hairs on midrib below; laterals impressed above; impressed midrib above without hairs. Shrub or small tree — ***Polyceratocarpus parviflorus*** [*WEDEABA-HOA*] 518
  - Finer veins not v. distinct; tiny white hairs on lamina; midrib channelled (near base) above, with hairs; small tree with short, regular horizontal boughs clustered at top of cylindrical bole; bark greenish black; slash (black line/) pink-brown or orange + fine network of lines, darkening — ***Piptostigma fasciculatum*** [*DUASIKA-FUFUO*] 508
- Lf + many laterals (*c.*20 or more) with visible scalariform venation; young lvs + long spreading hairs along midrib; yellow hairs on finer veins; **evergreen forest** — ***Piptostigma fugax*** 509

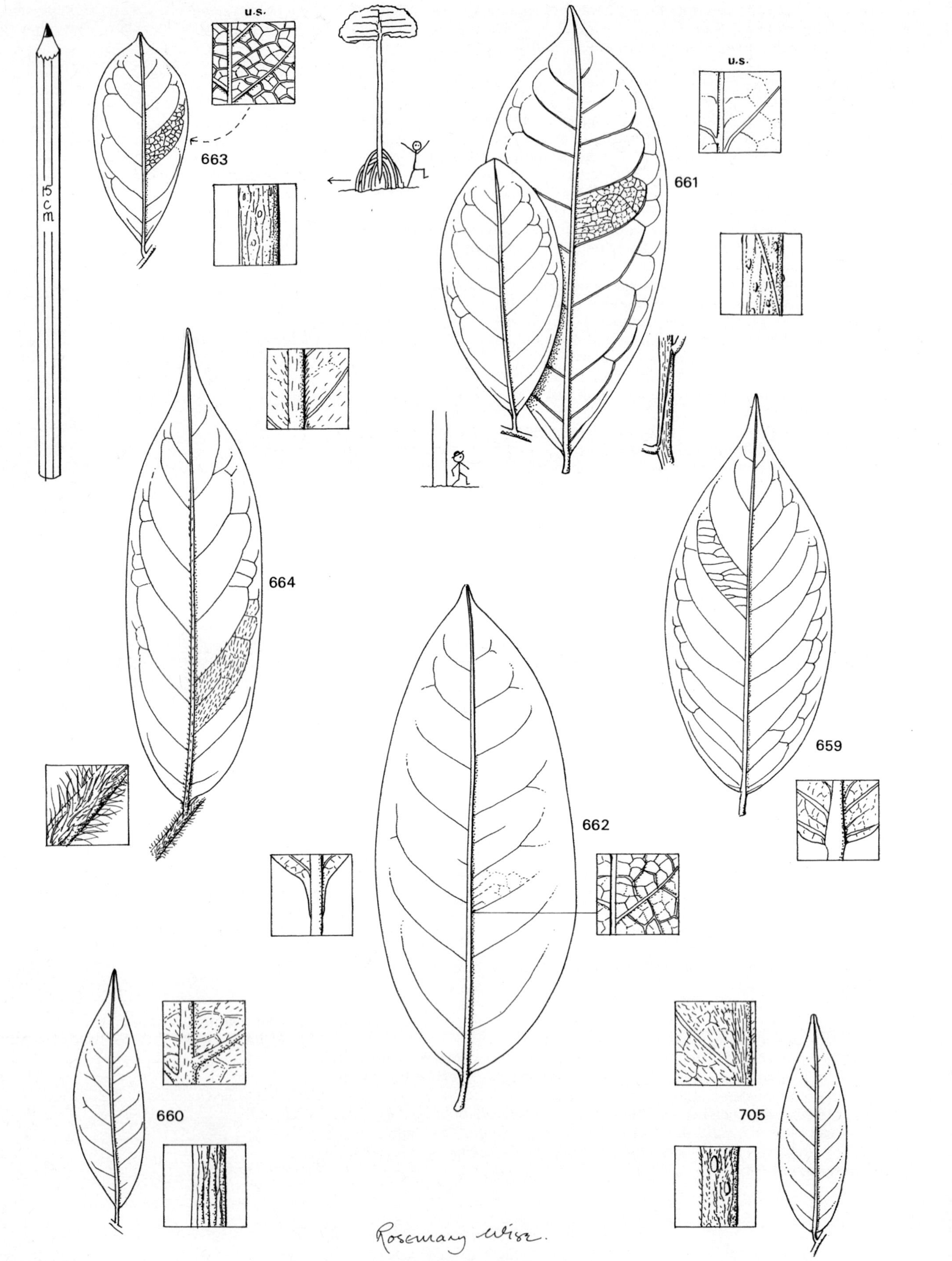
u.s.
663
15 cm
u.s.
661
664
659
662
660
705
Rosemary Wise

Group 12C: *Xylopia*, etc.
(Annonaceae: slender trees with ovate, obovate or lanceolate-elliptic lvs.)

Canopy lvs broadest above the middle and $\pm$rounded at apex; (shade leaves sometimes broadest below middle and acute or acuminate, but then lvs not glaucous nor very hairy); hairs, if present, sparse and inconspicuous; slash v. fibrous, yellow darkening to brown, and strongly scented; fts with arils

- Trees often with stilt roots; venation finely reticulate and prominent above;[6] twigs rough, with inconspicuous lenticels, and pale; fts 2-4 seeded with cupular arils around base; moister for. only (esp. **evergreen for.**) — ***Xylopia staudtii***[1] [*oBAA KoKo*] 663
- Trees without stilt roots (sometimes fluted); venation rather lax and not v. prominent above; twigs with conspicuous lenticels; young twigs often with two raised, sharp but small ridges running diagonally across from base of one petiole to base of next; fts with stringy arils surrounding whole seed — ***Xylopia quintasii*** [*oBAA*] 661

Lvs lanceolate (-elliptic), either hairy or glaucous[4]

- **Lvs + long, dense, obvious, orange-brown hairs on lower surface** (except when old); lvs lanceolate, NOT glaucous; twigs also v. hairy; fts hairy, 2 cm wide, ellipsoid, several seeded
  1) Laterals rather wavy, often forking before half-way to margin, and minutely prominent above; tree with short horizontal brs; bark quilted-rough; slash with fibrous and granular streaks, spongy, gritty and contoured with yellows and browns, with a pale inner bark, all darkening with time; sapwood almost white and wet — ***Xylopia villosa***[2] [*OBAA-FUFUO*] 664
  2) Laterals evenly curved up to nr. margin and sl. impressed above; branches slender and horizontal at first, but tree becoming large and irregular and **deeply fluted**; outer bark with narrow, oblong flakes; slash (with dark outer line) brown outer merging to yellow inner, all darkening, peeling in v. long strips, without granular sections; slash sometimes with slightly salty taste (as well as normal Annonaceae smell and taste); lvs drying orange-ish — ***Hexalobus crispiflorus*** [*DUABAHA*] (See Gp 12A)
  3) Laterals not as above: tree in wet places in western region – consider next species
- **Lvs not so hairy, but often glaucous**; usually in **dry or swampy forest**; fts narrow (<1.5 cm wide)
  - Lvs regularly >10 cm long and >3 cm wide; slightly leathery; elliptic-lanceolate; slash stringy
    - Lvs with finely reticulate, regular, raised venation below and above; (yng lvs v. hairy) lf base decurrent into raised edges of petiole channel; lvs drying a vivid red-orange colour below; twigs v. striate, with many lenticels; tree often **by rivers** and with **stilt-roots**, but neither is always the case — ***Xylopia rubescens*** 662
    - Lvs sometimes with finer venation raised and reticulate, but then not as regular little squares; often v. glaucous below and above; lf base often obtuse, with midrib above reddish especially at the base where it is broadened; tree in **dry forest or savanna** — ***Xylopia aethiopica***[5] [*HWENETIA*] 659
  - Lvs usually shorter than 10 cm, narrowly elliptic or lanceolate and rather thin; slash rather brittle[3]; trees often near rivers
    - Lf symmetrical, blue-glaucous, with rounded-acute apex and rounded-obtuse base; twig with young bark usually black, with fine striations and rounded lenticels; with many v. fine hairs, especially along midrib below, and in slight channel at base of midrib above; often by rivers; bole, slash, etc., like *X. villosa* — ***Xylopia parviflora***[3] [*'OBAA-HWO'*] 705
    - Lf $\pm$asymmetrical, not glaucous, with acuminate apex and cuneate base; twigs and young lvs with silky fine hairs of visible length — ***Xylopia elliotii***[3] 660

NOTES:
1) Sapling lvs of *X. staudtii* are longer. If twigs are black and lvs are long, consider *X. rubescens*, which has similar venation.
2) It is possible that more than one species in Ghana will key to this point: a plant with 4 cm long ovoid fruits, very dense, shiny orange hairs, which are beautifully silky-shiny on young trees, and are more persistent than in *X. villosa*, is known from **evergreen forest** regions. It appears to be closely related to *X. hypolampra*.
3) *Xylopia acutiflora* is a scrambling shrub (or small tree, at least in Nigeria) similar to these last two. *Xylopia* is in need of a taxonomic revision, so collection of fertile specimens is desirable.
4) A shrub or small tree of the **very driest, and disturbed forest**, with lvs elliptic or oblong, with rounded apex and rounded or obtuse base and with petiole <1 mm long is *Hexalobus monopetalus*
5) *X. aethiopica* fruits are *c.*4 cm long x 3 mm wide, slightly constricted between the seeds, and often sold in markets for their peppery, medicinal qualities. The tree is sometimes cultivated (?possibly introduced to Ghana), and best distinguished from *X. rubescens* on the basis of its preferred habitat.
6) If lf. being keyed at this point has v. fine reticulate venation, prominent above, consider *Mischogyne elliotiana* (Gp 12D) a **dry forest**, small tree without stilt roots.

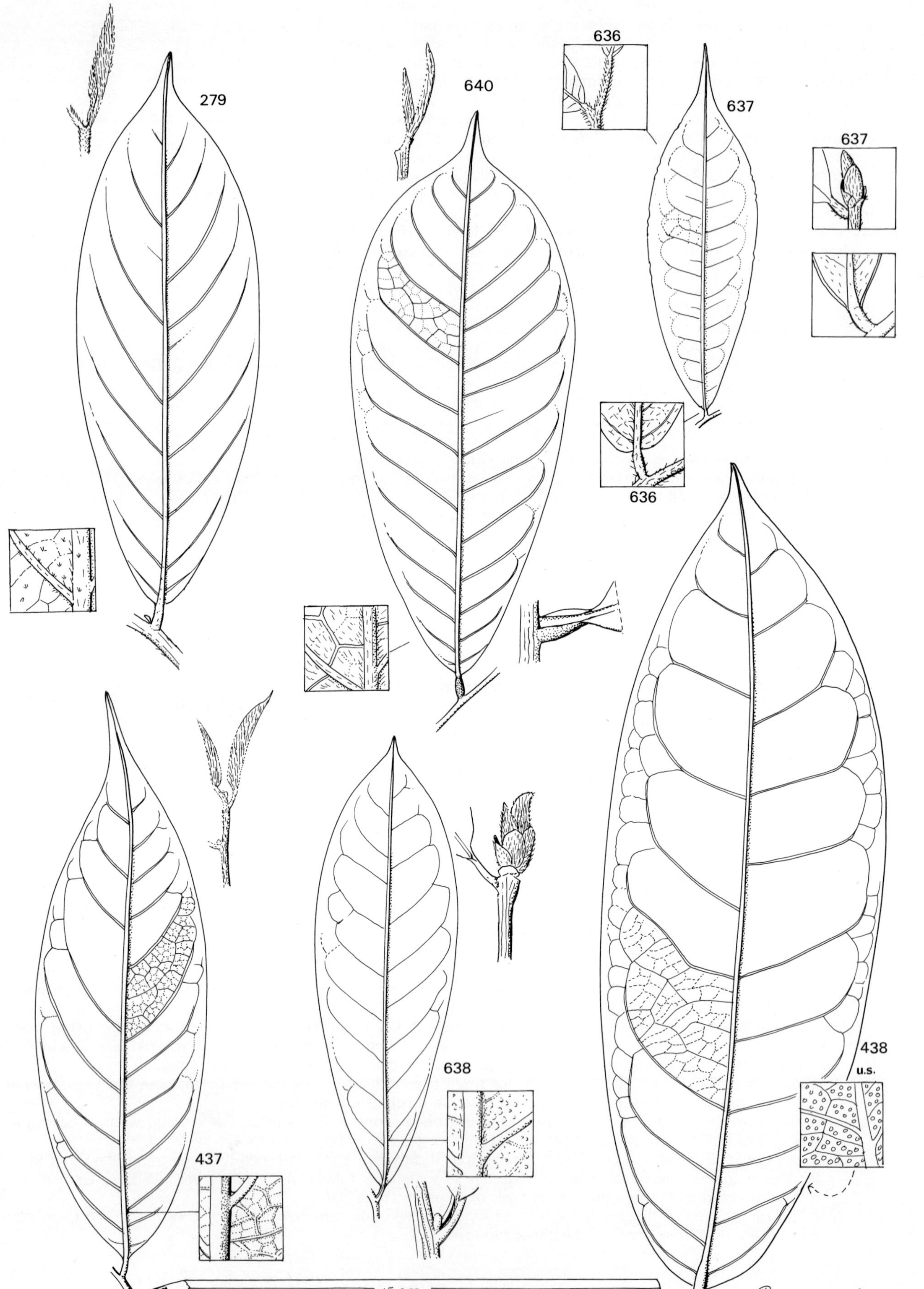
279
640
636
637
637
636
437
638
438
u.s.
15 cm
Rosemary Wise.

Group 12D
(Annonaceae: lvs oblanceolate and acuminate; base of leaf cuneate)

- Lvs with **long yellow** or brownish hairs on lower surface and often in midrib channel above (lens) OR laterals looping conspicuously and meeting ½ or ⅔ way to margin
  - Plant with (yellow) stellate hairs (mixed with simple ones) on lvs and twigs; venation below ±obscure; lower surface appearing v. bumpy and rough (lens); laterals slightly impressed, and midrib channel filled with hairs above; **slash bright, golden yellow**, contoured, easily peeled, with mealy (like crushed grain) areas and wide pores; crown with small horizontal boughs at top of cylindrical blackish stem; fts 1-seeded, ellipsoid, long-stalked many in 'heads' from each flower — ***Enantia polycarpa*** [*DUASIKA*] 279
  - Plant without stellate hairs; venation not obscure; slash not unusually golden; fts ±several-seeded
    - Lateral nerves not nearly joining; hairs below usually long and dense; young undeveloped leaves at apex of stem curled to one side and densely covered in yellow hairs; lvs often pale and shiny below when fresh; lvs v. variable; slash v. stringy — ***Hexalobus crispiflorus*** [*DUABAHA*] (See Gp 12A)
    - Lateral nerves nearly or actually joining; hairs rather sparse
      - Laterals (10-15 prs) meeting more than ⅔ way to margin; lvs usually >15 cm long; midrib channelled above; petiole smooth and swollen; young leaves at apex of leaf curled over to one side and densely covered in golden yellow hairs (like *Hexalobus*, above); undeveloped lvs hairy, usually only 2 visible, without any small, outer ones (bud scales); twigs corky; bole straight with rough and scaly bark; slash with corky outer layer, fibrous-leathery with pale brown and darker streaks; inner bark with net pattern; fts clustered, from older wood, 5 cm x 2 cm, hairy, several-seeded — ***Uvariodendron occidentale*** [*ESONOKWADU*] 640
      - Laterals (generally <10) arching over and nearly or actually meeting ½ to ⅔ way to margin;
        - Lvs <15 cm long (usually *c.*10 cm); margin and lamina wavy, with impressed laterals; trees with branches clustered at top and sometimes with fluted boles
          - Laterals prominent above; lf base cuneate; unfolding leaves with hairs rather flat, with a few small 'bud scales' around their base; bark rough, flaky; slash finely contoured, sometimes with a black outer layer, v. fibrous, pale pinkish-yellow turning dark brown, wet and gritty, with yellow or pinkish sapwood; fts smooth — ***Uvariastrum pierreanum*** [*OTWE-EHI*] 637
          - Laterals impressed; lf base very slightly rounded or obtuse; unfolding lvs and young twigs with long, spreading hairs (longer than width of midrib); **evergreen forest**; bark, etc., probably like previous species; fts with obvious raised reticulations — ***Uvariastrum insculptum*** 636
        - Lvs often >15 cm long, rather thick and shiny, not wavy; twigs with raised lines running between nodes: juvenile or shade lvs — ***Xylopia quintasii*** [*oBAA*] – (See Gp 12C)
- Lvs without hairs, or with a few on youngest lvs only OR gland-spotted, OR with reticulations raised above; fts on older wood[2]
  - Lvs with some hairs when young, particularly on unfolding lvs at twig tip; **dry forest** and not (recorded) in evergreen forest
    - Venation conspicuously fine-reticulate and prominent above; without glandular spots; petiole not v. swollen; venation v. fine; young twigs rather hairy; fts hairy, up to 4 cm long, on long slender pedicels — ***Mischogyne elliotiana*** 437
    - Venation not remarkably prominent and reticulate above; undersurface of lf + fine gland dots; fresh lvs v. scented when crushed; lvs + slight metallic colour below, often drying blue-grey; midrib prominent or flat above *c.*3 cm from base, usually hairy; unfolding lvs with small bud scales at base; dry petiole swollen and cracked; lats. 9-12 pairs; lvs usually <15cm long; slash becoming brown, with a darker inner bark; fts hairy on short pedicels — ***Uvariodendron angustifolium*** [*BOMMOFOKWADU*] 638
  - Lvs and twigs glabrous, even when very young, with obvious pellucid gland spots, and venation prominent and reticulate above; petiole swollen (2 mm wide); venation not dense (*c.*1 vein/ mm or fewer); **evergreen forest** — ***Monocyclanthus vignei*** 438

NOTES: 1) If laterals are v. wavy and the young twigs have dense, long hairs, check *Xylopia villosa* (page 75).
2) If the descriptions do not match your specimen at this point then reconsider Group 12F, which has generally smaller lvs, varied in shape on one twig etc. BUT with flwrs and fts amongst foliage.

Group 12E
(Annonaceae: leaves oblanceolate with base minutely or broadly cordate, or obtuse)
(Usually understorey spp. with lvs longer than 20 cm)

Lvs with stellate hairs, and minutely cordate base — See young *Pachypodanthium* (Gp 12A)

Lvs without stellate hairs

- Midrib + dense, long orange-brown hairs; lvs up to 35 cm long; laterals regular + joined; shrub or v. small tree with stringy pale red slash turning brown; young lvs pinkish and soft, drooping at the end of twigs — ***Uvariodendron calophyllum*** [*'ESONOKWADO-KoKoo'*] 639
- Midrib without dense hairs
  - **Evergreen forest** tree with dense gland dots below, and almost cuneate lf base – see *Monocyclanthus* (12D)
  - Semi-deciduous or dry forest trees, OR lvs not gland-dotted OR lf base cordate
    - Lf often with a few fine hairs; midrib with narrow channel above; slash thick fibrous, yellow-brown, darkening; terminal bud without overlapping scales
      - Veins clearly reticulate and visible; rather erratic; prominent below and above, especially along the line where lamina folds over into midrib channel; lf base minutely lobed; lvs normally with gland-dots visible below — ***Uvariopsis globiflora*** [*ASUMPA-AKOA*] 641
      - Veins, apart from a few of the largest ones, not v. clear, although laterals v. prominent; lvs often >20 cm long + fine hairs on midrib and often many on lamina as well (v. hairy when unfolding); crown deep and dense, even when shaded; bole straight; slash gritty, very thick, fibrous, soon turning very dark; fts up to 35 cm long on stout stalks from older wood, like a custard/sweet-apple (*Annona* spp.) — ***Anonidium mannii*** [*ASUMPA*] 82
    - Lf totally hairless; midrib with thickened channel or flat above; flowering when lfless; terminal buds with several overlapping scales; crown rather shallow and sparse if tree in shade; slash (black line over) pale orange-pink, merging with the orange inner bark, with bundled fibrous lines, peelable; fts spherical, ±woody, up to 20 cm diam — ***Monodora myristica***[1] [*WEDEABA*] 440

Group 12F
(Small trees; lvs usually <15 cm and narrow; several young lvs unfolding at once, v. varied, often drying black)

The v. young twigs of dried specimens have transverse creases; olden parts are black and smooth becoming corky at the end of the previous season's growth.

- Fresh lvs v. pale below; young twigs lenticellate, yellowish; apical bud with several overlapping, similar scales in rows; lvs v. variable: lanceolate to obovate; yngest twigs of dried specimens with conspicuous bulges, black; petiole >3 mm long; bark distinctive: grey with paler ridges; venation barely prominent above; flowers spectacular, on lfless branches; fts spherical, *c.*2 cm diam., + many seeds — ***Monodora tenuifolia*** [*MOTOKURADUA*][1] 444
- Fresh lvs dark green below; young twigs (and often older bark) black, with regular diamond-shaped narrow channels; flowers amongst new lvs; lvs often peppered with minute gland spots below (powerful lens often needed)
  - Midrib above fairly smooth, or raised ±a fine channel; terminal bud with only a few hairs, often in a thin line along one of the unfolding lvs; petiole not shorter than 2 mm; abrupt transition from youngest, v. black twigs to older, yellow-brown, glossy bark; slash creamy, turning steadily dark brown; flowers amongst new lvs, 1 cm or more long; fts large and conical, *c.*5 cm long, with bumps above the many seeds — ***Isolona campanulata***[2] [*DUAWISA*] 369
  - Midrib above + several fine raised dots nr. base; terminal bud of unfolding lvs v. hairy
    - Treelet of very **driest forest types**, often in rocky places; petiole usually >2 mm long, rather 'fat' at base, with transverse wrinkles like an earthworm; venation prominent above; flowers + fts cauliflorous — ***Dennettia tripetala*** 221
    - Treelet common and widespread in **moister forests**; petiole often <2 mm long and slender; flowers in lf axils hairy outside, 2-3 together with petals red towards base; crown dense, heavy, dark green; bark black with lenticels and girdled with raised horizontal rings; slash yellowish v. rapidly darkening, with many large pores; lvs markedly asymmetric; fts <1cm diameter, spherical, 2-seeded, red — ***Greenwayodendron oliveri*** [*DUABIRI*] 346

NOTES: 1) The name refers to the patterning of the bark, with pale and dark stripes like the coat of the squirrel-like *ABOTOKORA*: the bark seems to be the best means of separating species of this group when not fertile. *Monodora brevipes* is similar to *M. tenuifolia*, but is a shrub of **evergreen forest**, with broad inner petals, contrasted with the narrow inner petals of *M. tenuifolia* which have hairy structures on the middle of their inner face. See also NOTE 2, Group 1A.

2) There are other species in the genus *Isolona* in Ghana, which are shrubs or (v.) small trees. *I. cooperi* has large leathery leaves: *I. deightonii* has long orange hairs on young stem and midrib. 'DUAWISA' is often applied to *Pachypodanthium* as well.

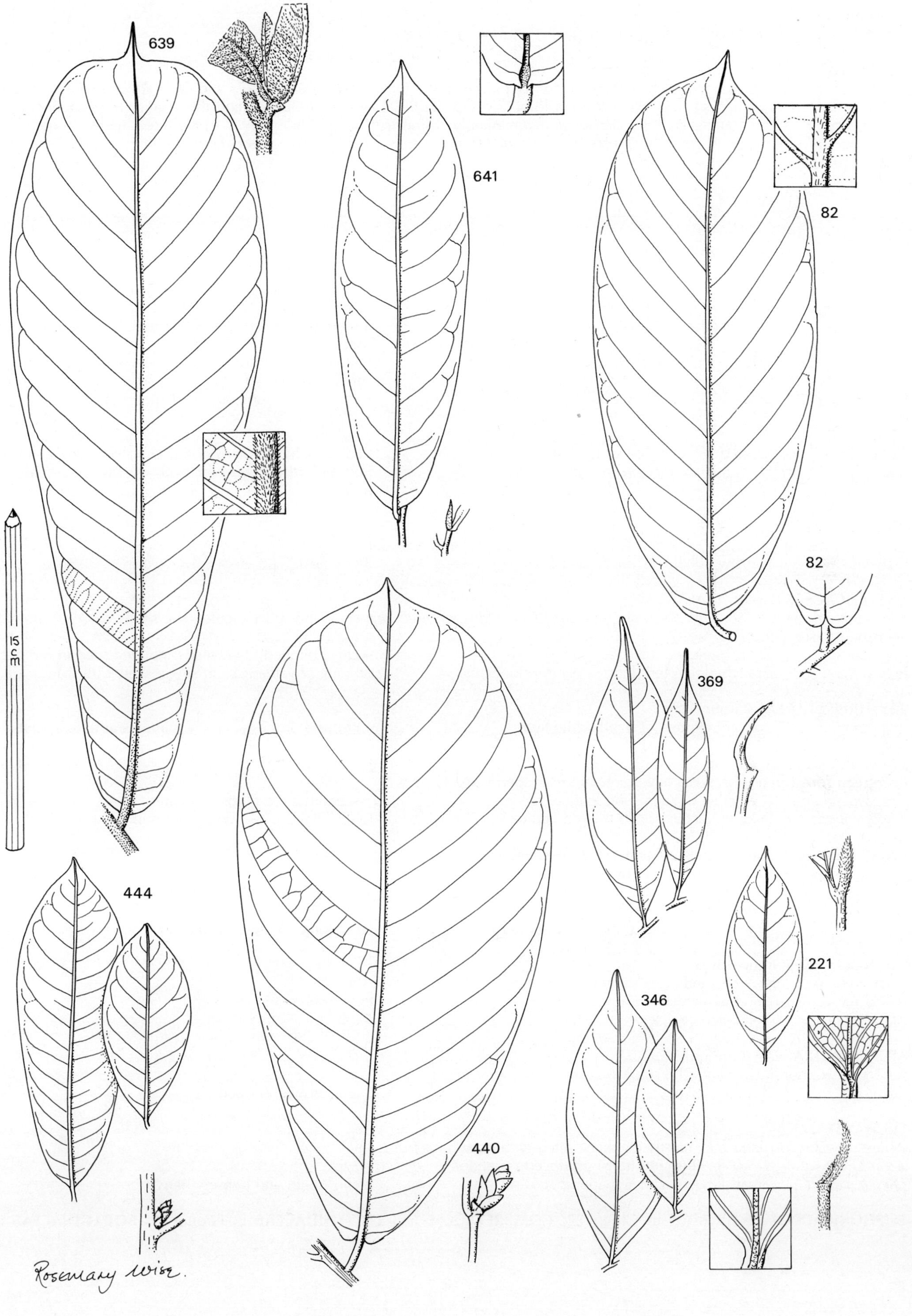
639
641
82
82
15 cm
369
444
221
346
440
Rosemary Wise.

GROUP 13: OLACACEAE – EUPHORBIACEAE (part), etc.
(Lvs simple, alternate, entire; slash not both fibrous and sweet scented)

Notes on families
Group 13 is a repository for an assortment of species in many families.

OLACACEAE
This family, includes *Octoknema* which was previously placed in the Octoknemataceae. *Coula* and *Ocktoknema* are exceptional in having stellate hairs and petioles often >2 cm long (see Gp 27B). *Strombosia, Ongokea* and the smaller trees *Aptandra* and *Heisteria* have yellow-orange, granular slashes and remarkably cylindrical boles. The slash is sometimes slightly scented. The calyx often develops around the fruits. *Ptychopetalum anceps*, a common small shrub in this family, has almost irridescent, small reddish fruits.

Genus	Flowers	Fruits
(Group 13A, 13B)		
Aptandra	Short axillary racemes of small flwrs	Calyx grows into 4 cm pink funnel around ellipsoid drupe
Heisteria	Small axillary clusters	Calyx becomes 4-5 lobed, red 4 cm around 1.5 cm drupe
Olax	Short axillary racemes	3-seeded, red, 1.5 cm with small 2 mm calyx cup at base
Ongokea	Small panicles of small flwrs.	Yellow 2cm drupe, inside 2-3-valved calyx, like a capsule
Strombosia	Small, red in axillary clusters	Purple 2 cm drupe, calyx remnants at top
(Group 27B)		
Octoknema	Small axillary spike-like panicles	Red, v. hairy drupes 1.5 cm, with tiny calyx at apex
Coula	Short axillary racemes of small flwrs	Red, edible drupe, 4 cm with minute, hairy calyx at base

ICACINACEAE (Group 13B)
Closely related to Olacaceae, a family of many climbers and the following tree:

Leptaulus	Narrow, tubular, clusters near axils	3 cm, fleshy, pointed; calyx at base

MYRISTICACEAE (Group 13A)
The 'nutmeg' family: the fruits resemble nutmegs. Trees with whorled branches and red exudate. The seeds have a 'ruminate endosperm' (see Group 12 notes)

Coelocaryon	Short branched infl of tiny flwrs	4 cm ±bilobed, 2-valved + pointed seed in red aril
Pycnanthus	Branched, twisted, loopy infls (Monoecious)	Oblong, 2-valved; 2 cm single seed with stringy aril

ERYTHROXYLACEAE (Group 13A)

Erythroxylum	Axillary clusters of small white flwrs	1 cm, red drupes (calyx + stamens at base at first)

SCYTOPETALACEAE (Group 13A, 13B)
Evergreen forest trees; the seeds have ruminate endosperm (see Gp 12)

Rhaptopetalum	3-petalled flwrs in short axillary racemes	1-seeded, 2 cm ellipsoid drupe
Scytopetalum	Axillary racemes with twisted pedicels	Red, 2 cm, ribbed, fleshy but dehiscent

CAPPARIDACEAE (Groups 27B, 31 and 13B respectively)
Three species with very different, but always glabrous foliage.

Bucholzia	1 cm brush of many stamens; in racemes	10 cm x 7 cm with 1.5 cm 'stipe'
Euadenia	2 linear petals 5 cm long; in terminal racemes	5 cm x 2 cm with rough surface
Maerua	On slender pedicels; many red stamens	6 cm long, many-seeded, rough

IRVINGIACEAE (Group 13C)
Large trees with large fruits and large stipules.

Irvingia	Short axillary racemes	Like small mangoes (5 cm)
Klainedoxa	Spike-like panicles	5-lobed, *c.*7 cm x 4 cm.

MEDUSANDRACEAE (Groups 13C, 13D)
Resemble Euphorbiaceae and Chrysobalanaceae.

Soyauxia	Axillary spikes	3 cm, brown, 3-lobed

DICHAPETALACEAE (Group 13D)
Related to Chrysobalanaceae; mostly climbers, often with poisonous fruits.

Tapura	Small flowers (and fruits) produced on petioles!	
Dichapetalum	Small flowers with Y-shaped petals	2 cm, rough and densely hairy

EUPHORBIACEAE are discussed in Group 22; **COMBRETACEAE** and **LECYTHIDACEAE** in Group 25; **BORAGINACEAE** in Group 26.

GROUP 13: Olacaceae – Euphorbiaceae, etc.: key

Nerve axils with domatia
- Pit domatia, without hairs
  - Petioles often >3 cm long and lvs clustered — Group 25
  - Slash red; lvs with glands — Group 14B (*Magnistipula zenkeri*)
- Tuft domatia
  - Petiole swollen at tip or lvs clearly trinerved — Group 21
  - Petiole not swollen at tip, or lvs barely or not at all trinerved — Group 13D

Leaves without domatia
- Petiole distinctly swollen at one end or articulated towards tip; often >1.5 cm long
  - Leaves, or young twigs with scurfy stellate hairs (petiole often >2 cm long) — Group 27 (esp. 27B)
  - Leaves without stellate hairs
    - Petiole v. clearly swollen; lvs hairy or slash fibrous + smell of green beans — Group 37A (*Baphia* spp.)
    - Petiole not particularly swollen, but discoloured or bent at tip, with fine channel; lf completely glabrous; slash granular, with gritty streaks — Group 13B (*Strombosia*)
- Petiole not swollen, or barely longer than broad
  - Leaves with silvery scales on lower surface or many, regular, tiny (lens) red spots (glands) below — Group 21
  - Leaves with no silver scales and not peppered with tiny red spots
    - **Evergreen forest or riverine** trees with glands or gland-like structures on base of leaf or petiole (REMINDER – if SLASH is RED and the leaf has basal glands see Gp 14)
      - Glands at the base of large-medium lvs, and ±brown hairs.
        - Outer bark black (even on twigs), very hard and brittle — See *Diospyros gabunensis* (Gp 11)
        - Outer bark not hard; slash fibrous — See *Strephonema*, etc. (Gp 21)
      - Fine thread-like structures on petiole — See *Didelotia* spp. (Gp 37A)
    - **Drier forest tree** OR petiole and base of leaf without such irregularities
      - *Mature leaves with hairs*, or twigs v. hairy
        - Base of leaf cordate; tree usually with red, watery exudate — See *Pycnanthus* (Gp 13A)
        - Base of lf not cordate: tree without red exudate (not *Pycnanthus*)
          - Branches whorled and leaves extremely clustered at twig ends, in layers — See Group 25
          - Branches not whorled OR lvs not clustered into 'pseudowhorls' — Group 13D
      - *Mature leaves (and usually twigs) hairless*
        - Venation between laterals (and a few larger veins) obscure below **OR tree with obviously red exudate**; lower surface often slightly glaucous or discolorous — Group 13A
        - Venation clearly visible between laterals
          - Leaves >10 cm long and rather leathery, with only a few of the larger veins visible between the lateral nerves; small trees in **wetter forests** — Group 13A
          - Leaves shorter, OR not leathery OR tree not in evergreen forest OR veins completely visible
            - Twigs with conspicuous stipules, or with obvious scars on nodes of young twigs where (large) stipules have fallen — Group 13C
            - Twigs without conspicuous stipules and without stipule scars at nodes; trees without large buttresses
              - APEX of lvs mucronate (i.e. with a tiny sharp projection at tip); small trees usually in **dry forest** understorey; slash sometimes scented — Group 13B
              - APEX of lvs not mucronate, OR tree in **evergreen forest**
                - Young twigs with two well-defined, opposite wings
                  - Slash fibrous; fragrant; twigs lenticellate — See *Xylopia* (Gp 12B)
                  - Slash not fragrant, granular; twigs not lenticellate — See *Heisteria* (Gp 13B)
                - Young twigs rounded, or sometimes flattened, wrinkled or many-winged, but not 2-winged
                  - *Twigs flattened, strap-like*
                    - Venation ±scalariform
                      - **Evergreen forest** tree; paired stipules at lf base — Check *Soyauxia grandifolia* (Gp 13C)
                      - Not *Soyauxia*; lf base extremely asymmetric; with many translucent small lines in lamina; (minute teeth often just visible with lens)
                        - In swamps or nr. sea in **evergreen for**. zone — ***Casearia barteri***
                        - Elsewhere, especially in forest in hilly places — ***Casearia calodendron*** (see 17D)
                    - Veins not at all scalariform; usually ±obscure — Group 13A
                  - *Twigs not strap-like*
                    - Margin typically undulate or (almost crenate)
                      - Finer venation markedly transverse and prominent above; slash yellowish, darkening (outer bark sometimes rather pale, soft and corky) — ***Diospyros viridicans*** (Gp 11)
                      - Finer venation not transverse-prominent; petiole often swollen at tip; leaves pustular below; slash soft with sweet watery exudate — ***Scottellia*** (Gp 17D)
                    - Margin not undulate, etc.
                      1) Two strong ascending nerves arising near lf base; [lf + fine white spots] — ***Holoptelea*** (Gp 18)
                      2) Lf base asymmetric; slash 'oPAHA'-scented — ***Drypetes*** spp. (Gp 17)
                      3) Neither of above — Group 13B

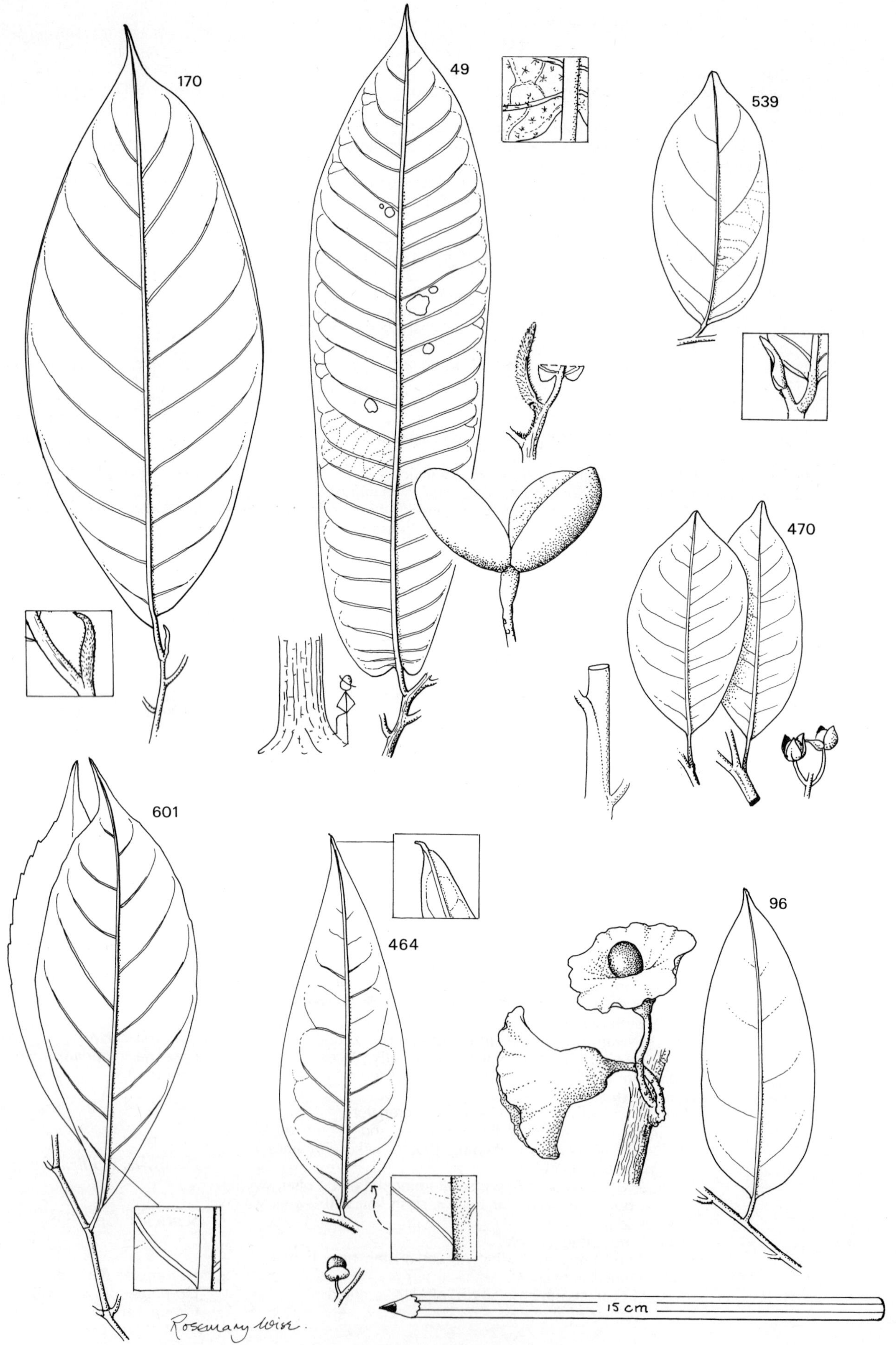

170
49
539
470
601
464
96
15 cm
Rosemary Wise.

Group 13A: Myristicaceae-Olacaceae, etc.
(Glabrous lvs with obscure venation OR tree with red exudate)

Leaves hairy, with cordate base, ±oblong, acuminate with >20 prs *c.*perpendicular laterals; lvs tattered with many insect bites; underside with rusty, stellate hairs; bole straight and cylindrical, without large buttresses; outer bark brown, becoming rough with many rectangular flakes; slash with watery red exudate, pink-brown fibrous and gritty; boughs ±horizontal, whorled, but drooping greatly towards tips in exposed trees; common in **disturbed forest** — ***Pycnanthus angolensis***[1] (MYRI) [*OTIE*] 49

Leaves not hairy, nor cordate; venation obscure

- **Not *Erythroxylum* (see bottom of page): leaf leathery, OR pointed OR >15 cm long OR twigs without stipules, etc.**
  - *Young twigs not flattened* (although often striate) OR tree with red exudate or lanceolate lvs
    - Leaves symmetric, often leathery or broadest above the middle; **evergreen or swampy forest** species
      - Slash with red or brownish watery exudate; lvs oblanceolate and rather rubbery to touch when fresh, with red and yellow translucent spots; youngest leaf always present, pointed and curved, *c.*1 cm long, at tip of twigs; midrib v. wrinkled on fallen lvs; common unbuttressed tree + whorled horizontal boughs within **evergreen forests** (also rivers in savanna areas) — ***Coelocaryon oxycarpum*** (MYRI) [*ABRUMA*] 170
      - Slash without reddish exudate (although slash red): not *ABRUMA* (not this combination)
        - Leaves (obovate to broadly oblanceolate), with rather rounded apex; ±obtuse base; lf base sometimes asymmetric, meeting petiole at different points on each side; apex of twigs sharp-pointed; lf underside slightly silvery; slash spongy, pink-brown; small tree in **swamps**, etc. — ***Rhaptopetalum beguei*** (SCYT) 539
        - Leaves (elliptic), with sharp (acute – acuminate) tip OR with cuneate base; slash reddish, granular, slightly turpentine scented — See *Amanoa* spp. (Gp 13C)
    - Leaves basically asymmetric, rather thin, broadest below middle, acute, usually much less than 15 cm long; (**OR found in non-swampy semideciduous or drier forest)**
      - Young twigs with rings at nodes, strongly 'zigzagged'; lvs sometimes slightly serrated or opposite; disturbed forest, mainly exposed to sun; slash with red or brownish watery exudate, yellow and granular; lvs ±papery — ***Tetrorchidium didymostemon*** (EUPH) [*ANENEDUA*] 601
      - Young twigs without nodal rings or not strongly zigzagged; lvs never serrated nor opposite; midrib channelled above; trees without red exudate
        - Lvs apparently almost stalkless ('subsessile') when viewed from above, petiole <2 mm long; midrib clearly channelled to apex; young twigs v. striate (with raised lines leading to lvs); lf tip usually mucronate; understorey tree often in shade; lvs rather brittle, sometimes with visible venation; slash yellowish, with a distinct, not v. pleasant smell ±like boiled meat — ***Olax subscorpioidea*** (OLAC) [*AHOoHENEDUA*] 464
        - Lvs clearly stalked from above, with petiole >2 mm long, not normally mucronate; lvs fleshy, often sl. leathery; twigs smooth, not v. grooved — ***Aptandra zenkeri*** (OLAC) [*AYENTUDUA*] 96
  - *Young twigs flattened* and striate, with short downward-running scars leading to edge of petiole channel (but no stipules); petiole sometimes slightly discoloured towards tip; lvs <15 cm long, variable, but usually with acute apex, slightly asymmetric; veins sometimes visible: laterals in steep angle to midrib; **bole very cylindrical to base; slash thick, yellow-brown, soft-granular, with** a blackish outer layer — ***Ongokea gore*** (OLAC) [*BoDWE*] 470
- **Leaf symmetric, obovate with rounded apex, <10 cm long, rather thin**, and glaucous purple to reddish below; stipules persistent, sharp triangles above petioles on rather flattened twigs; ±two vague basal nerves passing close to margin; slash pink to brown, fibrous, contoured, darkening; crown with stout, ascending, regularly forking branches — ***Erythroxylum mannii*** (see Gp 13C) (ERYT) [*PEPEA-NINI*] 287

NOTE: 1) *Pycnanthus* seedlings can be recognized by their whorled branches, with leaves bluish glaucous below and stellate, brown hairs (which rules out *Diospyros*). The juvenile leaves are often *NOT* cordate, and suffer less insect damage.

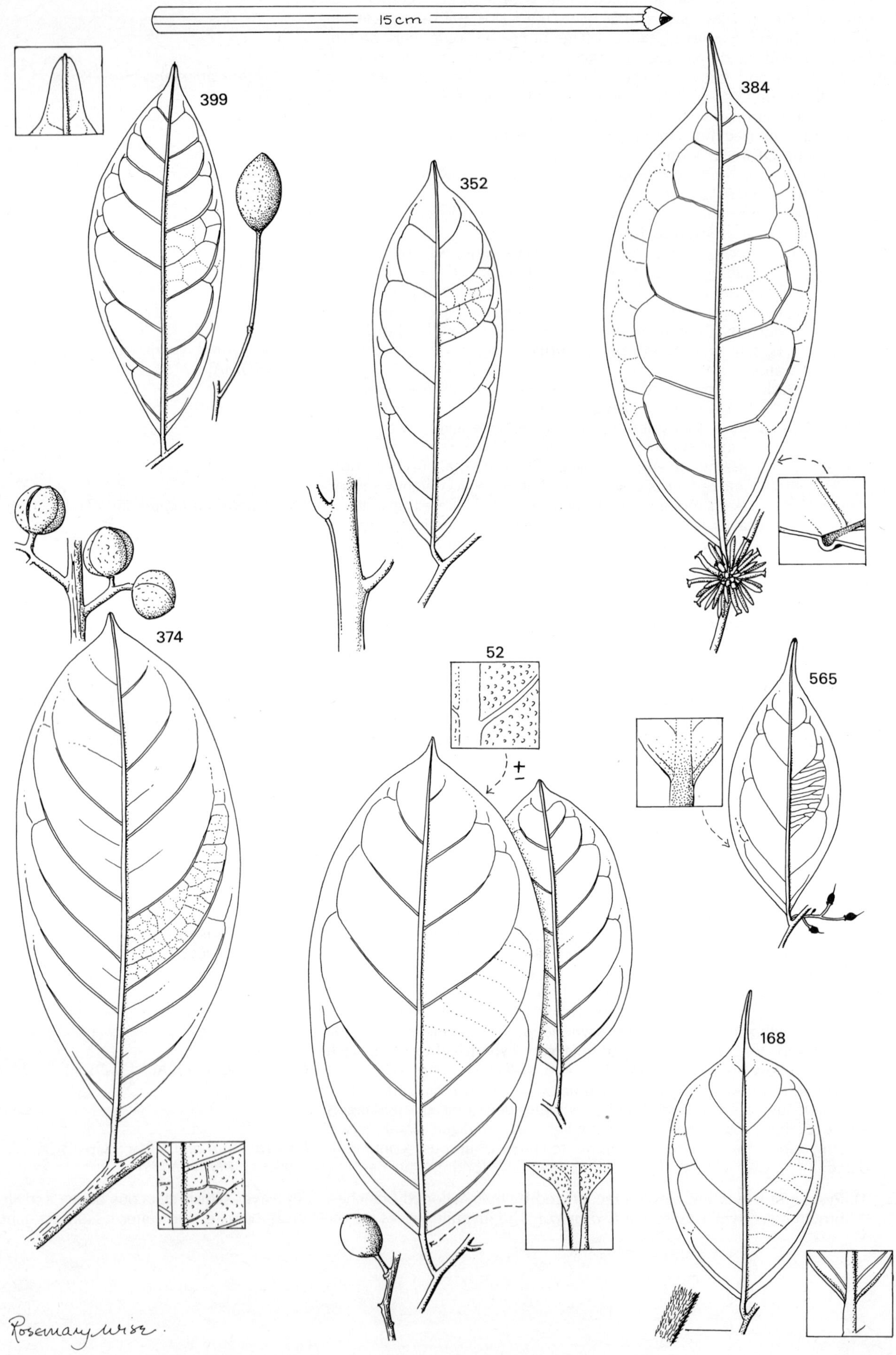
15cm
399
352
384
374
52
565
168
Rosemary Wise

Group 13B: OLACACEAE-EUPHORBIACEAE etc.
Glabrous lvs; veins visible

Families are Euphorbiaceae (see Gp 22) unless abbreviated otherwise.

- Apex of lvs mucronate (i.e. with a tiny sharp projection at tip); small trees especially in **dry forest** understorey
  - Lvs with basal glands — See *Dichapetalum barteri* (Gp 13D)
  - Lvs without basal glands
    - Petiole >3 mm long; lateral nerves clearly looping and joining; lvs glossy, elliptic, symmetrical; surface below slightly uneven or 'bumpy'; petiole channel wavy; slash bitter, granular, gritty, brown-orange with a smell of green beans — ***Maerua duchesnei*** (CAPP) [*KONINI-BERE*] 399
    - Petiole almost 3 mm long; lateral nerves not looping; lvs papery, lanceolate but asymmetric; slash white with pinkish marks, with yellow inner bark, fibrous, with foetid or garlic smell; fresh lvs also with unusual smell — ***Olax subscorpioidea*** (see 13A) (OLAC) [*AHOoHENEDUA*] 464
- Apex of lvs not mucronate, or slash not scented; (in or outside dry forest zone)
  - Young twigs with two well-defined, opposite wings; midrib ±flat or shallowly impressed; venation very prominent above and wavy or rippled, even over recurved margin; petiole often 1 cm or more long — ***Heisteria parvifolia*** (OLAC) [*SIKAKYIA*] 352
  - Young twigs rounded, or sometimes wrinkled or many-winged, but not strongly 2-winged (but see ***Margaritaria*** – below)
    - Midrib finely channelled above, but laterals ±raised above, looping *c.*⅔ way to margin; twigs not lenticellate; slash granular, cream with brown gritty streaks, unscented and without taste; lvs drying a golden-green colour below — ***Leptaulus daphnoides*** (ICAC) [*AFENA-AKOA*] 384
    - Midrib not finely channelled above, etc. (i.e. not *Leptaulus*)
      - **Leaves** papery and slightly glaucous with marginal nerve; yng twigs + stipules or their scars, sometimes with slight wings, often with hairs; slash pink to red, fibrous but crumbly or brittle, with orange gritty streaks; outer bark fissured and flaky; old stems ±sinewy — ***Margaritaria discoidea*** [*PEPEA*] (see 13C) 416
      - **Leaves** not glaucous OR leathery OR with drip tip; bark not of this distinctive type
        - LEAVES usually <10 cm long OR 'drip-tipped' (long acuminate); finer venation often transverse and parallel; petiole <7 mm long, **not obviously swollen-discoloured at tip**
          - *Branches strongly whorled* (small trees)[2]
            - Slash pinkish or yellow, darkening rapidly — See *Diospyros* spp. (Gp 11)
            - Slash pale yellow; lvs with v. looping laterals; young twigs several-sided or several-winged; old and new growth (separated by bud scales on twigs) markedly and abruptly different in colour and texture; slash yellowish, soft-fibrous — ***Napoleonaea vogelii*** ('leonensis') (see Gp 17D) [*oBUA*] 446
          - *Branches not whorled*, but sometimes clustered at top of stem[2]
            - Petiole colouration spreading beyond petiole to top of midrib or side of midrib on lower surface (see illustration)
              - **Slash** spongy-fibrous, with slightly unpleasant scent; petiole *c.*2 mm long; venation tending to be finely transverse; medium tree with all lvs apppearing small in canopy of **evergreen forest** — ***Scytopetalum tieghemii*** (SCYT) [*oPRIM*] 565
              - **Slash** red and granular: venation different — See *Licania elaeosperma*, etc. (Gp 14)
            - Petiole sharply distinct in colour from surrounding midrib; small trees usually in **wet places**
              - Lf without drip-tip, glabrous when young — ***Rhaptopetalum bequei*** (see 13A) 539
              - Lf with long drip-tip; slash reddish; lf with dense rusty hairs when young, becoming glabrous v. soon; lf margin rolled over at base; petiole 4–5 mm long and densely wrinkled or honeycombed; ±in **swampy places** — ***Cleistanthus polystachyus*** 168
        - LEAVES usually >10 cm long and not drip-tipped, OR petiole discoloured at tip; lvs often with fine raised spots ('pustulate'); **bark with rounded scales falling to leave patches of different colours**; medium-sized trees with v.straight, slender, unbuttressed boles; **slash creamy with orange gritty streaks**[1]
          1) VERY COMMON TREE; pustules mainly on upper surface OR absent; tip of petiole often slightly discoloured; petiole with fine channel; venation sometimes transverse; slash hard, granular over orange; normally v. straight, without buttresses — ***Strombosia glaucescens*** (OLAC) [*AFENA*] 52
          2) VERY RARE tree; pustules on both lf. surfaces; lf apex acute; twigs wrinkled or many-winged, lenticellate; bark with roundish scales — ***Keayodendron bridelioides*** (EUPH) 374
          3) Small tree of sunny places – check colour v. variable — See *Ehretia cymosa* (Gp 13D)

NOTES: 1) Unless the reader has strong reasons for doubt, trees keyed correctly to this point will almost certainly be *AFENA*, one of the commonest trees in Ghana, with very few invariant leaf characters. The other 2 species have been collected perhaps less than five times altogether in Ghana.

2) Lvs with glands: check ***Pteleopsis hylodendron*** (Gp 13D) which rapidly becomes glabrous.

Group 13C: EUPHORBIACEAE – IRVINGIACEAE etc.
(Lvs glabrous; twigs with conspicuous stipules or scars; veins visible)

Stipules not very large, but persistent at base of younger leaves; slash often reddish or soft fibrous		
Lf base cordate	See *Magnistipula* spp. (Gp 14)	
Lf base not cordate		
Stipules small, sharply triangular, persistent above nodes; leaf symmetric, obovate with rounded apex, <10 cm long, rather thin, and glaucous purple to reddish below; stipules persistent, sharp triangles above petioles on rather flattened twigs; ±two vague basal nerves passing close to margin; slash pink to brown, contoured, darkening; crown with stout, ascending, regularly forking branches	***Erythroxylum mannii*** (ERYT) [*PEPEA-NINI*]	287
Not *Erythroxylum*		
Stipules like threads; lvs ±ovate, <15 cm long	See *Licania* (Gp 14)	
Not *Licania*		
Stipules triangular, 2/node spreading away from stem; lvs acuminate; young twigs rounded, but with v. fine membranous wings most visible when fresh; lvs usually <10 cm long.	***Margaritaria discoidea*** (juvenile – see Gp 13B)	416
Stipules with rounded tips or twigs not finely winged: lvs >10 cm long; **evergreen forest**		
–Lvs oblong , with paired stipules *c.*1 cm long, 2 mm wide; venation conspicuous, almost scalariform, prominent below; **evergreen forest**		
Younger lvs without silvery scales (lens); without two short nerves nr base; midrib raised above, but channelled towards petiole	***Soyauxia grandifolia*** (MEDU) ABOTESIMA	571
Younger lvs with silvery scales; lvs often with two short, inconspicuous nerves just above base; margin recurved; tree with arching and rooting branches forming tangles in understorey of **evergreen forest**	See *Scaphopetalum amoenum* (Gp 20).	561
–Lvs not this shape OR venation clearly reticulate; stipules kidney shaped, a few mm wide; midrib with slightly raised crest nr. petiole; twigs ±conical buds. trees ±stilt roots in swamps in **evergreen forest**; slash red		
Flowers >1 cm apart along infl., sometimes apparently solitary in lf axils; lvs often >20 cm long, without sharply pointed apex	***Amanoa strobilacea*** (EUPH)	75
Flowers congested along infl., ±touching; lvs usually shorter with sharp acuminate apex; (known only from Ankasa)	***Amanoa bracteosa*** (EUPH)	74
Stipules large and conspicuous but confined to end of twigs; falling soon to leave rings at nodes, **OR TREES VERY LARGE** e.g. emergents of upper canopy; slash hard fibrous and granular, yellow-orange with pale orange, striate sapwood; venation reticulate; twigs v. longitudinally striate; trees ±large buttresses; fts *c.* 5 cm across; young lvs often pale blue-glaucous		
Stipules up to 2 cm long; veins prominent above, but not remarkably densely reticulate; laterals more conspicuous than veins; surface very glossy, especially above; midrib prominent above, disappearing into petiole channel; fruits like mangoes; slash with sweet watery exudate; inner bark with short, **dark vertical streaks**	***Irvingia gabonensis***[1] (IRVI) [*ABESEBUO*]	368
Stipules up to 7 cm long; venation extremely densely reticulate, regular and often prominent above in diamond-shaped or square cubicles (esp. dry lvs); young trees with spines on bole; old trees with v. large, tall and spreading buttresses presenting a flat wall and so often difficult to slash; fruits globose but in several sections	***Klainedoxa gabonensis***[2] (IRVI) [*KROMA*]	376

NOTES: 1) Aubréville (1959: vol 2, p.122) notes the existence of two 'varieties' of *Irvingia gabonensis* in Côte d'Ivoire. Okafor (*Bull. Jard. Bot. Nat. Belg.* 45: 211–221) describes var. *excelsa* in Nigeria as a buttressed, very large tree with spreading crown and inedible fruits, in contrast with var. *gabonensis*, which is smaller, has a cylindrical or fluted bole and narrower crown, and bears sweet, edible fruit. Var. *gabonensis* is semi-cultivated in Volta region (Hall & Swaine, 1981) and var. *excelsa* is more typical of wetter sites.

2) ***Pseudagrostistachys africana*** (22B) has a similar apical bud to *Klainedoxa*, but has basal glands, serrated margins and long petioles.

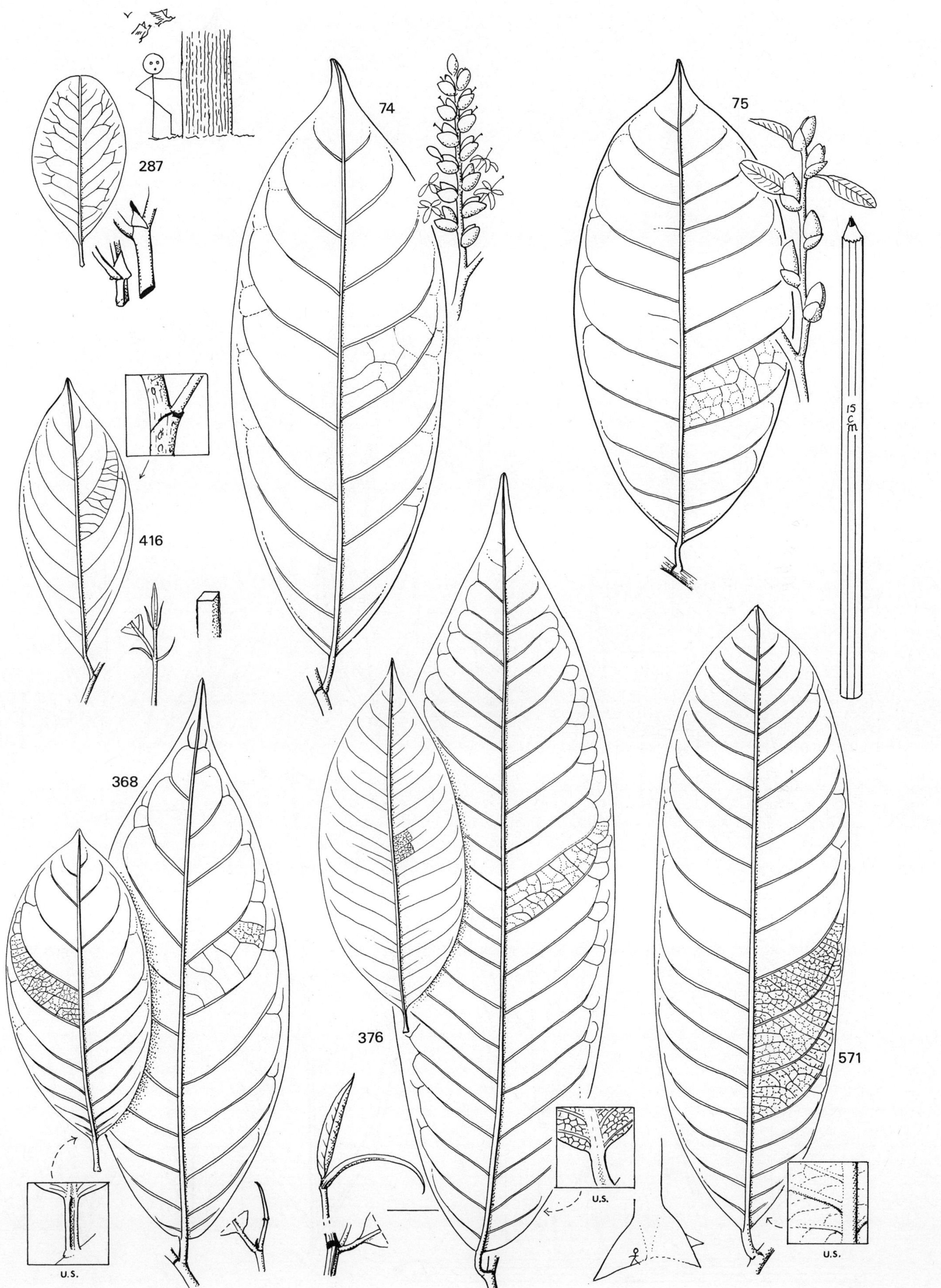
287
74
75
15 cm
416
368
376
571
U.S.
U.S.
U.S.

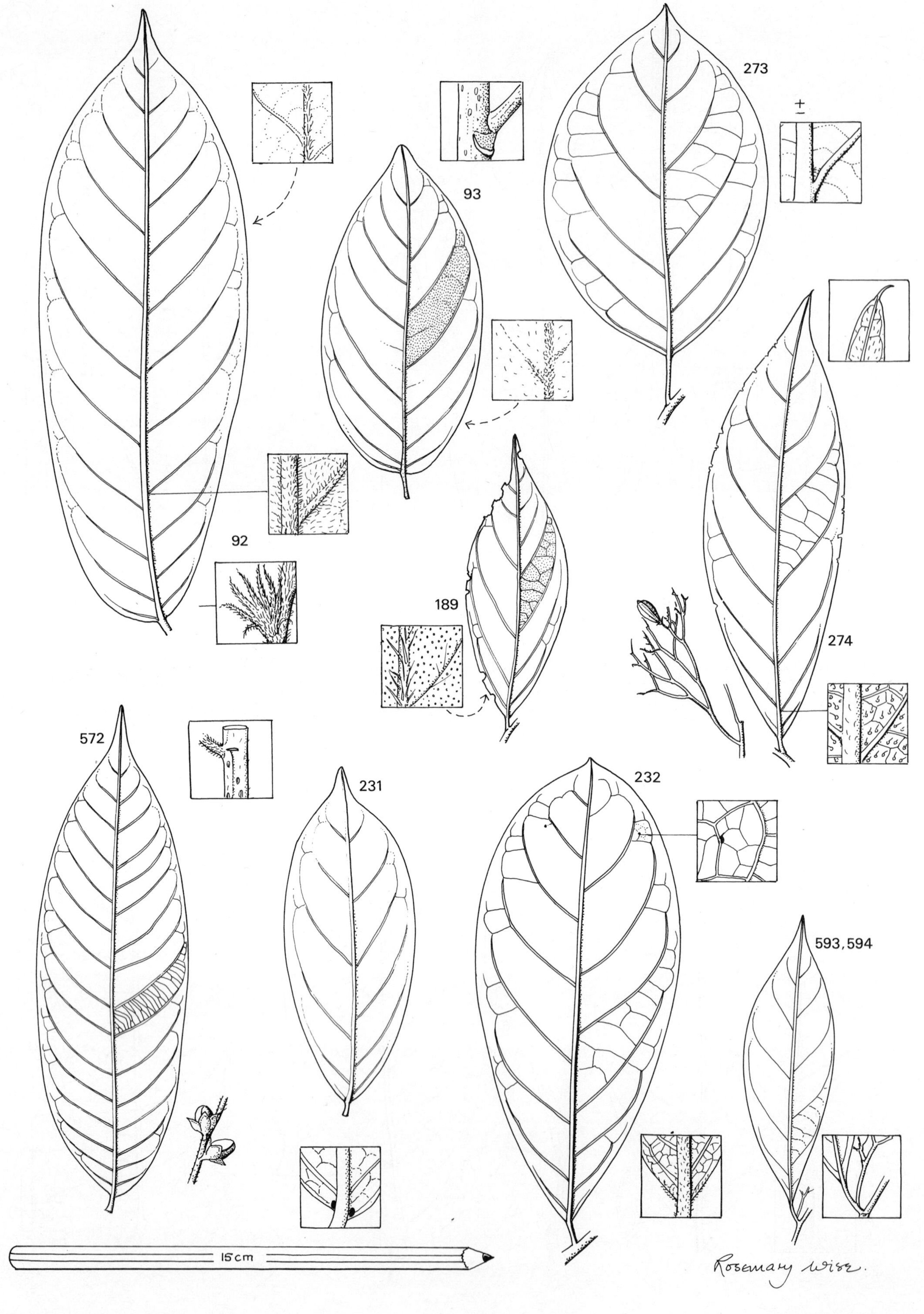
92
93
273
189
274
572
231
232
593,594
15cm
Rosemary Wise

Group 13D: EUPHORBIACEAE – DICHAPETALACEAE etc.
(Lvs simple, alternate, hairy, short-petioled, not trinerved, etc.)

Nerves reaching margin and fusing with marginal nerve; slash red-fibrous	See *Bridelia* spp. (Gp 15)	
Nerves not reaching margin		
Stipules persistent, and conspicuous on twigs for at least a few nodes; lvs sometimes cordate; slash reddish		
Stipules finely divided, like the legs of a hairy spider; hairs within midrib channel; venation lax-scalariform; slash pink over cream, fibrous-spongy, darkening	***Antidesma laciniatum*** (EUPH) [*FOTO-NINI*]	92
Stipules not divided		
Lvs with many white spots on top surface	See *Magnistipula* spp. (Gp 14)	
Lvs without white spots (lens); not *Magnistipula* spp.		
Stipules solitary, falling soon; twigs, lf undersurface and midrib above softly hairy; older twigs glabrous, lenticellate; venation lax-scalariform; slash pink over yellowish, fibrous-spongy, darkening;	***Antidesma membranaceum*** (EUPH) [*NUMANUMA-GYAMA*]	93
Stipules in pairs; twigs flattened; venation dense, almost scalariform; 10-15 prs. laterals hairs short away from veins but ±dense on them; tree often near water in W. region; slash red-brown and granular	***Soyauxia velutina*** (MEDU) [*ABOTESIMA-NUA*]	572
Stipules falling rapidly and not conspicuous		
Slash reddish over yellow at first; twigs lenticellate; lvs oblong, softly hairy	*See last two species.*[2]	
Slash without reddish outer layer; lvs not softly hairy		
Leaves with stiff or rough hairs (or hairs lying flat on lamina) often with white spots centred around base of hairs; lvs without knotted vein or petiole 'glands'; slash 'TWENEBOA'-like (cf. Gp 26)		
Surface below mature lvs very rough with long orange hairs; top lf surface with white spots; yng twigs densely hairy; weedy tree of **disturbed forest**	***Cordia vignei*** (see Gp 26) [*TWENEBOA-AKOA*]	189
Surface below not very rough OR without long orange hairs		
Scattered stiff hairs in nerve axils and on midrib; broadly elliptic	***Ehretia cymosa*** (see Gp 26, and below) (BORA)	273
Many white, stiff hairs lying flat against lower surface; medium tree of **evergreen** forest with elliptic to oblanceolate lvs. acuminate, often with sharp mucronate tip; bark smooth, pale; slash with bands, fibrous, darkening	***Ehretia trachyphylla*** (BORA) [*OKYINI*]	274
Leaves without stiff hairs and white spots; trees with flaky bark and slash darkening, but not through greenish shades OR lvs with (tuft) domatia		
Tree of **very dry forest or savanna**; lvs < 10 cm long, with translucent spots, and veins between ascending laterals dense, parallel and transverse; branches very slender, drooping, graceful and whip-like	***Anogeissus leiocarpus*** (COMB) (see Gp 25)	81
Not *KANE*; in **moist forest** or lvs different		
Veins not v. clear, or scalariform; lvs with 2 glands *c.*⅔ way to apex; lvs thin-papery with scattered long hairs; twigs and axillary buds with dense rusty hairs; tree with fibrous slash, darkening	***Pteleopsis hylodendron*** (Gp 25)	530
Veins visible and reticulate		
LVS broadly elliptic or ovate with rounded to acute apex (without glands); laterals ±impressed above; lvs with stiff hairs or glabrous, ±domatia in axils of nerves and/or outer venation; lf base ±asymmetric; bark lenticellate; slash pale, fibrous, darkening, with musty smell; small weedy tree of **disturbed forest**	***Ehretia cymosa*** (above)	273
LVS not broadly elliptic nor ovate; usually acuminate, sometimes glandular or with domatia		
–Petiole with flowers attached (or uneven as if glandular) OR lf + domatia OR thin papery		
Leaves v. thinly papery; base of lamina reaching midrib at slightly different places on each side (lens); hairs whitish, ±in tuft domatia on yng lvs; slash gritty, yellow→brown; **dry forests**	***Tapura fischeri***	593
evergreen forest small tree v. similar to previous sp.	***Tapura ivorensis*** (DICH)	594
–Leaves without domatia, and not particularly thin papery; petiole without fls or 'glands'		
Twigs with dense orange hairs contrasting with colour of midrib; glands ±visible on either surface towards lf apex as dips or bumps around 'knotted' venation; **small contorted, sinewy tree; bark with very papery flakes;** slash v. contoured, yellow-orange, fibrous →red-brown	***Dichapetalum madagascariense*** (DICH) [*ESONOWEDIE*][1]	232
Twigs (and lvs) without obvious orange hairs; lvs with **basal glands**; small tree of southern types of **dry forest**	***Dichapetalum barteri*** (DICH)	231

NOTES: 1) *D. madagascariense* used to be called *D. guineense*. The new conception of the species is extraordinarily broad.
2) If your specimen is very young, consider also *Cleistanthus* (Gp 13B).

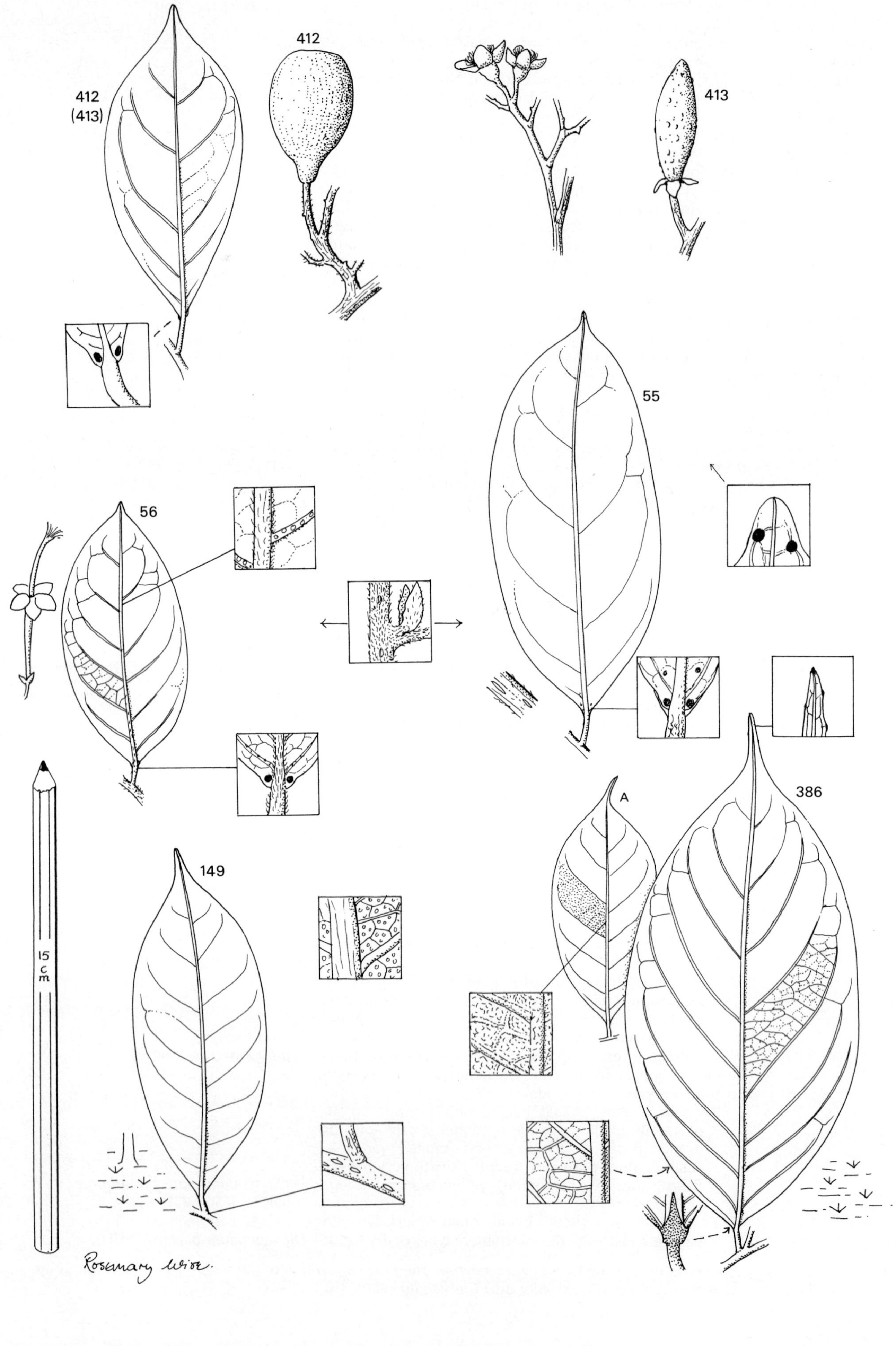
412
(413)
412
413
55
56
15
cm
A
386
149
Rosemary Wise.

GROUP 14: CHRYSOBALANACEAE
(Lvs simple, alternate, entire, not trinerved, ±basal glands and stipules; petioles <2 cm)
(Slash red to purple, brittle; bark v. lenticellate)

This is a group of small to large trees, growing, as far as is known, according to Troll's architectural model (Hallé *et al.* 1976). Most of the larger species have very spreading crowns. The bark is typically rough and very lenticellate. The slash, apart from being red to red brown, is either brittle-fibrous (esp. *P. excelsa*) or mostly granular (*Maranthes* spp.), but most granular-barked species still have some fibrous component in the slash. Often the slash is layered (i.e. contoured), typically with a hard yellow-orange sapwood, with a rapid watery exudate (which may turn red with time). Most species have basal glands on the leaf base or petiole. Stipules are usually persistent for a few nodes – sometimes they are attached to the base of the petiole rather than to the twigs. Basal glands are otherwise rare amongst species with short petioles and entire margins (but see *Diospyros gabunensis* (Gp 11), Dichapetalaceae (13D) and certain Euphorbiaceae (21) (and many with serrated lvs in 22,23). Several species with red slashes (but no basal glands) are listed in Groups 13 and 17 (see also 200 Main Species Key).

The flowers have many stamens, with the style attached to the base of the ovary. *Licania* has no petals, unlike the other species. In *Maranthes*, *Parinari* and *Dactyladenia* (=*Acioa*) the flowers are slightly, to very (*Dactyladenia*) asymmetric. The fruits are (often plum-like) drupes with one or two seeds. The genera have recently been reviewed (Prance & White, *Phil. Trans. Roy. Soc.* B 320, 1-184 (1988)).

Key to subgroups

Leaf margin entire, lf base not cordate AND leaf without dense hairs, but sometimes with hairs on twigs, petiole, midrib, etc.	Group 14A
Leaf margin with teeth OR base of leaf cordate, OR leaf discolorous with thick hairs	Group 14B

Group 14A
(Chrysobalanaceae with lvs not very hairy)

NOTE: In *Dactyladenia* (new name for *Acioa*) the stipules are typically attached to the base of the petiole.

The leaves of many of the hairy-leaved species lose their hairs with age; the following key only includes those species for which there is a likelihood that the hairy leaves (and discolorous crown) might not be noticed on the tree.

Glands raised and rounded, at top of petiole or at base of lamina; without glands at apex of leaf		
— Glands visible on upper surface; lvs without conspicuous spots along veins, ±glabrous; base of lf cuneate; red; stipules not v. persistent, nor attached to base of petiole; tall trees often with yellow exudate turning		
—— **Evergreen forest** tree (inflorescence branches hairy) (or swamps elsewhere); fruits very smooth	***Maranthes glabra*** [*AFAM-NINI*]	412
—— Tree by **rivers** (or planted) in drier zones (infl. glabrous); fruits with fine raised lines	***Maranthes kerstingii***	413
— Glands by midrib on lower surface; lvs with conspicuous spots along veins on both surfaces (x10 lens), and short hairs, e.g. along **midrib above**; base of leaf cuneate to obtuse; stipules attached to base of petiole; twigs without obvious lenticels; stipules persistent; **small tree in understorey or by rivers**	***Dactyladenia dinklagei***[1] [*ATWERE*]	56
Glands, if present, not large nor raised-rounded at top of petiole OR leaf with glands nr apex		
— Leaf with coarse, long hairs	See *Magnistipula* spp. (next Gp)	
— Leaf without coarse, long hairs		
—— Lvs ovate-elliptic with acute tip, usually with glands at apex of leaf; 2, 4 or more glands near to midrib towards base of lf, and often with glands at top of petiole; venation reticulate; twigs with conspicuous lenticels; small **understorey or riverside** tree; fts ovoid, *c.*4 cm long	***Dactyladenia barteri***[1]	55
—— Lvs broadly elliptic to oblanceolate, with acuminate tip; venation scalariform; **swamp or riverside** trees		
——— Twigs with many white lenticels; lvs drying reddish; glands in lamina nr base; laterals joining; midrib short and becoming corky; lf often with minute white spots between the finer veins; small tree of **swamps**	***Chrysobalanus icaco***[2]	149
——— Twigs without conspicuous lenticels; lvs drying yellow-orange; younger tree or lower lvs (A) densely hairy, + acuminate apex BUT lvs usually glabrous; glands small or absent; yng petiole with long hairs; top of older petioles smooth, with black colour which extends into midrib above, sometimes with ? glands on side of this part of petiole; **leaf acumen often with minute tooth-like glands**; venation clearly reticulate ±raised; crooked, sinewy, tree nr. **swamps or sea**	***Licania elaeosperma***[3,4]	386

NOTES: 1) Another *Dactyladenia* species has been discovered in **evergreen forest** (Hall and Abbiw GC44730): it has oblong lvs, stipules minutely leafy (with hairy midrib, venation etc.). This may be what they refer to as *'Acioa' hirsuta* in the Forestry Dept. list.

PTO for other notes.

Group 14B
(Chrysobalanaceae with discolorous, cordate or toothed lvs)

Leaf margin serrated (or, at least, with glandular teeth)

Leaf without two rounded basal glands; **swamps, riversides** — See *Licania* (above and Note 3)

Leaf with 2 basal glands; juvenile (? shade) lvs (A) + cuneate base, acuminate, with flat-ended cylindrical teeth and dense, short, orange cottony hairs on lower surface; lvs of mature trees ±cordate, with small serrations nr. apex; unfolding twigs and lvs enclosed in conspicuous buds of hairy overlapping scales; **evergreen forest** — ***Maranthes aubrevillei*** 410

Leaf margin entire

Base of leaf cordate (except on seedlings)

Leaves mostly >10 cm long; surface (esp. venation below) often with many fine pale spots (x10 lens); glands, if present, at base of leaf or top of petiole OR along lamina; stipules often persistent

Leaf with many long hairs below and especially along midrib channel above; base cordate to cuneate; top lf surface sl. bumpy: veins appearing 'lumpy'; twigs hairy; young undeveloped branches in lf axils with many overlapping, hairy bud scales; favouring **swampy areas in evergreen for.**; mature fruits hairy, the size and taste (Aubréville, 1958) of an apple — ***Magnistipula butayei*** 403

Leaf with, at most, a few hairs along midrib; base deeply cordate, sometimes auriculate; ±pocket domatia; ±several glands along lamina, close to midrib; flush of new lvs purple, turning metallic brown; by **rivers** and **swamps**, but rare; fts ovoid, 8 cm long, glabrous outside, with hairy seed — ***Magnistipula zenkeri*** 404

Leaves <10 cm long, elliptic or ovate; often discolorous due to dense hairs; laterals >2 per cm; glands *c.*½ way along petiole; dense-leaved, spreading, gregarious, **river-side** tree — ***Parinari congensis*** [*KOTOTRAMPO*] 487

Base of leaf not cordate: cuneate to obtuse (Leaves hairy below, often discolorous white or orange-cream)

Laterals many (>15 pairs), parallel, nearly perpendicular to midrib, prom. above; glands (if present) *c.*½ way down petiole; hair persistent, between veins; young twigs and old bark + obvious lenticels; crown golden; tree tall and straight; slash smelling like rotting sugar — ***Parinari excelsa*** [*AFAM*][3] 488

Laterals <15 pairs; glands, when present, not in middle of petiole

Glands not like two round, raised eyes; lf ovate-lanceolate and acute-tipped; tree of **riversides or swamps**, with discolorous *juvenile* foliage only. — See *Licania* (14A) and *P. congensis* (14B)

Glands like two raised eyes OR lvs not within this range of shape

Hairs fairly long and coarse; lf oblong to oblanceolate (slender, ±broader beyond middle) ±tiny spots (x10 lens). (See above) — ***Magnistipula butayei*** 403

Hairs, when present, v. short on lower lf surface, forming a dense felty mat; Leaf not slender, but broadly elliptic, oblong, ovate, etc.; apex abruptly acuminate or rounded; stipules not persistent, bud falling as bud scales on expanding shoots; fts ±smooth, ovoid, *c.*3 cm, sometimes with a ridge and shaped like an oven-ready chicken

Moister forest tree; hairs persistent even to mature or fallen lvs, forming a vivid orange lower surface soft to touch on young lvs; base of lf obtuse (rarely slightly cordate) to cuneate, but usually steeper than 45° to midrib; lf often pustulate on top; buds large and hairy; lenticels often elliptic on young twigs — ***Maranthes chrysophylla*** [*AFAM-KoKoo*] 411

Swamp forest, especially in drier forest zones; small tree; hairs only on younger lvs, never particularly soft to touch, and usually so sparse as to be dull-orange brown on lower surface; base of lf cuneate; last season's petioles and twigs v. corky; twigs often dark and peppered with raised, *rounded* lenticels; stipules (bud scales) falling at time of bud expansion — ***Maranthes robusta*** [*AFAM-BERE*] 414

NOTES (contd.) from 14A

2) This is *Chrysobalanus icaco* subspecies *atacorensis*. *C. icaco* ssp. *icaco* has obovate lvs with a rounded apex and basal glands near the midrib base below, and is found only **near the sea**.

3) Prance and White (see top of Gp.) note that *Licania elaeosperma* may be dispersed by water; *Maranthes* spp. fruits are eaten (in some areas at least) by primates; and *Parinari excelsa* fts are dispersed by bats, elephants and baboons. No doubt because of this, *P. excelsa* seedlings are commonest in Ghana in forests inhabited by elephants.

4) *Rhaptopetalum beguei* (group 13A) often has lvs slightly discoloured by the midrib at the base, vaguely suggestive of basal glands. As this species has a red slash, it could be confused here.

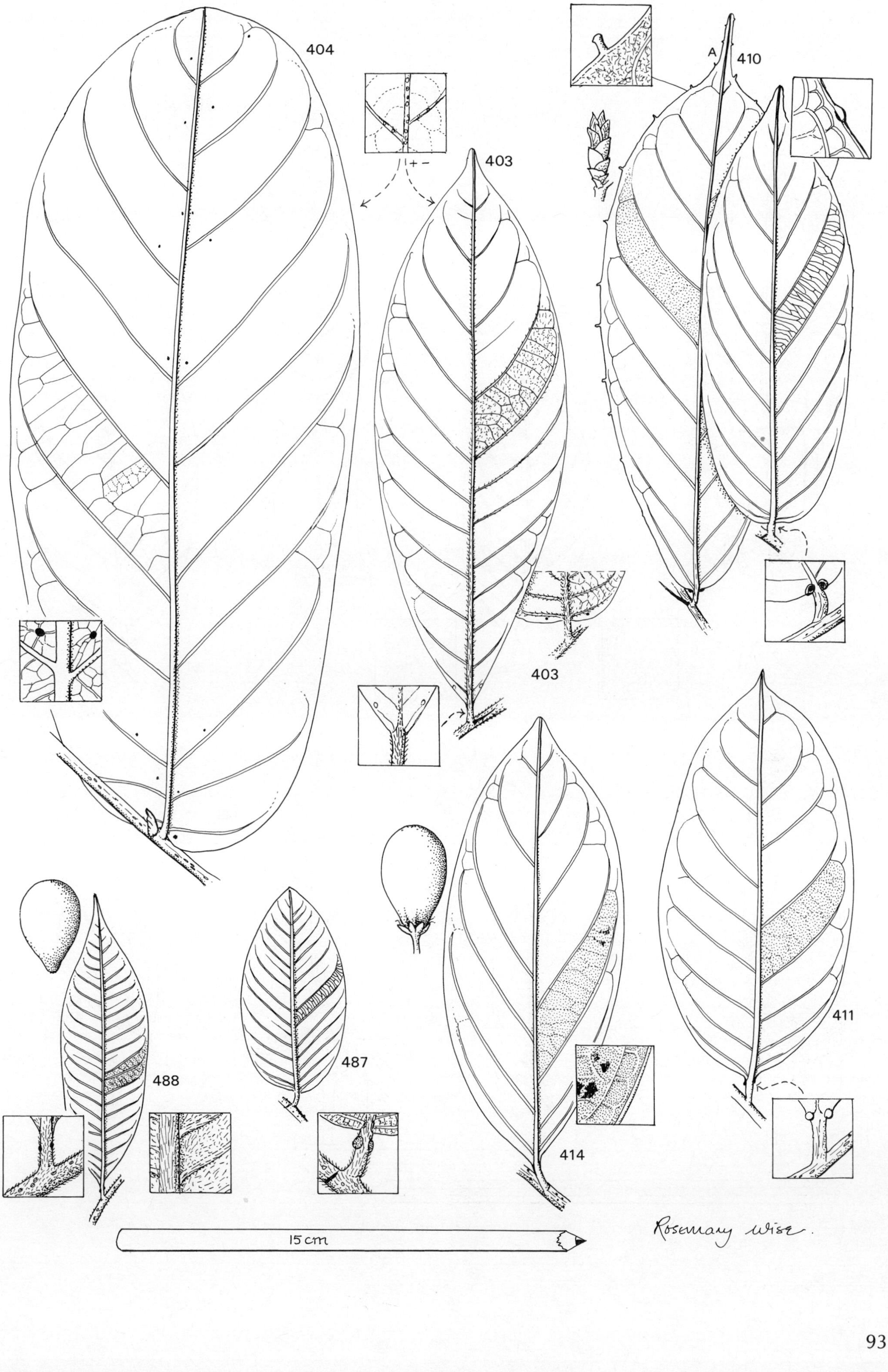
404
403
+ −
A
410
403
411
488
487
414
15 cm
Rosemary Wise

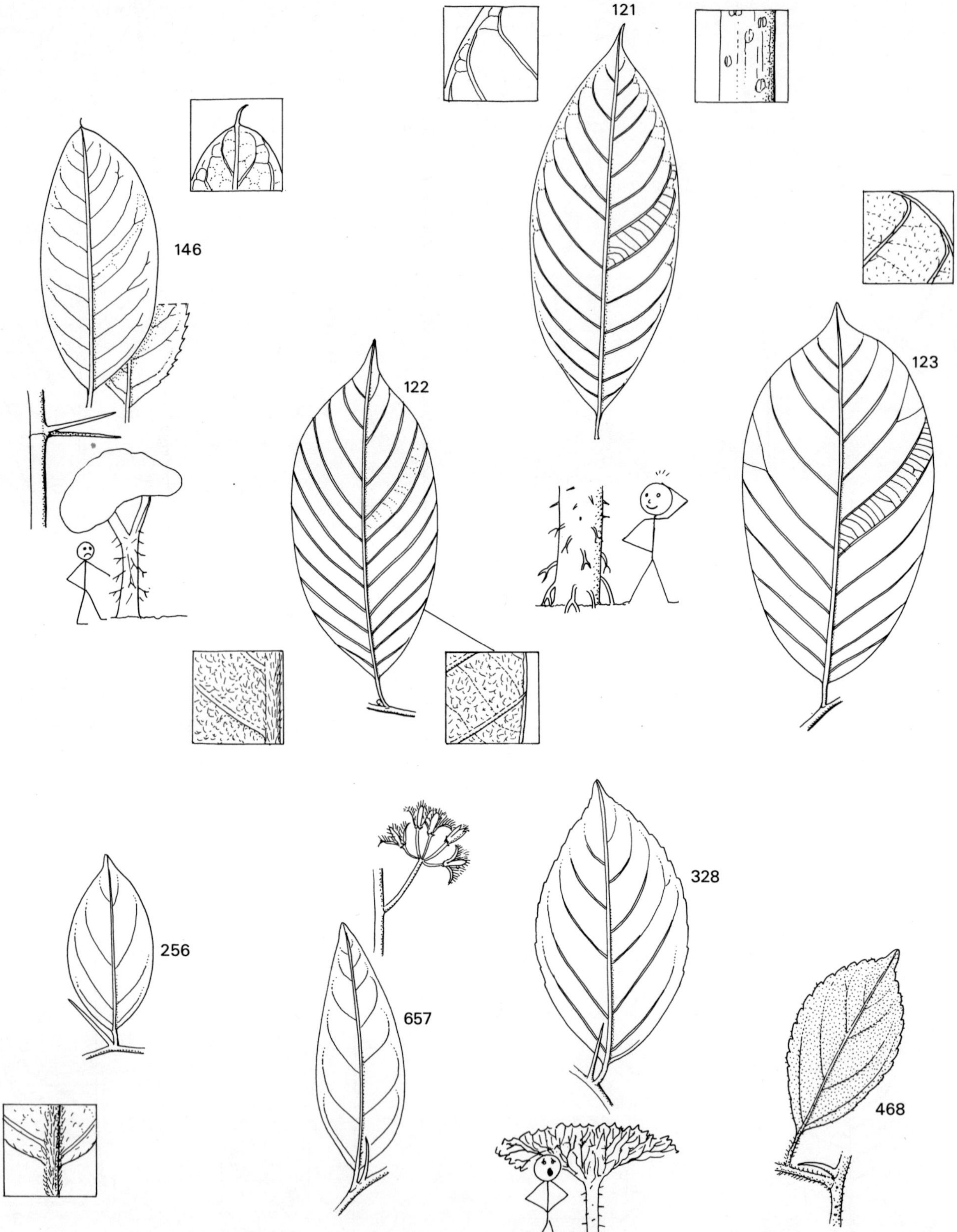
146
121
122
123
256
657
328
468
15 cm
Rosemary Wise

GROUP 15
(Spiny plants with short petioles and alternate lvs)

Margin entire, lf not gland-dotted		
LVS with tufts of hairs in domatia; outer bark, even on young twigs, shiny and orange-brown, peeling in papery flakes from older stems; nerves on lvs running over into petiole channel above	***Diospyros monbuttensis*** (See Gp 11) (EBEN)[1] [*ATWERE-NANTIN*]	
LVS without tufts of hairs in nerve axils		
Lvs with sharp, mucronate tip, usually asymmetric at base, with stipules at apex of twigs falling to leave scars at base of spines; lf sometimes with fine spots on upper surface; **dry forest**; small tree branching low; spines on bole often branched; slash yellowish	***Chaetachme aristata*** (ULM) [*ESONO-ANKAA*]	146
Lvs without these 'mucronate' tips		
Laterals reaching margin and fusing with marginal nerve; slash red and fibrous; tree often with basal, adventitious or prop roots as well		
Medium forest tree with hairs dense over lower leaf surface ±hiding finer veins; lvs usually narrowly elliptic; outer bark ±black rough; slash red, fibrous, with strong, sweet, camphore-like smell	***Bridelia grandis*** (EUPH) [*oPAM-KOTOKRODU-KESE*]	122
Small tree of v. disturbed forest; forest edge etc.; a few large hairs on nerves, otherwise hairs v. small and sparse on lower surface: not hiding nervation; lvs usually broadly elliptic, oblong or obovate	***Bridelia micrantha*** (EUPH) [*BADIE*]	123
Laterals not reaching margin		
Spines on older stems of young trees; not usually amongst lvs; twig ends (at least) with **stipules**; lvs sometimes bluish-glaucous below		
Venation scalariform; base of tree often with adventitious or prop roots as well as spines; twigs pale and lenticellate, with small stipule scars at nodes; slash red and fibrous.	***Bridelia atroviridis*** (EUPH) [*oPAM-KOTOKRODU*]	121
Venation not scalariform; bole without small adventitious roots at base		
Leaves ±10 cm long; spines *c.*1 cm long; slash fibrous; venation not close reticulate; youngest (green) twigs with fine wings and diverging pairs of stipules; not leaving complete ring scars, lenticellate	***Margaritaria discoidea*** (young) (See Gp 13B)	
Leaves and spines often larger (on these juvenile plants); venation close reticulate; twigs striate with ring scars where long terminal stipules have fallen; slash granular	***Klainedoxa gabonensis*** (young) (See Gp 13C)	
Spines even on younger twigs, amongst leaves, and often thick and sharp; on youngest parts even of mature trees; lvs never glaucous, but often hairy; **without stipules**		
Lvs ovate or oblong, with obtuse base; spines in lf axils more than ¼ as long as lf; **brown hairs** on midrib, etc. below	***Dovyalis zenkeri*** (FLAC)	256
Lvs lanceolate, slightly asymmetric; spines at base of lf <¼ length but some spines very long, themselves bearing leaves	***Ximenia americana*** (OLAC)	657
Margin serrated, or at least lf with glandular spots or pellucid points below		
Lvs without glandular spots; usually growing on **riverbanks**		
Lvs with few hairs; spines straight; finer veins prominent above; parallel and mainly transverse in mid-leaf		
Small tree or shrub (without petals on flowers); rarely in forest; lvs drying brown; teeth with only small thickened area	***Flacourtia flavescens*** (FLAC)	328
Medium sized tree (with large petals); whole tooth often thickened; lvs often drying green	***Oncoba spinosa*** (FLAC) [*ASRATOWADUA*]	
Lvs softly hairy when young; spines curved downwards; finer veins not parallel, laxly reticulate; spines recurved	***Oncoba brachyanthera*** (FLAC)	468
Lvs with glandular or pellucid spots	See *Aeglopsis, Afraegle* (RUTACEAE—Gp 31A)	

NOTE: 1) FAMILIES are abbreviated as follows: EUPH (Euphorbiaceae, see Gp 20); FLAC (Flacourtiaceae, see Gp 17); ULM (Ulmaceae, see Gp 18); OLAC (Olacaceae, see Gp 13); EBEN (Ebenaceae, see Gp 11).

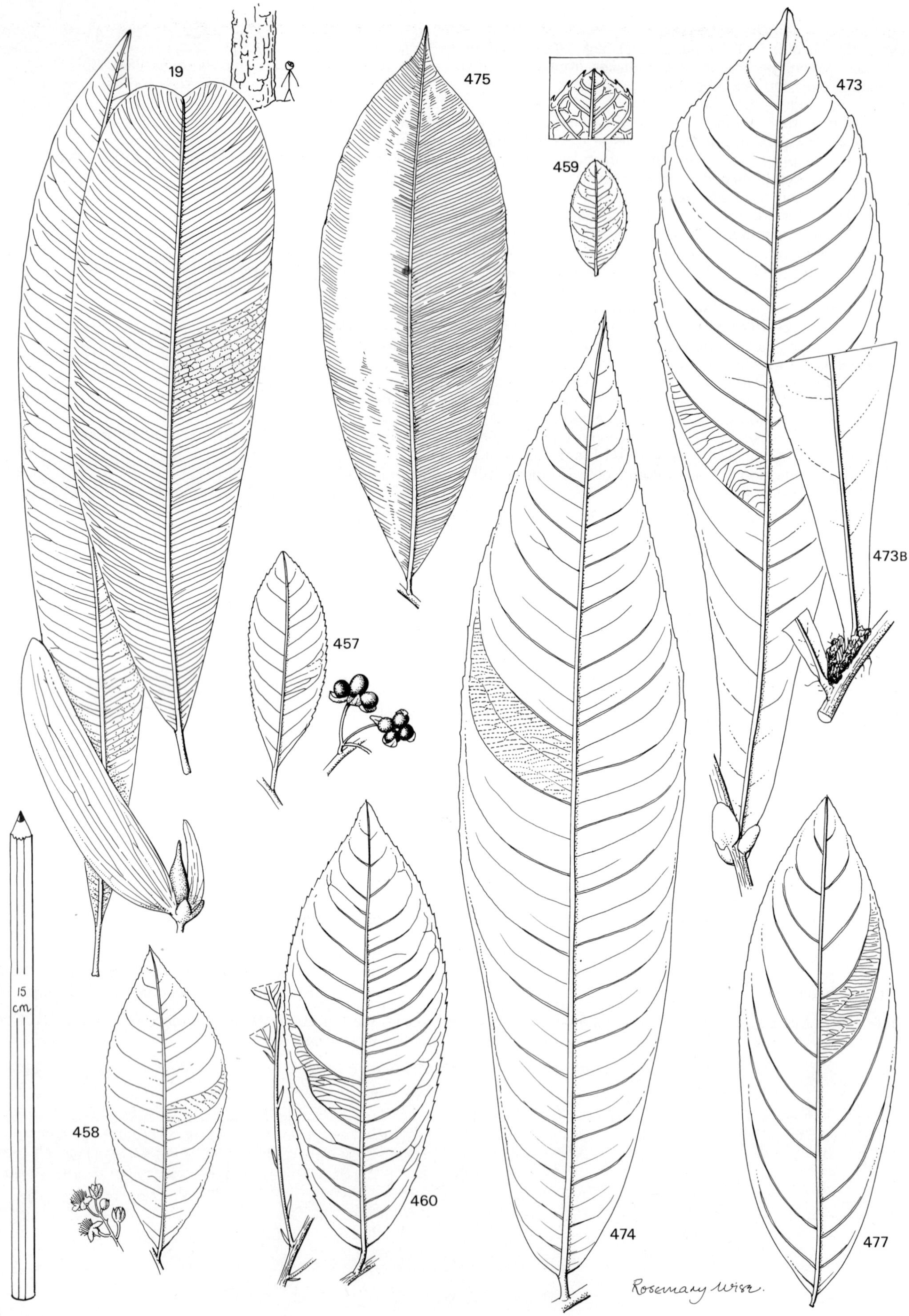
19
475
459
473
457
473B
458
460
474
477
15 cm
Rosemary Wise

GROUP 16: OCHNACEAE

Most species of this family are small trees or shrubs, but *Lophira* is a large tree (with very hard wood). The leaves of this family are always completely glabrous, and are often bright red when young. The twigs bear stipules, but these often fall very soon. The flowers are yellow, with five separate petals. *Ochna* and *Lophira* flowers have many stamens, whereas those of *Ouratea* species have only ten. The fruit of *Lophira* is a small pointed nut with one long wing (derived from sepals) 5 cm long and another shorter one. The fruits of *Ochna* and *Ouratea* are typically fleshy, often black carpels sitting on a swollen receptacle, with red sepals persistent at the base.

Leaves entire, with veins and laterals apparently closely parallel and finely transverse when held at arm's length		
Tall tree with leaves clustered at branch ends; some veins reticulate between laterals; apex of lf rounded or emarginate (except in young trees); upper and lower leaf surface v. similar; **evergreen forest** (or swamps elsewhere) tree with red new lvs, and all lvs clustered at twig tips; straight or fluted (not buttressed) bole + flaky bark; slash with bright yellow layer just under flaking outermost layer;	***Lophira alata***[1] [*KAKU,EKKI*]	19
Small to medium understorey tree; apex acuminate; lf underside discolorous; most veins parallel with laterals; bark thin and grey	***Ouratea calophylla*** [*OPUNINI*]	475
Leaves (almost always) serrated; venation varied, but not as above		
Leaf base deeply cordate or petiole absent or almost so; leaves usually >15 cm long, on little-branched treelets or shrubs in **evergreen forest**		
Lf base deeply cordate or with basal lobes; ±serrations; often in wet places; often 'filled' with leaf-litter around lf bases[2,3]	***Ouratea amplectens***	473
Lf base cuneate; lvs broadly oblanceolate with closely parallel veins; fine-serrate	***Ouratea duparquetiana***	
Leaf base not cordate; petiole clearly present and easily visible (or tree in v. dry forest)		
1) Lvs >20 cm long, leathery, with ±scalariform venation; conspicuous stipule scars; small tree by **rivers**	***Ouratea calantha***	474
2) Lvs <5 cm long, with proportionately short petioles; ovate with rounded apex and minute serrations, and small triangular stipules; in **v. dry southern forest**	***Ochna ovata***	459
3) Lvs of moderate length (usually 5-20 cm long)		
Twigs with conspicuous, persistent triangular stipules *c.*5 mm long; serrations sharp; venation prominent above, not obviously transverse nor densely reticulate; twigs NOT v. lenticellate; flowers *c.*4 cm across when stretched open	***Ochna staudtii***[4,5]	460
Twigs without such conspicuous stipules OR flowers <4 cm wide OR venation not so		
Venation between lateral nerves densely and regularly reticulate; **v. dry forests**		
Lvs rather papery; laterals much more conspicuous than intervening finer venation; twigs dark, not v. lenticellate; stipules ±thread-like and sometimes persistent; small tree normally in thickets (e.g. nr coast), with smooth bark falling off in large scales	***Ochna membranacea***	458
Lvs brittle (thicker than paper); laterals only a little more conspicuous than finer nerves; twigs dark with obvious white lenticels; stipules *c.*1 mm triangles, not v. persistent	***Ochna afzelii***	457
Venation between lateral nerves v. fine and ±transverse; lvs dark green and glossy above and paler below; twigs glossy, and not v. lenticellate	***Ouratea flava*** [*oWAN*]	477

NOTES:
1) *L. lanceolata*, a **savanna** tree related closely to *L. alata*, has petioles >2 cm long.
2) This distinctive 'litter bin' habit, with leaf litter collecting at the base of long, upward-pointing leaves, has arisen by parallel evolution in *Pycnocoma* spp. The plant probably gains some nutrients from the habit (nutrients are at a premium in evergreen forest where the soils are so poor): adventitious roots can often be found penetrating the litter.
3) ***Ouratea schoenliana***, another **evergreen forest** shrub with cordate leaves, has lvs <15 cm long; ***O. cameroniana* (473B)** is another 'litter bin' shrub from evergreen forest having barely cordate lvs >15 cm, but with lateral nerves impressed above.
4) ***O. staudtii*** used to be called ***O. kibbiensis***.
5) ***O. sulcata*** is a shrub with *c.*1 cm long peristent stipules and an unbranched infl. which will key here. There are many other *Ouratea* species of shrubs, which will rarely, if at all, reach 5 cm dbh. Three of the more distinctive of these are: ***O. morsonii***, which has peculiarly thread-like teeth, known from Ankasa f.r.; ***O. congesta*** which has the finer veins ± obscure but, when viewed x10, the veins can be seen as a fine ripple-like pattern; and ***O. glaberimma***, which has obscure venation, leathery leaves, and is known from **scrub forest** on white sands near the **sea**.

Notes on certain families in Group 17

As with Group 13, this group is a repository for species from many families. The Violaceae and Pandaceae are discussed in Group 17B and 17D respectively.

FLACOURTIACEAE

The Flacourtiaceae parallel the Euphorbiaceae to some extent, particularly in the wide range of leaf type. Previously some of the species (*Dissomeria, Homalium* and *Casearia*) were included in another family, the Samydaceae. Of these species, only *Ophiobotrys, Homalium* and *Scottellia* can (just) reach the forest canopy. These species have predominantly orange slashes, very gritty in *Homalium*, rather fine-granular and soft in *Ophiobotrys*, and more leathery-granular in *Scottellia*, which also is rather obviously scented, not unlike *Drypetes*. The crown of all these species is dense and narrow. The family is of very limited socio-economic importance, although some species have attractive flowers, and *Scottellia* has potential as a timber tree.

Genus	Group	Flower	Fruit
Flacourtia	15	No petals; small racemes	Globose-angled, 2.5 cm; edible + many seeds
Oncoba	15	Showy, single; 8–12 petals; many stamens	Spherical + raised lines; hard shell + pulp
Dovyalis	15	Males in spikes; female solitary	6 cm velvety, fleshy spheres
Dasylepis	17D	Sparse spike-like racemes of small flowers; many stamens	Ovoid and pointed, shell around seeds
Casearia	17D	Axillary clusters; no petals	3-valved capsules + seeds in red pulp
Scottellia	17D	Racemes or panicles; 5 stamens, 5 petals	Small, 3 valved capsules + 3 seeds
Homalium	17E	Racemes or panicles; 5-7 petals	Petals enlarge to form wings
Ophiobotrys	18A	In panicles; no petals	Small, 3-valved + 1 seed
Lindackeria	22A	Slender, hairy, sparse racemes of 1 cm white fls	Spiny capsules; black seeds + red aril
Caloncoba	23, 27B	Similar to *Oncoba*, but in clusters	Smooth or spiny capsules

The **IXONANTHACEAE** include only the two *Phyllocosmus* species listed in Group 17C, within Ghana. Their flowers are white, yellow or pink, in axillary inflorescences 6-20 cm long. The fruits are *c.* ½ cm long capsules with a sharp point. *Irvingia* and *Klainedoxa* are sometimes included in this family e.g. *Flora of Tropical East Africa*

The **HUMIRIACEAE** include only one *Sacoglottis* species in Ghana. The yellow flowers in large inflorescences develop into floating, ellipsoid, edible fruits up to 4 cm long with a thick woody pericarp.

The **PASSIFLORACEAE** or passion-fruit family, include in Ghana the three genera of trees or shrubs listed in Group 17C, and several climbers. The family is related to Flacourtiaceae. The flowers of all species have a 'corona' or ring-like structure between the petals and the stamens. In *Smeathmannia* rather spectacular flowers are produced singly in the leaf axils. They have white to orange-ish petals, many stamens and a brown hairy calyx. *Androsiphonia* produces panicles with leafy bracts whereas a similar arrangement in *Paropsia* is interpreted as flowers at the nodes of more or less leafy, spike-like twigs. The flowers in these last two genera are smaller than in *Smeathmannia*. The fruits are rather empty, brittle capsules, which in *Androsiphonia* are orange and very light.

GROUP 17: EUPHORBIACEAE (part) -FLACOURTIACEAE, etc.
(Leaves serrated, not 3-nerved, with short petiole)

An enumeration of smaller trees in any forest in Ghana will usually reveal an abundance of some of the trees in this Group. Unfortunately, although the key will bring you easily to this point, the path now becomes more difficult, because many of the following species are very similar.

Key to subgroups

Groups 17A and 17B can be recognized in most cases by the two last statements marked '**'. 17C includes all the species with glandular leaves or twigs. Group 17D and 17E have glandless leaves; those of *Homalium* (17E) are generally larger and more hairy, often on timber-sized trees. 17D species have smaller leaves or a reddish slash, like many of Group 17C.

The '**oPAHA**' scent, in the bark of *Drypetes* spp., is very distinctive: hot, somewhat sweet, like fresh oil paint mixed with pepper, or hot mustard. It has a more manufactured, unnatural smell than the fragrant but hot smell of Annonaceae. The bark of certain *Drypetes* spp. (especially *D. pellegrinii*) is used in herbal medicine. Small roots, which also have the distinctive taste, are used as chewsticks. *Scottellia* usually has a slash with a somewhat similar smell, but the slash is more leathery although still brittle.

Slash red to purple, or with reddish exudate	
Leaves or twigs with obvious glands, or peculiar teeth	Group 17C
Leaves and twigs glandless	
Leaves with domatia in nerve axils	See juvenile *Maesopsis*
Leaves without domatia	
Slash with v. strong oPAHA scent	Group 17A, especially *D. pellegrinii*
Slash not strongly oPAHA-scented	Group 17C, 17D (turn to 17C)
Slash yellow or orange or brown; without reddish exudate (OR SLASH NOT KNOWN to reader)	
Lvs with (tuft) domatia, but otherwise glabrous[1]	
Young branches with spines; lvs ±elliptic; venation scalariform	See *Maesopsis* (Gp 2)
Not *Maesopsis*	Group 17E
Lvs without domatia	
Large (timber-sized) trees; slash very gritty but NOT strongly scented	Group 17E
Lower storey trees or shrubs, OR slash strongly scented	
Leaves or twigs with obvious glands or peculiar teeth	Group 17C
Leaves and twigs glandless, and teeth not obviously abnormal	
Petiole with fine lateral wings or swollen or jointed	Group 17C, 17D (turn to 17C)
Petiole rounded, not swollen, often short or channelled, without (discoloured) joint	
Teeth very many, close, small and very regular; lvs glabrous; slash v. thin, not scented	See Gp 16 (OCHNACEAE)
Teeth not unusually dense and regular OR lvs hairy; (not matching descriptions in Group 16)	
Finer venation above transverse between laterals	Group 17D
Venation reticulate	
Venation obscure, or impressed above, with laterals in broad loops ⅔ of way to margin	Group 17C, 17D
Venation clearly visible, not impressed above nor in broad loops	
Leaves with pustules (lens), or easily-seen translucent spots and streaks	Group 17D
Leaves without pustules	
**Twigs with narrowly conical buds at tip which fall to leave (ring) scars at nodes; lf base usually symmetrical; slash unscented and dry; inflorescences long and branched; fruits ±dehiscent	Group 17B (*Rinorea* spp.)
**Twigs without pointed cones at tips (except *D. pellegrinii*) and without ring scars; leaf base usually v. asymmetric; lamina ±glabrous; lvs of some species entire; slash usually scented, often with exudate; flowers in fascicles; fruits indehiscent	Group 17A (*Drypetes* spp.)
Neither of above	See notes under Group 17B

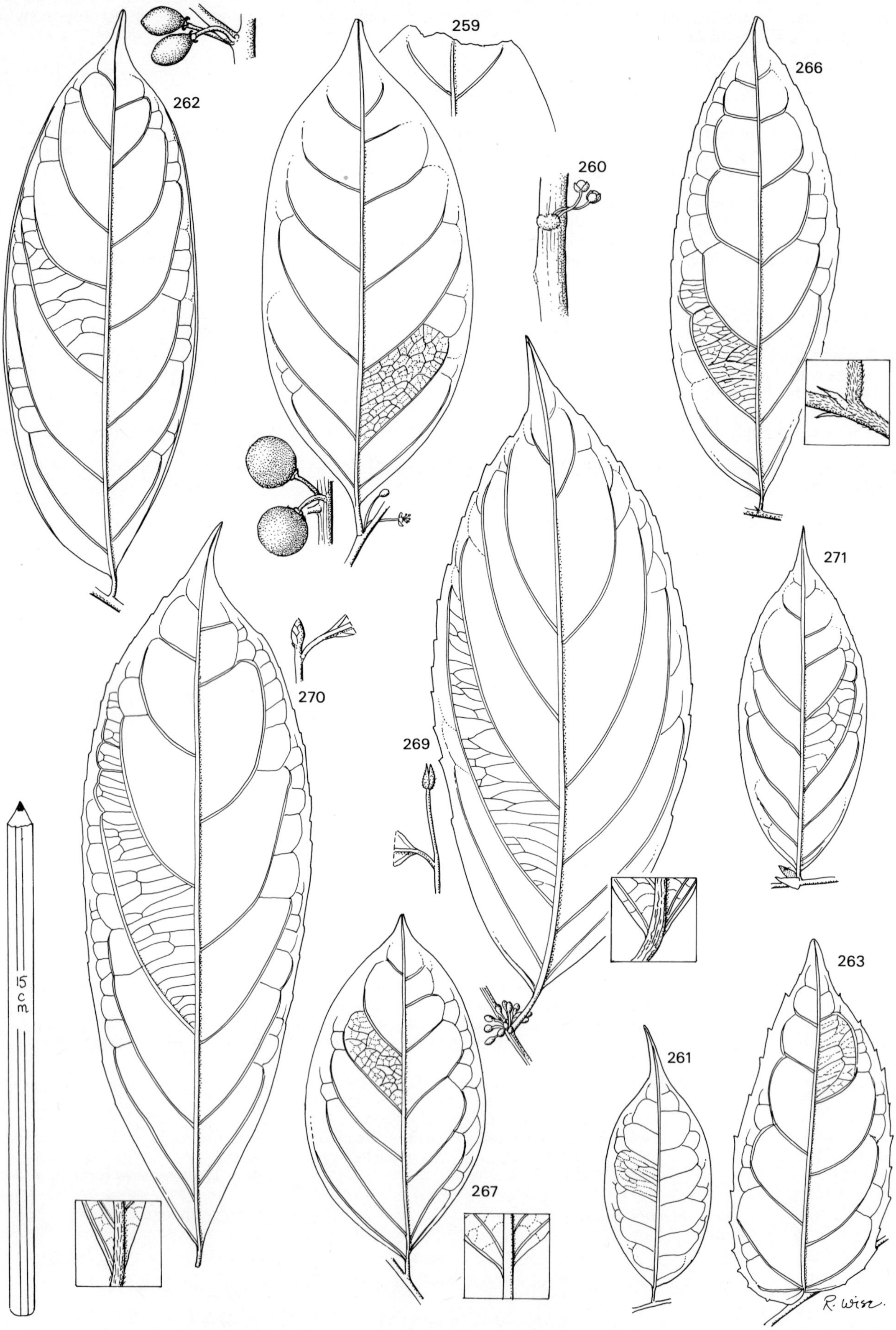

259
260
262
266
270
269
271
267
263
261
15 cm
R. Wise.

Group 17A: *Drypetes* (Euphorbiaceae)
(Simple, alternate, serrate, short petioled, not strongly trinerved; asymmetric; ±hot slash)

Leaf base extremely, and unequally cordate, often surrounding twig; twigs with short hairs; small tree with papery bark, or shrub	***Drypetes chevalieri*** [*KATRIKA-AKOA*]	263
Leaf base only v. slightly, or not at all cordate		
Young twigs with persistent stipules (uncommon trees)		
Stipules leafy, ovate, with nerves, etc.; old bark on twigs yellow, v. corky; young twigs v.smooth; **dry forest**	***Drypetes singroboensis*** [*oPAHA-AKOA*]	271
Stipules thread-like; lvs often with v. flat-ended tip; adventitious shoots with lvs broadly elliptic, almost circular with laterals looping *c.*½ way to margin, and with sharp teeth and long drip tip; other shoots with lvs ±oblong with extremely asymmetric base, with 1 side of midrib more than twice as wide as other side 2 cm above lf, base; slash yellow-orange, gritty-streaked, darkening; **evergreen forest**	***Drypetes ivorensis*** [*OPAHA-BERE*]	266
Young twigs without persistent stipules		
Lvs (shade lvs) extremely asymmetric at base, OR broadly elliptic with serrations at apex		
Slash reddish; Leaves broadly elliptic: not v. asymmetric	See *Scottellia* spp. (Gp 17D)	
Slash not reddish	See *D. ivorensis* (above)	
Not shade leaves of *Scottellia* or *D. ivorensis*		
Most leaves longer than 12 cm or leaves entire		
Margin entire OR serration extremely fine (x10 lens)		
a) Some lvs with >7 prs laterals; **tip of lf often eaten away**; venation very prominent-reticulate below; lamina creased beside midrib base, esp. when dry; dries yellowish; slash gritty, orange →brown, slightly scented	***Drypetes aframensis***[1] [*oPAHA-TENE*]	259
b) Lvs with 5-7 prs laterals looping together; venation lax (<8 veins/cm), not v.prominent below; margin recurved; lvs pale below, but drying brown; slash very pale yellow-orange, darkening to grey; scented; **evergreen forest**	***Drypetes aylmeri*** [*oPAHA-FUFUO*]	262
c) Lvs with 4-5 laterals	See *D. leonensis*	
Margin with clear (but ±blunt) serrations; scent often strong		
Slash revealing red inner bark; young shoots with pointed-ovoid buds, covered in dense yellow hairs; lvs often with a dense 'streak' of 2-3 veins running v. close to margin nr base or *c.*½ way along; base of midrib below wrinkled (on fallen lvs) and glabrous; slash (even roots and twigs) **very scented**; outer slash orange and gritty, inner bark pink to red ±white streaks; fts few, in leaf axils	***Drypetes pellegrinii*** [*oPAHA-KoKoo*]	269
Slash revealing brown to orange inner bark; twigs without such pronounced, pointed ginger buds at tips; leaves with no such 'marginal streak'; Base of midrib below finely striate, often with a few hairs, especially nr margin; serrations sometimes with tiny sharp points; slash slightly scented; fts velvety, hairy, often in clusters; also on older wood	***Drypetes principum*** [*oPAHA*]	270
Most leaves shorter than 12 cm		
Lvs entire, ±ovate, asymmetric, with 4-5(-6) laterals and 1-3 laterals arising nr base; laterals ±impressed above but finer veins prominent; fine hairs on petioles; petioles and twigs →pale, corky; habit and slash like *D. principum*; prefers **moister forests on hills**	***Drypetes leonensis*** [*oPAHA-NUA*]	267
Lvs slightly serrated		
Tip of leaf with long drip tip; **venation fine and parallel, apparently finely transverse**, but only 6-8 laterals joining in submarginal nerve; slender, buttressed tree with v. strong, hot taste and yellowish exudate in slash; **W. region forests**	***Drypetes aubrevillei*** [*oPAHA-NINI*]	261
Tip of lf merely acuminate; slash not extraordinarily scented; small or spreading trees/shrubs		
Lf margin meeting petiole at very different points, but minutely cordate, at least on one side; slash thin, brown, darkening; with sharp root processes nr base	***Drypetes gilgiana*** [*KATRIKA*]	265
Lf margin ±reaching petiole at equal points on each side; (mostly) small trees of **dry forest** close to savanna boundary		
Serrations not sharp pointed		
Lvs without 2-4 small 'basal nerves'	See *Microdesmis* 17D	
Lvs with 2-4 weak nerves arising nr lf base; dried midrib striate; usually a (large) shrub	***Drypetes parvifolia*** [*KATRIKA-BERE*]	268
Serrations v. small but v. sharp; midrib beneath v. grooved in dry lvs; often a small, crooked tree; slash cream or yellowish grey, darkening	***Drypetes floribunda*** [*BEDIBeSA*]	264

NOTE: 1) *D. afzelii* (**260**) is a rare treelet, with lvs often eaten away at tip like *D. aframensis*, but with larger flowers; when not eaten away the acumen is usually emarginate, or with several large nerves running to flat-ended tip.

GROUP 17B: Rinorea (part 1) (Violaceae)
(Simple, alternate, serrated, short petiole, not trinerved; stipule-cones at twig tips)

VIOLACEAE

The Violaceae discussed here are small trees or shrubs. *Rinorea oblongifolia* (Gp 22B) is an exceptionally common small tree or shrub, but the other species are less common, and often hard to tell apart. The family includes also *Allexis cauliflora*, which bears flowers and fruits on older wood, and *Decorsella*, which has fruits which are exceptional amongst flowering plants. Both genera are restricted to **evergreen forest**.

The flowers are almost regular with five overlapping petals, in long inflorescences (cymes). The fruits are capsules which split into three in *Rinorea*, but in *Decorsella* the seeds develop exposed on an opened placenta, which is star-like and woody at maturity.

Because the following species are barely trees, and very similar, **MPAWU-BERE** can be used as a local name for all species except the first.

- Margin with extremely sharp, long-pointed, spiny serrations (shrub) — ***Rinorea ilicifolia*** [*APOSE-NINI*] 545B
- Margin without sharp teeth[1]
  - Lamina thin and papery; lvs + twigs with long brown hairs scattered, or dense (on yng lvs), especially on upper surface or lower midrib; often with fine spots below (lens); teeth with minute sharp processes[2]
    - Leaf base obtuse or minutely cordate; nodal rings often not v. well defined on sl. wrinkly twigs — ***Rinorea kibbiensis*** 546
    - Leaf base cuneate; petiole up to 2 cm long; scars well-defined — ***Rinorea dentata*** 545
  - Lamina thicker than paper; ±cuneate at base; ±glabrous or, sometimes, with hairs on petiole[3]
    - Leaves not narrowly lanceolate, usually <15 cm long and with hairy petiole
      - Petiole and young twigs covered in dense yellowish hairs; (flowers with petals folded completely backwards) — ***Rinorea angustifolia***[3] 542
      - Petiole with hairs in channel (into which midrib disappears), but otherwise not densely hairy; petals not reflexed — ***Rinorea convallariflora*** 544
    - Leaves narrowly lanceolate–oblanceolate, often >15 cm long; petiole glabrous and channelled; venation reticulate and prominent above — ***Rinorea prasina***[4] 548

NOTES for Group 17B:

1) *Maytenus undata* is a rare tree of **dry forest** which will key here with sharp teeth, narrow petioles, without stipules; the flowers are in fascicles, and the fruits are capsules with red seeds.
2) *Rinorea welwitschii* is a variable shrub with many spots below, ±hairs, usually with thick leaves (see Gp 22B)
3) *Decorsella paradoxa* (=*Gymnorinorea abidjanensis*) is a small tree of **evergreen forest** with rather thick, fleshy, glabrous leaves which dry a pale orange-yellow, with pale twigs and small teeth.
4) See also *Homalium dewevrei* (17E) which sometimes lacks tuft domatia, and has stipule scars

---

NOTES for Group 17C (next text page)

1) The family Ixonanthaceae is discussed at the start of the Group.
2) The family Humiriaceae is discussed at the start of the Group.
3) The Euphorbiaceae have been discussed in Group 22. Other small shrubs or trees in the family which will key to this Group include *Alchornea floribunda*, which is a common shrub or small tree in the forests of Atewa range; it has long (>20 cm) extremely clustered lvs, and often has black spots on the lamina. The lvs are not long acuminate (see next sp.).
4) *Argomuellera macrophylla* [*PONKOHA*] is a similar locally dominant small shrub in the Euphorbiaceae; its lvs and twigs are densely hairy, soft and golden when young; mature lvs sometimes have black spots at the ends of the finest veins. The lvs have sharp teeth and long acumens.
5) *Pycnocoma macrophylla* [*KAFIE-KAFIE*], in the Euphorbiaceae, has v. long ±entire, sessile lvs, which are occasionally slightly serrated. It is a common small shrub with v. long thin, glossy leaves tufted at the top of the stem, and usually with leaf litter collecting at the bases of the leaves. The lamina is glabrous and pustulate. *P. cornuta* is similar, but is found in **drier forests** and has more markedly horned fruits.
6) The family Passifloraceae is discussed at the start of Group 17.

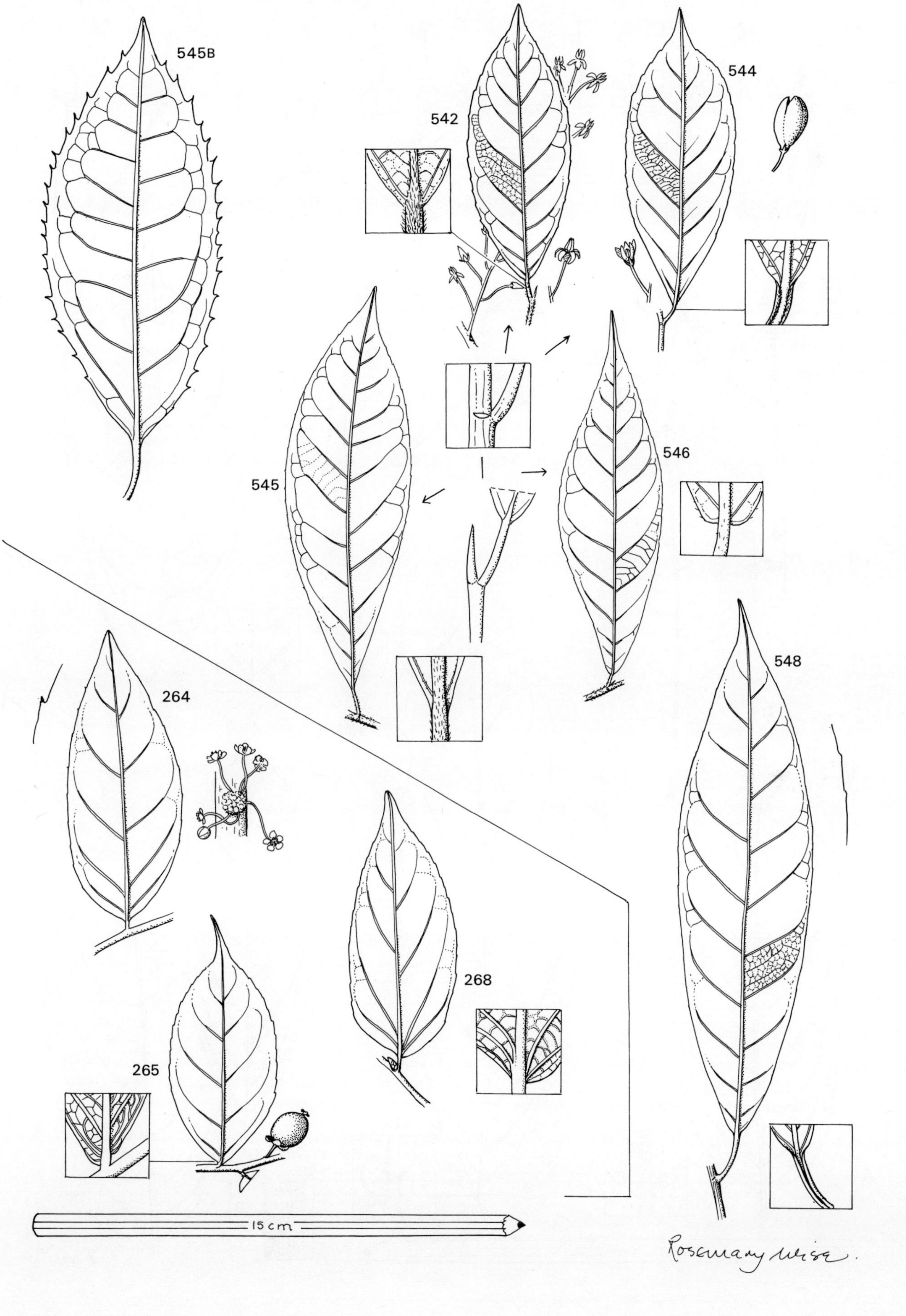

545B
542
544
545
546
548
264
268
265
15 cm
Rosemary Wise

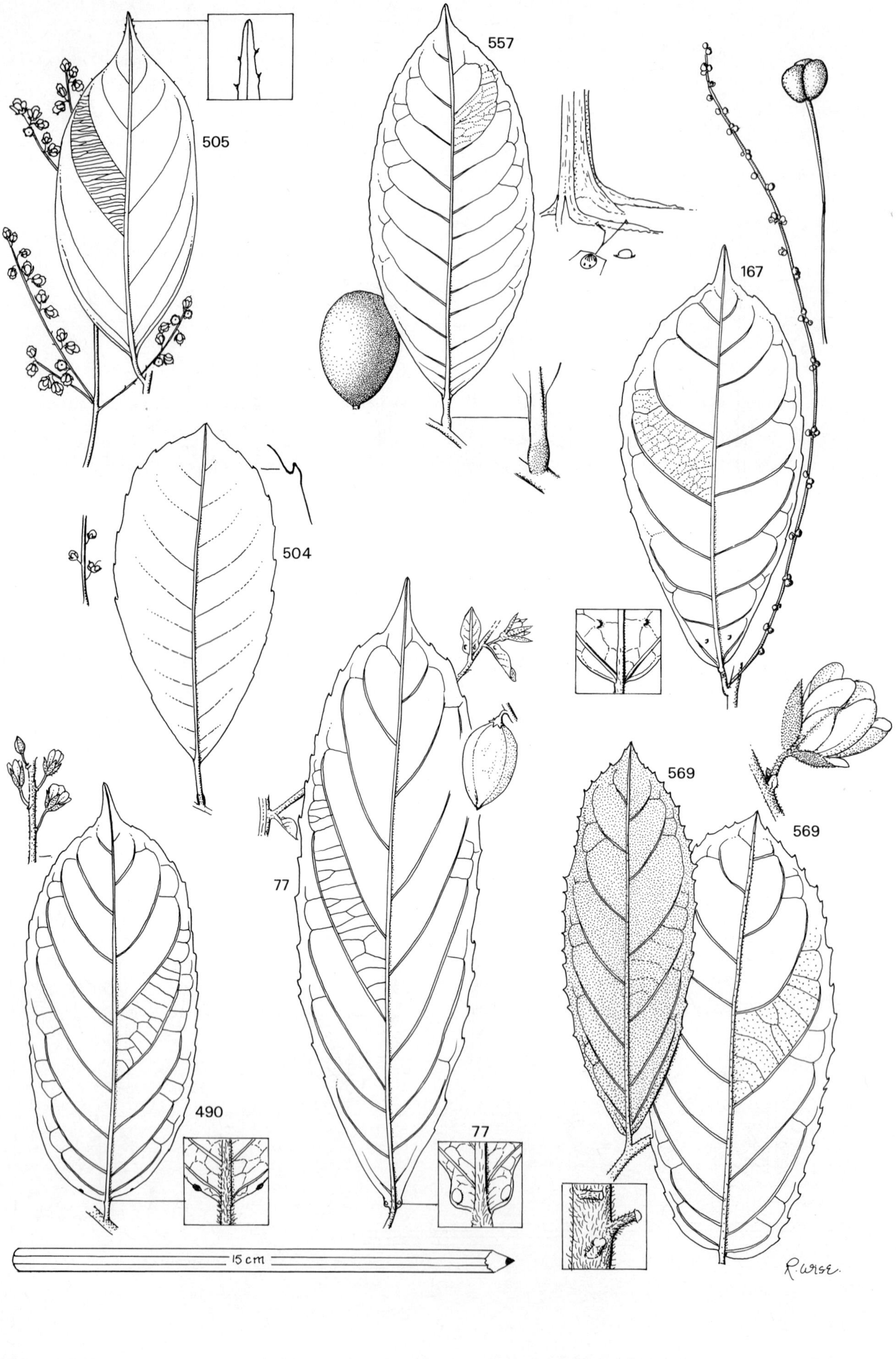
505
557
167
504
77
569
569
490
77
15 cm
R. Wise.

Group 17C: Passifloraceae-Ixonanthaceae
(Simple, alternate, serrated, short petiole, glandular, red slash)

For notes on this Group see below Group 17B, on page 102.

Finer venation obscure, leaf glabrous		
Lvs thinly papery with irregular margin; all veins obscure; reddish exudate	***Tetrorchidium didymostemon*** (see Gp 13A)	
Lvs thicker, with apex rounded or shortly acute; margin regular; only finer veins obscure; small tree in **evergreen forest** or **coastal scrub**	***Phyllocosmus sessiliflorus*** (IXON)[1]	504
Finer venation ±visible (or lvs hairy)		
Teeth, at least near apex, shaped either like minute tree stumps, or like tongues (slash red)		
Teeth crowded at acuminate tip, like small tongues; venation finely transverse between the laterals, but often not v. visible; tree often producing adventitious shoots nr base	***Phyllocosmus africanus*** (IXON)[1] [*AKoKORABEDITOA*]	505
Teeth all around leaf margin, shaped like tiny tree stumps (tapered cylinders with flat ends); leaf base slightly cordate, with two or more glands; twigs and leaves with long ginger, dense, cottony hairs	***Maranthes aubrevillei*** (see Gp 14B)	
Teeth not of a peculiar shape of the above types		
Petiole swollen or twisted or otherwise 'articulated' at top or (strongly so at) bottom		
Petiole (slightly) winged towards top (decurrent leaf base) and swollen near **base**; **evergreen forest zone only**; lf <15 cm long (except sapling *Sacoglottis*)		
Venation not prominent above; small tree with obovate lf with rounded apex; finer veins not clear	***Phyllocosmus sessiliflorus*** (above)	
Venation prominent above – clearly visible; lf margin with spots where veins emerge; highly fluted tree in **evergreen forest** with rough bark; slash red, fibrous + yellowish exudate; slash sometimes hissing	***Sacoglottis gabonensis***[2] (HUMI) [*FAWERE*]	557
Petiole ±articulated towards top, usually yellow-hairy		
Lvs without glands; venation usually markely transverse	See Group 17D	
Glands in lamina near base of midrib, marked by irregular venation (look at several lvs); leaves with inconspicuous, short basal nerves; laterals arching and joining finer venation ±reticulate; teeth **thickened, at the end of nerves, often with a fringe of hairs**; midrib prominent above; lvs yellowish below	***Cleidion gabonicum***[3-5] (EUPH) [*MPAWU*]	167
Petiole not articulated (often v. short)		
Leaves or twigs glandular, e.g. with basal glands or stalked glands on twigs; teeth usually thickened (?glandular); leaves usually hairy; small trees		
Basal glands visible as spots in the lamina **near (not on) the basal margin**		
Branches not whorled; young lvs hairy	See *Cleidion* (above)	
Branches whorled; (yng) lvs glabrous; petiole not swollen	See *Napoleonaea* (17D)	
Glands either on basal margin or on twigs		
Twigs without stalked glands, but leaves with glands on basal margin		
Basal glands much larger than, or otherwise unlike teeth		
Glands like paired ears on top of petiole; lf hairy →glabrous	See *M. aubrevillei* (Gp 14)	
Glands not on top of petiole		
Glands not on small basal lobes; leaves pustular/spotty below (lens); serrations well-defined; slash often with spots of latex	See *Sapium* spp. (Gp 19B)	
Glands on small basal lobes; lf base otherwise cuneate; **evergreen forest** small tree or shrub; apex acuminate or drip-tipped; twigs not v. lenticellate	***Androsiphonia adenostegia*** (PASS)[6]	77
Basal glands like slightly enlarged teeth, and in line with other teeth; **dry forest**, small-medium tree; fine venation v. distinct; outer bark brown and rough-flaky; slash contoured reddish and brown; twigs v. lenticellate	***Paropsia guineensis*** (PASS)	490
Twigs with stalked glands near nodes; lvs v. variable, but lvs and twigs normally with v. dense yellow hairs; saplings/shaded shoots with lvs ±oblanceolate, cuneate; but sun-leaves ±oblong and cordate; petiole <5 mm; slash pink-brown contoured	***Smeathmannia pubescens*** (PASS) [*TURUNNUA*]	569
Leaves and twigs not glandular, or (with the only 'glands' being) many obvious translucent spots	See next Group	

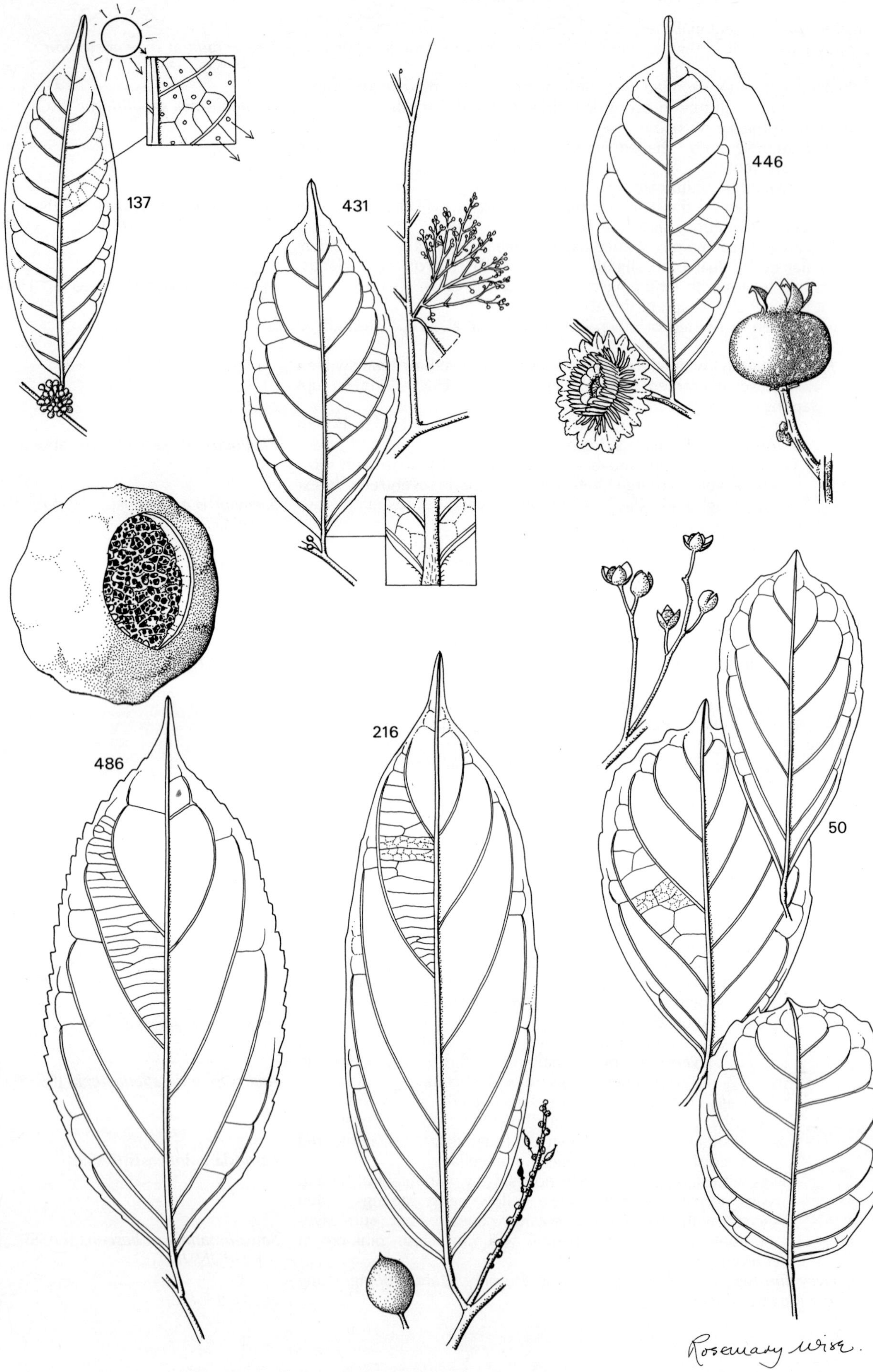
137
431
446
486
216
50
15 cm
Rosemary Wise

Group 17D: Pandaceae-Flacourtiaceae
(Simple, alternate, serrated, short petiole, glandless; slashes varied: red to yellow)

The Flacourtiaceae are discussed in Group 17E.

PANDACEAE

Previously, only *Panda* was placed in this family, but *Microdesmis* has recently been transferred from the Euphorbiaceae. *Microdesmis* has small flowers in axillary fascicles, thereby resembling *Drypetes,* although they are very often diseased and then look like much-branched panicles (as in illustration). *Microdesmis* fruits are small drupes or berries. *Panda*, whose ±asymmetric leaves could also be mistaken for a *Drypetes*, is dioecious: both sexes of tree have red flowers in long spike-like inflorescences, but the female flowers, and therefore the fruits, tend to be borne on older, stouter twigs. *Panda* fruits are *c.*5 cm ovoid drupe-like and eaten by elephants. The large, ±leathery cotyledons, shaped like a quarter moon with points away from the stem, are common in elephant dung, or on old logging roads in elephant areas.

Leaves glabrous, with many translucent spots[1]; teeth barely visible (lens); petiole >5 mm long; lf base extremely asymmetric (cuneate and obtuse); twigs zigzagged; small tree with narrow, ±horizontal, shallow crown; slash pale yellow-orange, v. soft granular, darkening soon to brown	***Casearia calodendron*** (FLAC) [*AKWANA-FUFUO*]	137
Without obvious translucent spots OR leaf with hairs OR symmetric		
— Petiole short (<1 cm) and slender, and not articulated; lvs with 5-8 laterals arching strongly and meeting *c.*⅔ way to margin; often <10 cm long		
— — Leaf and twigs ±glabrous; twigs with raised lines (esp. when dry); lf base ±symmetric; sometimes (rarely) with 2 basal glands, and sometimes (often) entire; apex usually + long drip tip, often broadened slightly at apex; branches strongly whorled; bark dark; slash whitish	***Napoleonaea vogelii***[2] (LECY) [*oBUA*]	446
— — Leaf and twigs generally with yellowish hairs (lens); leaf base v. asymmetric; twigs slightly swollen towards nodes; apex acuminate and often with a tiny mucronate point, but rarely drip-tipped; **brs not whorled**	***Microdesmis puberula*** (PAND) [*oFEMA*]	431
— Petiole robust, usually >1 cm long, often articulated; (flowers in spikes or racemes)		
— — Lvs glabrous; never with cordate base		
— — — Petiole not articulated; slash purplish, venation between laterals finely transverse; medium-sized tree, with very regularly branched boughs like large, compound lvs (Cook's model of Hallé *et al.*)	***Panda oleosa*** (PAND) [*KOKROBOBA*]	486
— — — Petiole usually articulated, or at least swollen or discoloured towards top; slash varied, often reddish brown, but never deep red-purple		
— — — — Leaf with well-defined, rather sharp teeth and finely transverse venation; with cuneate base; small stipule scar at base of petiole; slash slightly leathery, orange to red, not v. scented	***Dasylepis brevipedicellata***[3] (FLAC) [*ASRATOADUA-NINI*]	216
— — — — Leaf with rather rounded, blunt teeth or undulations; venation not usually finely transverse, but if so, leaf base rather obtuse; shaded leaves with obtuse apex, with teeth then sharper and ±restricted to apex; slash orange or reddish, gritty, darkening; with sweet exudate; smelling bitter and sweet (like almonds)	***Scottellia klaineana*** (FLAC) [*TIABUTUO*]	50
— — Lvs hairy or with cordate base	See next Group	

NOTES:

1) ***Suregada occidentale*** is a small shrub with extremely obvious translucent spots or raised pustules. It can be recognized by its ±sessile, symmetric leaves and almost four-sided twigs. If the slash is strongly scented check *Drypetes aubrevillei* (Gp 17A).
2) The family Lecythidaceae is discussed in Group 25. *Napolean vogelii* occurs in two forms in Ghana (see Hall and Swaine, 1981). The normal forest form is illustrated. Plants from riversides in savanna or dry forest have longer leaves without such drip tips. The leaves of both varieties are often ±entire (see Gp 13B).
3) ***Dasylepis assinensis*** is a rare shrub or small tree known only from Ankasa and Krokosua hills; it has smaller lvs (<15cms) and the venation is reticulate, but it otherwise resembles ***D. brevipedicellata***.

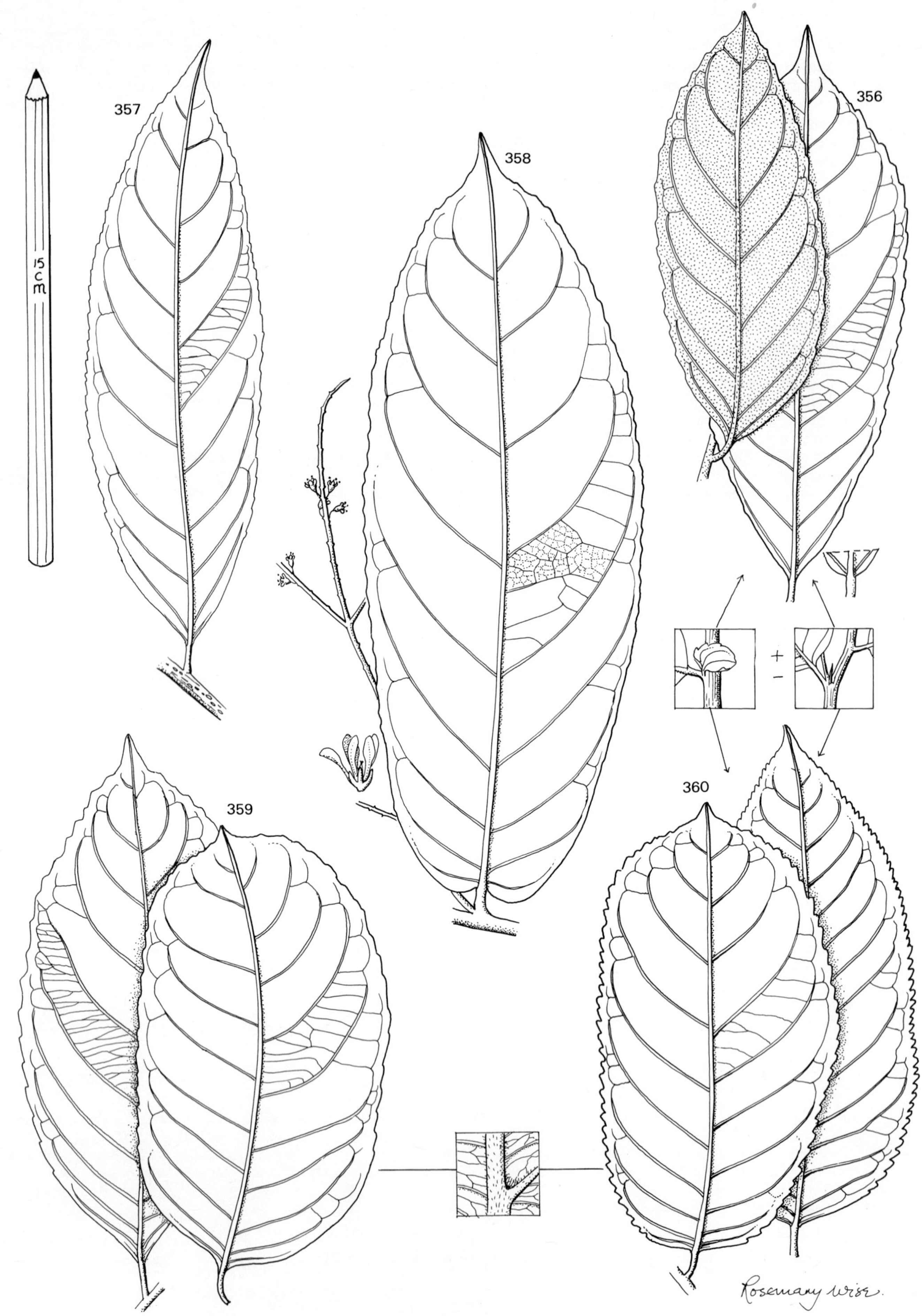
357
358
356
359
360
15 cm
+
−
Rosemary Wise

Group 17E: *Homalium* (Flacourtiaceae)

The Flacourtiaceae are discussed at the start of this group. *Homalium* spp. tend to have straight, cylindrical boles with small crowns of thin, drooping branches. The slash contain very conspicuous strands of orange grit, and the bark is often very lenticellate

Tree very deeply fluted, in evergreen forest	See *Sacoglottis* (17D)	
Tree only slightly, or not at all fluted, or outside evergreen forest		
Leaves and twigs glabrous AND deeply cordate, and usually >7 cm wide at middle; serrations well defined; often with several sets of laterals turned backwards into cordate base; tree cylindrical with rough outer bark (for slash see above); commonest *Homalium* in high forest.	***Homalium letestui*** [*ESONANKROMA*]	358
Leaves either with some hairs (although sometimes only in nerve axils) or without cordate base; (or if cordate,) then leaf usually <7 cm wide at middle; trees rare in 'ordinary' forest, but locally common on **steep slopes, swamps or riverbanks**		
Leaf base cordate-obtuse, or leaf densely hairy below; stipules often persistent and sometimes like small leaves themselves (but sometimes thread-like)		
Tree in **swamps or riversides**; leaf rather narrowly elliptic, and often softly hairy, but also sometimes glabrous; without a dense arrangement of laterals below; twigs glabrous, or with easily (x10 lens) visible, dense hairs, ±lenticels; slash said to smell of urine	***Homalium africanum***[1]	356
Tree not in wet place, or leaves with large, rather rounded, dense teeth; leaf usually rather broadly elliptic; usually with tuft domatia; base of leaf often cuneate, but when obtuse with conspicuous aggregation of 3-4 pairs of laterals in first cm at base; twigs not lenticellate, but with fine, dense hairs almost invisible individually with x10 lens	***Homalium stipulaceum***	360
Leaf base cuneate (rarely almost obtuse) and leaf with few hairs; tree slender, symmetrical; bole rough often with many small lenticels; slash finely, but densely gritty, yellow-orange with orange striate sapwood		
Nerve axils with tufts of hairs; twigs not obviously lenticellate		
Lf base obtuse to cuneate, often with hairs (e.g. on upper midrib); often with conspicuous lateral nerves	See *Homalium stipulaceum* (above)[2]	
Lf base always cuneate, sometimes with decurrent lf base; lf ±ovate-lanceolate with v. few hairs; even basal nerves ascending; finer veins usually rather parallel and raised above	***Homalium longistylum***[1]	359
Nerve axils without tuft domatia		
Leaf rather glossy; typically ovate with ±decurrent leaf base; twigs not very obviously lenticellate	***Homalium longistylum*** [*oWEBIRIBI*]	
Leaf papery; typically narrowly oblong-elliptic to oblanceolate; twigs dark, with conspicuous pale lenticels, even when viewed without lens; apparently rare in Ghana	***Homalium dewevrei***	357

NOTES: 1) These *Homalium* spp., as suggested by the key, can be very hard to tell apart. This is partly a consequence of the variability of the species, and it seems likely that hybrids (particularly with *H. africanum*) may be common. Fortunately the commonest species (*H. letestui*) can normally be reliably identified, but even identification of this is prone to error if immature (when it is sometimes not cordate). Because of these difficulties, *oWEBIRIBI* can be used as a local name for the last two species.

2) *Dissomeria crenata* is a similar species with domatia but with smaller leaves. (Fortunately) it is restricted to **riverine forest** in the savanna zone.

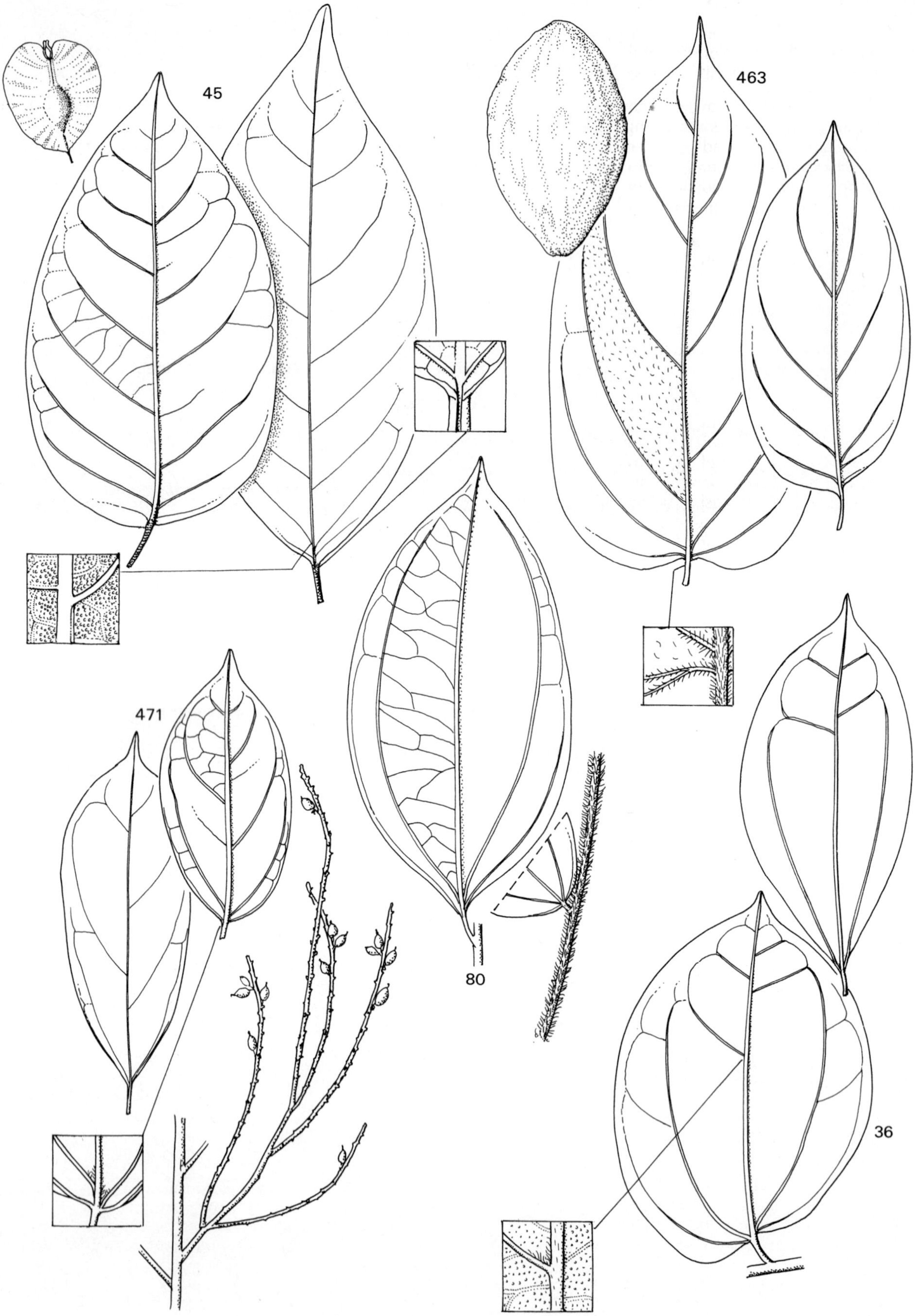
45
463
471
80
36
15cm
Rosemary Wise.

GROUP 18: ULMACEAE, etc.
(Lvs strongly 3- or more nerved at base; without scales; no stellate hairs; petiole short)

As well as *Celtis* spp., which are often the commonest trees in the semi-deciduous forests, the Ulmaceae include *Holoptelea*, *Trema* and *Chaetachme* (the latter is spiny, and has therefore been placed in Gp 15).

The larger *Celtis* spp. are recognizable immediately by their chocolate and cream-coloured slashes (see Plate 3), but *C. africana*, *C. wightii* and other members of the family do not share this feature. All species have paired stipules, but these usually fall very rapidly, and are usually seen only on seedlings. The crown is normally rounded, with a regular 'cloud' of foliage.

The Ulmaceae have small flowers with no petals. The flowers are arranged in small clusters. The fruits are two-winged in *Holoptelea*, with two style arms remaining at the apex. The other species produce small drupes (fleshy fruits, with a single stone (seed with hard 'endocarp')) and are often reddish or black when ripe (small red drupes are often bird dispersed). The stone is ridged or otherwise ornamented. That of *C. adolfi-friderici* is the largest, at *c.*2 cm long, with a 1 cm white, spherical endocarp which is deeply pitted. The endocarps of the other species are similar, but less than 1cm in diameter. *Trema* produces even smaller black drupes, a few millimetres in diameter. The ellipsoid drupes of *Chaetachme* have two long, hairy style arms persistent at the apex.

Okoubaka[1] (Santalaceae – previously in Octoknemataceae) produces long panicles on older wood, and *c.*7 cm ellipsoid drupes. *Anisophyllea* (Anisophylleaceae – previously in Rhizophoracaeae) produces small spikes of 4-petalled flowers, and *c.*2 cm ellipsoid drupes. *Ophiobotrys* is mentioned with the rest of the Flacourtiaceae at the start of Group 17.

Key to subgroups

Margin entire	
Lvs with (1-4) *vague*, unequal basal nerves	
Slash red-fibrous; basal nerves several, but not well-defined; petioles often with thread-like stipels; **evergreen forest**, esp. by rivers	***Didelotia*** spp. (Group 37A)
Slash orange yellow, or granular	
Lvs ±lanceolate; slash fibrous, darkening rapidly; **dry forests**	***Diospyros abyssinica*** (Group 11)
Lvs broadly elliptic to ovate; slash not *Diospyros*-like	
Lvs with few no or few hairs; discolorous; twigs winged; slash mustard-scented (like *oPAHA*)	***Drypetes leonensis*** (Group 17A)
Lvs hairy or with white spots; slash not scented of mustard, although possibly peculiarly iodine-scented	Group 18A
Lvs with very well-defined basal nerves	Group 18A
Margin serrated	
Lvs with 2-4 basal nerves, not strongly ascending	***D. parvifolia*** (17A), ***Cleidion*** (17C).
Lvs with very clear, strongly ascending basal nerves	Group 18B

GROUP 18A
(Lvs trinerved, entire, with short petioles)

Basal nerves not quite arising from base, but arising from widened part of midrib channel just above base; lf surface with many white spots; slash yellow-brown, brittle, with conspicuous green outer layer and slightly sweet iodine-like smell; twigs and bark with conspicuous lenticels	***Holoptelea grandis*** [*NAKWA*]	45
Basal nerves arising precisely at base		
Lvs or twigs with long or dense (use lens) yellow or brown hairs; often broadest below the middle		
Venation not scalariform; hairs long, yellow and weak; venation not v. close; lvs ±symmetric		
Basal nerves ±reaching apex; slash crunchy, orange-brown with vertical lines over hard orange, striate sapwood; **evergreen forest**	***Anisophyllea meniaudii***[2] (ANIS) [*KoKoTE-AKOA*]	80
Basal nerves not even approaching apex; venation rather obscure; base obtuse or cordate; slash yellow-orange, brittle-fibrous in layers; older trees with stunted forest around them, due to parasitic roots	***Okoubaka aubrevillei***[1] (SANT) [*ODII*]	463
Venation scalariform; slash brown and yellow	See *Celtis zenkeri* (18B)	
Lvs with short, inconspicuous, or no hairs; most lvs broadest above the middle		
Lvs with tufts of hairs in pit domatia of axils of upper laterals; very asymmetric; basal nerves (almost) reaching apex; slash with chocolate brown blotches in a cream background, not darkening rapidly; crown dark green	***Celtis adolfi-friderici*** [*ESAKOSUA*]	36
Lvs with no tufts of hairs in axils (except sometimes in basal axils) 1 or 2 weak basal nerves outside main ones, running very close to, or obscured from below by, recurved lower margin; slash yellow-orange, thin, brittle and gritty	***Ophiobotrys zenkeri*** (FLAC) [*AKWANA*]	471
Only 1 pair of basal nerves, well away from lower margin; bark often very silvery; slash with close, ±continuous bands of brown and yellow	***Celtis mildbraedii*** [*ESA*] – see 18B	

For NOTES see next page.

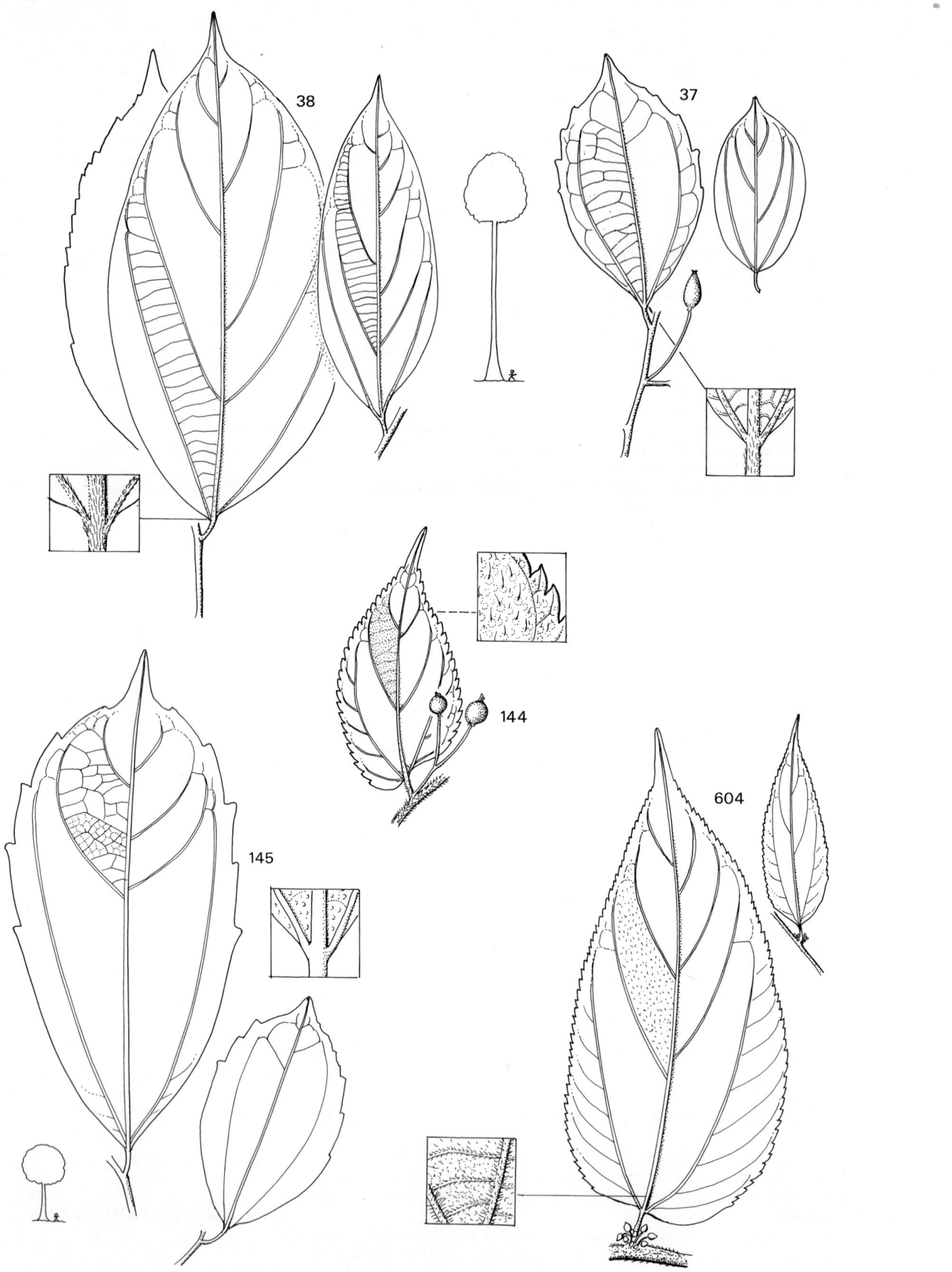
38
37
144
145
604
15 cm
Rosemary Wise.

Group 18B: Ulmaceae
(Margin serrated)

Leaves with strongly scalariform venation and dense hairs (shade leaf or sapling – **mature lvs ±entire**); hairs short, dense and orange-brown; lf base asymmetric; crown pale, yellow green; bark often rather yellow; **slash with broken black-brown bands, or spots in a cream background, turning markedly browner in the first 30 seconds**	***Celtis zenkeri***[1,2] [*ESAKoKo*]	38
Leaves without strongly scalariform venation OR glabrous		
Lvs not very rough above		
Leaf base not cordate; lf not very hairy (often completely glabrous)		
Serrations rather shallow (<1 mm steps), often in upper third of lf; finer venation conspicuous and prominent; often with v. fine orange brown hairs on petiole and young stem; Slender, tall tree with narrow crown; slash with close brown and cream bands, darkening only **a little in the first 30 seconds**	***Celtis milbraedii***[1,2] [*ESA*]	37
Serrations usually pronounced (>1 mm steps), often on end half or more of lf; petiole never with hairs; small tree with relatively spreading crown and slash yellowish, darkening rapidly	***Celtis wightii*** [*PREMPRESA*]	145
Leaf base cordate, or lf very hairy	(Consider saplings of Gp 19)	
Lvs rough above; often v. asymmetric		
Teeth not sharp, but margin irregularly crenate; yellowish exudate	See *Ficus exasperata* (Gp 19).	
Teeth well-defined, leaves acuminate or with silver, silky hairs.		
Lateral nerves with tuft domatia. Tree of **dry forests** only	***Celtis africana***	144
Lateral nerves without tuft domatia; hairs below v. fine and silvery; small weedy tree common in very disturbed forest	***Trema orientalis***[1] [*SESEA*]	604

NOTES: 1) *Celtis* seedlings have broad cotyledons. *Trema* seedlings have very narrow cotyledons. The higher seedling leaves of *C. zenkeri* are not as markedly scalariform as the lvs of the adults, and are hard to distinguish from those of *C. mildbraedii*. Young and shaded leaves of both these *Celtis* spp. are serrated, (more persistently so in *C. mildbraedii*), and become entire with age and exposure to the sun. Seedlings are normally abundant in sunny areas near to parent trees.

2) *Celtis zenkeri* and *C. mildbraedii* are often confused if only slash and general form are considered. Fallen leaves are almost always abundant, and the venation pattern then serves to distinguish them. Both are very common, and when growing side by side they can be distinguished by their crown colour, which is oranger in *C. zenkeri* because of the hairy leaves. *C. zenkeri* is slightly more common in dry forests.

NOTES FOR GROUP 18 AND 18A:

1) *ODII* is considered a powerful fetish tree. The leaflets of *Balanites wilsoniana* (which has similarly large fruits), are often remarkably similar to the leaves of the *ODII* tree, and are sometimes named by tree spotters as a female version of it. *Balanites* has one pair of leaflets in a compound leaf, is spiny, often fluted, and does not diminish the stature of the forest around it. Sapling *Okoubaka* have unusual velvety green stems.

2) *Anisophyllea*, in fact, produces opposite leaves, but one in each pair fails to develop.

FLOWERS AND FRUITS OF MORACEAE

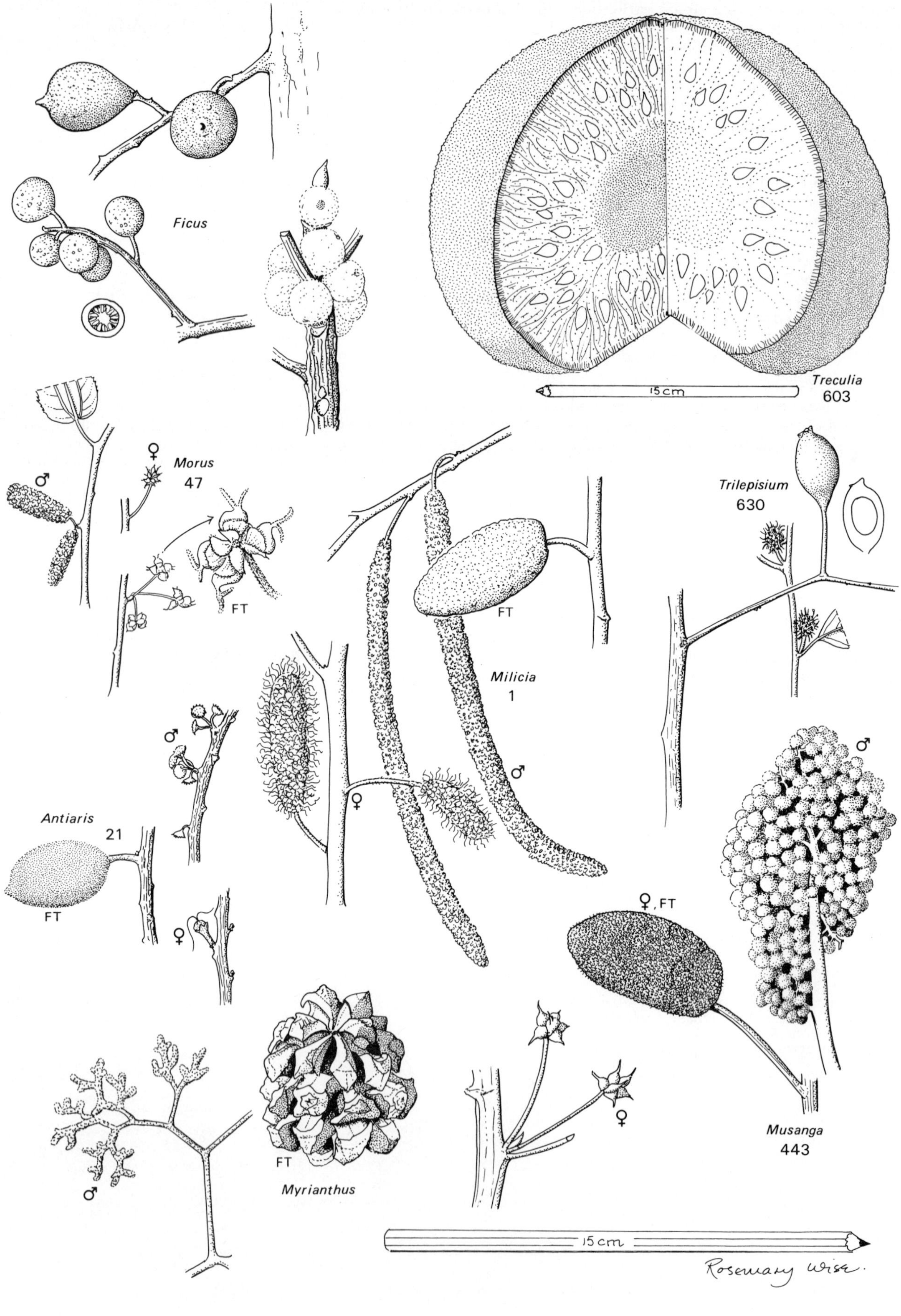

GROUP 19: MORACEAE (part 1)
(Lvs simple, alternate; slash with milky or brownish latex)
(Stipules often large or persistent; leaving round scars at nodes)

For part 2 see Group 28A.

The majority of the species in this family in Ghana are 'figs'; species of *Ficus*, which are of little economic importance and are often found as epiphytes or stranglers, growing even from seed on other trees (Gp 19C). By contrast, the *ODUM* tree (two species of *Milicia*, previously, and more widely known as *Chlorophora*) provides an exceptionally useful and important timber known more widely as 'IROKO'. Ecologically, the species apart from the strangling figs are common as pioneers in very disturbed forest. This is particularly true of *Musanga* and *Ficus exasperata*, but applies also to *ODUM*. For all of these, the many small seeds in the fruits clearly improve the chances that the species 'find' newly exposed gaps in the forest.

The family is divided into two major Groups in this guide. The species in this Group (19) produce a milky latex, whereas species like *Musanga* in Group 28A do not (their leaves are also lobed or compound). The bark of the family tends to be very fibrous, sometimes gritty or granular in outer layers, often with conspicuous lenticels. A cloth made from the bark of *Antiaris toxicaria* was once an important produce of Ghana's forest zone, and *Broussonetia papyrifera* – the paper mulberry tree which has escaped from cultivation in some forest reserves in Ghana – has a bark which was similarly used for fibre in Polynesia. Species of trees which produce latex and which also have alternate, simple leaves are found in Sapotaceae (Gp 10) and some Euphorbiaceae (see below). All latex-bearing species with serrated or long-petioled or trinerved leaves are keyed in this 'Moraceae' Group.

One of the most distinctive vegetative characters of the Moraceae concerns the apical bud, which tends to be conical and pointed, derived from wrapped-up leafy stipules, which normally soon fall to leave ring scars at the nodes in much the same way as *Klainedoxa*, etc. (Gp 13) and *Rinorea* (Gp 17). This characteristic is particularly obvious in some *Ficus* spp. and *Musanga*.

The flowers of Moraceae are generally tiny and lack petals, but are usually aggregated into caterpillar-shaped ('catkins'), or spherical inflorescences. Individual flowers are always of one sex. In *Trilepisium* and *Ficus* male and female flowers are arranged in a single structure, whereas in the other genera they are on separate structures. Indeed in *Milicia*, *Morus* and sometimes *Treculia* the male and female inflorescences are on different trees (i.e. the species is dioecious). There are differences at least in *Milicia*, between the appearance of the crown of male and female trees.

Genus	Notes on arrangement of flowers
	(Monoecious unless stated otherwise)
Ficus	Male and female enclosed in a globose structure ('syconium' or 'fig') with a hole for pollinators at apex
Trilepisium	Male surrounding female on small catkin
Antiaris	Male disc-like; females single with 2 styles; sexes together in lf axils
Milicia	Dioecious. Male catkins long, females shorter with many protruding styles
Morus	Dioecious. Both sexes of catkin much smaller than *Milicia*
Treculia	Dioecious or monoecious; m. infls. in lf axils: f. on larger branches and trunk. Both with minute circular scales between which stamens or styles protrude
Myrianthus	Male infls branched; females at end of unbranched stalk
Musanga	Male infls branched, in dense heads; females in paired globose heads in axils

The fruits are important sources of food for the many birds and mammals which disperse the seeds. In many cases the fruits are little different from the female inflorescences from which they develop. In some cases they are fleshy and produce a strong fruity-acid or fermenting smell when ripe (or over-ripe), e.g. the 'figs' of *Ficus* species, and the immense football-sized fruits of *Treculia africana* (from the crashing to the ground of which the onomatopoeic *Twi* name '*BReBReTIM !*' is said to be derived). Similar, immense fruits are produced by the cultivated *Artocarpus* spp., jack-fruit (*A. heterophyllus* with small, ±entire lvs and cauliflorous fruits) and bread-fruit (*A. communis* with sharply lobed lvs). The species with many small seeds are not found as seedlings in the forest shade, unlike the first two species below which produce fruits with single large 'stones' and which are often abundant in the shade near parent trees.

Genus	Notes on fruits
Trilepisium	Dark, shiny drupe (with single stone surrounded by fleshy layer
Antiaris	Reddish, softly hairy drupe
Ficus	Many small seeds in syconium (see flower description, Gp 19C)
Treculia	Firm, immense, with many medium-sized seeds in fibrous-fleshy mass which produces strong fermenting smell
Morus	Small fruits loosely aggregated, with conspicuous sepals
Milicia	Fleshy, with strong fermenting smell; many small seeds
Musanga	V. many tiny drupes in fleshy structure 10 cm long
Myrianthus	Rather woody or slightly fleshy cone- or crystal-like structure

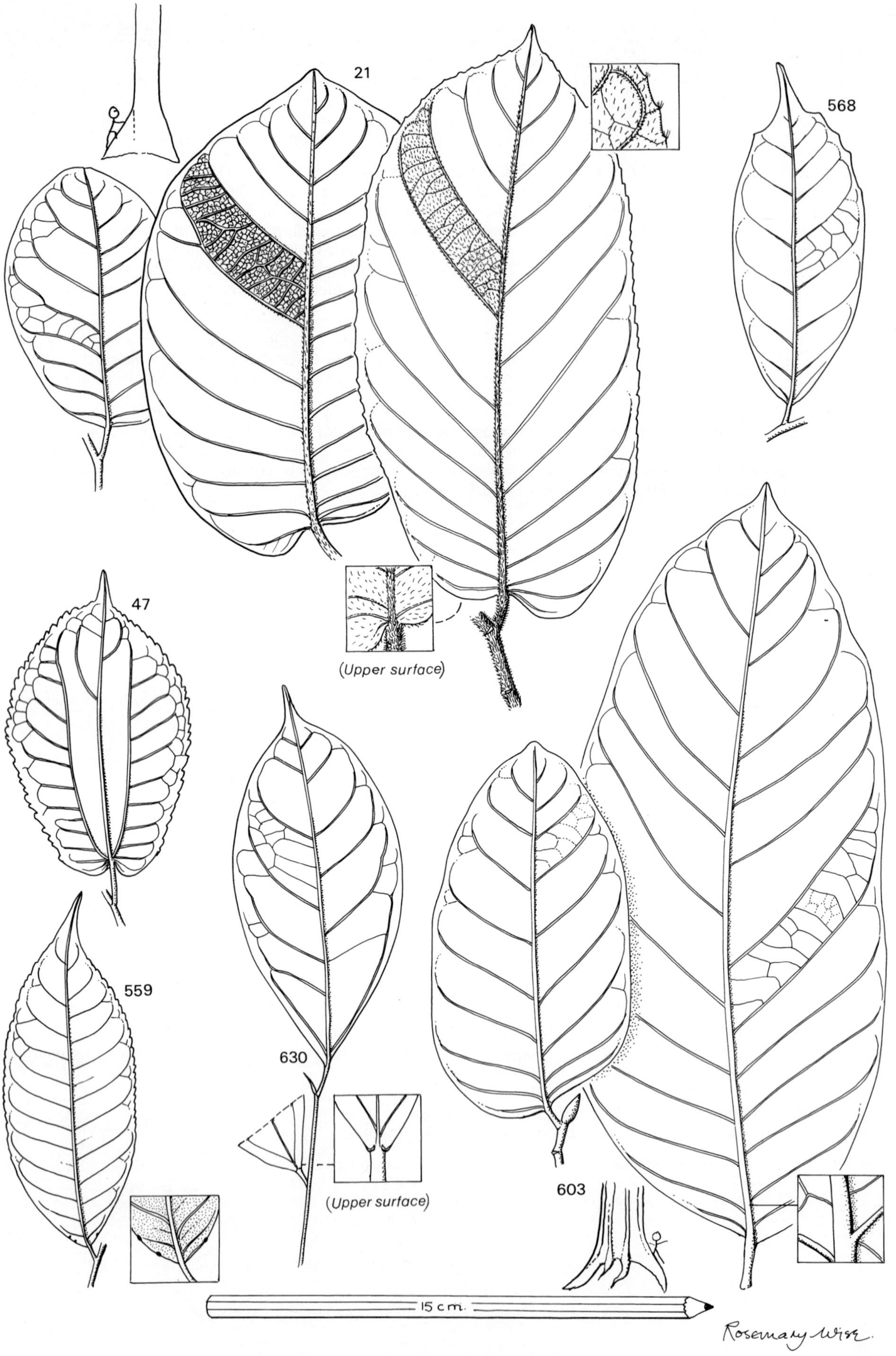
21
568
47
(Upper surface)
559
630
(Upper surface)
603
15 cm.
Rosemary Wise.

Group 19: Moraceae (part 1): Key to subgroups

Trees with 'strangling' aerial roots; often epiphytic on other tree species; stipules cone-like at apex of twigs, leaving ring scars	***Ficus*** spp. [*AMANGYEDUA*] Group 19C	
Trees not 'strangling figs'		
Leaves and twigs v.succulent; lvs clustered at br. ends; stems with small spines; branches clustered at top of stem	***Elaeophorbia grandifolia*** (EUPH) (See Group 22)	
Leaves and twigs not remarkably succulent		
–Leaves with rusty stellate hairs and scalariform venation – consider juvenile *Coula* (27B) or a detached leaflet from a compound leaf		
–Leaves without stellate hairs (or simple, and tree **outside evergreen forest**)		
Leaves with short petioles (<2 cm)	Group 19A	
Leaves with long petioles (>2 cm)	Group 19B	

Group 19A
(Moraceae (etc.) with short petioles)

Leaves serrated		
Leaves not trinerved		
Lvs with two glands on margin at base; scaly or pustular below; latex often only spots; lvs almost always broadest below half way; teeth fine		
Leaves broader than 3 cm (except saplings); veins not v. prominent below; prominent above; **evergreen forest**	***Sapium aubrevillei*** (EUPH) [*KETEBONTORE-NUA*]	559
Leaves narrow; veins prominent above and below; **dry forests**	***Sapium ellipticum*** (EUPH) [*KETEBONTORE*]	560
Leaves without glands, often coarsely hairy, with cordate base: saplings[1,3]		
Lvs ±oblong, with long chocolate brown hairs; brs ±whorled	Juvenile *Antiaris* (below)	
Lvs ±ovate-lanceolate, with yellowish fine hairs	Juvenile *Milicia* (19B)	
Leaves v. strongly trinerved[1], papery; crown dense, dark green; bole cylindrical; bark with large lenticels in conspicuous corky lines; slash cream-yellow, thick, soft-fibrous with copious latex	***Morus mesozygia*** [*WONTON*]	47
Leaf margin entire		
Lvs without ascending basal nerves; asymmetric, and ±cordate or obtuse base, at least on one side (if symmetric and not rough consider Group 10)		
Bole straight and cylindrical, sometimes with small buttresses; crown symmetrical and dark with horizontal lower boughs; leaves **<15 cm** long, generally **elliptic**; lvs sometimes v. rough-hairy (in 'var. *africana*') and sometimes hairless (in var. *welwitschii*) with white spots above; slash fibrous, slightly granular outer layer, yellow to brown with orange lines and copious, watery, milky-brown latex	***Antiaris toxicaria***[1,2] [*KYEN-KYEN*]	21
Bole fluted; crown rather untidy and inconsistent; lvs often **>15 cm** long; ovate to **lanceolate**, never roughly hairy; slash slightly fibrous, but brittle and gritty, yellow to brown, darkening with copious, watery, cream (turning brown) latex, ±green outer layer	***Treculia africana*** [*BREBRETIM*]	603
Lvs trinerved, or with a pair of ascending nerves arising v. close to base of midrib		
Leaves smooth, glabrous, with tiny folded-back 'lobes' at base, above which the 'basal' nerves arise; apex acuminate; crown dark and dense; bark smooth; base sometimes fluted or buttressed; bark easy to tear into long strips; slash yellow, v. rapidly darkening to brown with similarly darkening latex	***Trilepisium madagascariense*** [*OKURE*]	630

NOTES:
1) *Ficus exasperata* has very variable and rather irregular leaves and might key here. It has no latex, but yellow exudate (see Group 28A). The leaves are extremely rough like **sandpaper**; uneven at base and tip, often one nerve arising slightly above the other.
2) The two varieties mentioned here were treated previously as two species. The smooth-leaved var. *welwitschii* is found in **moister forests**, and var. *africana* with the rough lvs in **dry forest**.
3) ***Sloetiopsis usambarensis*** (568) is a small, often arching (and rooting at br. tips) treelet, with serrated or entire, glabrous lvs with conspicuous, pale reticulate venation below, rather like a *Ficus*, and with the laterals meeting in a strong, wavy, sub-marginal nerve. ***Dorstenia (Craterogyne) kameruniana*** is a similar shrub in this family, with large persistent stipules and ±hairy leaves more usually serrated, sometimes with large teeth confined to an almost flat-ended apex (like a small *Sterculia oblonga*). It is often found **near rivers**.

Group 19B
(Moraceae with long petioles and simple leaves)

Leaves with two and often more basal nerves besides midrib; bole appearing unusually smooth and pale, except sometimes for crescent scars (old lf scars); slash v. fibrous, sometimes with green outer layer, darkening with latex		
–Nerves reaching almost to apex; lvs serrated and glabrous; petiole rarely much more than 3 cm long	See *Morus* (19A)	
–Nerves not reaching so far, or leaves v. hairy		
Lvs not densely hairy nor v. rough; margin with rounded undulations or steps; ±symmetrical; petiole often <3.5 cm long; white spots on surface of young lvs; nodes with rings of hairs; medium or small, cauliflorous tree in **secondary forest**	***Ficus sur*** [*DOMINI*][1]	300
Lvs v. hairy or rough to touch (on upper surface), sometimes v. asymmetric.		
Lvs extraordinarily rough to touch; exudate, at least initially, yellow and watery (but sometimes with milky latex later in slash or in cut petioles); lf often v. asymmetric	(See *Ficus* spp. in Gp 28A)	
Lvs not so extraordinarily rough, but with long dense hairs over surface		
Twigs with hairs dense all over, with paired persistent stipules at young nodes	***Ficus vogeliana***[1]	327
Twigs with conspicuous long fringes of hairs at nodes, and with stipules falling rapidly; veins v. prominent below with long , soft hairs on veins; a large, canopy tree	***Ficus mucuso***[1,2]	314
Leaves not strongly trinerved; basal nerves, if any, almost perpendicular to midrib (bark not smooth)		
–Petioles always <3 cm long; mature lvs strongly discoloured, or not cordate	Try SAPOTACEAE (Gp 10)	
–Petiole often >3 cm long; leaf base slightly cordate; tall to very tall trees with straight, cylindrical boles; crown with outward sweeping layers of dark foliage; bark rough, with reddish-brown conspicuous lenticels; slash v. gritty with rapid, watery latex; surface roots often conspicuous (+ orange brown lenticels)		
Lateral nerves in >10 pairs[3]; venation on lower surface looking like a sponge, with the gaps between the veins as conspicuous as the broad, fibrous-looking veins themselves; all forest types	***Milicia excelsa***[3,4] [*ODUM*]	1
Lateral nerves in 6-10 pairs[3]; venation on lower surface arranged in neat, sharp little squares, v.regular, with the veins as fine lines between these green 'islands'; **moister forests** (W. region)	***Milicia regia*** [*ODUM-NUA*] (***Milicia = Chlorophora***)	2

NOTES: 1) All these *Ficus* spp. with their exceptionally smooth, yellowish stems, are regularly confused under the local names *DOMINI*, *DOMA* or (from a distance) *NYANKYERENE*. The name *AMANGYEDUA*, on the other hand, seems more often to be used for dark crowned, spreading, often strangling *Ficus* spp. '*ANOMANI*', a local name from which scientists christened *F.anomani*, appears to have a similar linguistic root to *DOMINI*. At any rate, the application of local names to *Ficus* spp. is far from species-specific.

2) The introduced tree *Broussonetia papyrifera* – the 'Paper Mulberry' tree from the Pacific region – has been planted in some areas. It is 'escaping' into disturbed parts of Afram Headwaters forest.

3) *Milicia* species are not normally distinguished by the timber trade, as the trees and timber are so similar. Sapling leaves of both species have shorter petioles than the mature trees, and sharp serrations. Those of *M. regia* have up to 12 pairs of laterals, but have v. few hairs below, except on the veins, whereas those of *M. excelsa* are normally very densely covered in fine, pale hairs.

4) The genus name *Milicia* has unfortunately replaced *Chlorophora*, which has also in recent times been changed to *Maclura*.

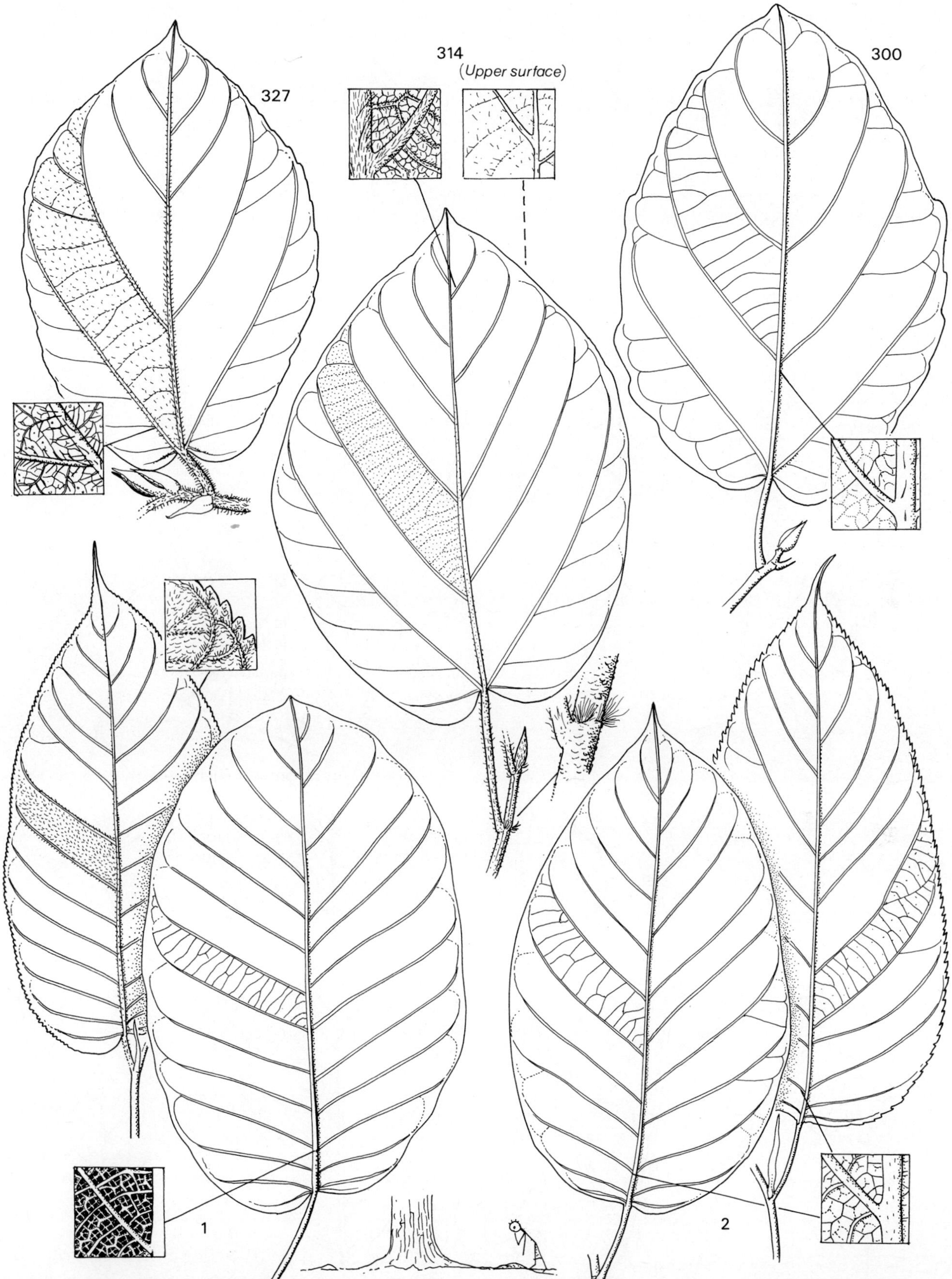
327
314
(Upper surface)
300
1
2
15cm
Rosemary Wise

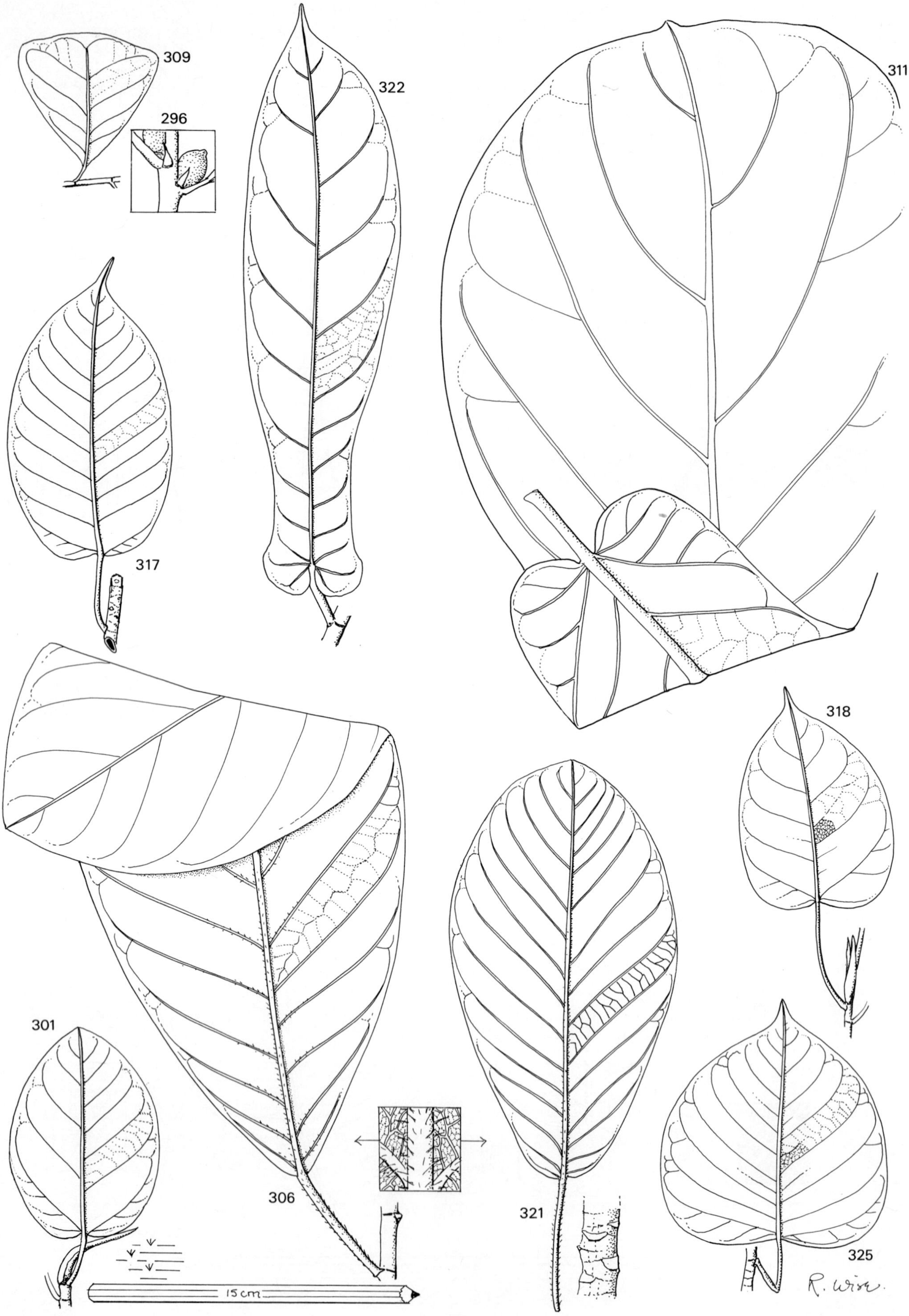
309
296
322
311
317
318
321
301
306
325
15cm
R. Wise.

GROUP 19C: *Ficus* (MORACEAE)
(Strangling figs)

Ficus is the largest genus of forest trees in Ghana, although most species are rare, and several barely qualify as trees because of their 'strangling' habit. Some species are common in towns, on buildings and as, or on, cultivated trees. Most species are called *AMANGYEDUA*, or *GYEDUA*. The leaves often have a distinctive venation of tiny, very regular reticulations. Several species have very pustulate leaves. Other species of *Ficus*, which are never stranglers or do not produce latex, are listed elsewhere (Gps 19B, 28A). The species listed here are sometimes found as 'normal' trees in swamps, along rivers, or planted in villages.

'Fig' (='syconium') is also a term for the specialized inflorescence and fruit of the genus. The minute flowers or seeds are on the inside of the modified receptacle, which has a small hole at the tip to allow the passage of pollinating wasps which develop with the fruit. In some species (*F. polita*, *F. umbellata*, *F. ottoniifolia* and *F. artocarpoides*) the figs develop on small 'pegs' on the older branches and main trunk, but most species bear figs in the leaf axils.

Several *Ficus* spp. have recently changed names. These are summarized in the index.

Leaves of the following very distinctive shapes (see illustrations)		
1) Lf ±triangular, with a flat, straight apex		
Stipules not persistent; figs <1 cm diameter	***Ficus leprieurii***	309
Stipules peristent; figs >1 cm diameter	***Ficus craterostoma***	296
2) Lf sagitate	***Ficus sagittifolia***	322
3) Lf lyrate	***Ficus lyrata***	312
Leaves not one of these distinctive shapes		
Leaves broadly ovate; strongly trinerved		
Tree in **very swampy** places with large (*c.*4 cm long) terminal bud and thick twigs; (>3 mm, usually >5 mm wide); figs pale with short hairs, on stalk <1 cm long	***Ficus trichopoda***	301
Tree in drier places, or without large terminal bud		
Most lvs with >9 laterals; **dry forest** species; some lvs ±lanceolate; figs sessile on young twigs, *c.*4 cm diam.,with (pairs of) young figs enclosed by the pointed, hat-like bracts	***Ficus ovata***[1]	317
Most lvs with <10 strong laterals; trees often planted in villages, deeply cordate; apex acuminate; figs on short woody shoots from older wood		
Twigs stout; apical buds often >4 cm long; petiole >2 mm wide; often with *c.*9 laterals; venation between laterals clearly subdivided by pale 'inter-lateral nerves'; lf base normally deeply cordate, or even v. slightly peltate; typically >15 cm long; figs on stalks <1 cm long	***Ficus umbellata***[1]	325
Twigs slender; apical buds usually <4 cm long; petiole <2 mm wide; venation between laterals almost entirely divided by equal-sized minute reticulations; lf base often obtuse; figs on stalks >1 cm long	***Ficus polita***	318
Leaves not markedly ovate (although sometimes trinerved)		
Venation finely transverse; lf broadly elliptic; lf margin folded over to meet edges of petiole channel on top of petiole; figs in pairs on a thick stalk 1 cm long, with 2-3 bracts *c.*5 mm long at base	***Ficus elasticoides***	304
Venation not so finely transverse, or lf not this shape		
Lvs long (>20 cm) and broad (often >15 cm); long spreading hairs on (young) midrib or petiole		
Lf base ±cordate or lf with with rounded serrations (on juvenile lvs); apical bud >5 cm long; venation rather scalariform with many 'rungs' much larger than the finest veins; figs in pairs in the leaf axils, on 1-4 cm long stalks	***Ficus recurvata***	321
Lf base not cordate; venation with pronounced wavy line running between laterals, hence venation not v. scalariform. In **swamps**	***Ficus saussureana***[2]	306
Lvs smaller, or at least completely glabrous... (*continued overleaf*)		

NOTES: 1) ***Ficus populifolia*** has similar leaves to ***F. umbellata***, but the figs are not borne on pegs. It is a fig tree of dry areas, but is also sometimes planted in villages. ***Ficus bubu*** is also similar to ***F. umbellata***, but has 7-12 pairs of laterals, and the bark of young twigs scales off when dry. It seems rare in Ghana, but is possibly more typical of **evergreen forest**. ***F. bubu*** also has more elliptic lvs with a barely cordate, or obtuse base (Berg, *Flore Du Gabon* vol. 26, p. 212).

2) ***Ficus vogelii***, a common strangler of buildings in towns, will key to *F. saussureana* which has dense woolly hairs and leafy stipules on young parts, and is more typical of swamps.

Lvs medium-sized or glabrous, (not of shape listed on previous page)

Lvs with v. conspicuous pale (white) reticulate veins below, and often with twigs covered in the persistent leafy stipules;

Lvs narrowly oblong-elliptic, midrib sharply impressed or channelled above, often >15 pairs laterals meeting in submarginal nerves; stipules not persistent, nor leafy; many pustules above; figs solitary on stalks in leaf axils — ***F. barteri*** 297

Not *F. barteri*: lvs not oblong, or midrib not impressed or lvs not v. pustulate, or twigs with persistent stipules

Stipules not extremely persistent, nor dense, or else lvs not obovate;

Venation without a prominent vein in the finer venation returning to the nerve axil; stipules sometimes large and persistent; petiole >2 cm long; lf not markedly pustulate — ***Ficus conraui***[2]

Venation with a prominent 'returning vein' (see illustration) like ***Sorindeia*** (Gp 35B); lateral nerves *c.*6 in broad arches on elliptic leaf; figs ±sessile; lvs with fine white pustules

Leaf broadly elliptic – abruptly acuminate; petiole often 4 or more cm long; **drier forests** — ***Ficus tessellata***[2] 323

Leaf narrowly elliptic; petiole <3 cm long; figs 1-3 in lf axils + short hairs; **evergreen forests** — ***Ficus ardisioides***[2] 299

Stipules leafy and persistent for many nodes, conspicuously clothing twigs; lvs ±obovate; figs on stalks up 2 cm long, 1-2 in axils, with two basal bracts — ***Ficus cyathistipula***[2] 303

Lvs with venation below not pale; twigs without persistent stipules

Venation between laterals appearing almost parallel to laterals from a distance; yng twigs with brown hairs; lvs ±oblanceolate with rounded-acute apex; lf base never cordate; lvs *c.*10 cm long or less; figs small (<1 cm) and sessile in lf axils — ***Ficus kamerunensis*** 308

Lateral nerves very distinct from finer ±square reticulations lvs trinerved or cordate on petioles up to 5 cm long; lvs v. pustulate above

–lf base often minutely cordate, and basal nerves not well developed; figs ±sessile in lf axils; **dry forest** — ***Ficus thonningii*** 324

–lf base ±cuneate, OR with well-marked ascending basal nerves;

Lf base cuneate; figs smooth, ellipsoid, on 1-2 cm stalks in lf axils or short woody pegs from older wood — ***Ficus ottoniifolia***[1] 316

Lf base cordate or the majority of lvs >15 cm long

Lf lanceolate to ovate, or even sometimes oblong, with a smooth petiole; dry forest species — See *Ficus ovata* (previous page)

Lf more nearly oblanceolate-elliptic, with stout, rough petiole with waxy flakes; lvs often >15 cm long, often with long hairs — Consider *F. vogelii* (notes, previous page).

NOTES: 1) ***Ficus artocarpoides*** is similar to ***F. ottonifolia***, but differs in having 10-16 pairs of laterals, often without well-defined basal nerves.

2) ***Ficus conraui*** has rather variable leaves, but is not apparently very common. Fertile material may be the only means to obtain a satisfactory name for these species. (*F. conraui* to *F. cyathistipula*)

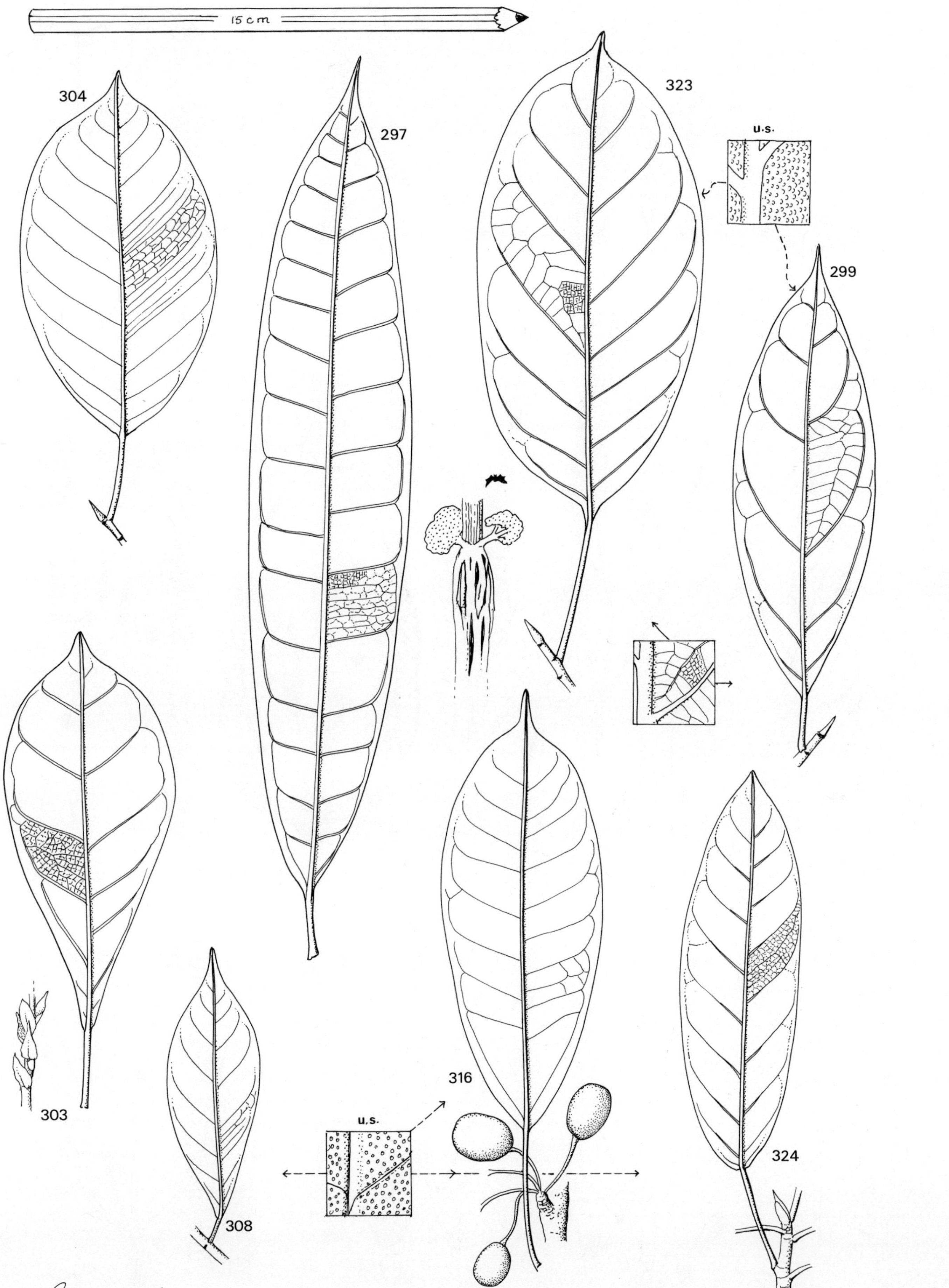
15 cm
304
297
323
u.s.
299
303
308
u.s.
316
324
Rosemary Wise

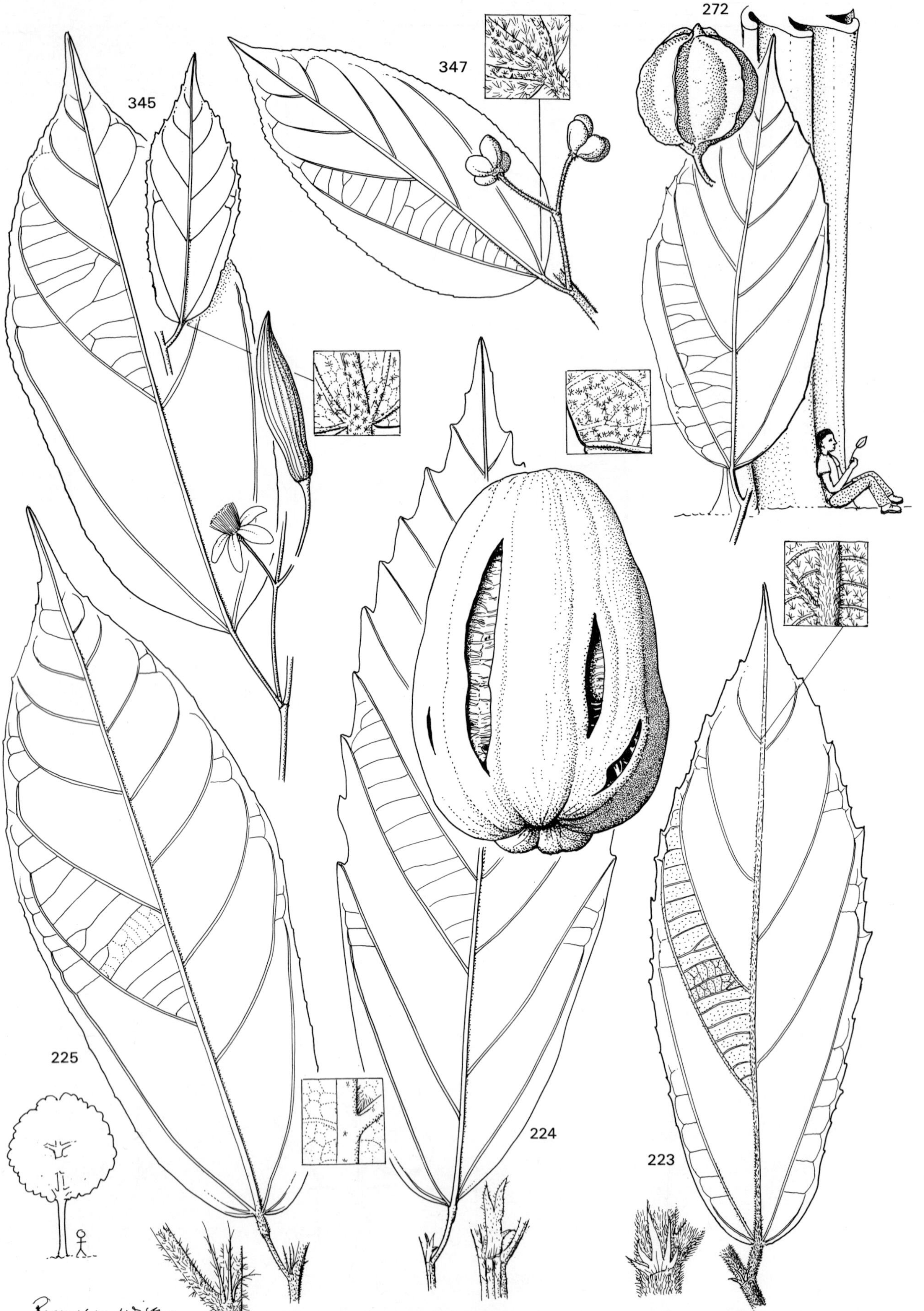
345
347
272
225
224
223
15 cm
Rosemary Wise

GROUP 20: TILIACEAE

(Lvs simple, alternate, serrated; with stellate hairs, 3 ± nerved at base, not v. long petioles)
(Midrib prominent above; lf base usually cordate or asymmetric)

A family, closely allied to the Sterculiaceae (see Gp 27), which includes few Ghanaian trees. As well as the following species, of which only *Duboscia* is an upper canopy tree, the family includes *Christiana* (Gp 27C), which has a long petiole. Otherwise, the intermediate (not particularly short nor long) petiole length of Tiliaceae is a useful distinguishing feature amongst species with stellate hairs, although it is important to check several leaves. The family includes several sun-loving herbs and straggling shrubs (e.g. species of *Corchorus*, *Grewia* and *Triumfetta*) which are common in abandoned farms and other open sites. Several of these 'weeds' have very fibrous stems and are used in the production of string, baskets, etc.

Mature *Duboscia* trees are deeply fluted, usually with the narrow ridges curving to one side (e.g. clockwise). The slash (of all the tree species) is fibrous, usually in layers (contoured pattern), but thick and spongy in *Desplatzia*, and yellowish or red brown, with dilatation bands, darkening. *Desplatzia* species have whorled boughs and rather flat layers of foliage at first. Leaves from the layered branches have more asymmetric bases than leaves from the vertical shoots.

The flowers are regular, yellow (*Glyphaea*, sometimes *D. dewevrei*) or white sometimes tinged with pink (*Christiana*, *G. pubescens*, *Duboscia*, *Desplatzia* spp.). In *Duboscia* two or three flowers are enclosed together in a single envelope of bracts. *D. chrysochlamys* has very hairy clusters of flowers in the leaf axils, whereas in the other two *Desplatsia* species the flowers are in more open axillary or terminal inflorescences.

The fruits are usually indehiscent. Those of *Desplatzia* species, which are obviously relished by elephants as seedlings are common in elephant dung, are very distinctive fibrous structures. When fresh, *Desplatsia* fruits contain a gum that stains (hands etc.) like ink. *Christiana* fruits are hairy follicles (*c.* 1 cm diam.) splitting to reveal brown-white marbled seeds.

Leaves usually <15 cm long; **stipules not branched nor divided and rarely persistent, except on saplings**		
– Leaves with dense pale hairs below; whole surface often white-discolorous		
– – Hairs rather coarse and woolly; surface soft to touch; midrib similarly clothed; upper surface often rough; small tree of **disturbed forest**	***Grewia mollis***[1] [*KYAPOTORO*]	347
– – Hairs v. fine: surface not soft to touch; midrib with stellate hairs; young stems v. white-hairy; sometimes with paired stipules persistent for a few nodes; tall tree with spiral fluting; slash pale red-brown and brown contoured, fibrous and brittle, darkening + white sapwood turning yellow to reddish, + **distinct scent of sweet vegetables, like boiled fresh sweet corn or roast plantain**	***Duboscia viridiflora*** [*AKOKORAGYEHINI*]	272
– Leaves without dense pale hairs (but nevertheless usually with stellate hairs v. visible with lens); lf ± symmetrical; sometimes v. narrow; twigs not v. hairy and stipules falling very early; v. often bearing yellow flowers; small spreading treelet	***Glyphaea brevis*** [*FoTo*]	345
Leaves >15 cm long; **stipules persistent and branched or several-fingered**; small trees with whorled branches and thick, spongy-fibrous yellow slash		
– Lvs not densely clothed in rough or soft orange stellate hairs; lvs papery or very highly asymmetric; tufts of orange hairs in nerve axils		
– – Lf long acuminate; stipules often <8 mm long, linear 'fingers', divided to base; dense short ginger hairs on twigs and stems; lf thinly papery, ± symmetric. serrations v. fine (not deep)	***Desplatsia subericarpa*** [*eSONOWIESAMFIE-BERE*]	225
– – Lf shortly acuminate and v. asymmetric at base, with robust serrations; stipules ± strap-like fingers, sometimes hairless, >8 mm long; twigs with v. few or no hairs	***Desplatsia dewevrei*** [*eSONOWIESAMFIE-NINI*]	224
– Lvs, stems, stipules, midrib on upper surface etc. densely covered in long, coarse orange hairs (stellate on lower side of lf); stipules stiff, flattened, deeply divided; lf base only slightly asymmetric; **evergreen forest**	***Desplatsia chrysochlamys*** [*eSONOWIESAMFIE(-NWI)*]	223

NOTE: 1) Specimens previously named *Grewia pubescens* in Ghana have been re-annotated *G. mollis* (by J. Hall). *Grewia mollis*, in the narrower sense seems to be a hairier, **savanna** species which has yellow flowers (*Flora of West Tropical Africa*), but it is very variable. A taxonomic revision of *Grewia* is needed to clarify the situation.

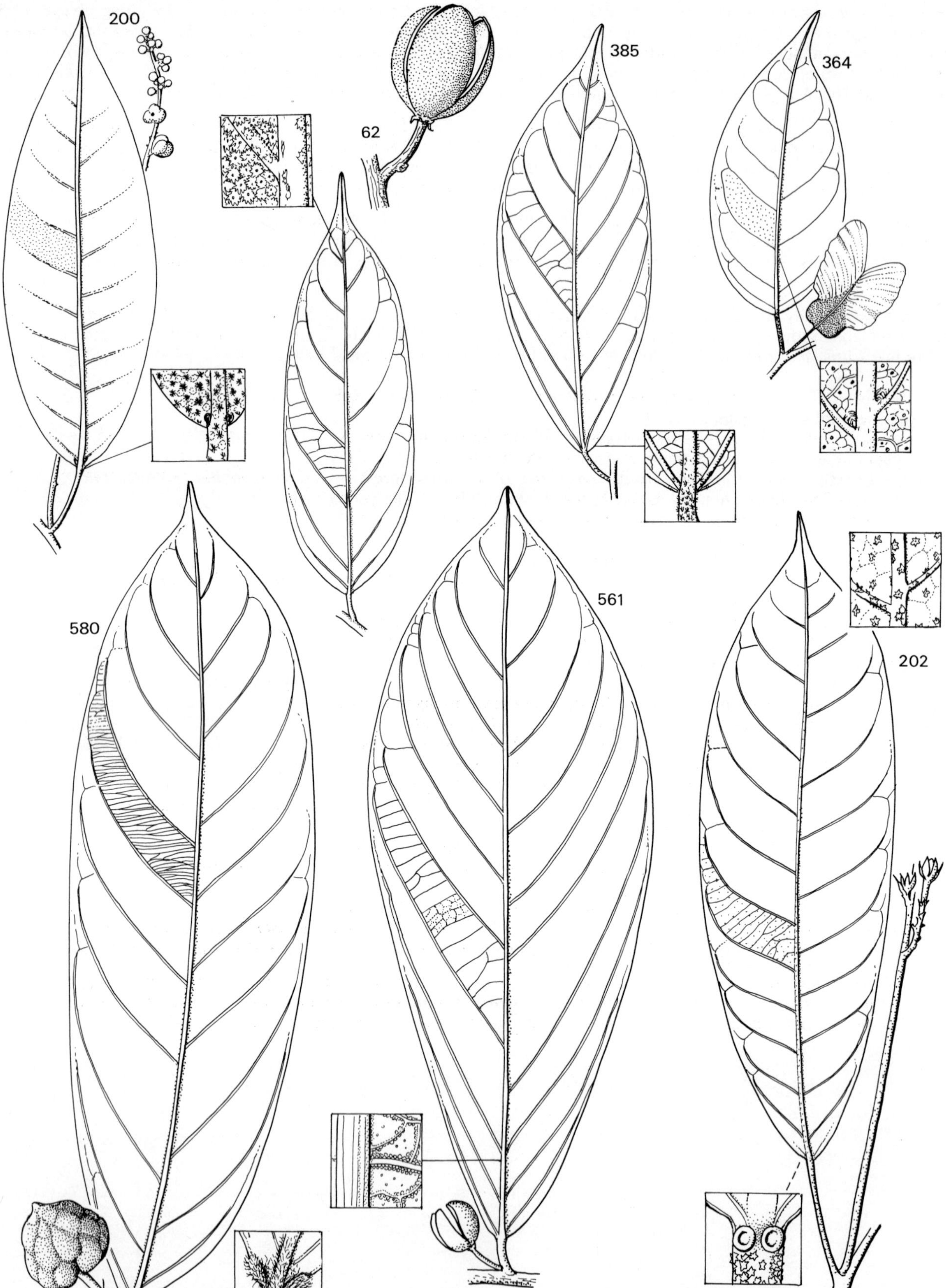
200
62
385
364
580
561
202
(Upper surface)
15 cm
R. Wise.

GROUP 21
(Species with basal glands, stellate hairs or scales)
(Sometimes trinerved or long-petioled, but not both)

Species with distinctive leaves, which fit better in this 'odds and ends' group than elsewhere. Included are the few short-petioled Sterculiaceae. For notes on Sterculiaceae see Group 27. For notes on Euphorbiaceae see Group 22.

Leaves with **extremely dense silvery or golden scales** on lower surface of mature lvs; crown therefore discoloured		
Dry forest tree **with basal glands**; petiole long; lf narrowly oblong, <2 cm wide; scales silvery; bark rough; sometimes planted for shade	***Croton zambesicus***[1] (EUPH) *[DODWATU]*	200
Evergreen forest trees without basal glands		
Leaves with short petiole and drip tip; slash, etc. smelling strongly of **onions**; scales golden brown	***Afrostyrax lepidophyllus***[2] (HUA) *[DUAGYENNE]*	62
Leaves with long petiole and without drip tip (Sapling lvs or lvs of flowering branches: **lvs otherwise digitate**); slash fibrous, reddish; tree with stilt-buttresses	***Heritiera utilis*** (STER) See Group 28B	
Leaves without very dense silver or golden scales on mature lvs; sometimes with scattered scales		
Leaves without glands at base (OR tree in **drier forests**)		
Lvs without domatia in nerve axils; small understorey trees		
Leaves long, not strongly trinerved (lamina shape and size of cocoa lf) with a few silvery scales, especially nr. midrib below; **forming tangles in understorey by rooting** from trailing branches; **evergreen forest**	***Scaphopetalum amoenum*** (STER) *[NSOTO]*	561
Leaves <15 cm long, **trinerved**, with stellate hairs; parts of slash becoming pink, then brown with time; not forming tangles and not restricted to evergreen forest	***Leptonychia pubescens***[3] (STER) *[FoTo-NUA]*	385
Lvs with tufts of hairs in axils; medium to large trees (lvs trinerved)		
Leaves with stellate hairs, broadest above or below middle; petiole swollen at top; slash pink on young trees; stipules pointed and clustered at top of young stems	***Nesogordonia papaverifera***[3] (STER) (See Group 27A)	
Leaves without stellate hairs[4]; covered in tiny red dots; broadest below middle; fruits two-winged; **dry forests**	***Hymenocardia lyrata***) (HYME)[4]	364
Leaves with basal glands; trees in, or near **evergreen forest**		
Petiole short; rusty brown hairs besides midrib; small straight tree in **evergreen forest**; slash pinkish brown, soft, brittle and gritty with large pores, becoming darker, with gummy exudate	***Strephonema pseudocola***[5] (COMB) *[AWURUKU]*	580
Petiole long; small **evergreen forest** tree with v. conspicuous, but scattered scurfy scales on lvs and young stems; twig tips with persistent, broad-based stipules	***Crotonogyne manniana*** (EUPH)[6]	202

NOTES:
1) *Croton aubrevillei* is apparently (Aubréville, 1959, vol. 2, p. 90) a similar species to *C. zambesicus*, with trinerved, scaly leaves and thick granular bark and dirty orange slash.
2) *Afrostyrax* is the only species in the Huaceae in Ghana. Previously it was in Styracaceae.
3) *Leptonychia* sometimes has tuft domatia; the lvs of *Leptonychia* can then be distinguished from those of *Nesogordonia* by being slightly asymmetric, often broadest below the middle, and by the stouter, non-swollen petiole. The (invented) local name reflects tree spotters' common confusion of this species with *Antidesma* spp. (Gp 13D) and *Glyphaea* (Gp 20).
4) *Hymenocardia* was previously in Euphorbiaceae. Now it has been segregated in Hymenocardiaceae.
4) *Diospyros monbuttensis* will key to this point if the spines are not observed.
5) *Diospyros gabunensis* may key to this point if its characteristic black bark has not been observed.
6) *Crotonogyne chevalieri*, a similar species with lvs normally serrated, is mentioned in Group 22.

FLOWERS AND FRUITS OF THE EUPHORBIACEAE

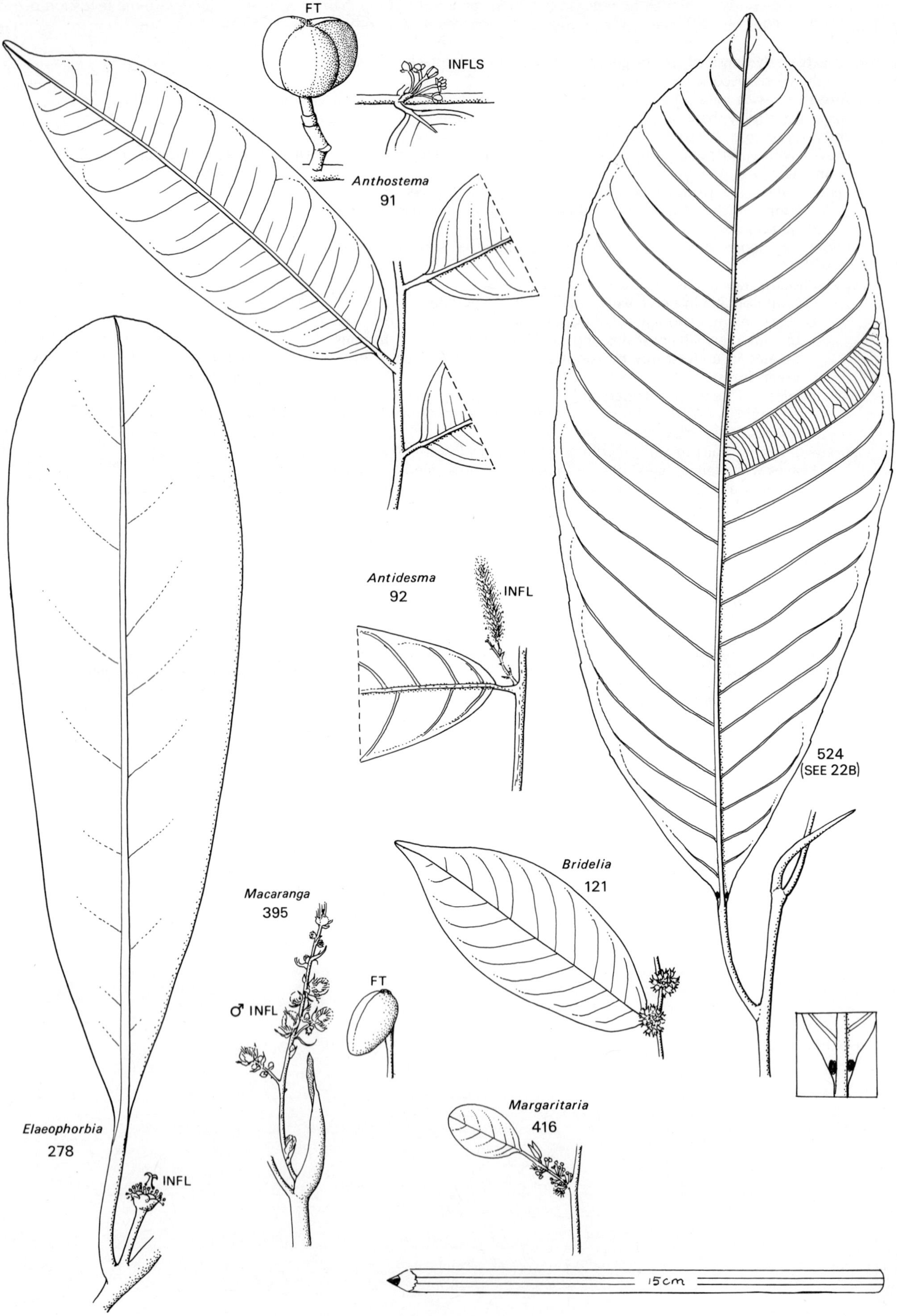

GROUP 22: EUPHORBIACEAE (part)

This family is unusually varied, and so has been split up into several Groups in this guide. Most species have simple leaves, and can be found in Groups 10-24 (see below). Only in *Mallotus oppositifolius* and *Tetrorchidium* are the leaves opposite. *Ricinodendron* is exceptional in having digitate (compound) leaves; the introduced para rubber tree, *Hevea brasiliensis,* has trifoliate leaves. Other commonly cultivated trees in this family include cassava (*Manihot esculenta*), castor (*Ricinus communis*) and the sandbox tree (*Hura crepitans*) which has prickly stems and explosive fruits.

In many species, the leaves have glands, stellate hairs, serrations or basal nerves. Certain species produce white latex (cassava, *Hevea, Elaeophorbia, Sapium* spp. and *Anthostema* (see Gps 10, 19)). Most species are small trees. Larger trees often have stilt roots of various sizes (*Bridelia grandis* (Gp 15) *Uapaca* spp. and *Protomegabaria* spp. (Gp 24)). Larger trees that do not produce stilt roots, i.e. *Margaritaria* (Gp 13) and *Ricinodendron* (Gp 28), have red or partially red slashes. Many smaller trees, e.g. *Antidesma* and *Amanoa* spp. (Gp 13), have red slashes as well, but a similar number of these small tree species (e.g. most *Drypetes*) have orange-brown slashes. Different members of the family resemble in their field characteristics members of a variety of other families, like *Chrysobalanaceae* (red slash and basal glands) on *Sterculiaceae* and *Tiliaceae* (stellate hairs, basal nerves). *Phyllocosmus, Soyauxia* and *Panda* represent small families with some affinity with the Euphorbiaceae. Indeed, *Microdesmis* has been recently transferred from Euphorbiaceae to Pandaceae. *Hymenocardia* (Gp 21) has also been recently removed from the Euphorbiaceae.

The flowers are always unisexual, and usually individually small and clustered on long inflorescences. The details of flower structure are in some cases complicated. What appears to be a single flower in *Elaeophorbia, Anthostema* and *Uapaca* is in fact derived from several very modified flowers. Sometimes single inflorescences include flowers of both sexes, but several species are dioecious. The fruits are very often two or three-lobed capsules, with the style persistent on top, although some species produce drupes (seeds in a stone, surrounded by soft tissue). In the table below the inflorescence and fruit types are summarized. Spikes (individual flwrs without stalk) and racemes (flwrs with stalk), which are inflorescences with a single unbranched axis, are often very similar, but are clearly different from panicles, which have branched axes, and clusters (also called 'fascicles'), which have no (or very short) axes.

Genus	Group	Flowers	Fruits (+ approx. max. width)
Specialized inflorescence resembling single flowers (see above)			**(dehiscent) capsules unless stated**
Elaeophorbia	22	Heads (3) in axils	2 cm, hard drupe with one 3-grooved stone
Uapaca	24	Heads (1-few) in axils + petal-like bracts	2 cm+, rounded (edible drupe + *c.* 3 seeds
Anthostema	22	Heads (many) in axils	1.5 cm. 3-lobed capsule
Clusters of flowers on twigs, branches or bole			
Margaritaria	13B	Clusters. Male on ±lfless brs; f.±below lvs	1 cm, 3-lobed with 6 blue-black seeds
Keayodendron	13B	Clusters in lf axils	2.5 cm, ovoid (±7 sided)+1 hard seed; indehiscent
Bridelia	15	Clusters in lf axils	1 cm, drupes
Drypetes	17A	Clusters in axils or cauliflorous	1.5 cm, ovoid-globose. 1-few seeds; indehiscent
Unbranched inflorescences with regular brush (catkin) of flowers			
Tetrorchidium	13A	Dense spikes; female opposite the lvs	½ cm, 3 red seeds
Antidesma	13D	Dense spike-like racemes	¼ cm, drupes, red
Sapium	19	±Dense spike-like racemes	1.5 cm, 2-lobed + coiled styles
Croton	21,22A	±Dioecious + dense spike-like racemes; m. with petals	1 cm+, 3-lobed
Alchornea	22A	±Dioecious. Female spikes + short white hairs	½ cm, 2-3 lobed + 2 long styles
Unbranched inflorescences with scattered (clusters of) flowers			
Martretia	(13)	Dioecious. Racemes glabrous	4-lobed + long styles
Cleistanthus	13B	Dioecious. Small brown-hairy racemes + petal-like sepals	1 cm, 3-lobed, 6-seeded
Cleidion	17C	Dioecious. Male in sparse spikes; f. solitary	1 cm, 3-lobed, + long styles
Maesobotrya	22B	±Sparse racemes. Spike-like on stem knobs	1 cm, globose, edible, red drupe
Mareya	22B	Sparse spikes + short hairs	¼ cm, deeply 3-lobed
Crotonogyne	21	Sparse spikes + scales. M + petals. f. ±in panicle	3-lobed
Necepsia	(22B)	Sparse spikes (+ small bracts); dense short hairs	1 cm, 3-lobed
Pseudagrost.	22B	Racemes hairy + hairy basal bracts + petals	Velvety 1-seeded
Disclaoxylon	22B	Sparse raceme long (15 cm+) and hairy	1.5 cm, rounded velvety
Mallotus	22A	Dioecious. Racemes with scales or stellate hairs	1 cm, 3-lobed
Protomegabaria	24	Sparse racemes	3 cm, globose or 3-lobed
Unbranched inflorescences with relatively large bracts			
Amanoa	13C	Spikes with large overlapping bracts. Yellow petals	3 cm, capsule globose
Pycnocoma	(17C)	Spike-like racemes, with conspicuous bracts	2.5 cm, 3-lobed + 'horns'
Macaranga	23	±Dioecious. Female in racemes + small bracts	Round or 2-lobed capsules
Branched inflorescences			
Macaranga	23	±Dioecious. Male often panicles + large bracts	
Discoglypremna	22A	Dioecious. Racemes/panicles cream-coloured	1 cm, 3-lobed black seeds + red coat
Alchornea	22A	±Dioecious. Male panicles + short white hairs	
Grossera	22B	Panicles cream, + petals	3-lobed capsule
Spondianthus	24	Panicles cream, males + petals	1 cm, ±ovoid + red seeds
Ricinodendron	28B	Dioecious. Panicles	2.5 cm, 2-3 seeded, 2-3 lobed; indehiscent

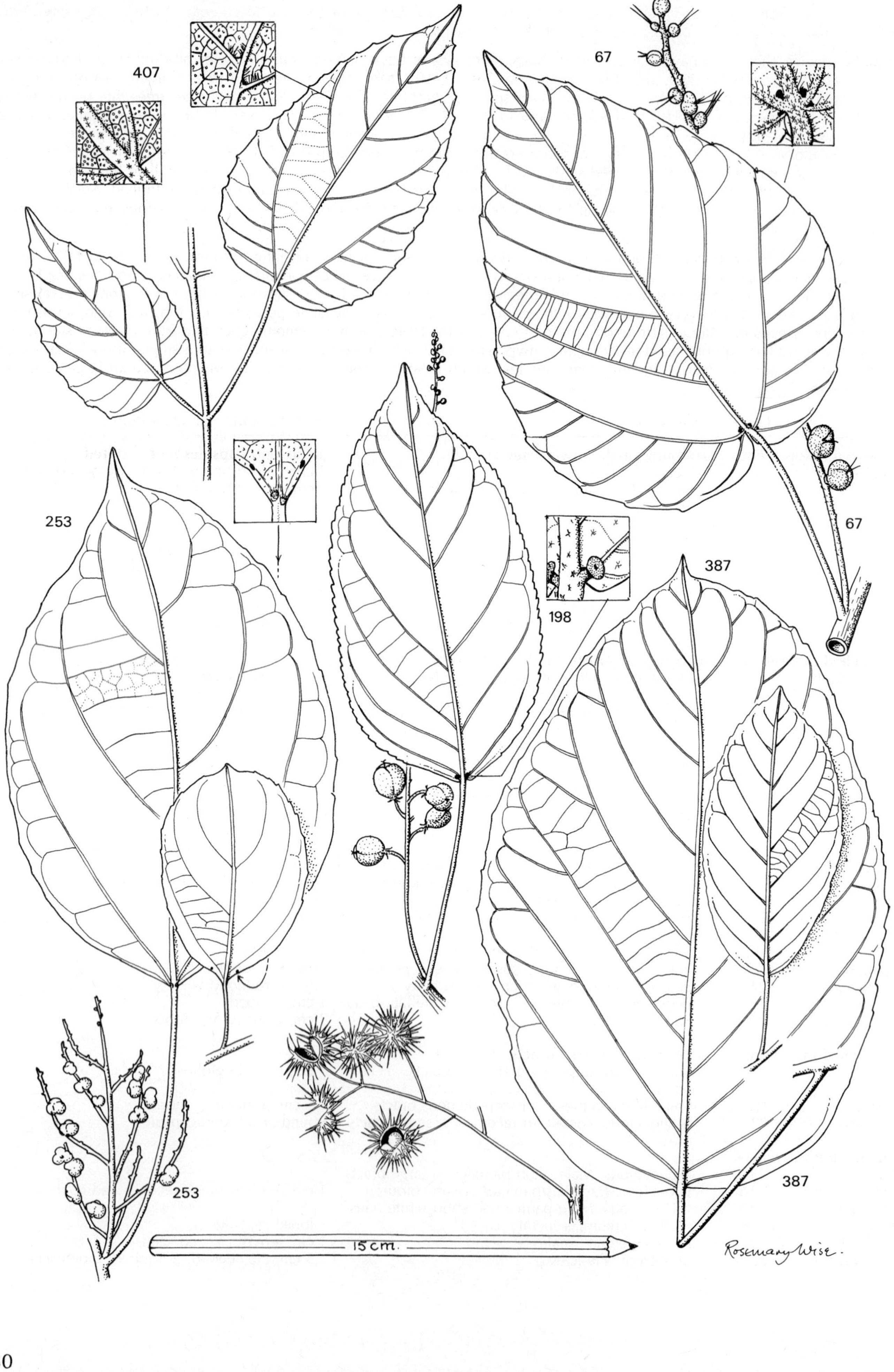

407
67
253
67
387
198
253
387
15 cm.
Rosemary Wise

GROUP 22: EUPHORBIACEAE (part) (contd.)
(Lvs with long petioles, serrated, often with glands)

Key to subgroups

Leaves completely white below (but prominent veins visible) and strongly tri-nerved; some lvs trilobed; most lvs broadest at middle; see ***Myrianthus libericus*** or sapling ***Musanga*** — Group 28

No lvs trilobed; fine white hairs absent or only on larger veins, etc.
- Lvs with many stellate hairs below, $\pm$asymmetric at base, with petiole <3 cm — Group 20
- Petiole >3 cm (or not Tiliaceae for other reasons)
  - Leaf lamina broadest below the middle, with strong basal nerves, with base cordate or obtuse — Group 22A[1]
  - Leaf lamina broadest around or above middle, or without strong basal nerves — Group 22B

Group 22A: Euphorbiaceae
(Leaves broadest below middle or with strong basal nerves)

Lvs opposite, one lf smaller than the other; lf with minute scales; shrub or small tree — ***Mallotus oppositifolius***[2] [*NYANYANFOROWA*] 407

Lvs alternate
- Base of lf cordate, with deep lobes; nerves reaching margin; hairs in axils; darker glandular cushion-like spots around base of basal nerves; small tree or bush, in **disturbed**[2] **or swampy forest** — ***Alchornea cordifolia*** [*GYAMA*] 67
- Base of lf not cordate OR without cushion-glands
  - Lvs with glands at base
    - Small spiny trees normally with stilt roots — See *Macaranga* spp. (Gp 23)
    - Tall trees without spines
      - Basal glands as bumps nr base of lamina, by petiole, often in two pairs, those closest to the petiole being v. small; small tree very common in **v. disturbed forest**, or larger tree slightly to very fluted at base; bark smooth; slash granular, green outer layer, orange-ish brown (or pinkish) with white speckles, sometimes + vertical ginger bands, gritty, with fruity-acid taste — ***Discoglypremna caloneura*** [*FeTeFRe*] 253
      - Basal glands on stalks at top of petiole; lf with numerous white spots due to stellate hairs (x10 lens); **dry forest** (or slopes) tree with yellow-brown, peppery, gritty and brittle slash; not (or only locally) common — ***Croton penduliflorus***[4] 198
  - Lvs without glands at base; without stellate hairs or scales[3]
    - Petiole usually >3 cm long; lvs with fine white hairs on veins etc.; small tree, slash (thin), orange and scented — ***Lindackeria dentata*** (FLAC) [*SOFO-SE*] 387
    - Petioles barely even exceeding 2 cm; sometimes swollen at tip
      - Leaves with pustular surface; margin more wavy than serrated — ***Scottelia klaineana*** (FLAC) See Group 17D
      - Leaves with smooth surface; venation fine, $\pm$transverse and prominent above — ***Dasylepis brevipedicellata*** (FLAC) See Group 17D

NOTES:
1) *Maesa lanceolata* (Myrsinaceae), known from 700 m on the Togo plateau, has ovate leaves without basal nerves: they are fleshy and young parts have brown stellate hairs and simple hairs. The upper surface has small spots and the venation is almost obscure.
2) *Mallotus subulatus* is similar to *M. oppositifolius*, but has almost entire lvs with v. fine teeth. The lf surface is not v. hairy, but the petiole bears long hairs. It has stout, $\pm$hollow twigs.
3) *Acalypha neptunica* is a **dry forest** shrub (rarely treelet) like this with many raised spots above, usually with hairy, slightly cordate leaves, but without basal glands.
4) *Croton sylvaticus* is a similar species, fertile as a shrub and found in v. **disturbed dry forest**. See also notes on *C. aubrevillei* (Gp 21).

Group 22B: Euphorbiaceae
(Lvs broadest above middle or not trinerved; small understorey trees)

NOTE – when tasting the slash, be careful not to swallow, nor to sample more than a small crumb of bark, as it may be poisonous in large doses.

Slash bitter, or sweet, sometimes soft; **leaves often with raised pale spots or bumps peppered over surface OR basal glands**; or bole very uneven with large bumps		
Base of lamina folded up around midrib into petiole channel; bole with raised, irregular knobs or bumps where flowers arise; teeth with v. sharp, curved **endings (x10), ±small hair tufts as well; petiole with orange** hairs; lvs drying bluish black to green; slash soft, pale orange, with bittersweet taste, like fresh groundnut	***Maesobotrya barteri*** *[APOTREWA]*	401
Midrib base not channelled, but raised, guttered OR with glandular bumps; leaves usually peppered with raised pale spots		
Lf with large terminal buds, and conspicuous ring scars at nodes (like *Ficus*); venation markedly scalariform; lvs with two basal glands by side of midrib on (decurrent) lf base; slash hard, red; tree **gregarious in swamps of evergreen forest zone, rivers etc.**	***Pseudagrostistachys africana*** *[SUKROMA]*	524
Lf without such buds (or tree in drier places)		
Young twigs hollow, pale yellow with raised lenticels; leaf scars clearly visible even on larger branches or trunk (as in papaya tree); teeth unusually large and curvaceous, arching towards the apex; midrib flat or guttered at base, with petiole slender and channelled; bark easily peeled, with conspicuous green outer layer; dry lvs v. wrinkly	***Discoclaoxylon hexandrum*** *[DUBRAFO-NINI]*	252
Young twigs not hollow; teeth not curvaceous; leaf scars not so persistent; white spots above often abundant		
Petiole/midrib junction with 2-3 bumps amongst red-brown hairs, resembling spider eyes; teeth usually sharp, with hairs; lf ± two ('knotted vein') glandular spots nr leaf base; lower surface sometimes v. discolorous; twigs pale, with a few lenticels similarly coloured; midrib hairy, even above; slash orange-brown, brittle-fibrous, bitter, with watery exudate	***Mareya micrantha***[1,2] *[DUBRAFO]*	415
Petiole without 'spider-eye' bumps at top, channelled or grooved; teeth terminating in small, blunt, rounded knobs (x10), petiole often with yellowish hairs, sometimes thickened at top; midrib usually guttered at base, usually without hairs; twigs hairy, dark, + raised pale lenticels; slash sometimes with spots of **red exudate**, sweet, cream to red-brown; lvs dry purple-brownish green	***Grossera vignei*** *[MPEDURO]*	348
Slash with no obvious taste, dry, granular orange and brown contoured; lvs not covered in small spots, without glands; bole smooth, but youngest twigs with Y-shaped scars where bud scales have fallen	***Rinorea oblongifolia***[3,4,5] (VIOL) *[MPAWU-NTUNTUM]*	547

NOTES:
1) ***Crotonogyne chevalieri*** is a small tree of **evergreen forest** which may key here. Some lvs have conspicuous basal glands; stipules are persistent; from a distance finer venation appears parallel; close inspection of lower surface (lens) reveals curious fine raised short veins and scattered knotted vein glands.
2) ***Necepsia afzelii*** is an uncommon **evergreen forest** tree with lvs of this type, with basal glands and finer venation also approximately parallel, yet without the curious raised lines mentioned under 1). The twigs have many stipules clustered at the tips, each with a white ±ciliate margin.
3) ***Cleidion*** (*MPAWU*) and various species of *Rinorea* are often confused under the same name. *Cleidion* has short petioles.
4) ***Rinorea brachypetala***, ***R. subintegrifolia*** and ***R. welwitschii*** are uncommon species which will probably key to this point. They have minute raised spots on the leaf, unlike the very common *R. oblongifolia*, but possess the stipules/scars typical of their genus. Fertile material is important for accurate identifications of *Rinorea* spp.
5) The family Violaceae is discussed in Group 17B.

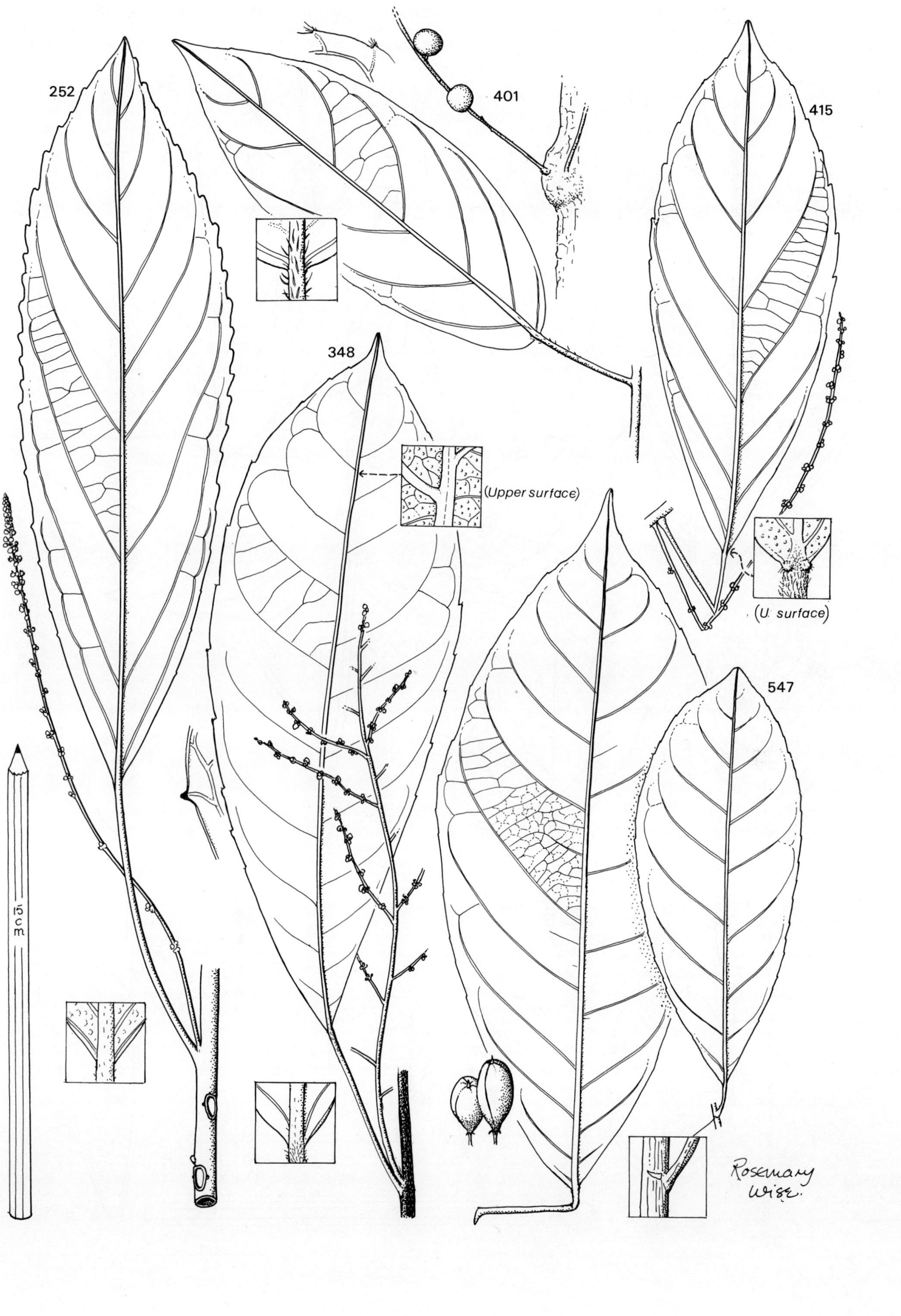
252
401
415
348
(Upper surface)
(U. surface)
547
15 cm
Rosemary Wise.

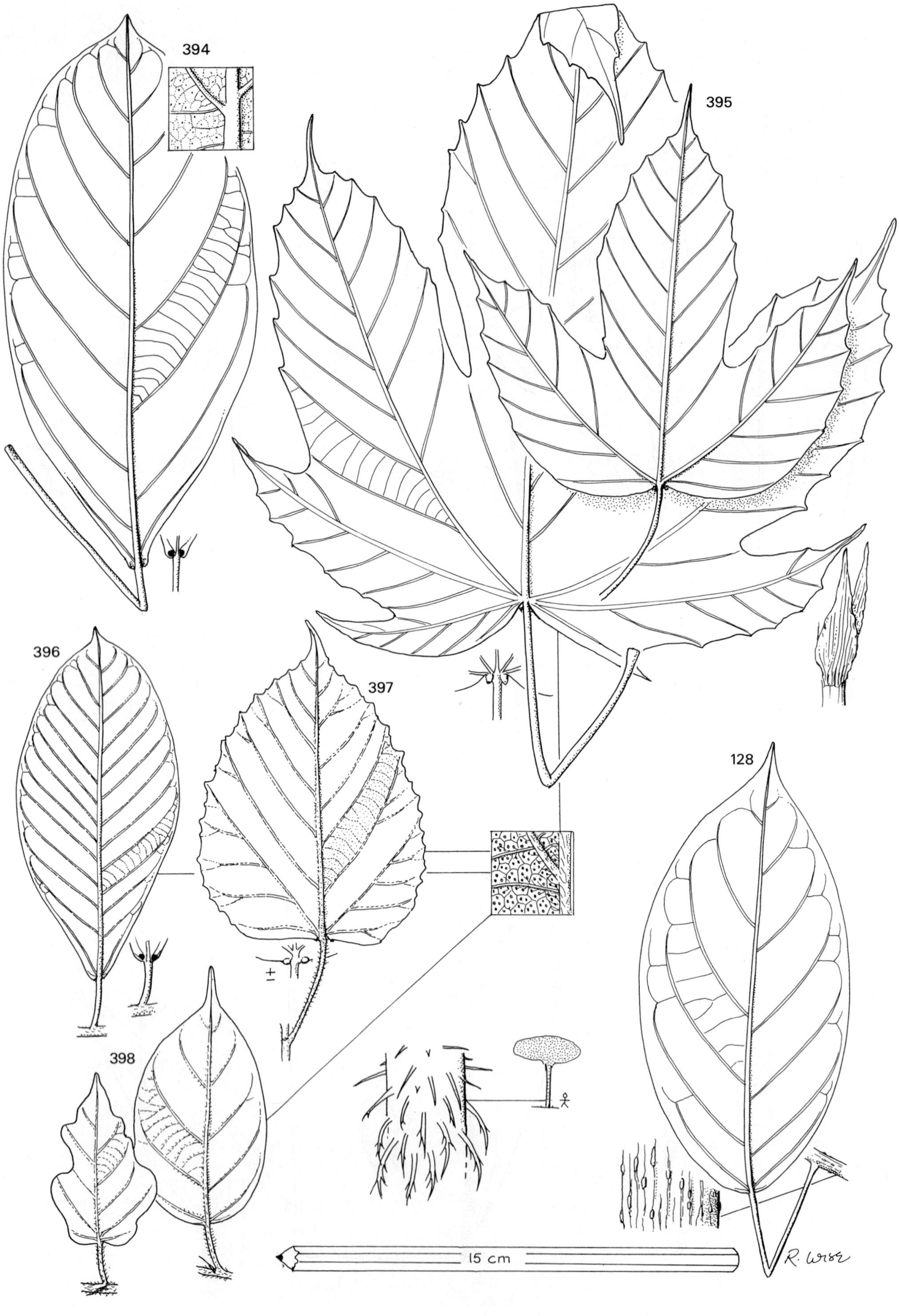
394
395
396
397
398
128
15 cm
R. Wise

GROUP 23: *Macaranga* spp. (EUPHORBIACEAE part)
(Long petioles and spines, often small stilt roots as well)
(Lvs often with basal glands, and/or silvery scales below)

The family Euphorbiaceae is discussed in Group 22.

Macaranga trees resemble *Bridelia* (Gp 15) trees, with spines which, on the lower part of the stem are intermixed with many slender stilt-root-like outgrowths. Both are often crudely named 'OPAM', although the leaves, with long petioles and glands in *Macaranga* spp., enable easy, precise identification.

The species are typical of **swamps or other more open or disturbed forest patches**. The slash is white or red-brown, sometimes contoured, darkening. The leaves usually have scales and basal glands. Stipules are often conspicuous at the base of the mature leaves. Leafy structures (bracts) surround the male flowers on the inflorescences, although they are inconspicuous in *M. spinosa* and *M. heudelotii*. The rounded or 2-lobed fruits are often covered in yellow glands.

- Leaves lobed, often with long drip tip
  - Lvs irregularly lobed, teeth not well-defined, with long white hairs below — ***Macaranga spinosa*** [*OPAM-NUA*] 398
  - Lvs with large regular lobes with clear teeth; with large stipules
    - Lvs 3-7 lobed; long (>2.5 cm) leafy stipules at base of petiole — ***Macaranga heterophylla*** [*oPAM-KoKoo*] 395
    - Lvs (0-)3 lobed
      - Stipules >2 cm long; lobes, if present, deep — ***Macaranga heterophylla*** 395
      - Stipules lanceolate, <2 cm long; shallowly lobed (v. small tree) — ***Macaranga beillei***
- Leaves not lobed
  - Lvs regularly serrated, broadly ovate
    - Lf with long hairs on petiole and midrib; always unlobed — ***Macaranga hurifolia*** [*oPAM-FUFUO*] 397
    - Lf without long hairs; large persistent stipules; lobed lvs often on tree as well — ***Macaranga heterophylla*** 395
  - Lvs not regularly serrated, nor cordate: elliptic, oblanceolate
    - Lvs with glands at base, sometimes on two small lobes; plant often with stilt rts
      - Tree of **swamps and riversides**; petioles usually <3 cm long; 10-13 prs lateral nerves; short spines on young branches/twigs — ***Macaranga heudelotii*** [*AWORA-oPAM*] 396
      - Tree of **disturbed forest**; petioles usually >3 cm long; 6-10 prs laterals; spines on older branches — ***Macaranga barteri*** [*oPAM*] 394
    - Lvs without glands at base
      - Short spines even on young twigs; petiole and stem with long hairs; 5-10 prs laterals with scalariform venation — ***Macaranga spinosa*** 398
      - Spines normally on older brs; leaf without obvious hairs with short acumen or acute apex; 5-7 prs laterals with open reticulate venation; without stilt roots — ***Caloncoba gilgiana***[1] 128

NOTE: 1) *Caloncoba gilgiana* is in the Flacourtiaceae (see Gp 17), but the tree looks superficially similar to *Macaranga*. The leaves are glandless, without stipules. The flowers could hardly be more different from the typically euphorbiaceaeous flowers of *Macaranga*, as they are very showy, 10 cm diameter, fragrant and white. The fruit has the approximate dimensions of an egg, but has 5-6 grooves along which it splits at maturity.

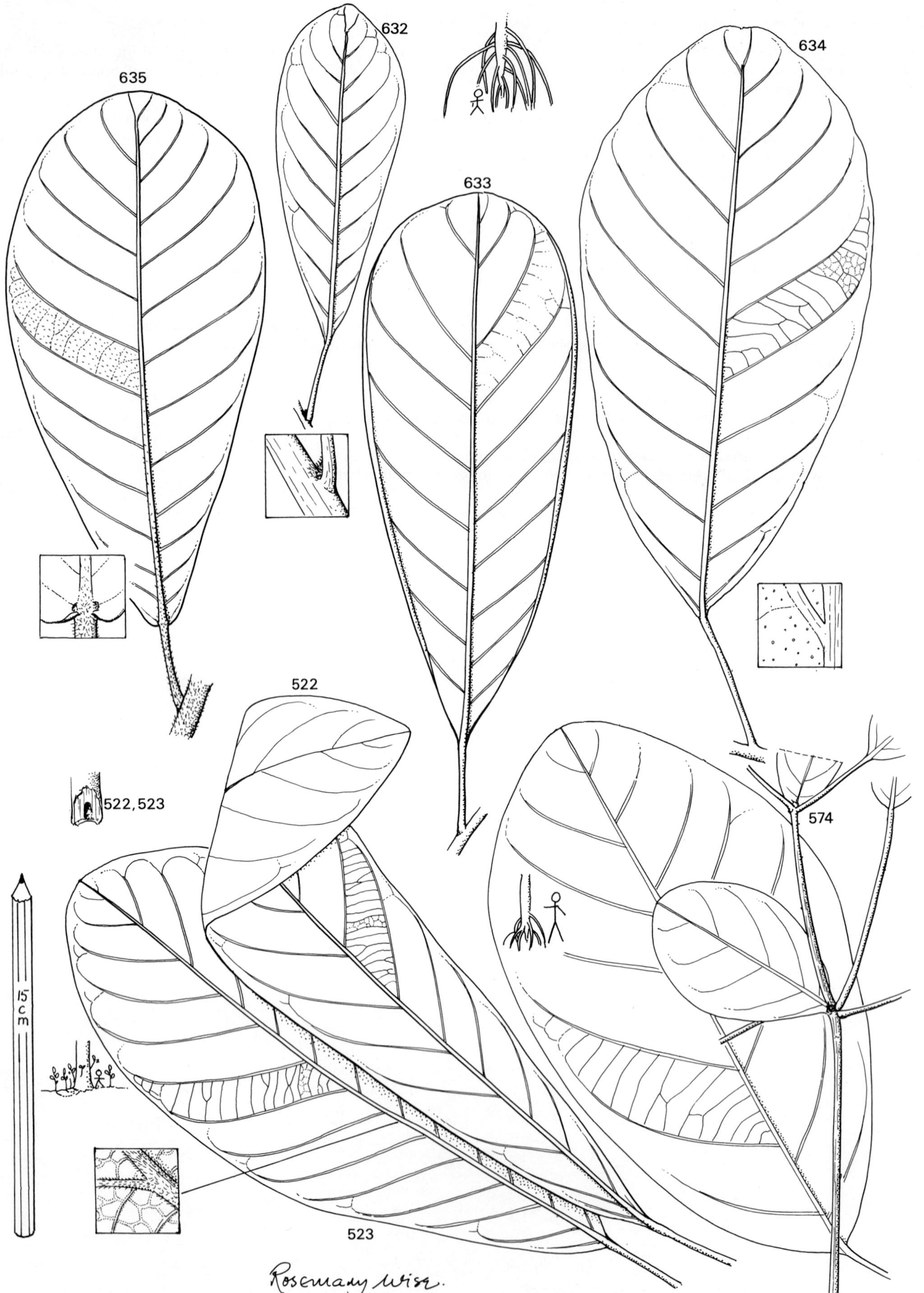
632
635
634
633
522
522, 523
574
15 cm
523
Rosemary Wise.

GROUP 24: *Uapaca* spp., etc.: (EUPHORBIACEAE part)
(Medium to tall trees with lvs v. markedly clustered at branch ends; lvs with long or variable petioles; with blunt apices; without stellate hairs)

The family Euphorbiaceae is discussed in Group 22.

This group of Euphorbiaceae (see Gp 22) is typical of, but not exclusively found in **swamps**. They have strongly clustered leaves with rather rounded apices and long petioles. Stilt roots, which are very large in *Uapaca*, are sometimes very small or absent from *Protomegabaria* and *Spondianthus* trees.

Stilt roots and large, clustered leaves are produced also by *Anthocleista* species (Gp 4: opposite leaves and prickles) , *Dracaena* spp. (Gp 39: main nerves all running along leaf) and *Heretiera* (Gp 28B: golden scales and crown, with rather flattened buttress-like stilts).

The slash, or cut twigs of *Spondianthus* and *Uapaca* spp. sometimes produce a sticky, red exudate. *Uapaca* fruits are edible.

Lvs rounded or elliptic, with rounded apex and base (+ basal gland-like spots); a wide range of sizes of lamina and petiole clustered at tips (yngest red); slash red-brown, hard, fibrous, + red exudate; ±small stilt roots; fts small, capsules with red seeds and persistent sepals; in **swamps** — ***Spondianthus preussii*** [*AWORATWEANKA*] 574

Lvs oblanceolate usually >15 cm long, with gradually tapering base; often with large stilt or adventitious roots

- **Trees with v. pronounced stilt roots** arching out several metres from the tree; petiole swollen at top and bottom; lvs ±gland-spotted; slash sometimes gritty or darkening; fts with edible pulp
  - Lvs without hairs below; stipules thin and falling rapidly; fts ovoid or globose, often warted, *c.*3 cm diameter
    - Petiole axils with tufts of conspicuous red hairs; **wet places** — ***Uapaca heudelotii*** [*KONTAN-AKOA*] 632
    - Petiole axils without conspicuous red tufts
      - Lvs thin and papery; margin v. undulate; tree often away from water; upper surface with small white scales — ***Uapaca guineensis*** [*KONTAN*] 634
      - Lvs leathery, margin not normally wavy; **evergreen forest**, not strongly favouring swamps, etc. — ***Uapaca corbisieri*** [*KONTANMIRI*] 633
  - Lvs with hairs on nerves, etc. below; stipules broad and persistent; **swamps** in **evergreen forest**; fts 3 cm long, ovoid — ***Uapaca paludosa*** [*AWORA-KONTAN*] 635

- **Trees without large stilt roots**, but often buttressed or with adventitious roots or shoots; petiole swollen at top; twigs hollow; lvs slightly silvery below; slash with dull orange, waxy fibres + a **remarkable hissing noise when slashed**; fts 4 cm across, hard, slightly 3-lobed like those of rubber trees, curling up when ripe
  - Leaves with hairs on midrib and nerves; base of lamina abrupt, not long-cuneate, sometimes minutely rounded, the margin ending abruptly at top of the petiole channel; gregarious in **swampy or hilly areas**; (male racemes *c.*7 cm long) — ***Protomegabaria stapfiana***[1] [*AGYAHERE*] 523
  - Leaves glabrous; base of leaf narrowly cuneate: margin of leaf gradually decurrent into the closely joined edges of the petiole channel, along the centre of the petiole; rare tree in **evergreen forest**; (male racemes >10 cm long) — ***Protomegabaria macrophylla*** [*AGYAHERE-NUA*] 522

NOTE: 1) Doubt has often been expressed about the separate identity of these two species of *Protomegabaria*. More fertile material, particularly of the latter species, would help sort out whether they are indeed two species.

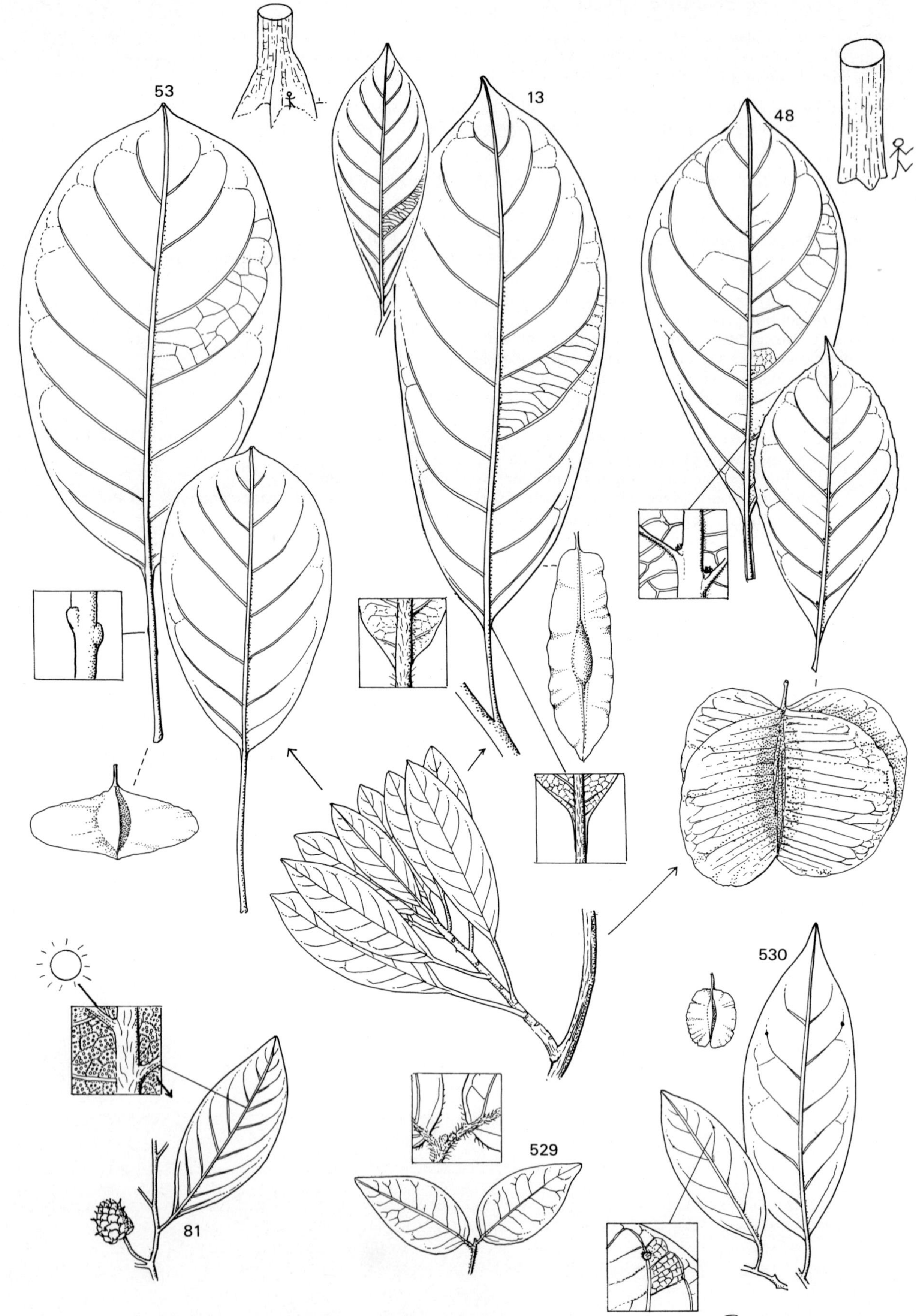
53
13
48
530
529
81
15 cm
Rosemary Wise

GROUP 25: *Terminalia, Petersianthus*
(Long petioles; lvs clustered; branches whorled)

This group includes species from two related families, the Lecythidaceae and the Combretaceae. *Petersianthus*, in the Lecythidaceae, used to be called *Combretodendron*, an indication of the similarities between the two families. The Lecythidaceae includes also *Napoleonaea*, but this variable species never has long petioles and so is keyed out elsewhere (Gp 13B & 17D). Similarly, as well as two *Terminalia* spp., the Combretaceae includes *Pteleopsis* spp. (Gp 5 & 13D) and *Anogeissus* (Gp 13D), with short petioles, and the rather more divergent mangrove, *Laguncularia* (see Gp 3) and *Strephonema* (Gp 21). In this introductory section the flowers and fruits of these families will be summarized, after mention of the vegetative features of the species keying to this Group. The *Terminalia* spp. and *Petersianthus* are important timber species which thrive in disturbed forest.

Terminalia, Petersianthus (and to a lesser extent *Pteleopsis*) have very clustered leaves, with the twigs often looping between the clusters. This, together with the strongly whorled branches, which is also a very marked feature of *Napoleonaea*, give the crowns a distinctive appearance, particularly on younger trees. (The details of the growth pattern are in fact not the same in *Terminalia* (Aubreville's model) and *Petersianthus* (Koriba's model) – see Hallé *et al.* (1978)). The slash (details in the key) is fibrous and often yellowish in all the species.

The flowers of all these species are bisexual. In *Petersianthus* they are about 1 cm across, with 4 rounded white petals, 4 rounded sepals and many stamens, smelling rancid. They are arranged in panicles or racemes. *Napoleonaea* has very unusual, beautiful flowers, with a thin white and purple jelly-fish-like (in appearance) corolla, sometimes growing directly from older wood. *Pteleopsis* (*Strephonema* and *Combretum*) flowers have petals, whereas *Terminalia* and *Anogeissus* (and *Laguncularia*) flowers do not. *Terminalia* flowers are arranged in spike-like racemes; *Anogeissus* produces globose heads; *Pteleopsis* produces small umbrella-like clusters.

Apart from *Napoleonaea* (which has hard-skinned orange berries), *Laguncularia* and *Strephonema*, the species produce papery-winged fruits. In *Terminalia* there are two wings, whereas in *Petersianthus* (and in lianes and savanna trees of the genus *Combretum*) there are four, usually golden yellow wings. *Pteleopsis* fruits have 2 or 3 wings. In *Anogeissus* the winged fruits are very small and clustered into a cone-like structure.

Other species with similarly-winged fruits are *Hymenocardia* (Gp 21), and *Holoptelea* (2-winged, Gp 18). Other types of winged fruit are produced by *Pterocarpus* (Gp 37), various Sterculiaceae (see Gp 27), *Lophira* (Gp 16), and *Homalium* (Gp 17E). Winged seeds are produced inside the dehiscent fruits of Meliaceae (see Gp 34), Mimosaceae (see Gp 37); Sterculiaceae (Gp 27); *Anopyxis* (Gp 5); Bignoniaceae (Gp 31) and *Schrebera* (Gp 5).

Leaves always with short petiole	See Group 5 or 13D	
1) **Savanna tree** with alternate lvs	***Anogeissus leiocarpus***	81
2) **V. dry, rocky forest**; twisted tree with alternate or opposite, small lvs	***Pteleopsis habeensis***	529
3) **Moister forest**, straight tree with glands ⅔ way along lamina	***Pteleopsis hylodendron***	530
Leaves with long petiole[1]		
Leaves without pit domatia in nerve axils (entire); laterals not usually meeting; venation often reddish; slash fibrous dark yellow; fruits 2-winged.		
Leaves often with two sub-opposite glands within wing of petiole; without hairs along midrib; venation lax; lf ±many fine translucent points within lax venation when held up to light; lower boughs of older trees almost horizontal; with high plank buttresses; **bark with large, silvery, thick scales; slash moderately yellow;** fruits broader than long	***Terminalia superba***[2] [*oFRAM*]	53
Leaves never with two glands beside rounded, sometimes rather short petiole; often with orange-brown hairs beside midrib; venation very fine; younger trees with horizontal boughs, but older trees with slightly ascending boughs and graceful pale green foliage; buttresses rather small and thick; bole scaly at first, **becoming black and fissured with age; slash very vivid, bright yellow;** fts longer than broad	***Terminalia ivorensis*** [*EMIRE*]	13
Leaves with pit domatia, sometimes fringed by hairs, in the axils of nerves; margin rarely perfectly entire, and sometimes serrated; laterals meeting in submarginal nerve; old leaves, on ground or in crown, conspicuously reddish; fts 4-winged; outer bark deeply fissured; bole straight and unbuttressed; slash yellow-orange to pink-brown, v. fibrous-spongy, thick, bitter, with **an unpleasant, foetid, smell**	***Petersianthus macrocarpus*** [*ESIA*]	48

NOTES: 1) *Caloncoba gilgiana* (Gp 23) is a small straggly tree, normally spiny, with ±glabrous lvs and long petiole.
2) The first two seedling leaves of *T. ivorensis* are alternate, but opposite in *T. superba*. The leaves of *Petersianthus* seedlings and saplings are serrated.

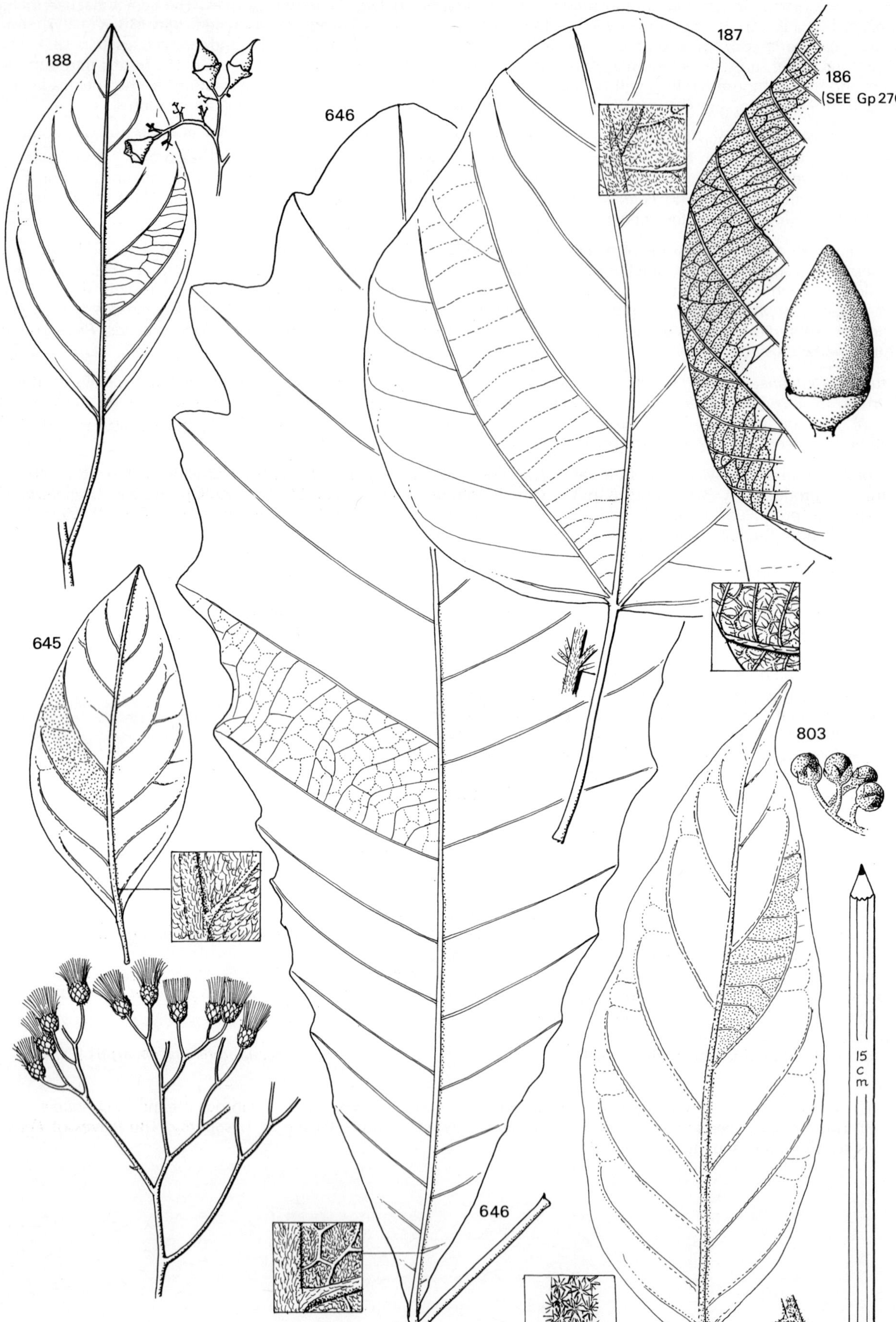
188
646
187
186
(SEE Gp 27C)
645
803
15 cm
646
Rosemary Wise

GROUP 26: BORAGINACEAE, SOLANACEAE, COMPOSITAE
(Lvs simple, alternate, with long petiole; irregular or entire margin)

This group includes three related families of sun-loving trees. They are, in fact, a little out of place amongst the other families of Groups 13-27, as they are more closely related to families such as the Rubiaceae (Gp 1), Loganiaceae (Gp 2) and Apocynaceae (Gp 9) (which have simple, opposite lvs) on the one hand, and to Verbenaceae (Gp 29) and Bignoniaceae (Gp 30) (compound opposite lvs) on the other. Their simple, alternate, often clustered leaves with long petioles, however, necessitate their mention here.

Only *Cordia millenii* and *C. platythyrsa* become large trees. Species of *Cordia* and *Ehretia* (see Gp 13D: Boraginaceae with short petioles) have a distinctive ('TWENEBOA') slash:

***TWENEBOA* slash**: thick, soft-fibrous, spongy, with large pores, and peelable with brownish exudate; white to creamy yellow or pink **darkening through dirty greenish shades** to brown; often contoured or with streaks; distinctively but not strongly earthy-aromatic, somewhat like mushrooms. *Vernonia* species do not develop such a soft-fibrous bark but the slash darkens similarly; *Solanum* tends to be white to cream with pale orange strips, and is also slightly fibrous. Many species of the related families (especially Verbenaceae, Bignoniaceae and some Rubiaceae) mentioned above resemble these species in their slash.

The flowers of these species have petals fused together to form a ±tubular corolla. In *Vernonia* (as in the many, herbaceous weeds of the Compositae) many tiny, bluish flowers are aggregated into a 'capitulum', which resembles a single flower, which is itself borne in a panicle; the small fruits have a plume of cottony hairs for wind dispersal. *Cordia*'s white flowers are borne in panicles; the fruits are ovoid drupes with a saucer-like calyx tube attached at the base. *Ehretia* flowers are like those of *Cordia*, but *Cordia* has a twice-forked style, whereas that of *Ehretia* is once-forked. *Solanum* flowers are white to purple; the fruits are (poisonous) berries.

Vernonia species are used as chew-sticks and in medicine and cooking. *Cordia* spp. can be used for timber (and making drums). *Solanum* is an exceptionally common forest-zone weed.

Lvs very rough, like sandpaper	See *Ficus exasperata* (Gp 28A)	
Lvs not so rough		
Lvs almost trinerved OR glabrous; medium to large trees with *Tweneboa* slash (above); margin sometimes slightly uneven		
Leaves without basal nerves, thin, glossy and completely hairless; lamina normally <10 cm long, elliptic; clustered at branch tips; margin ±uneven, appearing irregularly serrulate, bright green drying black	***Cordia senegalensis*** [*OKOSU*]	188
Leaves with at least one strong basal nerve; usually with others not quite opposite at the unequal-sided leaf base; lvs with (dense) brown hairs; large trees with cylindrical boles, thick bark and short, thick (or without) buttresses		
Leaves usually >15 cm long, rather rounded, with projections of the laterals at the margin; hairs often coarse or long, and soft to touch; basal nerves often 3 or more; top surface with many, scattered, white spots	***Cordia millenii***[1] see Group 27 [*TWENEBOA-(NINI)*]	186
Leaves, when mature, usually <15 cm long, often ovate, with **very short cream-coffee hairs only just noticeable to touch (lens);** without projecting nerves; without long hairs and without white spots above; basal nerves usually not quite opposite; (young trees in **disturbed forest** with v. smooth bark, flat-topped, with hairs v. fine or absent)	***Cordia platythyrsa***[1] [*TWENEBOA-BERE*]	187
Leaves never trinerved, hairy or v. large , often with wavy margin; small, very weedy and sun-loving trees rarely much over 5 cm dbh.		
Leaves >15 cm; often >25 cm long; margin with deep undulations, but not quite lobed or serrated; hairs below short, not covering veins	***Vernonia conferta***[4] [*OFENA, OWUDIFOKETE*]	646
Leaves <20 cm long; hairs long and obvious		
Hairs not branched NOR stellate; seeds with a cottony-plume	***Vernonia colorata***[2] [*AWONWENE*]	645
Hairs stellate or branched; much smaller on young stem; Fts berries	***Solanum erianthum***[3] [*PEPEDIAWUO*]	803

NOTES:
1) These hairy *Cordia* spp. ('TWENEBOA') are not distinguished by the timber trade, nor by most tree-spotters.
2) Other sun-loving, weak-wooded *Vernonia* spp. key here as well. *V. amygdalina* is a savanna treelet with (more sharply) serrated leaves.
3) *Solanum erianthum* used to have the name *S. verbascifolium*. Other *Solanum* spp. are herbs or shrubs, often with spines on the leaves and twigs or deeply lobed leaves.
4) *Vernonia titanophylla*, which occurs in the Atewa, and possibly other evergreen forests, is similar to *V. conferta*. It differs in having leaves with dense, white, woolly hairs below and a line of peculiar, linear structures a few mm long at the nodes and along the base of the petiole.

FRUITS OF THE STERCULIACEAE

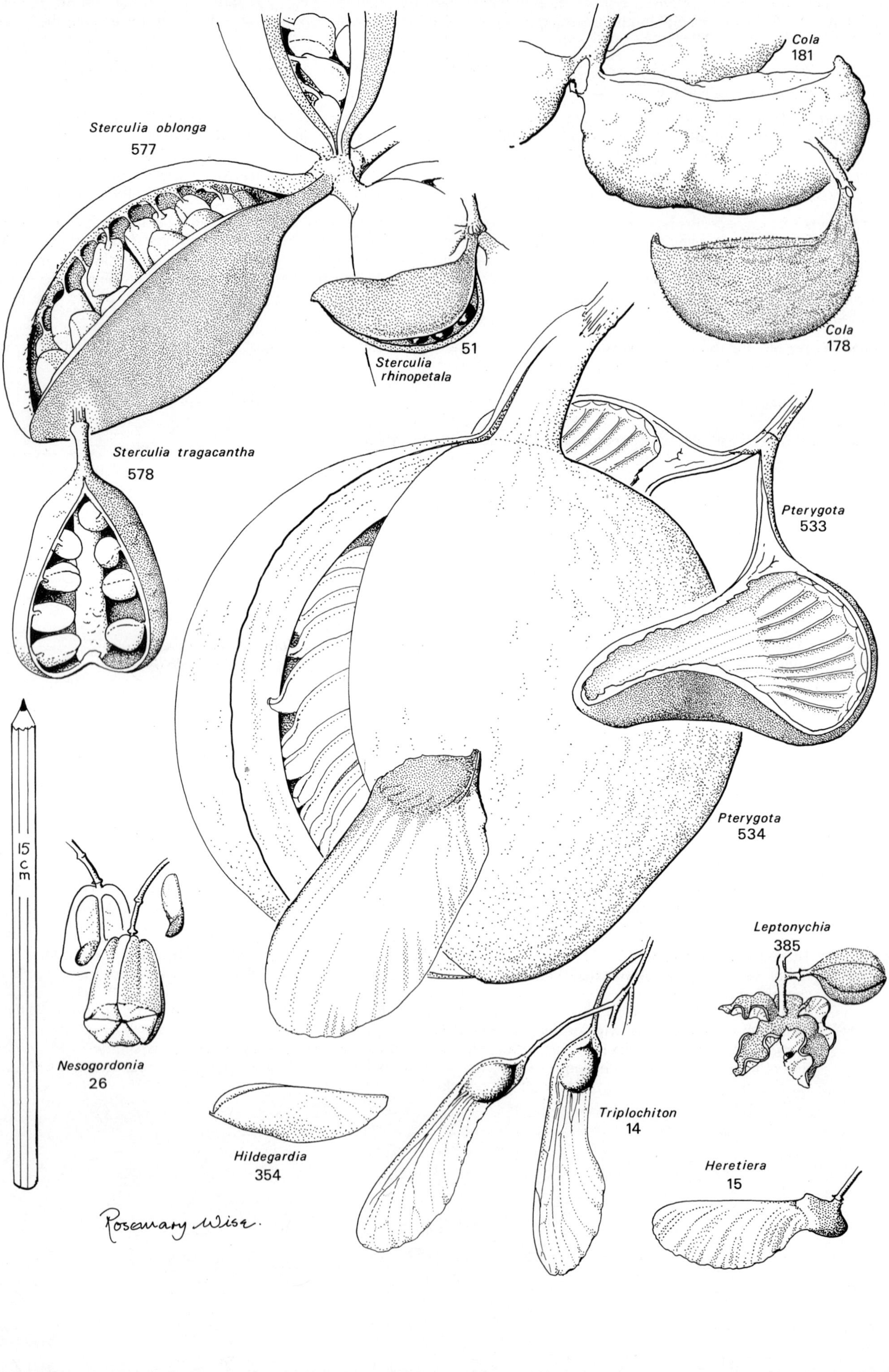

GROUP 27: STERCULIACEAE, etc.

(Leaves simple, alternate (often v.clustered), with long petiole; often with stellate hairs: without stilt roots)

Key to subgroups

Leaves without basal nerves, or with weak basal nerves lacking strong side branches	See Group 27A
Leaves with 2 basal nerves with strong branches or with >2 basal nerves	
Leaves not lobed, or densely hairy on lamina	
Leaves with <6 basal nerves, or fine teeth where laterals reach margin	
Leaves rather rounded, with 6 or more basal nerves	Group 27C
Leaves lobed; hairs NOT usually very conspicuous on lamina, or lamina v. rough	Group 27D
Leaves ±smooth on upper surface	Group 27D
Leaves very rough with coarse hairs on upper surface	Group 28A

STERCULIACEAE

The family Sterculiaceae is of exceptional economic and social importance in Ghana. Apart from the timber trees like *Triplochiton* [*WAWA*], *Heritiera* and *Mansonia* [*OPRONO*], the family includes also the understorey species *Cola nitida* and *C. verticillata* which produce Kola 'nuts', and the introduced *Theobroma cacao*, the Cocoa tree. Apart from the species mentioned below, the family includes *Leptonychia* and *Scaphopetalum* (with short petioles) in Group 21 and three *Cola* species and *Heritiera* with digitately compound leaves in Group 28B. Savanna trees in the genus *Dombeya* are also in this family. The family is closely related to Tiliaceae (Gp 21) and Bombacaceae (Gp 28C), and many species closely resemble certain Euphorbiaceae. In common with many such species, stellate hairs are common on the leaves and flowers of the Sterculiaceae, and the bark is typically fibrous, with a yellow or reddish slash, often with vertical bands of dilatation tissue and usually darkening rapidly on exposure. Most Sterculiaceae in Group 27 grow rhythmically, with whorled boughs on the ('monopodial') main stem, and with leaves clustered at the end of twigs (Rauh's model of Hallé *et al.* 1978).

The flowers are regular, with a superior ovary, usually in axillary inflorescences, but they are otherwise varied.

Genus	Notes on flowers
Sterculia	No petals, but calyx cream to pinkish, often brown hairy (lobes remain joined in *S. tragacantha*)
Cola	No petals, but calyx often coloured; flowers unisexual
Octolobus	No petals, hairy calyx 8-lobed; solitary and unisexual
Hildegardia	No petals, but calyx bright red, narrow tubular, *c.*2 cm long
Pterygota	No petals, but calyx brown hairy outside with reddish centres; united stamens
Heritiera	No petals; calyx tube small, with sharp lobes, in many-flowered infls
Nesogordonia	Fragrant; 5 cream, narrow petals *c.*1 cm long; 5 sepals; free bundles of stamens
Triplochiton	White, hairy petals with reddish base *c.*1 cm wide nr. tip; calyx 5-lobed; 30+ stamens
Mansonia	Fragrant; 5 white petals; *c.*1.5 cm wide nr tip; calyx splitting down one side; 10 stamens
Scaphopetalum	Petals with boat-shaped section at tip, yellowish with purple lines
Leptonychia	Inconspicuous, with long hairs on margin of petals

There is a wide range of fruit type, showing some relation to the ecology of the species, with most of the larger, often emergent trees bearing winged fruits or seeds, and with most smaller trees without such 'flying seeds' (with exceptions in *Cola gigantea* and *Sterculia* spp. – larger trees with bulky seeds). It is important to note that whereas *Mansonia* and *Triplochiton* produce 1-seeded winged fruits (directly on a stalk-like *Terminalia*), *Pterygota* and *Nesogordonia* produce many winged seeds within a single fruit, like *Khaya* and *Entandrophragma*.

	Winged seeds or fruits (wind or water-dispersed)
Mansonia	Whole 1-seeded fruit with one wing; indehiscent
Triplochiton	Whole 1-seeded fruit with one wing; indehiscent
Heretiera	Whole 1-seeded fruit with one wing; indehiscent
Hildegardia	Whole 1-seeeded fruit with inflated wing; indehiscent
Nesogordonia	5-locular capsule splitting to release small 1-winged seeds
Pterygota	Large *c.*woody 'follicle' splitting along 1 edge to release large, thick, 1-winged seeds

	Seeds and fruits without wings (probably animal-dispersed)	
Sterculia	Clustered follicles opening to reveal:	
S. oblonga	yellow (arillate) seeds	
S. rhinopetala	red (arillate) seeds	
S. tragacantha	grey (non-arillate) seeds	
Leptonychia	Capsules splitting to reveal black seeds with red arils	(See Gp 21)
Scaphopetalum	Capsules splitting; seeds without arils	(See Gp 21)
Cola	Clustered follicles eventually opening; with large (non-arillate) seeds	
Octolobus	Clustered follicles eventually opening; with large (non-arillate) seeds	

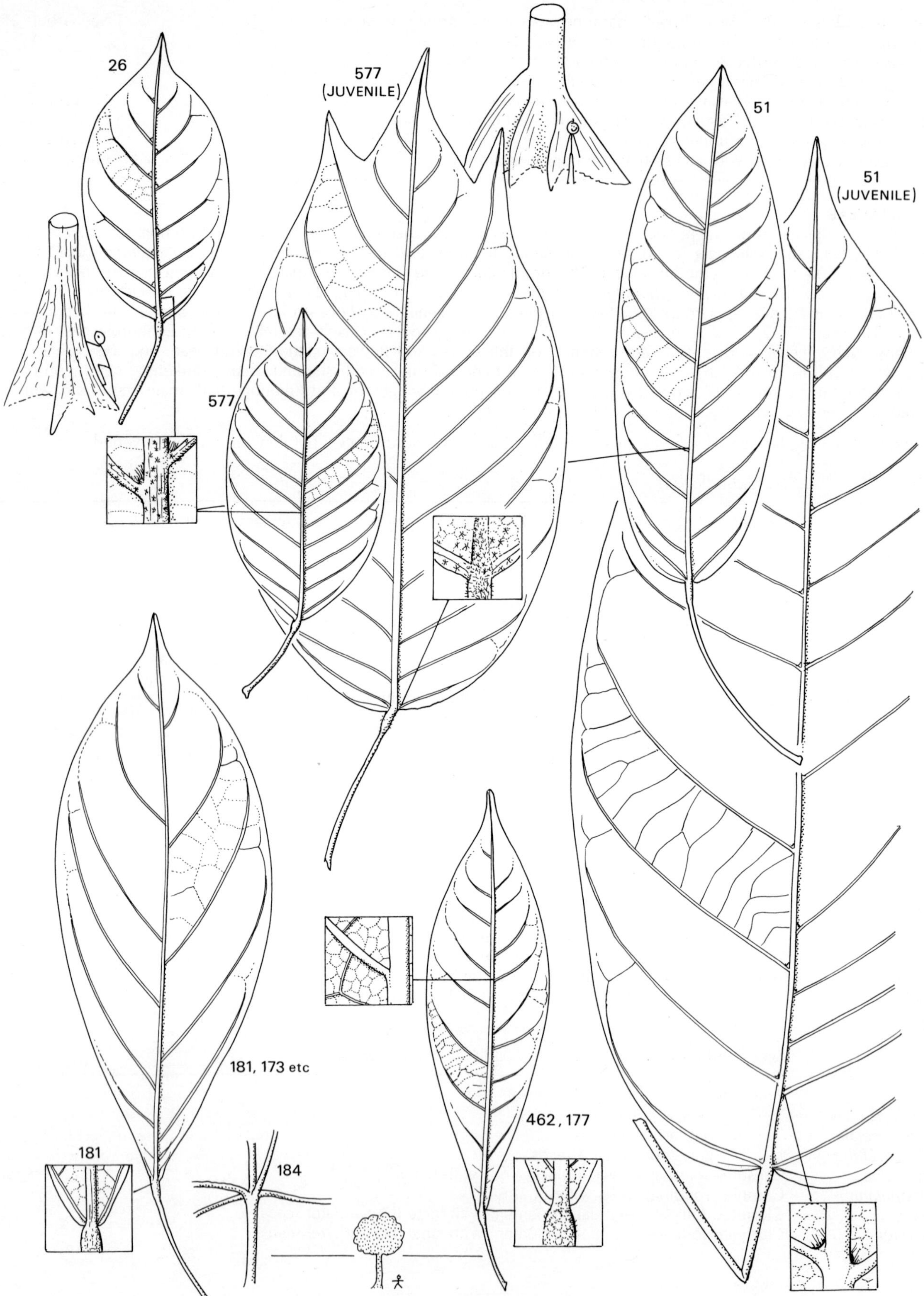
26
577
(JUVENILE)
51
51
(JUVENILE)
577
181, 173 etc
462, 177
181
184
R. Wise.
15 cm

Group 27A: Sterculiaceae

(Leaves entire, not cordate; long petiole; often with (inconspicuous) stellate hairs)

NB: some species of Group 23, for instance *Calancoba gilgiana*, will key to this Group when spines are not observed. If the leaf is NOT acuminate, check Group 24, especially *Protomegabaria* and *Spondianthus* which are often found without. Species of Group 24 lack domatia, however.

Axils of lateral nerves with tuft domatia; often stellate-hairy elsewhere; timber-sized trees, often with high, narrow buttresses.[1]

- Lvs usually with <10 prs laterals, and <3.5 cm wide, markedly acuminate; many stellate hairs below; petiole slender; young twigs with linear paired stipules crowded at tips; outer bark often flaky; **slash soft-fibrous, pink or brown becoming darker;** sapwood with fine ripple marks — ***Nesogordonia papaverifera*** [*DANTA*] 26
- Lvs usually with 10 or more prs laterals, >3.5 cm wide; branches whorled.
  - Lvs often barely twice as long as wide, ovate, sometimes ±3-nerved; **midrib channel above filled with brownish hairs**; stellate hair ±dense below; laterals ±impressed above; apex acute (3-pointed on saplings); **slash white to yellow, fibrous, with orange gritty streaks, ±dilatation bands and slight musty, distinctive smell** — ***Sterculia oblonga***[1] [*oHAA*] 577
  - Lvs oblong, lanceolate or oblanceolate (often like cocoa leaves); **midrib ±prominent and glabrous above**; only long acuminate when juvenile, and only new lvs have many stellate hairs; apex normally blunt; **slash red with cream lines darkening, 'chunky', bitter** — ***Sterculia rhinopetala***[1] [*WAWABIMA*] 51

Axils of nerves without tufts of hairs; lvs glabrous or with scurfy hairs; small-medium-sized, understorey trees.

- **Lateral nerves not joining into a clear loopy sub-marginal nerve,** or decreasing in width considerably before meeting adjacent laterals; lvs with or without hairs
  - **Lvs** strongly clustered at branch tips, the lowest (largest) in each cluster markedly broadest above middle: oblanceolate; almost glabrous (some hairs usually visible on young petioles with lens); slash of larger trees ±gritty.
    - *Leaves trinerved* at base, or nearly so; young twigs stout
      - Lvs alternate, although v. clustered; margin recurved; bark thick, rough, dark; slash pale pink with pale orange, gritty streaks, soft-fibrous, darkening, with a slightly soapy effect in mouth — ***Cola nitida*** [*BESE*][3] 181
      - Lvs in whorls, often of 4 — ***Cola verticillata*** [*BESE-TORO*] 184
    - *Leaves not trinerved* at base[2,3]
      - Stipules (/bud scales) linear *c.*1 cm long , densely aggregated at twig tips; sometimes with v. hairy twigs; hairs on young petiole swelling dense, brown, v. short (stellate), with 'arms' of hairs ±invisible without x10 lens
        - Treelet common in **some dry forests** — ***Octolobus spectabilis***[2,3] [*AFINAFI*] 462
        - Treelet in **moister forests** only — ***Cola flavovelutina***[2] 177
      - Stipules <1 cm long, partly or v. membranous[2]
        - *Stipules* and apical bud sharply pointed, with stellate hairs; petiole on flowering brs often <1 cm long; treelet of moist places — ***Cola reticulata***[2] 173
        - *Stipules* and opical bud broadly triangular to rounded, ±glaborous; stipules rather chaffy, like rice husks, and persistent on young twigs; Hairs few; if visible on petiole, then large and pale with the 'arms' clearly visible (x10 lens); common in upland for. (e.g. nr Nkawkaw) — ***Cola boxiana***[2] 173
  - **Lvs** evenly dispersed along twigs; often broadest around, or even below, middle; oblong or elliptic; leaves and twigs with 'scurfy' (like dandruff or flaking rust), red-brown hairs, or venation v. scalariform; petioles often <3 cm; slash thick and spongy fibrous (becoming darker with exudate), without gritty bits — See Group 27B
- **Laterals looping abruptly to form a well-defined sub-marginal nerve;** lvs absolutely glabrous — See Group 27B

NOTES: 1) *Nesogordonia* seedlings have serrated lvs at first. Saplings very soon develop leaves like the parents, with conspicuous aggregations of stipules at the end of the twigs. In *S. oblonga* the first pair of lvs are alternate, whereas in *S. rhinopetala* they are opposite. Sapling lvs are illustrated. Saplings of these species have obvious **whorled branches**, and are common in the understorey of **semideciduous forest**.

NOTES 2) and 3) are under Group 27B, overleaf.

Group 27B: (Miscellaneous families resembling Sterculiaceae)
(No tuft domatia; laterals looping)

Four species with leaves which can be easily confused with members of Group 27A, but which are not Sterculiaceae; abbreviations are for Olacaceae, Capparaceae and Flacourtiaceae (see Gp 13, 17). *Octoknema* was previously in the Octoknemataceae.

Laterals not forming well-defined sub-marginal nerve; young lvs + twigs ±hairy;		
Lvs evenly dispersed along twigs; leaves and twigs with 'scurfy' (like dandruff or flaking rust), red-brown hairs, or **venation v. scalariform**; petioles often <3 cm; slash thick and spongy fibrous (darkening with exudate), without gritty bits[1]		
Lateral nerves not arching close to margin; veins not closely parallel; lvs often slight asymmetric-oblong, acuminate; scurfy hairs very persistent; not only in **evergreen** forest; slash with a sweet, fruity taste; crown dense, with a subdued metallic reddish hue	***Octoknema borealis*** (OLAC) [*WISUBONI*]	461
Lateral nerves arching *very* close to (recurved) margin; veins usually closely parallel and scalariform (like ripples on water, spreading from nerve axils); laterals impressed above; lf often elliptic with a long drip tip; older lvs without scurfy hairs; lower surface with scattered, raised dots	***Coula edulis***[1,2] (OLAC) [*HWEREME-DUA, BODWUE*]	191
Laterals looping abruptly to form a well-defined sub-marginal nerve; lvs absolutely glabrous		
Petioles often >3 cm long; 8-10 pairs laterals; basal nerves absent; twig v. winged; lf pale green below; slash reddish brown, bitter, slightly hot	***Buchholzia coriacea***[3] (CAPP) [*KONINI, ESONO-BESE*]	125
Petioles <3 cm long; 6-8 pairs laterals; basal nerves ±present; twig rounded, with lenticels; small tree or shrub with bark rarely thick enough to slash	***Caloncoba echinata***[4] (FLAC) [*AWIEWU-NUA*]	127

NOTES: 1) The slash of younger *Coula* trees sometimes exudes a white latex.
2) The name '*BODWUE*' should not be confused with that of *Ongokea gore* ['*BODWE*'].
3) *Bucholzia* fruits are dispersed by elephants.
4) *Caloncoba glauca* has been recorded in Ghana: Nigerian specimens have ±lanceolate-acuminate lvs with many white scales below, with a petiole often >3 cm long. *C. gilgiana* will key to this point when the small twig-like spines on the older branches are overlooked – the latter species has minutely cordate lves.

NOTES for Group 27A (contd.):

2) The following four species are very hard to differentiate when sterile. If flowering use the following key:

Flowers 8-lobed (>1 cm long).	***Octolobus***
Flowers 5-lobed	
Flwr bud (mature, but unopened) <0.5 cm wide, v. slender, on 1cm stalks, ±glabrous	***Cola boxiana***
Flwr bud >0.5 cm wide, or at least hairy or sessile	
Flwr stalk <5 mm long - covered in red spots (glands)	***Cola reticulata***
Flwrs clearly pedicellate, densely covered in orange hairs	***Cola flavo-velutina***

3) Two species of small shrubs with similar lvs are *Penianthus zenkeri* (Menispermaceae) with a deeply grooved petiole, and *Cola heterophylla* (**evergreen for.**) with v. corky, hollow twigs and papery lvs with irregular margins.

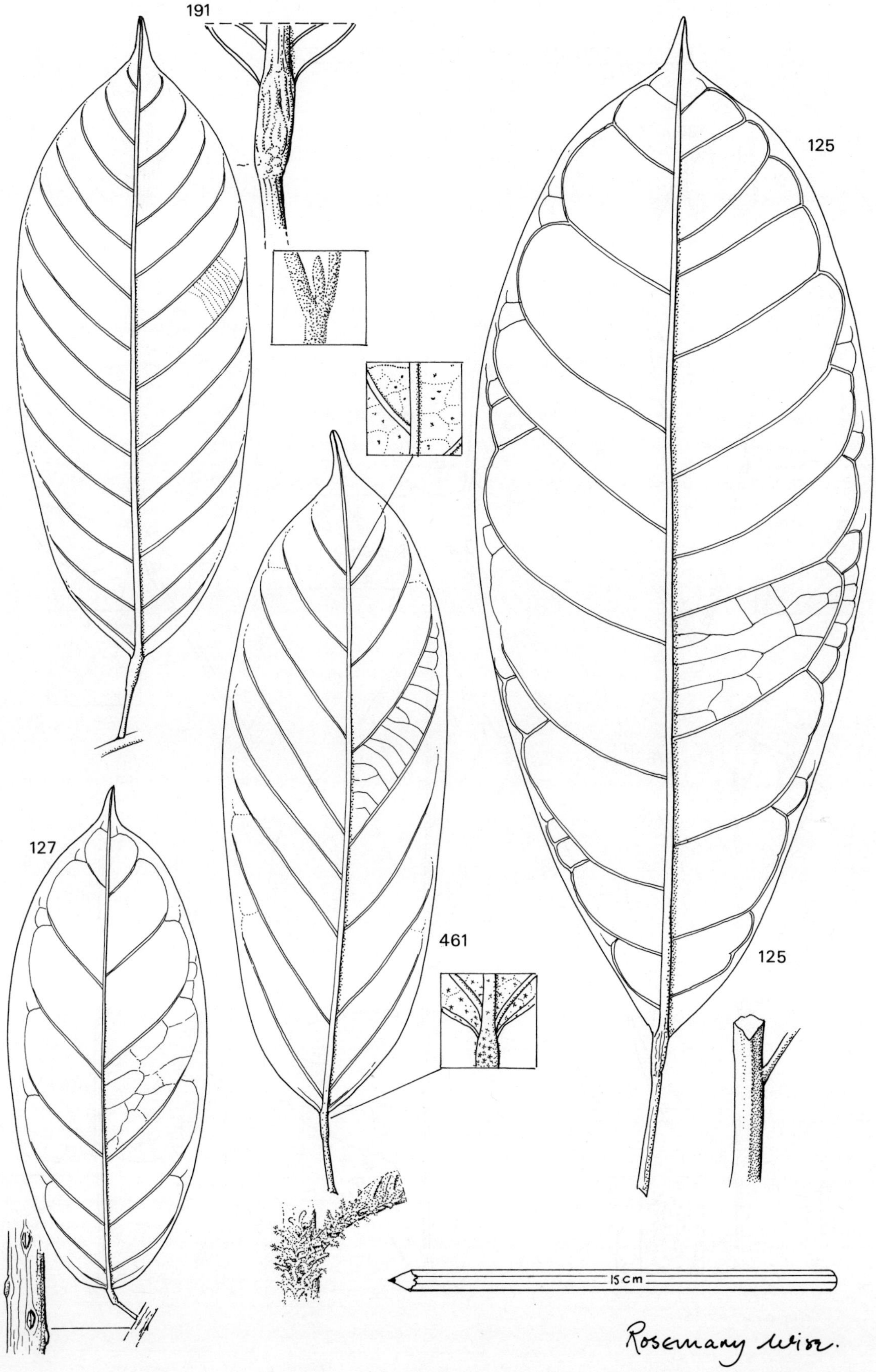

Rosemary Wise.

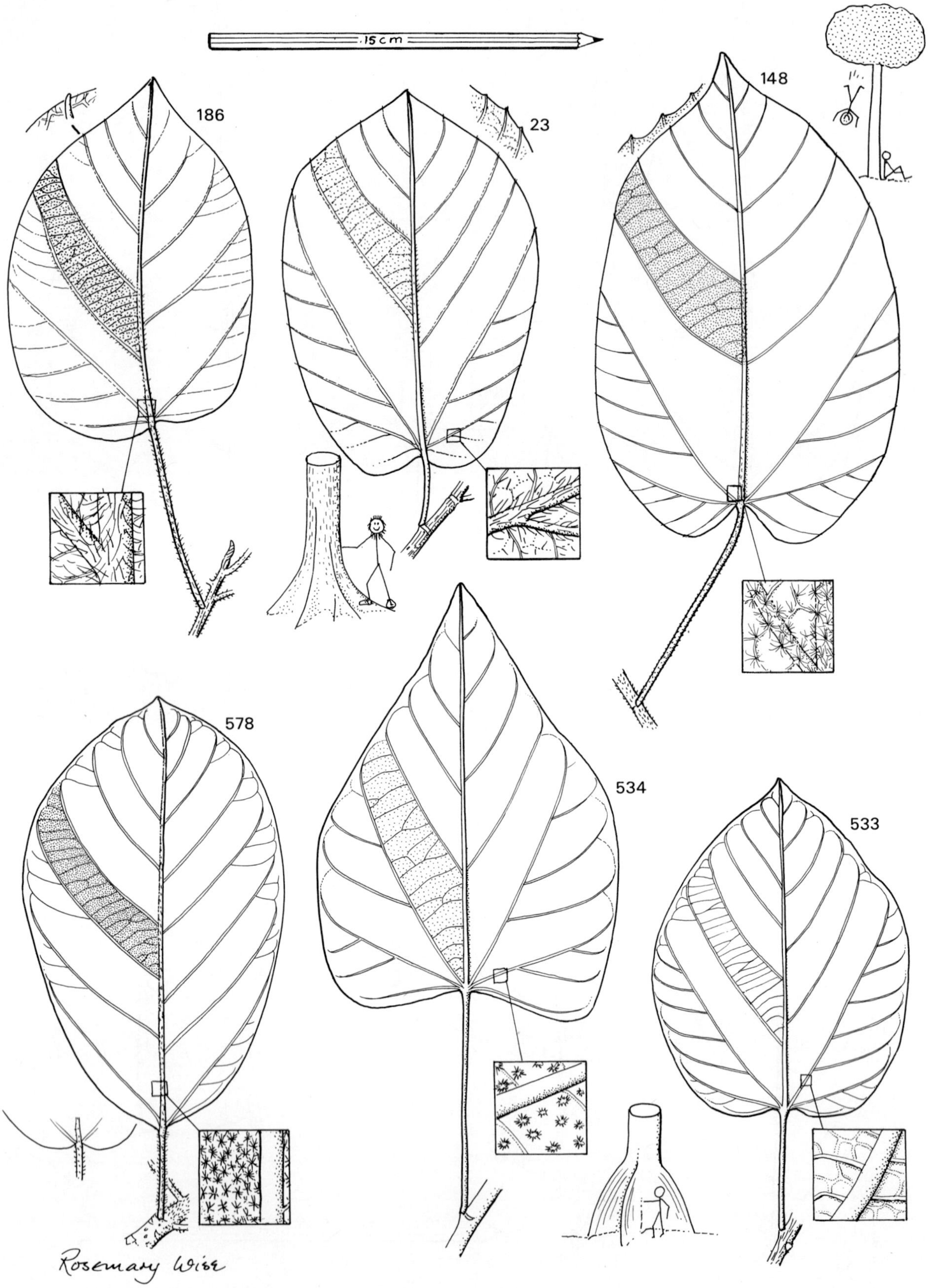
15 cm
186
23
148
578
534
533
Rosemary Wise

Group 27C: Sterculiaceae
(Leaves entire or with small teeth; if entire then with basal nerves well developed)
(Base usually cordate or obtuse; usually large lvs)

- Latex-producing tree with fringes of hairs at nodes OR lvs very rough, like sandpaper and tree with yellowish clear exudate; bole usually extremely smooth, grey to yellowish — See *Ficus* spp. (Gp 19B)
- No latex; lvs not rough like sandpaper
  - Leaves with thread-like extensions of laterals at margin, or teeth, with dense hairs even on mature leaves; often v. asymmetric at base.[1]
    - Leaves without stellate hairs; no stipules; 1 pair basal nerves reaching halfway up lamina, ±one other pair; not v. cordate; petiole grooved above; slash contoured, fibrous, darkening through greenish shades, with brown watery exudate — ***Cordia millenii*** (See Group 26) (BORA) [*TWENEBOA*] 186
    - Leaves with stellate hairs, especially visible on top surface or young twigs; petiole not grooved; young twig tips, at least, with stipules; slash fibrous darkening rapidly, sometimes contoured, but NOT darkening through greenish shades; outer bark often with diamond-shaped markings *c.* 2 cm wide
      - Twigs with ring scars at young nodes where stipules have fallen; hairs long, erect, especially conspicuous on main nerves, etc. below; most lateral nerves ending in small thickened teeth;[2] base of lf usually asymmetric; apex very obtuse; often broadest above the middle; tall tree with narrow crown of ±horizontal boughs; slash yellowish, fibrous, + vertical bands, darkening rapidly; sapwood with ripple marks — ***Mansonia altissima*** [*OPRONO*] 23
      - Twigs with linear, not broad-based stipules: without ring scars; hairs on lower surface all clearly stellate (x10), with outstretched arms, not erect, not especially long on nerves, etc.; few or no nerves ending in points, teeth sometimes large, lf almost lobed; apex gradually tapered to a point from *c.* half-way; base usually symmetric; apex sometimes 3-lobed; medium-sized tree gregarious, especially but not entirely in **wet places**; slash yellowish or pink-orange contoured, darkening, ±bands, v. fibrous — ***Christiana africana*** (TILI) [*SUPRONO*] 148
  - Leaves without 'teeth'; lvs with (stellate) hairs dense at first, but disappearing with age; without dense and soft covering of simple hairs as well
    - Lvs oblong-elliptic, or slightly broader above the middle; 2 strong basal nerves; base sl. cordate or ±obtuse; bark ±rough; slash thick, fibrous, brownish pink, darkening with gummy exudate, contoured scented; branches whorled ±horizontal and lvs v. clustered — ***Sterculia tragacantha*** [*SOFO*] 578
    - Lvs varied, but not oblong nor elliptic; usually tapering with almost straight edges from half-way to the apex; branches not whorled; crown rounded with (large) lvs not v. clustered
      - Slash reddish, fibrous slightly sticky; some lvs lobed — See *Cola gigantea* (Gp 27D)
      - Slash soft, peelable pithy-fibrous, white, pale orange (or rarely pink) with vertical streaks of orange gritty flecks; mature trees without lobed lvs.[3]
        - Petiole usually <5 cm long; only 2-4 basal nerves; young base sometimes obtuse, otherwise cordate; young shoots covered in dense brown hairs; old leaves glabrous; rare tree apparently absent from drier forests; slash smells of green beans — ***Pterygota bequaertii*** [*KYEReYe-BERE*] 533
        - Petiole often much longer than 5 cm; lvs often v. large, deeply cordate; distinct in having a relatively parallel-sided lower section as if a heart-shaped leaf had been stretched out in the middle; usually with 6 basal nerves; hairs on young stem short, but sometimes dense; buttresses large and thin, smooth, lenticellate, often with many parallel raised lines ('crease marks') at right angles to buttress edge; slash soft, pale yellow (see above) — ***Pterygota macrocarpa*** [*KYEReYe*] 534

NOTES: 1) Leaves with v. few hairs, on a small tree – see *Lindackeria dentata* [*SOFOSE*] (Gp 22A).
2) Juvenile *Mansonia* leaves have very pronounced serrations.
3) *P. macrocarpa* seedlings have oblong, sessile cotyledons, barely-lobed young leaves, grey-green older stems and not v. distinct veins, whereas *Cola gigantea* seedlings have petiolate, auriculate cotyledons and deeply-lobed young leaves which are reddish at first (Taylor).

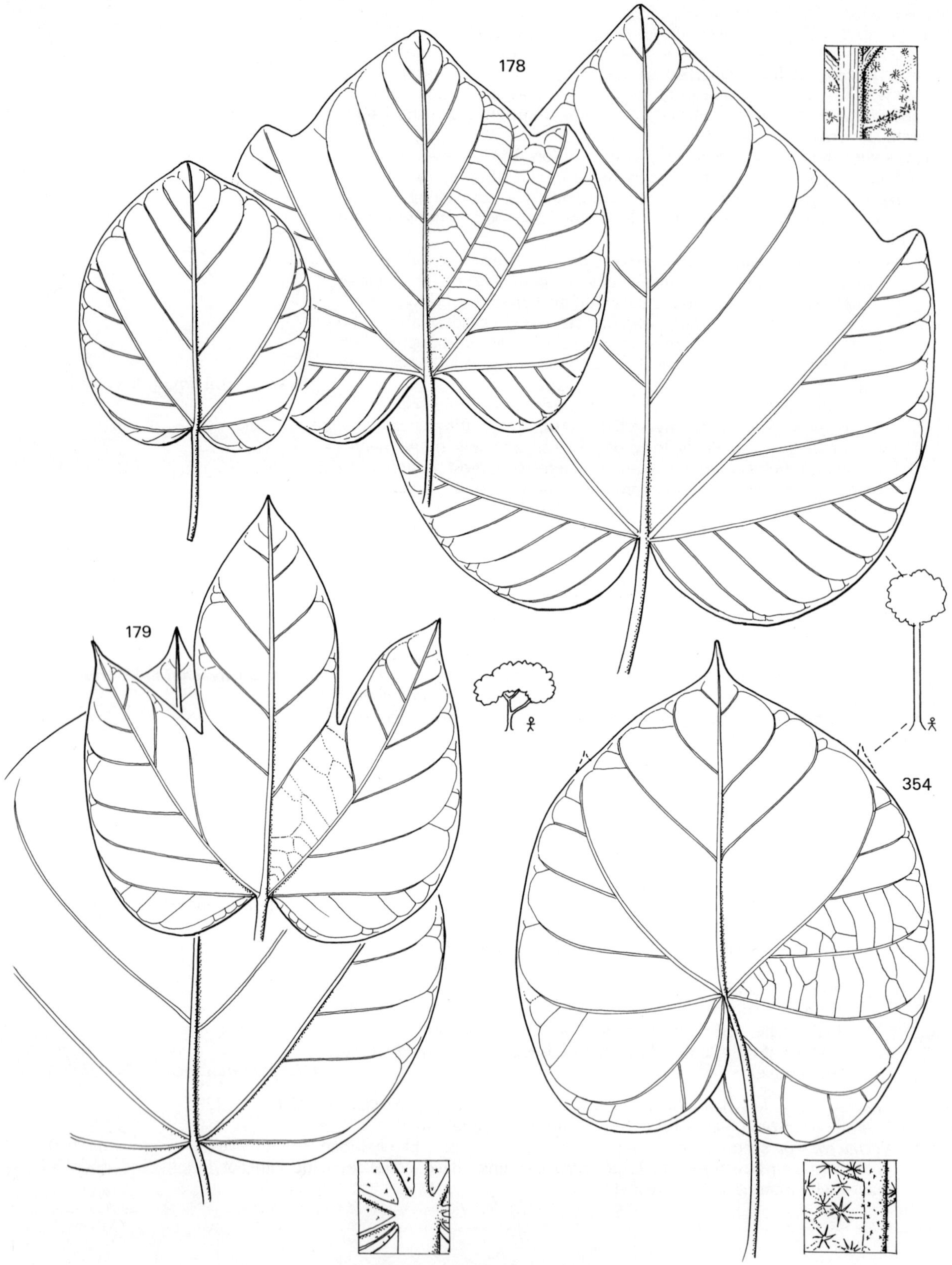
178
179
354
15cm
R. Wise

Group 27D
(Leaves lobed or rounded, many basal nerves, with few hairs or glabrous)

Leaves not densely hairy; lvs usually lobed or rounded + very few (stellate) hairs below or glabrous; some lvs without clearly-dissected lobes; lvs usually large

- Lvs v. rounded but with a cordate base and acute apex; sometimes with a slight acute 'outgrowth' of margin where basal nerves approach it; lamina rather thin and pale; tree of **dry forest, shallow soils, rocky areas**, etc.; older trees sometimes with high buttresses; young trees + whorled boughs; with v. smooth, greenish lenticellate bark and soft fibrous-spongy, white slash with strong scent and vertical bands, darkening — ***Hildegardia barteri*** [*AKYERe-KYEWEWA*] 354
- Lvs not smoothly rounded, often broadly lobed; sometimes rounded at apex or not cordate at base; crown dark, and dense; slash fibrous, with sticky exudate, not obviously scented, reddish or orange with bands of different colour
  - Lvs never highly dissected like a papaya, NOR digitately lobed (sometimes unlobed on sun leaves); without long hairs; young lvs red; slash pink or orange-ish with a sticky exudate
    - Apices of lobes of leaf (+lf apex) distinctly acuminate or, if not, base of leaf obtuse or only very mildly cordate; small spreading, often crooked tree; never emergent; stellate hairs below, sparse except around base of lamina; laterals not v. much more prominent than the barely raised lateral nerves; undersurface not particularly rough — ***Cola lateritia***[1] [*WATAPUO-BERE*] 179
    - Apices of lobes of leaf (+ lf apex) rounded-off; base of leaf usually very markedly cordate; stellate hairs below normally abundant on lateral nerves and veins; fine veins rather prominent with laterals much more: the undersurface therefore distinctly rough or abrasive to touch; tree **with a long ±straight bole,** ±high buttresses, unbranched below the rounded crown; often emergent — ***Cola gigantea***[1,2] [*WATAPUO*] 178
  - **Lvs all conspicuously lobed**.... (continued overleaf)

NOTES: 1) In practice, these two species are most readily distinguished by overall appearance and habit; it may not be realistic to separate saplings with this key, as younger *C. gigantea* have more pointed lobes.
2) For notes on *Cola gigantea* seedlings see Gp 27C. *C. lateritia* seedlings are probably indistinguishable.

Lvs all conspicuously lobed; lobes clearly defined on all leaves, even in sun; usually dissected to ½ or more of length of main nerves or, if not, leaves <15 cm broad; If lvs large (on shade or sapling branches) then the lobes usually themselves v. lobed[3]

- Lobes usually narrowed slightly at base; all basal nerves arising from the perimeter of the top of the petiole
  - Lvs deeply dissected like papaya lvs — ***Cola caricifolia*** (shade) [*ANANSE-AYA*] 174
  - Lvs with lobes never themselves sharply lobed
    - Lvs always with more than 3 lobes
      - Tips of lobes usually pointed; hairs on base of main nerves below, and on top of petiole as long or longer than the petiole width; small, little-branched trees — ***Cola caricifolia*** (sun)
      - Tips of lobes rather rounded, with 'nipple-like' tips; hairs on base of main nerves etc. short and dense, <1 mm and <width of petiole; outer bark ±smooth, lenticellate; tree becoming medium-sized and well-balanced with ±cylindrical bole; slash pale yellow, soft with narrow, vertical, ladder-like strands, darkening — ***Cola millenii*** (shade, ?moist areas)
    - Lvs with 3 lobes; hairs as described above — ***Cola millenii*** (sun, ?dry conditions) [*ANANSE-DODOWA*] 180
- Lobes usually widest at their base and narrowing in a smooth curve to an apical 'nipple' shape; outer basal nerves arising from a short branch extending from the top of the petiole round the basal margin; normally 5 lobes[4]; twigs + rings at nodes; often v. large trees with tall, narrow buttresses; bark on young trees smooth, grey with many lenticels, but becoming rough and scaly with age; slash thick, fibrous (esp. inner bark), grey-yellow with browner vertical bands, darkening; with ripple marks — ***Triplochiton scleroxylon*** [*WAWA*] 14

NOTES: 3) Saplings of *Sterculia oblonga* have an unusual, sharp pair of lobes next to the acuminate apex, making the leaf 3-pronged like a fork. *Cola heterophylla* is also sometimes lobed. See Group 27A for notes on these.

4) *Triplochiton* seedling lvs are similar to mature ones, but with only 3 lobes. The leaves of middle-aged trees have much larger lvs than canopy trees, sometimes with 7 lobes.

15cm
174
174
180
180
14
Rosemary Wise.

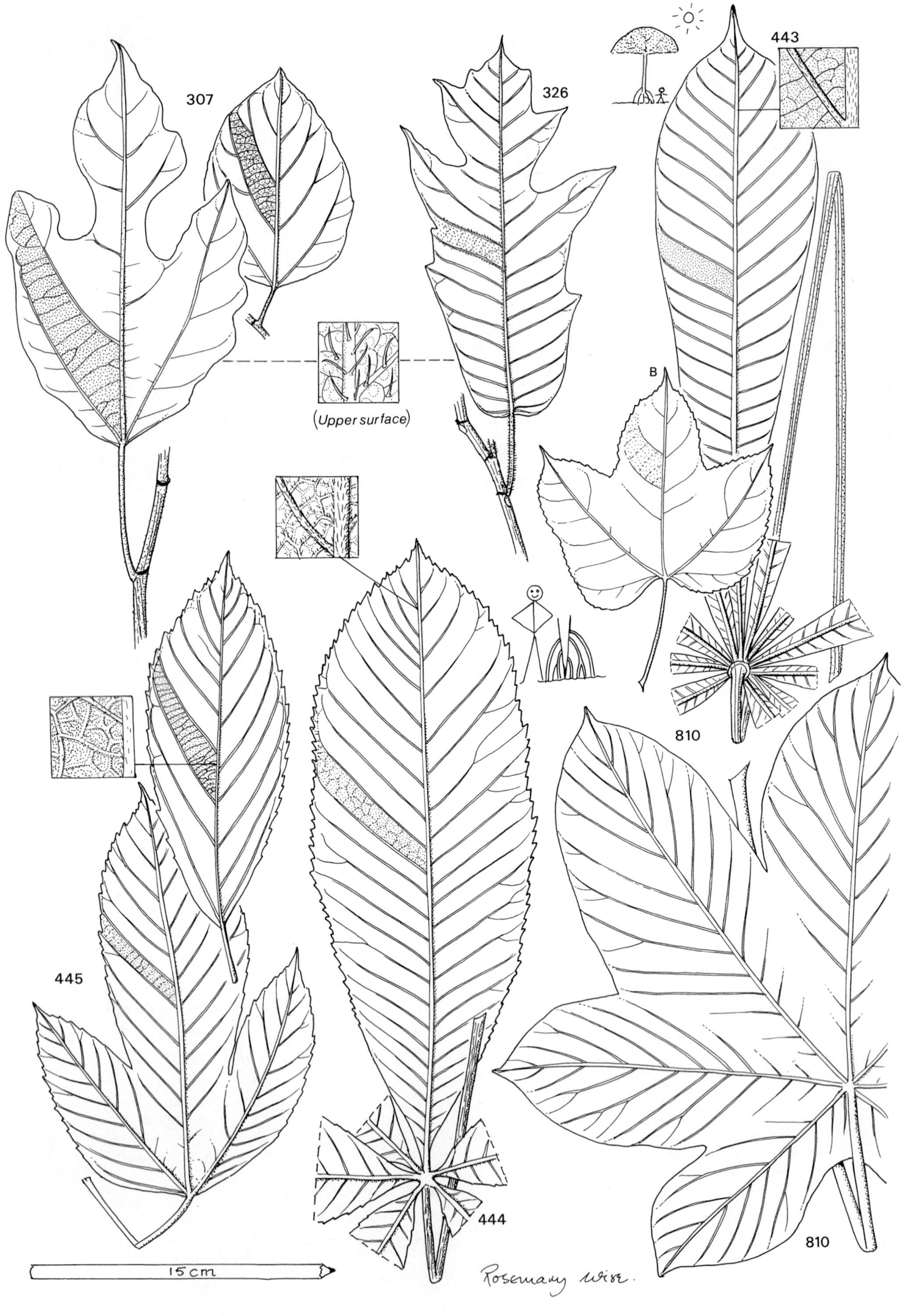
307
326
443
(Upper surface)
B
810
445
444
810
15 cm
Rosemary Wise

GROUP 28
(Moraceae (part) to Bombacaceae)

Key to subgroups

Leaves whitish or silvery below with cottony hairs between veins	***Myrianthus*** etc. (Gp 28A)
Leaves sometimes golden, with scales, but usually hairless and not discolorous	
LEAVES LOBED (simple)	
Lvs with many reddish glands below, and drip-tipped; (tree with spines)	***Macaranga*** (Gp 23)
Lvs without glands, drip tips and spines; with yellowish exudate	***Ficus*** spp. (Gp 28A)
LEAVES DIGITATEly compound	
Lvs often with more than ten leaflets arranged in a circular fashion	***Musanga***[1] (Gp 28A)
Lvs with less than 10, normally gradually acuminate leaflets	
Evergreen forest trees lvs with golden scales or translucent spots	Group 28B
Lvs without scales and spots	
Small, unarmed trees, OR large lvs (try first if in doubt)	Group 28B
Large or armed trees	BOMBACACEAE (Gp 28C)

Group 28A: Moraceae (part 2)
(Lvs lobed and serrated or digitate: tree without latex, often with stilt roots)

For notes on the Moraceae, see Group 19.

Leaves simple, (1-) 3-lobed[2,4]		
Lvs with long hairs, papery and v. rough; veins not finely raised and reticulate below; nodes with ring scars; bole unusually smooth green or yellowish, but often with large expanded nodal scars; slash fibrous, not contoured, with yellowish watery exudate; common in **dry forests (sapling or sucker shoots)**; (sometimes with latex, especially in young parts)		
Lobes not themselves sharply lobed; <7 prs laterals; common	***Ficus exasperata*** [*NYANKYERENe*]	307
Lobes sometimes lobed; lvs often >7 prs laterals; lvs drying yellowish	***Ficus variifolia*** [*NYANKERENe-NINI*]	326
Leaves white below, with short hairs; not coarsely rough on top, glossy surface; with fine, ±square raised reticulation below; slash orange-brown, fibrous-brittle, usually with a clear, watery exudate	***Myrianthus libericus***[1] [*NYANKUMA-NINI*]	445
Leaves compound, with 5 or more leaflets; usually **stilt-rooted**		
Leaflets (3-)5-7; serrated or with glandular teeth on margin		
–Margin with glands, not serrations; with stellate hairs.	See *Ricinodendron* (Gp 28B)	
–Margin serrated; lvs with many parallel laterals; venation and hairs as in *M. libericus* (above); not stellate	***Myrianthus arboreus***[1,3] [*NYANKUMA-BERE*]	444
Leaflets more than 8, entire, arranged in a circle around the top of petiole, like an umbrella; tree as a whole also umbrella-like; spreading, with few, stout whorled branches supporting a dark green dome of lvs; fallen stipules like red, papery banana skins; tree v. common in **secondary forest**, except in **drier forests**	***Musanga cecropioides***[1, 2] [*oDWUMA*]	443

NOTES: 1) *Musanga* seedlings (**443B**) up to a few feet tall, have 0,3 or 5 lobes (increasing with age) on thin papery lvs with serrations and dense white hairs below.

2) ***Cecropia peltata*** (**810**) [*'FRENCH oDWUMA'*] is a tree similar in habit to *Musanga*, but has broad, 3 to 7-lobed pellate lvs with white hairs, rather like sapling *Musanga* or *Myrianthus*, but on fully mature trees: it has been planted in Côte d'Ivoire and is now 'escaping' into the western region of Ghana. There are tufts of hairs on the underside of the petiole at the base.

3) *Ceiba* sapling leaves (28C) are, unlike the adults, digitate *and serrated*.

4) *Ricinus communis*, the castor oil plant, is a common weedy shrub which has many-lobed, serrated leaves.

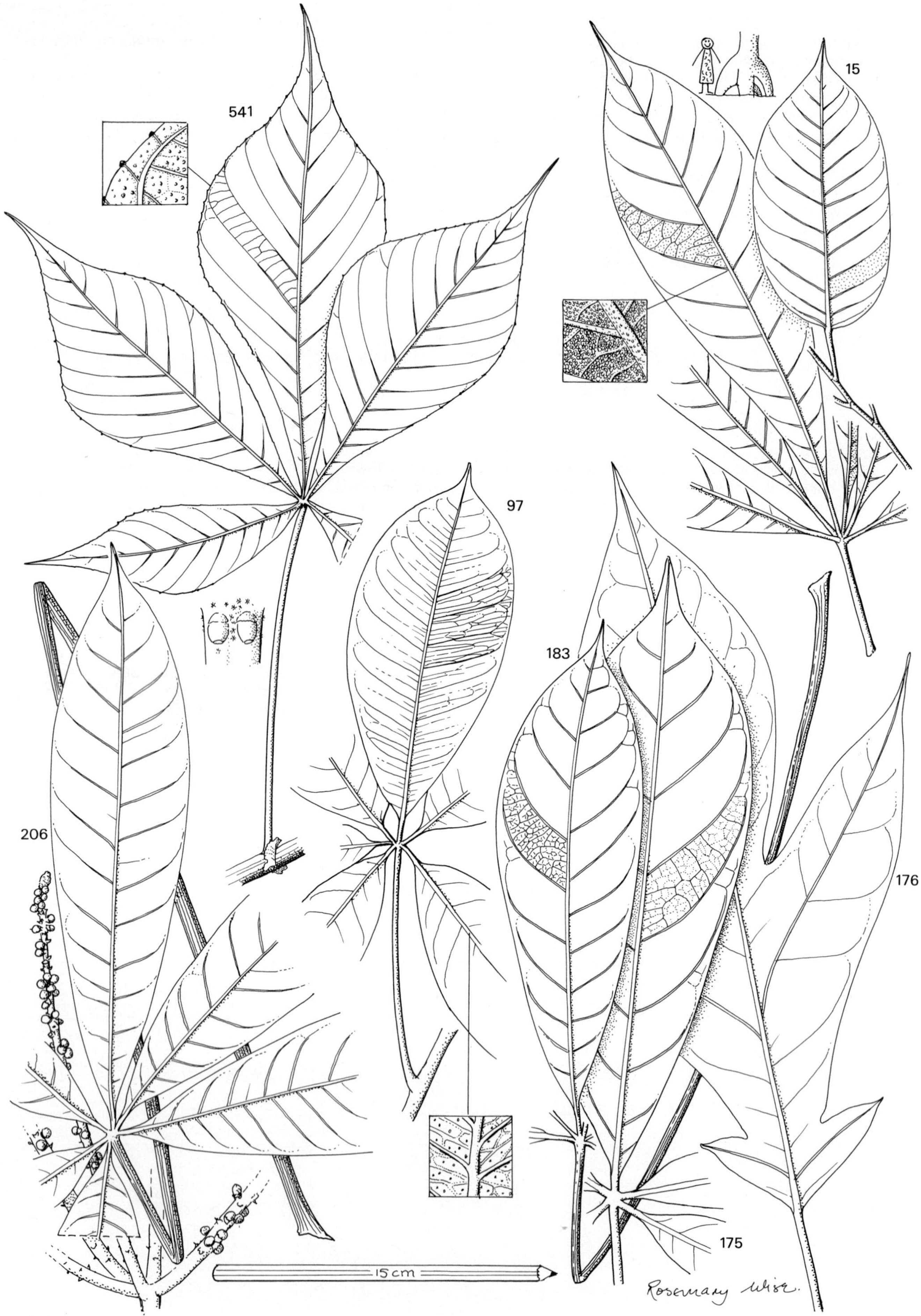
541
15
97
183
206
176
175
15 cm
Rosemary Wise.

Group 28B: (Sterculiaceae, etc.)
(Lvs digitate, alternate)

Several families are reprsented here. The Euphorbiaceae are discussed in Group 22; the Sterculiaceae in Group 27; and the Rutaceae in Group 31. *Cussonia* is in the Araliaceae and has flowers in long, foetid spikes clustered at the end of short twigs with small fleshy fruits.

Margin with small glandular teeth; leaflets almost 4-sided; venation ±scalariform; lf with small glands, or stellate hairs, especially when young; paired glands at base or top of petiole; twigs with large, leafy stipules; tree common in **secondary forest**, with whorled, horizontal boughs; slash orange to pink, stringy, peelable, with streaks of orange grit, and reddish, watery exudate; sometimes savoury-scented (like raw bacon or other lightly smoked meat)	***Ricinodendron heudelotii***[1] (EUPH) [*WAMA*]	541
Margin without gland-teeth; not *WAMA*		
— Lvs with scales or raised or translucent spots below; **evergreen forest** trees with hot and bitter taste **OR** with stilt buttresses		
— — Lvs with golden scales; (0)-7 leaflets; apex of lflets acute to acuminate; golden-brown colouration of scales visible even in crown as a whole; tree often with stilt-buttresses; slash v. fibrous, stringy, with visible sponge-like pores, yellow to pink, not strong tasting; **evergreen forest**	***Heritiera utilis*** (STER) [*NYANKON*]	15
— — Lvs and crown without golden tint: lvs without scales, but with fine translucent spots; usually 5 leaflets with rounded apex; with many close laterals; trees without stilt-buttresses; slash gritty, + bitter and hot taste, yellow turning brown, increasingly fibrous towards inner bark; **evergreen forest**	***Araliopsis soyauxii*** (RUTACEAE) [*MEAWERE*]	97
— Lvs without scales and translucent spots (or trees found outside **evergreen forest**)		
— — –Unarmed trees with long petioles		
— — — Large tree, like Papaya when young; petioles long with **pair of stipules at base**; lvs (+infl. spikes) clustered at ends of thick twigs; crudely-branched tree with thick-fibrous and fissured bark; slash soft and yellowish; leaves rather thin and papery; uncommon, but widespread	***Cussonia bancoensis*** (ARALIACEAE) [*KWAE-BRoFRE*][4,5]	206
— — — Small (-medium) cauliflorous treelets of **evergreen forest**, with fruits borne on stem; lvs clustered at top of stem; petiole often >20 cm long; venation close-reticulate		
— — — — Leaflets not lobed		
— — — — — Leaflets papery with finer veins raised (lens); sometimes without petiolule; petiole/midrib sometimes with dense, brown, v. short, stellate hairs; flowers stalkless (sessile); fruits densely hairy	***Cola umbratilis***[2,3] (STER) [*TANANFRE-BERE*]	183
— — — — — Leaflets rather brittle and glossy with recurved margin; finer veins not prominent; lvs and flowers always with stalks; fruits almost hairless	***Cola chlamydantha*** (STER) [*TANA-NFRe*]	175
— — — — Leaflets lobed: incised almost to midrib	***Cola digitata*** (STER)	176
— — –Prickles on tree or petiole <15 cm long; trees becoming very large; young trees with boughs clearly whorled, but lvs not strongly clustered around top of stem or branches; lvs not papery when mature; seeds dispersed in cottony kapok; sapwood usually with ripple marks	Group 28C (BOMBACACEAE)	

NOTES:
1) *Ricinodendron* has large (5 cm wide) ovate cotyledons with dense white hairs below and with glands at the top of the petiole. Later leaves become 3-lobed and serrated.
2) These *Cola* spp. can not be easily separated when sterile, so both might reasonably be called *TANA-NFRE*. In practice, the last species, which is easily distinguished as shown, is often given this same local name as well.
3) *Cola umbratilis* was previously included in *C. buntingii*, a species from more western parts of W. Africa.
4) *Polyscias fulva* (ARALIACEAE) is a rare tree of dry forest with distinctive, dense stellate hairs below and *Canarium*-like lvs. The dried lvs have a slightly sweet, peculiar smell.
5) *Schefflera barteri* (ARALIACEAE) is a large, strangling climber with large, intrapetiolar stipules and petiolules, many cm long, on palmate leaves. It is restricted to evergreen forest.

Group 28C: Bombacaceae
(Small-medium digitate lvs; emergent, often armed, trees; boughs whorled)

Three large to extremely large, deciduous trees, including the well known *Ceiba* tree, which is very widespread and very common. All have branches arising in whorls on the stem (following Massart's architectural model of Halle' *et al.* (1978)), which is particularly obvious on the younger tree. A similar foliage pattern is created in *Terminalia*, Sapotaceae and Apocynaceae species which have whorled or clustered **simple** leaves on whorled branches (and follow different architectural models). It is interesting that several very common pioneer species look superficially similar in this respect.

The slash of the species is similar to many Sterculiaceae; fibrous, with vertical bands, and darkening. The details are, however, different in each case (see below).

The large, bright red flowers of *B. buonopozense* are conspicuous on leafless trees in the dry season. The **savanna** species *Bombax rubrocostatum* is very similar in this and other respects to *B. buonopozense*. The other two species have white, also notably large flowers. There are many stamens, except in *Ceiba* flowers which only have five.

All species produce capsules with seeds enclosed in dense, cotton wool-like 'kapok', which is red in *Rhodognaphalon brevicuspe*[1] and white in the other species.

The amazingly 'fat' baobab tree, also in Bombacaceae, has been planted in many villages in the forest zone, although it is a savanna species. Younger trees of this species can be deceptively 'slim'.

Trees armed with prickles, lfts much more than twice as long as wide, usually with clearly visible venation; laterals distinct

- Leaves slender, tapering gradually to tip; midrib ±red; stipules not conspicuous; bark smooth except for prickles (which are not densely crowded) on younger parts, metallic grey, or green on saplings; bole on older trees slightly bulging or irregular, and rarely perfectly cylindrical; base sometimes with huge convex buttresses; slash **hard**-fibrous, yellow or pink with pinkish vertical bands, darkening; tree extremely common in **secondary forest**; flwrs white; kapok white — ***Ceiba pentandra***[2] [*ONYINA*] 143
- Leaves broadest close to tip and ending **abruptly** in a short triangular tip; oblanceolate; stipules paired, persistent and conspicuous on young twigs; spines sharply black-tipped, with rings on older trees; densely crowded on young trees, sometimes in vertical rows and falling to leave reddish marks; bole straight and cylindrical ±small convex butts; slash red to pink, with white streaks, **soft**, fibrous, (±gritty) darkening; flwrs red; kapok white — ***Bombax buonopozense*** [*AKONKODIE,AKATA*] 119

Trees unarmed; lvs ±obovate; *c.*2 times as long as broad; nerves not v. distinct and finer venation often obscure; with an abrupt apex and reddish midrib; young petioles hairy; bark smooth or very rough and dark with age; bole cylindrical (±small convex butts.); slash pink, soft fibrous, with many bright white bands, very rapidly darkening; **ripple marks very broad and conspicuous;** flwrs white; kapok orange-ish — ***Rhodognaphalon brevicuspe***[1] [*ONYINAKOBEN*] 118

NOTES: 1) *Rhodognaphalon brevicuspe* was previously called *Bombax brevicuspe.*

2) The sapling and seedling leaflets of *Ceiba* (**143B**) are, unlike those of the other two species, serrated. Those of *B. buonopozenze* have the conspicuous paired stipules; *Rhodognaphalon* seedlings are unarmed and have ring scars at the nodes.

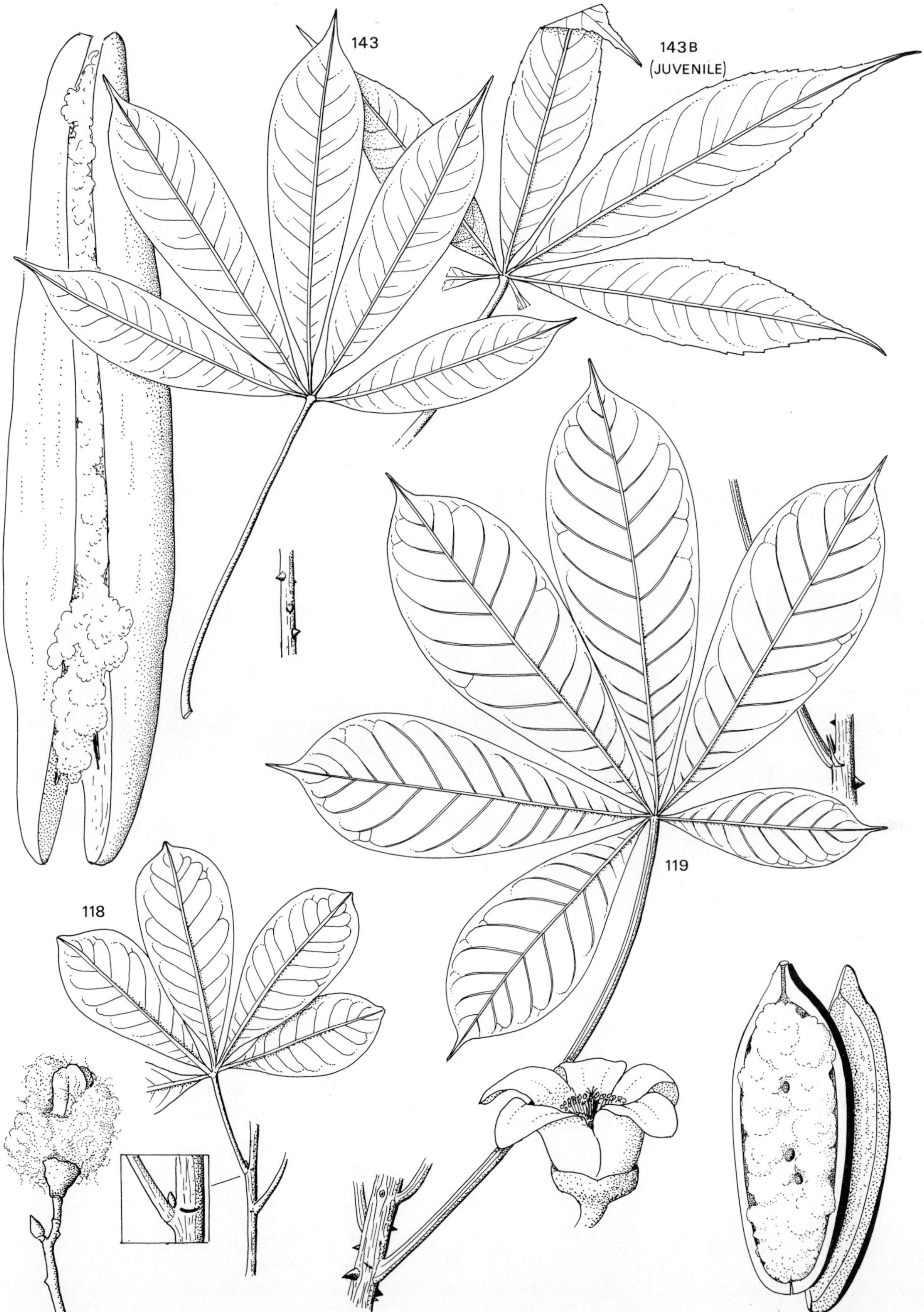
143
143B
(JUVENILE)
119
118
15 cm
Rosemary Wise

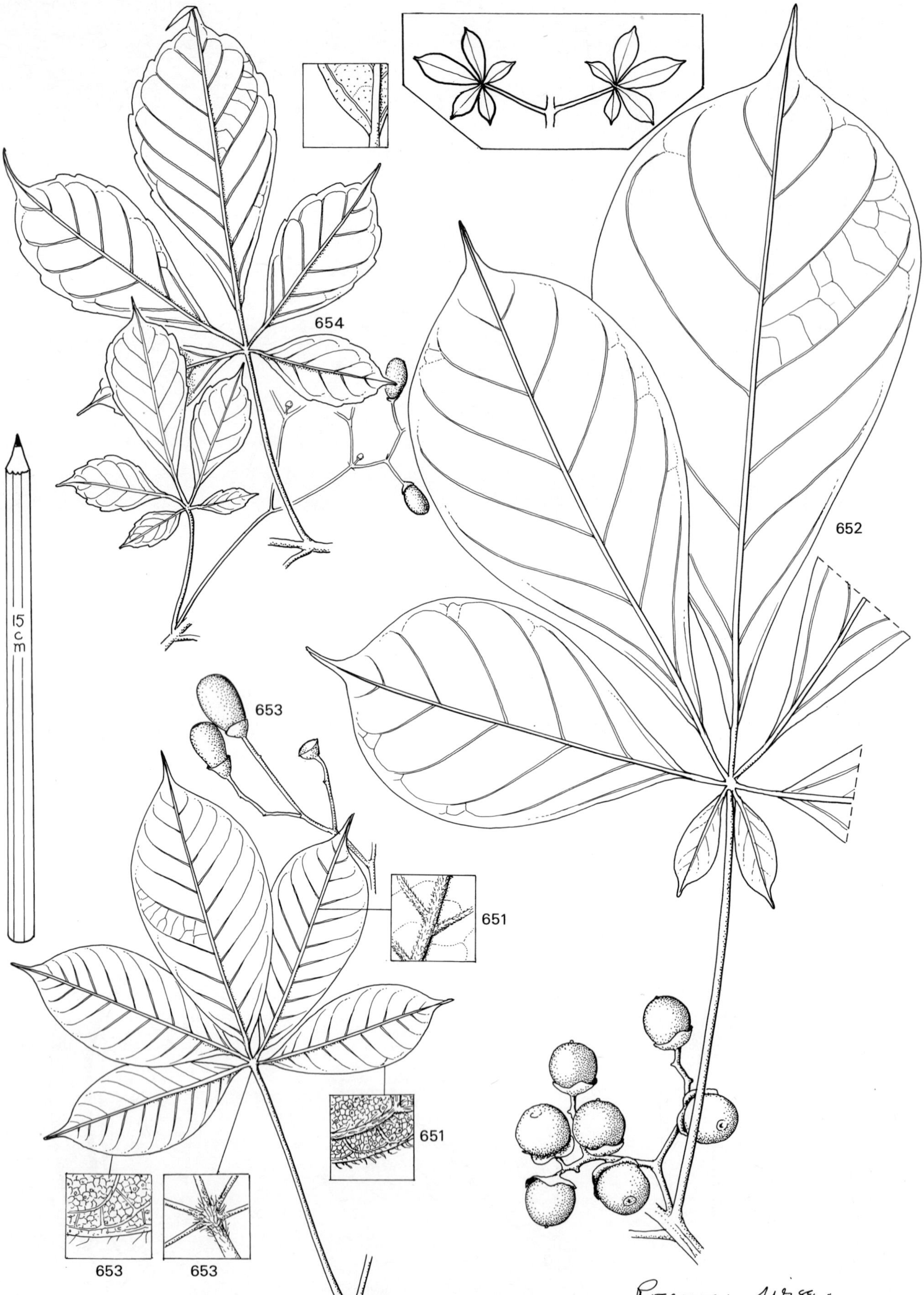
654
652
15 cm
653
651
651
653
653
Rosemary Wise

GROUP 29: Vitex (VERBENACEAE)
(Lvs digitate and opposite)

Vitex spp. are medium-sized trees with a characteristic soft-fibrous bark which, when slashed, darkens from cream, through a dirty green-grey colour, to brown-green then brown. The outer bark is usually vertically fine-fissured, or furrowed, and is often of a colour and texture that suggests from a distance that the tree is dead. Larger trees are, as a rule, slightly fluted.

The flowers are very asymmetric, usually pink to purple, tubular at the base with 4 stamens attached to the base and 5 lobes at the apex, of which one is larger than the rest (similar to, but smaller than Bignoniaceae). They are arranged in much-branched inflorescences. The fruits are fleshy (1-stoned), edible drupes, sometimes with several seeds inside the stone. The calyx remains, like a saucer or egg cup, around the base of the fruits (like Boraginaceae – Gp 26).

Two introduced species of verbenaceous tree are very widely cultivated; *Gmelina arborea* and *Tectona grandis* (teak). These have simple, opposite leaves. *Gmelina* has trinerved, ovate leaves with 2-4 glands at the base and short hairs below; it produces orange flowers with dark streaks and 2cm long drupes, in narrow panicles. *Tectona* has very large leaves *c.*30 cm long which are rough above, and stalkless, and with branched laterals, on square twigs. Its flowers are white, in large panicles, with the dry fruits covered in the inflated, leafy, persistent calyx.

Leaflets with raised amber-like (yellow-orange, translucent) spots and normally-spaced laterals; not large.		
Leaflets small, often <6 cm long (≪10 cm), glabrous; pale-discolorous or shiny below; usually serrated; venation reticulate; ±pit domatia; inflorescences few-(white-)flowered, even smaller than the lvs	***Vitex micrantha*** [*OTWENTOROWA-NINI*]	654
Leaflets, at least the terminal one >6 cm long, with hairs; venation ±scalariform; not normally serrated, except on shade leaves		
Hairs on top of petiole, on nerves, etc. longer than width of midribs and major nerves; spots (x10) golden; inflorescences v. many flowered, almost as long as the leaves; flwrs white	***Vitex rivularis***[3] [*OTWENTOROWA-BERE*]	653
Hairs mostly shorter than width of midrib, densely appressed; glands often hard to see, red-orange; inflorescences compact, much smaller than the leaves; flwrs mostly blue	***Vitex ferruginea***[3]	651
Leaves without spots; large and leathery; lflets often >8 cm broad; petiole often >20cms; small trees; venation lax; apex rounded or abruptly acuminate; infl. compact, of yellow and purple flwrs	***Vitex grandifoliola*** [*SUPOWA*][1,2]	652

NOTES:

1) **Vitex doniana** is very similar to *V. grandifoliola*, but is found only in farms, thickets, extremely disturbed forest, etc. It has longer petiolules (1-2.5 cm long) than its forest understorey relative.
2) **Oldfieldia africana** (Euphorbiaceae) has not been found in Ghana, but may be discovered, particularly in western region, because it occurs in Côte d'Ivoire. It is a large tree, with opposite digitate leaves, and many slightly raised, fine lateral nerves; densely ferruginous petioles (to 8 cm long), but glabrous, glossy leaflets, dark above and pale below. The fruit is a 3-valved capsule.
3) These two species are very similar, and barely distinguishable in the field. It may therefore be more realistic to record the local name of both species as [*OTWENTOROWA*].

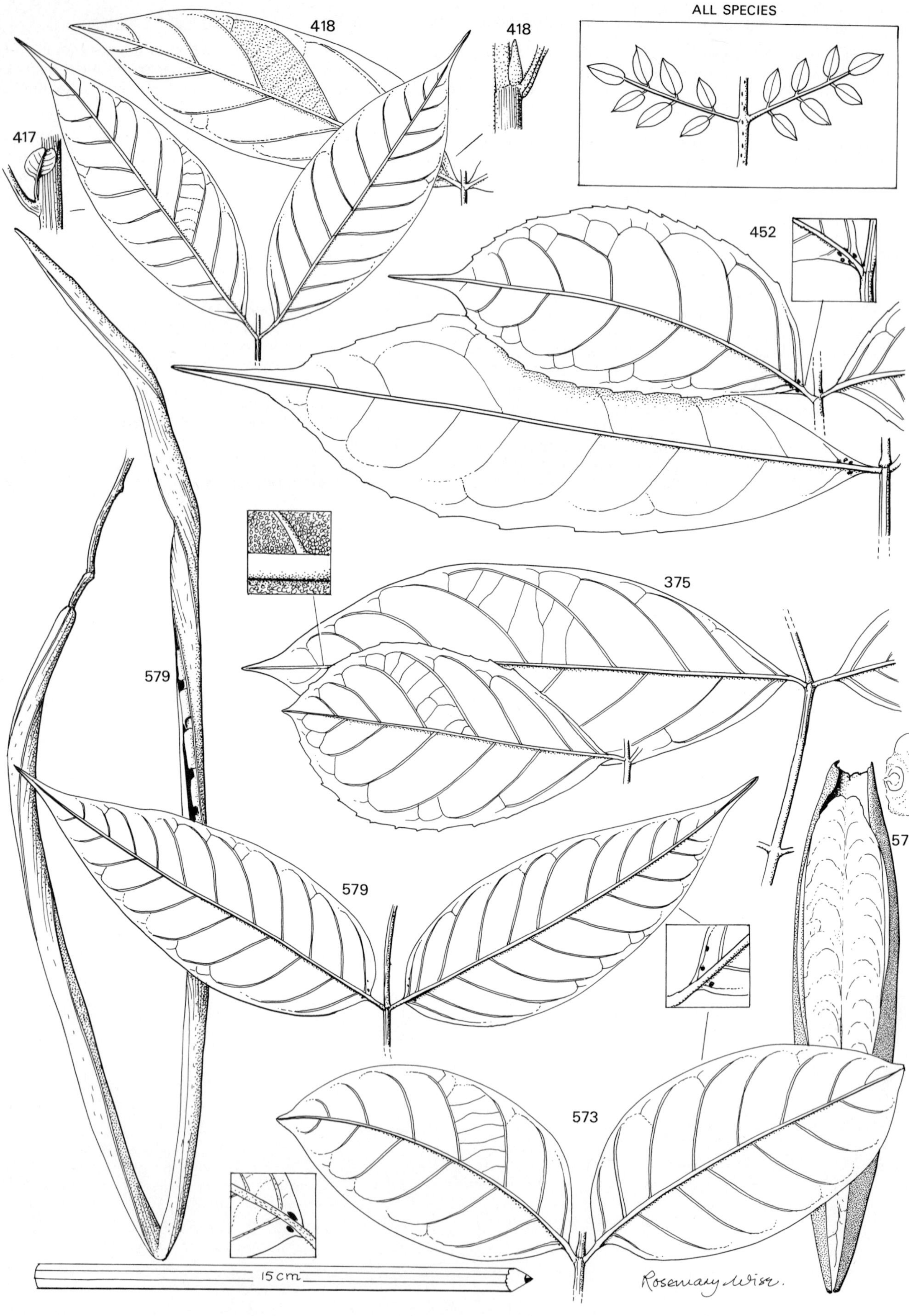
ALL SPECIES
418
418
417
452
375
579
579
573
573
15cm
Rosemary Wise.

GROUP 30: BIGNONIACEAE
(Imparipinnate, opposite leaves)

This small family of sun-loving trees includes the only Ghanaian tree species to have opposite, pinnate leaves. The leaflets are normally paired, with an odd leaflet at the end. In some species the leaflets have basal glands, near the midrib. The rachis is never broadly winged, but usually rather smooth. Often, there is a small discoloured or constricted length of rachis at the base of the leaflets. The small 'calabash tree', *Crescentia cujete*, commonly planted in villages, is in this family (but this species is unusual in having simple, alternate, clustered leaves). *Millingtonia hortensis*, with white fragrant flowers, is fairly commonly planted in Accra. *Stereospermum kunthianum* is a common **savanna** tree. The wood of this family is usually white, light and rots quickly, but the species have many uses in local medicine.

The flowers are tubular and conspicuous (similar in 'design' to Verbenaceae (Gp 29) or Boraginaceae) often in large panicles. There are five lobes at the end of the flower tube (corolla lobes), and four stamens inside it.

Genus	Flower colour	Fruit
Kigelia	Mottled purple to yellow	Massive 30 cm pendulous 'sausages'
Markhamia	Yellow with purple lines in throat	Pendulous, 50 cm, strap-like falcate capsules
Stereospermum	Purple-pink with darker streaks	Pendulous 80 cm cylindrical capsules
Newbouldia	Purple-pink with darker streaks	Pendulous 25 cm capsules
Spathodea	Flaming scarlet with yellow-edges	Upright 20 cm capsules

Spathodea trees (Flame trees) in flower are unmissable, particularly as this happens in the dry season when the tree is leafless (see also *Rhodognaphalon* – Gp 28). The fruits (except *Kigelia*) are pod-like capsules, opening along 1 (*Spathodea*) or 2 lines, and with winged seeds (*Spathodea* winged all-round; other spp. winged at each end). The fruits could be confused from a distance with those of the Apocynaceae (Gp 9) or (*Kigelia*) those of *Allanblackia* (Gp 8).

Leaflets with serrated margin[1]		
Venation not finely reticulate, nor prominent; no basal glands; laterals not joining; lflets ±metallic lustre underneath; **often in wet places**; (margin usually entire, see below); slash fibrous grey-green, darkening through greenish shades	***Kigelia africana*** [*NUFUTEN*]	375
Venation prominent and markedly reticulate; often with 2 or more dark glandular spots at side of base of midrib; laterals strongly joined in sub-marginal nerve; ±pronounced drip tip; (margin rarely entire); minute glands barely visible also on lower lft surface (lens); small-medium tree; bark smooth and pale; branches markedly ascending; slash cream over orange to white, soft, chewy, with orange grit lines	***Newbouldia laevis***[2] [*SESEMASA*]	452
Leaflets with entire margin OR lflets v. hairy		
Laterals strongly arching and joining; lft glabrous and glandular (see above)	***Newbouldia laevis***	
Laterals not obviously joining, or lf hairy		
Lflets hairless, or hairs short, inconspicuous and not in domatia		
Leaflets with two small glands at base, nr petiolule; slash v. bitter, like quinine		
Rachis and rest of leaves completely hairless; venation lax-reticulate; prefers rocky soils; bole ±cylindrical; slash pale pink to white with thin, brown, gritty streaks, soft-fibrous, contoured chunky and crumbly	***Stereospermum acuminatissimum*** [*ESONO-TOKWAKOFUO*]	579
Rachis and midrib above with very short, brown hairs; venation scalariform; bole v.fluted, twisted or with high-reaching buttresses; bark rough with vertical lines of reddish lenticels; twigs also v. lenticellate; slash fibrous creamy, darkening to grey-green, with thin streaks; (flowering when lfless, with bright red flowers)	***Spathodea campanulata*** [*AKUAKUO-NINSUO*]	573
Leaflets without basal glands and hairs (but with minute translucent glands peppered over lower surface, ±visible); underside of lflet with metallic lustre; **wet places** (rarely serrated – see above)	***Kigelia africana***	375
Lflets with obvious hairs, or with hairs in nerve axils; **savanna, dry forest** or **riversides**; bark flaky or rough; slash yellowish, hard-fibrous, peelable; *lvs ±pseudostipules* at base, with minute glandular bumps (lens) over surface		
Hairs dense and soft below lflet; laterals ascending; venation scalariform and v. prominent; pseudostipules small and hairy; twigs with dense yellow hairs; capsules softly hairy	***Markhamia tomentosa*** [*oBOGYANEBOo-NINI*]	418
Hairs only in nerve axils, (±midrib, etc.) but dense on rachis; pseudostipules, when present, leafy, rounded, not v. hairy; venation lax, and not prominent; twigs scaly, soon glabrous; capsules not softly hairy; **often by rivers**	***Markhamia lutea*** [*oBOGYANEBOo-BERE*]	417

NOTES: 1) Sapling/sucker lvs of the entire leaved species, particularly *Markhamia* spp., are sometimes serrated.
2) *Newbouldia* is easily propagated with cuttings, and widely used in Ghana as a living fence, and for medicine. Aubreville (1958) has commented that this is likely to have encouraged wide dispersal of the species into 'natural' forest.

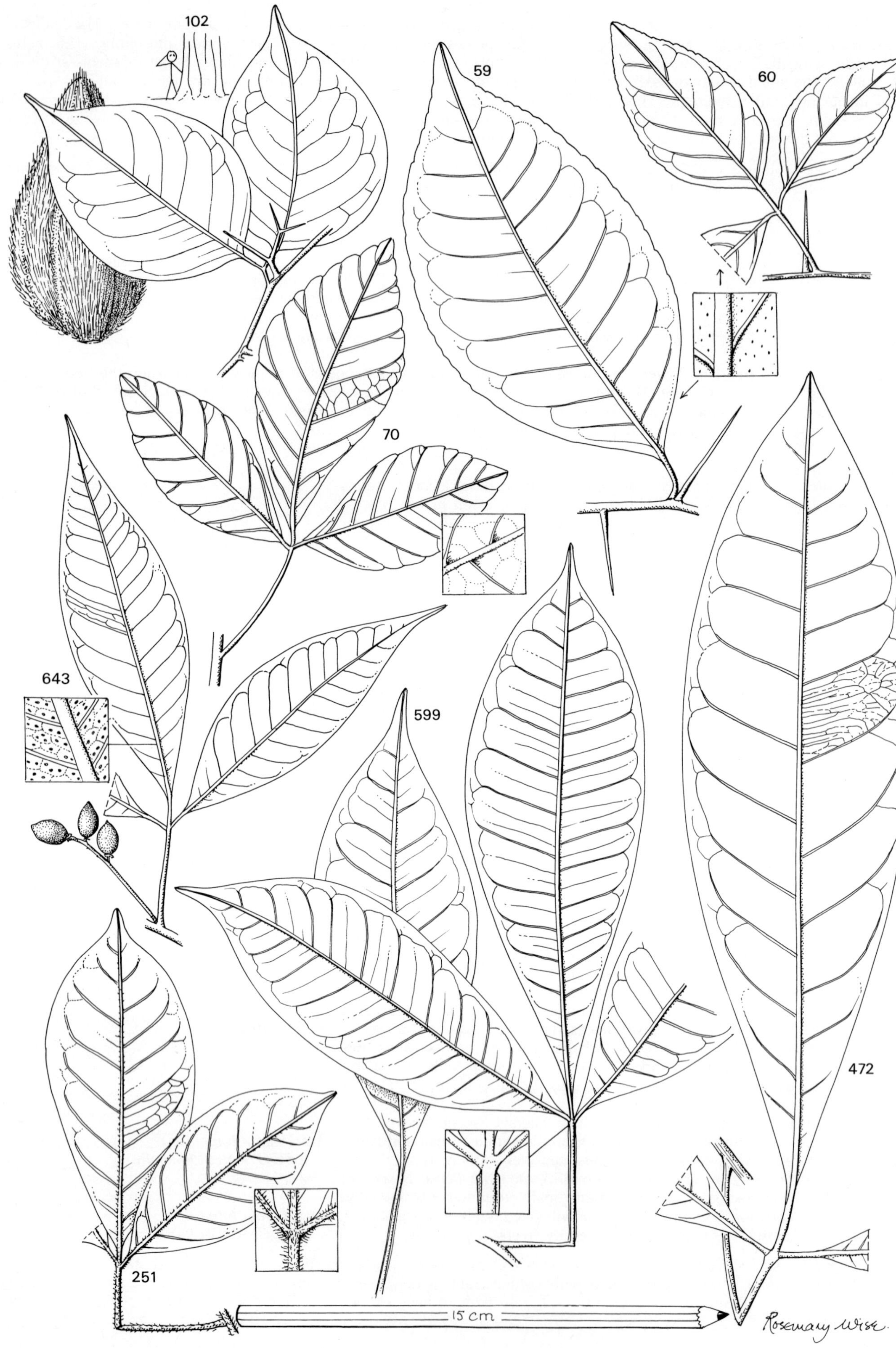
102
59
60
70
643
599
472
251
15 cm
Rosemary Wise

GROUP 31: RUTACEAE, etc.
(Trifoliate lvs; or pinnate leaves and armed or broadly winged)

Key to subgroups (and miscellaneous species)

Leaves with 2 opposite ovate to broadly elliptic leaflets; tree with spines often branched; **bole irregular, usually very fluted**; slash orange-cream, hard, gritty, with copious, scented, gummy exudate	***Balanites wilsoniana*** (BALAN) [*KROBODUA*]	102
Leaves with more than 2 leaflets, at least on most leaves		
Leaves trifoliate (with 3 leaflets at end of petiole)[2]		
Leaves ±serrated, without spines, prickles or gland dots; often with tuft domatia (in nerve axils); lamina papery; small tree of **disturbed forest**; (v. variable)	***Allophylus africanus*** (SAPINDACEAE)	70
Not *Allophylus*		
Leaves without large glands at base of petiolules, and without conical prickles, although sometimes with spines		
–Plants with spines, or lflets hairy or gland-dotted	Group 31A (*Rutaceae* (part))	
–Plants unarmed, **without gland spots and hairs**, with v. asymmetric side lflets		
Savanna (**dry forest**) tree; lflets <15 cm long, shorter than petiole; veins ±visible; slash cream, thick, gritty	***Crateva adansonii*** (CAPPA)	
Understorey little-branched shrub; lflets often >15 cm; twigs hollow; venation usually obscure	***Euadenia trifoliolata*** (See Gp 13) (CAPPA)	
Leaves with large glands[2]; tree with broad-based prickles; slash smelling of green beans	Group 31C (*Erythrina*)	
Leaves pinnate, often with leaves tufted at branch ends		
Rachis broadly winged	Group 31B (*Bersama etc.*)	
Rachis not broadly winged		
Tree unarmed; lvs thin, v.fragrant and v.hairy when young	***Clausena anisata*** (RUTA)	
Tree armed with prickles	Group 31D (*Zanthoxylum*)[1]	

NOTES: 1) Prickles are produced by the bipinnate-leaved trees *Cylicodiscus* and *Cathormion* (Gp 38).
2) *Hevea brasiliensis*, the introduced rubber tree, has trifoliate leaves with a gland at the base and, of course, copious **latex**.

Group 31A: Rutaceae

The Rutaceae include *Clausena* (above), *Zanthoxylum* (Gp 31D), *Araliopsis* (Gp 28B) and the cultivated citrus trees (orange, lime, etc.), as well as the current group. The most characteristic feature of these trees is the abundance of oil-producing glands, visible, for instance, as many translucent spots in the lvs which makes the lvs fragrant when crushed. Flowers, fruits and bark are normally also scented. *Afraegle* and *Aeglopis* (keyed from Gp 15) are like the *Citrus* spp. and produce spines, simple lvs and even fruits like that genus (but with a woody 'skin'). *Citropsis* (31B) is spiny, but has compound, v. winged lvs and smaller fruits. The other species, including the digitate-leaved *Araliopsis*, produce fruits with one or more seeds in a fleshy coat (drupes or berries).

Small *Citrus*-like tree with spines and some simple leaves (from Gp 15)		
–Lf always simple, serrated; laterals ±joining; veins ±obscure; glands as black spots on lower leaf surface; treelet often in forest undestorey	***Aeglopsis chevalieri*** [*KWAE-AKENKA*]	59
–Trees usually with trifoliate lvs as well; medium tree with translucent glands; petiole tips swollen or articulated; laterals irregular, forking at varied heights and often not joining; tree of villages, rarely recorded in forest; fts like hard oranges	***Afraegle paniculata*** [*oBUOBI*]	60
Trees without spines; majority of leaves trifoliate, except on saplings		
Leaves with conspicuous orange hairs along midrib above, lower veins etc.; twigs densely hairy as well; medium-sized tree	***Diphasia angolensis*** [*AMUDURO*]	251
Leaves without conspicuous hairs		
Dry forest trees with rather finely transverse venation (>1 lateral/cm) on acuminate lflets		
–Petiole often slightly winged; lvs sometimes opposite; lflets <15 cm long, narrowly elliptic to oblanceolate, acuminate; medium tree in **dry forest**	***Teclea verdoorniana***	599
–Petiole not winged; lflets 1-3, lanceolate and long acuminate with very conspicuous black gland spots below; in Shai Hills, (& ?similar) **(southern) dry forest**	***Vepris heterophylla***	643
Evergreen forest tree with normal density of laterals (just <1/cm); lvs broadly oblanceolate, and barely acuminate; petiole not winged; lflets sometimes 2, often >15 cm long; slash fibrous, scented, aromatic with gritty streaks	***Oricia suaveolens***	472

Group 31B
(Lvs imparipinnate with broad wings)

Species of small tree from related families with similar, unusual leaves. *B. abyssinica* is the only species in its family in Ghana.

Plant unarmed; lflet **sometimes serrated**; slash yellow and crumbly with watery exudate; lvs extraordinarily variable in size; small tree in wetter areas, particularly the edge of **evergreen forest**; fts globose 4-valved velvety capsules 2 cm across; red seeds + yellow arils	***Bersama abyssinica*** (MELIANTHACEAE) [*ESONODUA*]	113
Plant with spines or prickles e.g. paired at base of leaf; lflet always serrated		
Lvs gland-dotted; treelets		
Lflets ±acuminate; petiole wing very narrow (c.1mm); uncommon tree ? restricted to **evergreen** forest	***Citropsis gabonensis***	165
Lflets not v. acuminate, spines and wing >5 mm long/wide; in **dry forests**	***Citropsis articulata*** (RUTACEAE-above) KWAA-ANKAA	164
Lvs not gland-dotted; scrambling treelet of **dry forest**, etc.; ft 1-seeded	***Harrisonia abyssinica*** (SIMAROUBACEAE) (See Gp 32)	

Group 31C: *Erythrina* spp. (Papilionaceae)
(Spiny trees; trifoliate leaves with large gland at base of leaflets)

Trees in this and the next Group are normally armed with (conical) prickles, modest (but vicious) on the petioles and rachis, but becoming small woody pyramids on older bark. In *Erythrina* the pyramids seem more inclined to join up on the older bark into crocodile-tail-like ridges. In any case, the **slash** of *Erythrina* is easily distinguished by its smell of green beans and the conspicuous inner-bark with v.large pores, like cloth with an open weave.

The slash scent is not coincidental: this genus is in the bean family (PAPILIONACEAE – see Gp 37), and produce bright red or pink flowers shaped vaguely like those of peas or beans, in upright inforescences (racemes). The pods, however, are more like strings of beads, strongly constricted between the seeds (**See Gp 37**).

The 3 species found in forests are v. similar, but can usually be separated as follows:

Leaflets almost triangular, (terminal lflet) often broader than long, with an almost straight base and v. scattered hairs below; small tree with red flowers; pods leathery	***Erythrina addisoniae*** [*oSOROWA-NINI*]	283
Leaflets rounded to ovate; terminal lflet longer than broad; pod velvety, twisted		
Rachis grooved, sometimes with prickles; prickles remaining rather separate on older bark; small tree (<15 m) with red flowers; seeds <1 cm long when mature	***Eythrina vogelii*** [*oSORE*]	285
Rachis slightly or not wrinkled; without prickles; prickles forming into lines like crocodile tails on older bark; large tree with pink to mauve flowers; mature seeds c.2 cm long	***Erythrina mildbraedii*** [*oSOROWA*]	284

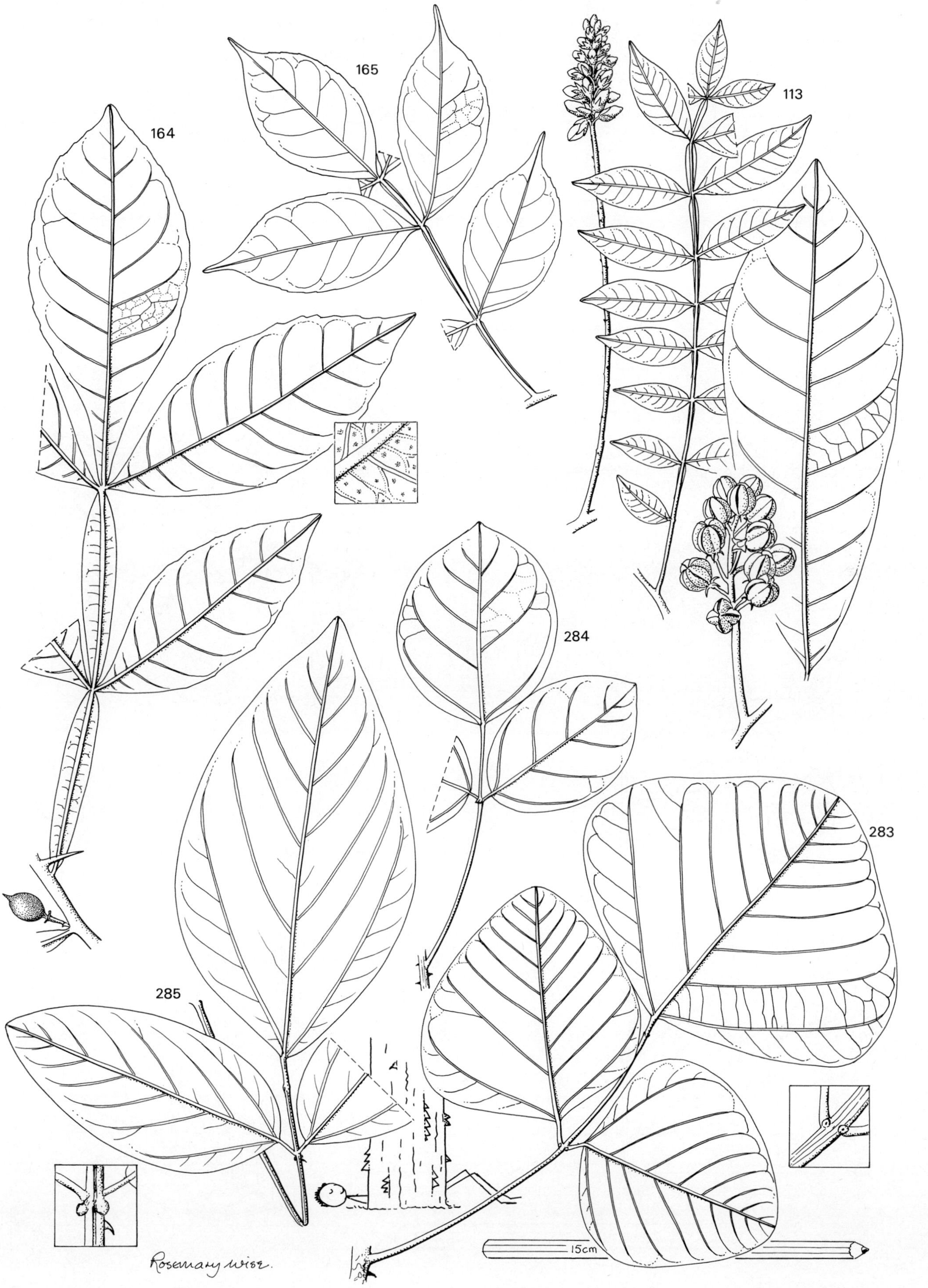
165
164
113
284
283
285
15cm
Rosemary Wise.

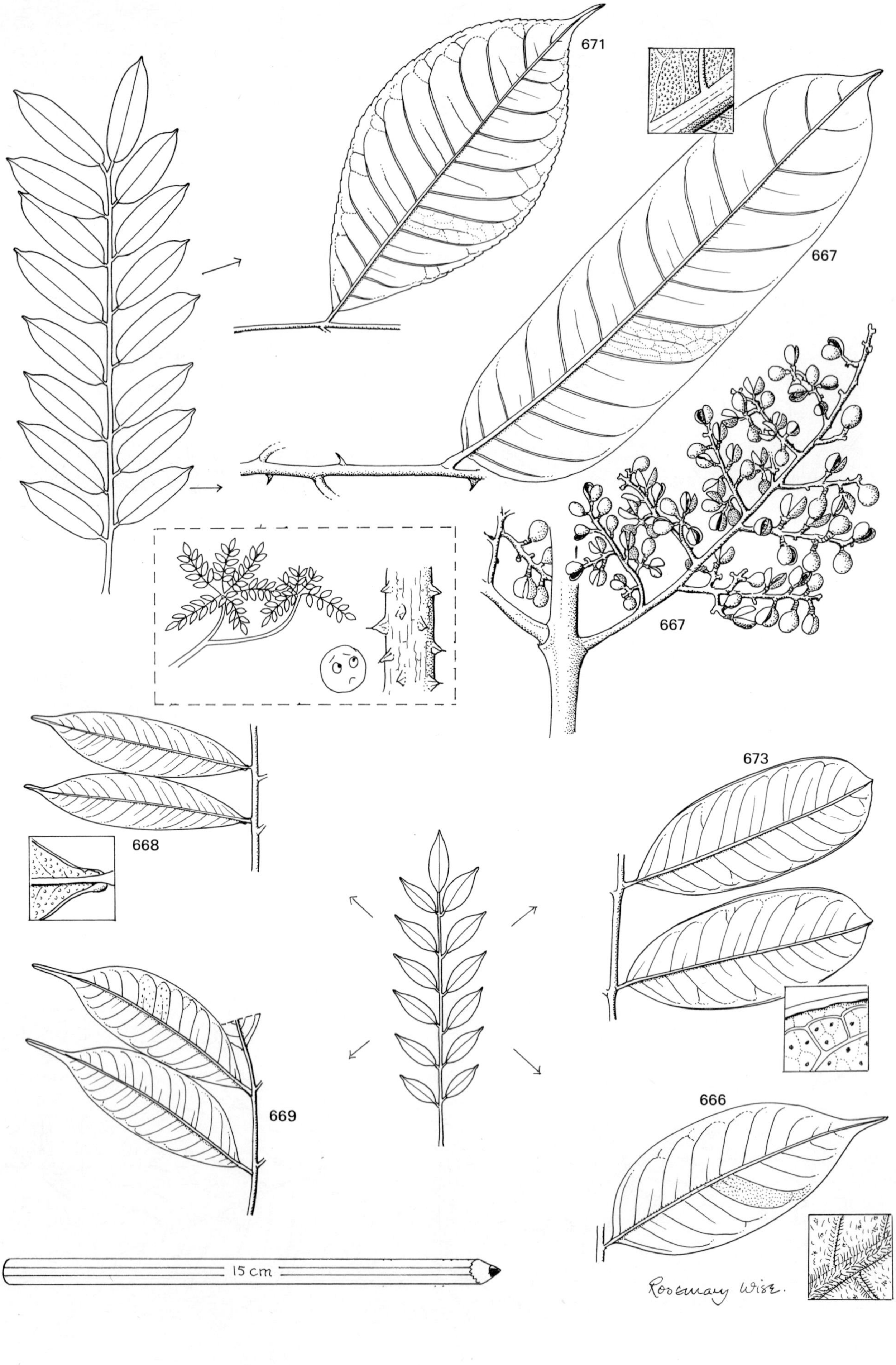
671
667
667
668
673
669
666
15 cm
Rosemary Wise.

Group 31D: *Zanthoxylum* spp. (Rutaceae)
(Pinnate, thorny trees)

Zanthoxylum spp. are easily recognized from their bark and slash alone. The bark is covered in broad-based woody prickles, often surmounted by a sharp prickle, similar to that of *Erythrina* spp. (Gp 31C). The slash, though, is yellow to orange-brown, very gritty, fibrous and/or brittle and has an unmistakeable, very fruity-acid taste which rapidly becomes hot, and later leaves a cooling, refreshed, or at least lip-tingling, sometimes pepperminty aftertaste (In *Z. lemairei*, particularly, the hot component is not pleasant). The yellow colour darkens with time.

Z. gillettii is perhaps the commonest species, and has a crown (even saplings) like *Entandrophragma* spp. The smaller-leaved species *Z. lemairei* looks like the Neem tree. The toothed or gland-dotted leaves and prickles of *Zanthoxylum*, though, should easily prevent confusion with Meliaceae.

The flowers are small, with whitish petals, clustered into panicles; the fruits are capsules *c.*1 cm or less in diameter, which split along a rib to expose a single black or bluish seed.

Spondias mombin (35B), which also has crown of tufted, pinnate leaves, is sometimes spiny.

Leaves v. large, 20 cm to 1 m long; lflets usually >10 cm long and >3 cm wide, often with prickles on midrib below; medium-sized straight tree with large *Entandrophragma*-like crown; common in **secondary forest**; (slash fibrous, gritty, hot – see above)	***Zanthoxylum gilletii*** [*OKUO*]	667
Leaves smaller; or without prickles on midribs, or leaflets (almost) opposite		
Leaflets not obovate-elliptic; with (just-visible) teeth, OR with drip tip OR papery and hairy OR in moist forest		
Leaflets >3 cm broad or >12 cm long, especially towards lf tip, papery, with rachis often >30 cm long; ±symmetric at base, but often falcate; lflets often almost opposite; lvs without prickles; small tree, not particularly common	***Zanthoxylum rubescens***	671
Leaflets smaller, or v. asymmetric		
Leaflets glossy and glabrous when mature; medium-sized trees		
Laterals many and parallel; lflet base v.asymmetric **with a tiny pouch on one side of top of petiolule**; prickles along rachis; lflet without drip tip	***Zanthoxylum lemairei*** [*OKUO-NINI*]	668
Laterals not unusually dense; lflets v.thin, with gland spots clearly visible; **with long drip tip**; ±prickles on lvs only nr base	***Zanthoxylum leprieurii***[1] [*OYAA*]	669
Leaflets thin-papery, hairy; hairs persistent and dense below, and on nerves, etc. above; small tree; twigs with 1cm prickles	***Zanthoxylum chevalieri*** [*OYAA-BERE*]	666
Leaflets obovate-elliptic, brittle and glossy, with laterals often forked; margin recurved; apex sometimes notched; lf glabrous; **savanna or v. disturbed forest**	***Zanthoxylum zanthoxyloides*** [*KANTO*]	673

NOTE: 1) *Z. viride* is typically a straggling shrub, with leaves which will key to *Z. leprieurii*.

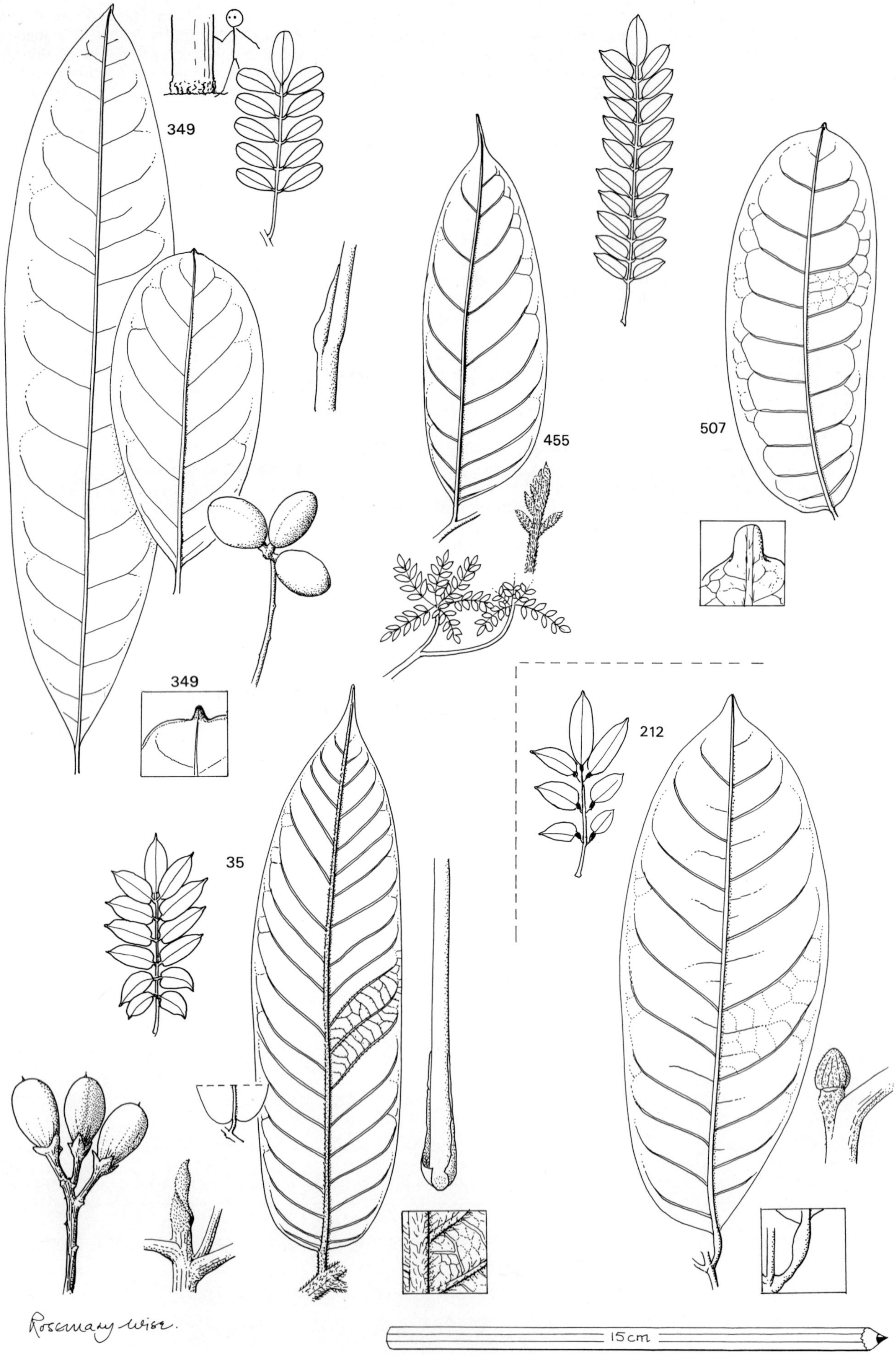
349
455
507
349
212
35
Rosemary Wise.
15cm

GROUP 32: SIMAROUBACEAE
(Imparipinnate; lflets often emarginate or apiculate; slash pale, fibrous)

A small family including *Hannoa klaineana*, a common pioneer tree in abandoned farms, with an *Entandrophragma*-like crown. The slash is thick-fibrous and bitter, but normally v. pale. The family also includes the spiny *Harrisonia* (Gp 31). The leaflets are normally opposite with an odd terminal one. The flowers are small, regular, in panicles, often with hairs around the base of the stamens. The fruits (drupe-like) are fleshy, with a central stone; in *Harrisonia* they are small and lobed; in *Pierreodendron* they resemble small mangoes; in *Hannoa* they are ovoid, *c.*2.5 cm long; in *Nothospondias*, a dioecious species, they are yellowish, slightly asymmetric, *c.*2 cm long.

Lflets and young twigs glabrous; veins between lateral nerves not v. clear, ±impressed below, but ±obscure (with metallic colour below on juvenile leaves); apex of lflet emarginate or thickened or both; petiole markedly winged nr base; bud at twig apex ±glossy, pointed; lflets 3-4(-7) pairs, +1 terminal; medium-large, sun-loving tree with *Entandrophragma*-like crown; bole cylindrical, often with bark flaking and appearing rotten at base; fruits ellipsoid drupes 2.5 cm long; slash pale yellow-brown, fibrous and ±gritty — ***Hannoa klaineana***[1] [*FOTIE*] 349

Lflets or twigs with hairs, especially when young and on midrib and nerves; rare small-medium trees with thick (±hollow) twigs and long compound leaves very clustered at twig ends (more like *Carapa* than *Entandrophragma*);

- Leaflet base not extraordinarily asymmetric; venation lax-reticulate, strongly **impressed below**, like cracks on leather or parchment; lflets with a v. conspicuous thickened process at end of midrib; rachis sometimes with undeveloped leaflet at tip; small tree with rough bark in ±square scales, slash like *Hannoa* (above) — ***Pierreodendron kerstingii***[2] [*FOTIE-NUA*] 507
- Leaflet base extremely unequal, ±obtuse-cordate on one side and cuneate on the other; veins clearly visible below; lflets ±lanceolate and acuminate (or drip-tipped); in 10-20 pairs; lvs v. large, v. clustered; unfolding lflets with dense brown hairs; older twigs 2 cm or more wide, v. hollow; petiolule (of lflets) deeply channelled; bark smooth; slash brown (rarely + reddish exudate) — ***Nothospondias staudtii*** 455

NOTES: 1) In saplings, the leaflets of *Hannoa* tend to be mucronate to drip-tipped, or even merely acuminate. At this stage the tree can be recognized from the obscure-veined, discolorous lower surface, and the smooth, cylindrical, yellowish, brittle rachis, sometimes constricted where the uppermost leaflets join, and with the lflet base cuneate-asymmetric. This could be confused with *Guarea thompsonii* (with latex and grooved, less brittle petiole) or *Trichilia prieuriana* (with fewer lflets (< 5prs) and not v. pale-discolorous lower surface).

2) *Brucea guineensis* is an uncommon, unbranched treelet of areas of more open canopy. The four most basal leaflets sometimes arise at the same part of the rachis. It has v. thick twigs 1-2 cm wide, and the brittle, sometimes hollow rachis is deeply constricted at the base of the leaflets, and at the base of the petiole. The lflet base is obtuse-cordate with petiolules *c.*1 cm long. The lfets are often pustulate, with spots particularly at the base of hairs.

GROUP 33: BURSERACEAE
(Imparipinnate; leaflets cordate–obtuse or petiolules jointed; slash aromatic)

The following two species have panicles of small flowers, **with their parts in 3s**. Both have sweetly-scented slashes (a sweet-incense rather than a cedar smell). *Dacryodes* is unusual in having stellate hairs (cf. *Lannea* in Gp 35). Both species produce edible fruits (berries) which are 4x2 cm ellipsoid and 1-seeded in *Canarium*, and *c.*1 cm and turpentine- or mango-scented in *Dacryodes*. **Both species have twigs and rachises finely striate, especially when dry.**

Petiolule not obviously swollen at tip; base of leaflet obtuse to cordate; leaflets narrow (< 4 cm), lanceolate or oblong and acuminate, v. hairy; finer venation prominent and visible; apical bud and twigs thick; bud ±ensheathed by leafy covering; rachis normally > 2 mm wide at middle; bole ±cylindrical; slightly fissured → roughly scaly on old trees; foliage strongly clustered at branch tips in crown; saplings also *Entandrophragma*-like; slash strongly fragrant, hard fibrous, yellowish to slightly pinkish, with resinous gritty spots; tall straight tree of **secondary forest** — ***Canarium schweinfurthii***[1] [*BEDIWONUA*] 35

Petiolules swollen at tip; base of leaflet cuneate to obtuse[2]; lf only hairy (with stellate hairs) when young; younger lvs with rusty stellate hairs, (spots under lens); newer leaves reddish in the crown; rachis ±slender (2 mm wide at middle); nerves above with a darker line on each side; bole usually slightly irregular, with smooth greyish bark with small scales; slash pink-brown, fibro-granular with slight gummy exudate and sweet scent; medium, irregular tree of shady (when young), **usually damp places** — ***Dacryodes klaineana***[1,3] [*ADWEA*] 212

NOTES: 1) The lvs of *Polyscias fulva* (Gp 28B (notes)) are similar to those of *Canarium*, and bear stellate hairs. The seedlings of these 'Gp 33' trees have very distinctive 'many-fingered' cotyledons *c.*2 cm long in *Dacryodes* and *c.*5 cm long in *Canarium*. The leaves of older seedlings/saplings of *Canarium* (always unshaded), although not always cordate, are pale and coarsely hairy, and the bark or outer stem is distinctively scented from an early age.

2) *Santiria trimera* (also Burseraceae) has not been recorded in Ghana, but occurs in Côte d'Ivoire and may be found in **evergreen forests**. It has stilt roots, compound lvs and floral parts in 3's.

3) The introduced plantation tree, *Aucoumea klaineana* from Gabon, has petiolules several cm long, but is otherwise similar to *Dacryodes*.

7
6
8
12
15cm
Rosemary Wise.

GROUP 34: MELIACEAE (MAHOGANY FAMILY)

This is the 'mahogany' family, which includes many of Ghana's most valuable timber trees, but also species quite different from the *Khayas* and *Entandrophragmas*, some of which are common understorey trees. Although the leaves of most species are rather clustered towards the ends of twigs, in *Entandrophragma* this is particularly marked, and in *Trichilia* and *Guarea* species this 'clustered compound leaf' feature of the crown is often not noticeable. All the larger tree species grow according to Rauh's architectural model (Halle' *et al.*, 1978). The crowns of the common plantation species *Cedrela* and *Azadirachta* (neem) are of a similar type. The slash of the family is very varied, though often reddish, bitter and/or sweetly-scented. The family also includes (simple-leaved) *Turraea*, which in Ghana includes only shrubs.

The flowers are small, in branched inflorescences, regular and appearing hermaphrodite (but often functionally dioecious), with a tube ('corona') supporting the anthers which in some *Trichilia* spp. are divided into deep lobes. Two main types of fruit are produced; woody capsules with winged seeds by Groups 33A and 33B (as well as the savanna species *Pseudocedrela kotschyi*) and more fleshy or leathery capsules, without winged seeds, by the rest. *Ekebergia* is exceptional in producing fleshy drupes (which do not split open at maturity). All capsules have a central column. The capsules are globose in *Khaya* and elongated in *Entandrophragma* and *Lovoa*. There is a similar trend to that noted in Apocynaceae (Gp 9), whereby canopy species are wind-dispersed, whilst the lower-storey species are probably animal-dispersed, except that *Guarea* spp. and some species of *Trichilia*, with seeds probably dispersed by birds, especially hornbills, eventually attain the canopy.

Key to subgroups

Leaves paripinnate; fruits dry, woody capsules with winged seeds
  Leaves totally glabrous, lflets usually in 2-7 pairs; slash distinctively scented (almost like rose-water) and capsules spherical or shortly cylindrical; lflets usually drip-tipped or many-nerved — 34A
  Lflets with some hairs, mostly >7 pairs; capsules cigar-shaped, >5 cm long; (long) lvs clustered at twig tips — 34B
  (Lflets glabrous, with apical process; lvs often >0.5 m long; seeds not winged) — 34C
Leaves imparipinnate; fruits (fleshy or leathery) without winged seeds
  Tree without latex etc.; lflets often falcate — 34C
  Tree with latex; lflets paired except for terminal odd one — 34D

Group 34A: *Khaya-Lovoa* (Meliaceae)

The *Khaya* scent in the slash of the following species is distinctive – between rosewater and the 'cedar' of other Meliaceae. The slash tastes bitter, and is at least partly red. *Khaya* seeds are winged round the edge whereas *Entandrophragma* and *Lovoa* seeds have an elongated wing at the base. The species are listed below in order of preference for wetness of forest type.

Lateral nerves normally <15 pairs; lvs markedly acuminate; fruits spherical; petiole and rachis smooth and rounded above abrupt basal swelling
  Leaflets in 2-4 (-5) pairs (saplings often with 4-5) and broadly elliptic or ovate; petiolule often >6 mm long; laterals not obviously impressed above[1]
    Lflets (3-5 prs) sometimes with 12-15 pairs of well-defined lateral nerves; larger inter-lateral veins slightly prominent below, like threads; bole normally twisted or leaning nr top; bark rough, scaly and pitted; **found only in drier forests** or v. hilly areas; slash red with white streaks, scented + viscous exudate; fts with 5 thick (*c.*1 cm) valves — ***Khaya grandifoliola*** [*KRUBA*][1] 7
    Lflets (2-4 prs) with 6-8 pairs of lateral nerves; veins poorly-defined; lower surface of dry lflets with appearance of leather, with ±impressed larger veins; bole ±straight; **bark normally smooth and pale; slash red-orange,** soft fibrous, gummy; fts with 4-5 thin (½ cm) valves — ***Khaya anthotheca*** [*KRUMBEN*][1] 6
  Leaflets in 4-7 pairs (saplings with 5-7) and oblong or oblong-elliptic + pronounced drip tips; with 5-9 pairs of lateral nerves, slightly impressed above; petiolule <6 mm long; bole long and cylindrical; buttresses becoming large, esp. on one side of tree; **older bark thick, dark, with deep pits where scales have fallen; slash deep red over** paler red, scented, fibrous and extremely bitter; tree ±absent from driest forests; fts + 5 thin (½ cm) valves — ***Khaya ivorensis*** [*DUBINI*][1] 8
Lateral nerves normally >15 pairs; lflets (4-7 prs) not strongly acuminate; visible venation tending to be parallel and transverse; rachis slightly winged, and stout (strongly winged in seedlings); petiole winged; bole straight, with no or only small buttresses; **bark dark**, smooth + many lenticels, becoming rough with age; slash red with white streaks very sweetly scented; in **moist** to **evergreen forest**; capsule *c.*5 cm long — ***Lovoa trichilioides***[2] [*DUBINIBIRI*] 12

NOTES: 1) *Khaya* seedlings have long drip tips on simple leaves. In *K. anthotheca* the juvenile leaflets are broadly ovate, but in *K. ivorensis* they are more narrowly oblong to oblanceolate, with exceedingly slender drip tips, *c.*2 cm long and <1 mm wide. Aubréville (1959) comments that *K.i* trees often occur in strips bearing NE-SW, due to predominant wind-dispersal.

2) The seedlings of *Lovoa* are readily identified from the alternate (entire) lflets, silvery below, with only *c.*8 laterals, and a markedly winged rachis. Often many are found in the vicinity of parent trees. Trade name =African Walnut.

Group 34B: *Entandrophragma*

Entandrophragma trees[1-5] can be recognized when mature by their straight boles and long compound leaves clustered at twig ends. See notes on saplings below. *Zanthoxylum* (Gp 31D), *Hannoa* (Gp 32), *Canarium* (Gp 33), *Antrocaryon* (Gp 35B) and some legumes show some similarities, however, particularly as regards long pinnate lvs clustered at twig ends (see '*Utile*-crown key' in the '200 Main Species' section).

Leaflets with >16 pairs of very prominent (below) and close laterals (*c.*2 per cm on lvs <5 cm wide); lvs often >5 cm wide; midrib finely and deeply impressed above; laterals also impressed; (some lvs with tuft domatia); yng twigs, etc. with dk brown, velvety hairs; crown dark, wide-spreading; bark → rough scales and small pits where scales fall; **slash *not* fibrous, pink with orange grit, bitter but not scented**; capsule opening at tip, *c.*15 cm long; peeling backwards from tip, with a stalked column — ***Entandrophragma candollei***[2] PENKWA-AKOA 16

Leaflets without so many densely-arranged laterals (<2/cm on lvs *c.*5 cm wide); slash fibrous, sometimes scented; crown NOT v. broadly spreading

- Leaflets never with apex folded up; often with tuft domatia in axils of nerves, but without a fringe of long hairs besides midrib; rachis ±hairy; bark rough or with regular vertical fissures; slash *not* soft, corned-beef-like nor chunky
  - Many lateral nerves forking *c.*⅔ of way to margin, and often with **tuft domatia in these forks**; laterals usually <12; veins finely prominent above; young twigs v. grooved or many-sided; bark becoming v. rough and flaky; **slash fibrous + part granular, red → rusty brown, with ripple marks in sapwood and a strong scent**; capsules only *c.*10 cm long, dark, smooth, opening at apex (and base) but with thin valves ±peeling backwards slightly at the apex — ***Entandrophragma cylindricum***[3] (SAPELE) PENKWA 4
  - Lateral nerves forking, if at all, c.¾ way to margin, and lflts with tuft domatia only in midrib-nerve axils; often >12 pairs laterals; veins below sometimes obscure; bark with pronounced, **regular vertical fissures; slash fibrous, not strongly scented, pink ±white lines with browner outer layer; with poorly-defined or no ripple marks**; capsule *c.*15 cm long, brown with pale, ±corky spots where broadest, and opening at apex, thick valves NOT peeling — ***Entandrophragma utile***[4] EFOoBRODEDWO 5
- Leaflets often with folded-up apex; without tuft domatia, but often with a fringe of long hairs alongside midrib; veins not finely prominent; rachis glabrous; bark not v. rough, although ±flakes or knobs; slash **very 'chunky', soft, red + white streaks; no ripple marks; weak scent**; ±large buttresses; capsule smooth, 15 cm + long, opening at base (stalk end) first, with all the valves falling off together — ***Entandrophragma angolense***[5] EDINAM 3

NOTES:
1) Saplings of *Entandrophragma* are distinctive, with long lvs clustered on top of unbranched stems. They are often seen in this stage in the semi-shade of forest understorey.
2) In ***E. candollei*** the characteristic nervation develops on the crinkly, dark green papery lflets at a young age, and the stems have dense brown hairs.
3) The scent of the scratched bark of ***E. cylindricum*** is also obvious in young plants, which also have brown hairs on the stem but glossy leaves. Saplings of the other species do not have scented slashes (but c.f. *Canarium* (Gp 33)).
4) Young ***E. utile*** has hairy, dull green lvs, and a hairy shoot.
5) ***E. angolense*** seedlings/saplings have v. glossy and glabrous lflets and shoots, with drip tips and asymmetric base, and with rough-corky purplish petiolules.

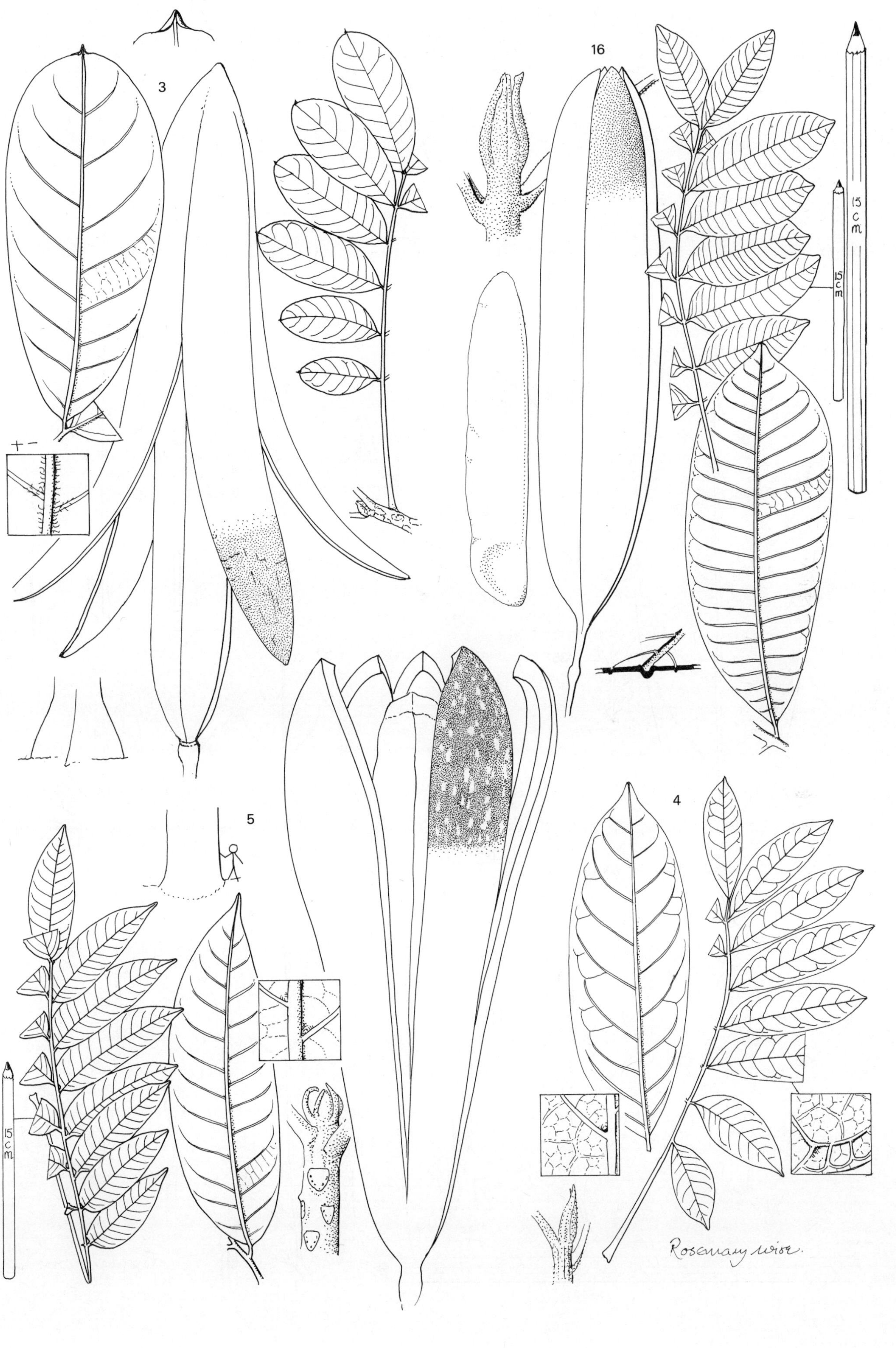
3
16
15 cm
15 cm
5
15 cm
4
Rosemary Wise

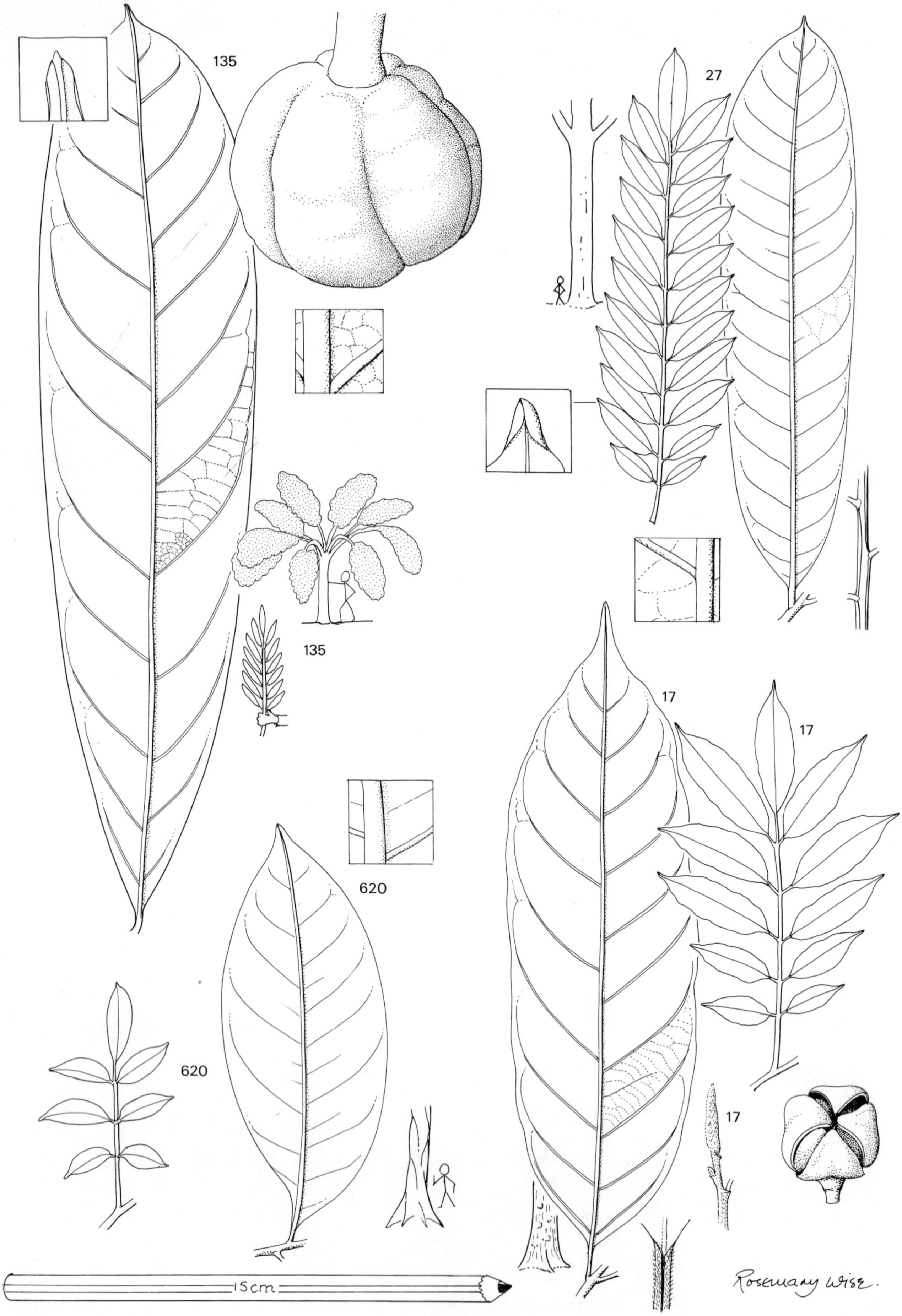
135
27
135
17
17
620
620
17
15cm
Rosemary Wise.

Group 34C: other Meliaceae

If your tree does not match the following species, having keyed to this point, then continue with Group 34D which normally *but not at all times of the year* produce latex.

Leaves large; most with >11, ±oblong, NOT strongly asymmetrical leaflets; lflets often with an abrupt, folded drip tip or apical projection		
Lflets often folded up at tip, or drip-tipped; leaves to 30 cm long with 5-15 prs oblong leaflets; lflets often alternate; petiole grooved, rachis slightly winged; medium-sized, rather inelegant tree with stout, low boughs and heavy crown; bark pale, shallowly fissured, ±pits on older trees; **slash yellowish, with orange speckles, scented**	***Turraeanthus africanus***[1] [*AVODIRE, APAPAYE*]	27
Lflets sometimes + apical point (protruding midrib), but never folded up; leaves up to a metre in length, with up to 20 prs opposite lflets; petiole not winged; laterals raised above; fine, square reticulation ±visible below; slash reddish; young lvs red; small tree often sprawling in swamps with v. long lvs clustered at twig ends	***Carapa procera*** [*KWAKUO-BESE*]	135
Leaves with <12 leaflets, or if more than 12, then leaflets highly asymmetric		
Leaflets always <12; usually some around 10 cm or longer[2]; lvs not v. clustered in crown		
Crooked, highly sinewy, or irregularly fluted tree; lflets *c.*5-9[2]; veins not always visible, but margin usually fairly regular; outer bark pale grey-brown, with many v. fine vertical striations , ±flaky; slash stringy-fibrous, contoured, with slight cedar scent	***Trichilia prieuriana***[2] [*KAKADIKURO*]	620
Tall, straight, cylindrical tree; all parts of outer stem, even on saplings, with v. strong, pleasant, bitter-sweet scent between apple and cedar scent; lflets papery, long acuminate, with wavy, recurved margins and slightly asymmetric; petiole with inrolled wings and dense, short, yellow hairs; large tree; flush of new lvs reddish; **bark often with sea-shell-shaped pits where scales fall; slash pale orange or slightly reddish, fibrous with strong, sweet scent, soon turning rusty brown**; capsules reddish, 4-valved with orange pulp inside	***Guarea cedrata***[3] [*KWABOHORO*]	17
Leaflets often >10, mostly alternate, all similarly-sized and usually <10 cm long; slash white-cream, ±camphor smell; veins and laterals prominent	See *Majidea fosteri* (SAPINDACEAE – Gp 36B)	

NOTES: 1) *Turraeanthus* seedlings have simple, v. long drip-tipped lvs, broadest at tip; they can be distinguished from *Khaya* seedlings by the ±obscure veins and short hairs on the terminal bud.

2) *Ekebergia senegalensis* is a tree of **savanna riversides** and certain other types of **dry forest**, but is apparently rare in Ghana. The leaves and the fluted bole resemble greatly those of *Trichilia prieuriana*. The fruits of *Ekebergia* are fleshy drupes (+ seeds without arils), whereas *Trichilia prieuriana* produces arillate seeds in leathery capsules. A tentative couplet to separate this rare species from the very common *T. prieuriana* is:

Lflets normally 9-13, on lvs very clustered at twig ends; veins normally visible, ? more translucent than rest of lf; lower surface of peculiar matt texture between veins; **margin irregularly, finely wavy, often appearing slightly serrated from a distance**; bark v. lenticellate on the stout twigs, becoming v. rough in square scales on older trees; slash pink-red with white streaks; tree of open vegetation or **v. dry forest** only; with fleshy drupaceous fruits ***Ekebergia senegalensis***

3) The stems and leaves of the seedlings of *Guarea cedrata* resemble at a very young age those of the parent, particularly as regards the very characteristic apple-cedar scent when the outer layers of the (hairy greenish) stems are scratched. The first leaves above the cotyledons are normally trifoliate, but sometimes simple.

Group 34D: *Trichilia-Guarea* (Meliaceae)
(White or brownish latex, often cedar-scented, lflts normally symmetrical)

Leaflets without dense, long hairs

Leaves glabrous, with lanceolate leaflets or broadly-winged petiole: with no hairs in midrib channel, nor on young rachis; venation often not v. clear

Petiole channel with fine inrolled wings on each side; edge of lflts very untidy, with irregularly recurved margin; laterals minutely wavy with few intermediate veins visible on plastic or plaster-like surface, drying metallic yellowish; leaves terminating regularly in either a pair or a single leaflet, with other (up to 12) lflets paired; bark dark, smooth, ±small pits; slash ±granular, pale brown or yellow, not v. scented — ***Guarea thompsonii***[1,2] KWABOHORO-NINI 18

Petiole without wing-edged channel; petiolule often v. slender, usually >6 mm long but relatively thin; leaves mostly broadest well below middle, <3 cm wide: lanceolate but asymmetric at base; midrib flat above; up to 18 lflets per leaf; bark with regular, oblong scales between vertical fissures; slash with unpleasant smell, red-orange, darkening, with drops of latex — ***Trichilia martineaui***[2] [*TANURO-NUA*] 616

Leaves (+lflets) with at least a few hairs (especially in midrib channel above), mostly broadest above middle, without broadly-winged petiole

Hairs conspicuously emerging above midrib channel; lvs with up to 21 (large) leaflets; **evergreen forest** tree, v. similar to next species — ***Trichilia ornithothera*** [*TANURO-BERE*] 619

Hairs of midrib channel above inconspicuous from a distance: level or below the level of the top of midrib channel OR tree of **dry forest**; veins usually indistinct on lvs of mature trees

Mature lvs without almost-coarse hairs; petiolules of (lower) lflets normally slender, <1 mm wide but >5 mm long; *mature* rachis without v. dense hairs; very common small-medium tree with cylindrical bole; crown dark, except for paler clusters of flushing lvs; slash pinkish, sweet-scented, darkening on exposure; *petals c.*1 cm long or less; fts + dense, v. short hairs — ***Trichilia monadelpha***[3] [*TANURO*] 618

Mature lvs with almost-coarse hairs giving larger veins, rachis, etc. a 'fuzzy' appearance when examined with naked eye; petiolules not slender: if >5 mm long then 1 mm or more wide; tall, straight tree of disturbed, especially drier forest; nerves of younger lvs+ fine long ginger hairs; not v. common; slash as *TANURO*; flwrs large petals *c.*1.5 cm or more long; fts softly hairy — ***Trichilia megalantha***[3] 617

Leaflets coarsely velvety with dense, long, soft orange hairs; midrib and lateral nerves impressed above; laterals often >20 pairs; large tree with dark crown and cylindrical bole; slash pinkish white, soft, fibrous, darkening, with drops of yellowish latex; flwrs >1 cm long; fts with short and long hairs — ***Trichilia tessmannii***[3] TANURONINI 621

NOTES: 1) Young *Guarea thompsonii* is similar to young *Hannoa* (see Gp 32), with rachis constricted near petiolules, etc.

2) *Trichilia* is very close to *Guarea*. The key difference is in the flowers; in *Guarea* the anthers are hidden inside the staminal tube, whereas in *Trichilia* the anthers are visible.

3) *T. monadelpha* (previously *T. heudelotii*) cannot be reliably separated from *T. megalantha* without flowers. The foliage of both species varies widely. Although very often *T. monadelpha* has leaves with more slender petiolules and less coarse hairs than the rarer *T. megalantha* this is hard to quantify. The lvs of *T. megalantha* sometimes resemble those of *T. tessmannii* (previously *T. lanata*) more closely than suggested by this oversimplified key. Some of the variants of these *Trichilia* spp. suggest hybridization or other genetic complications.

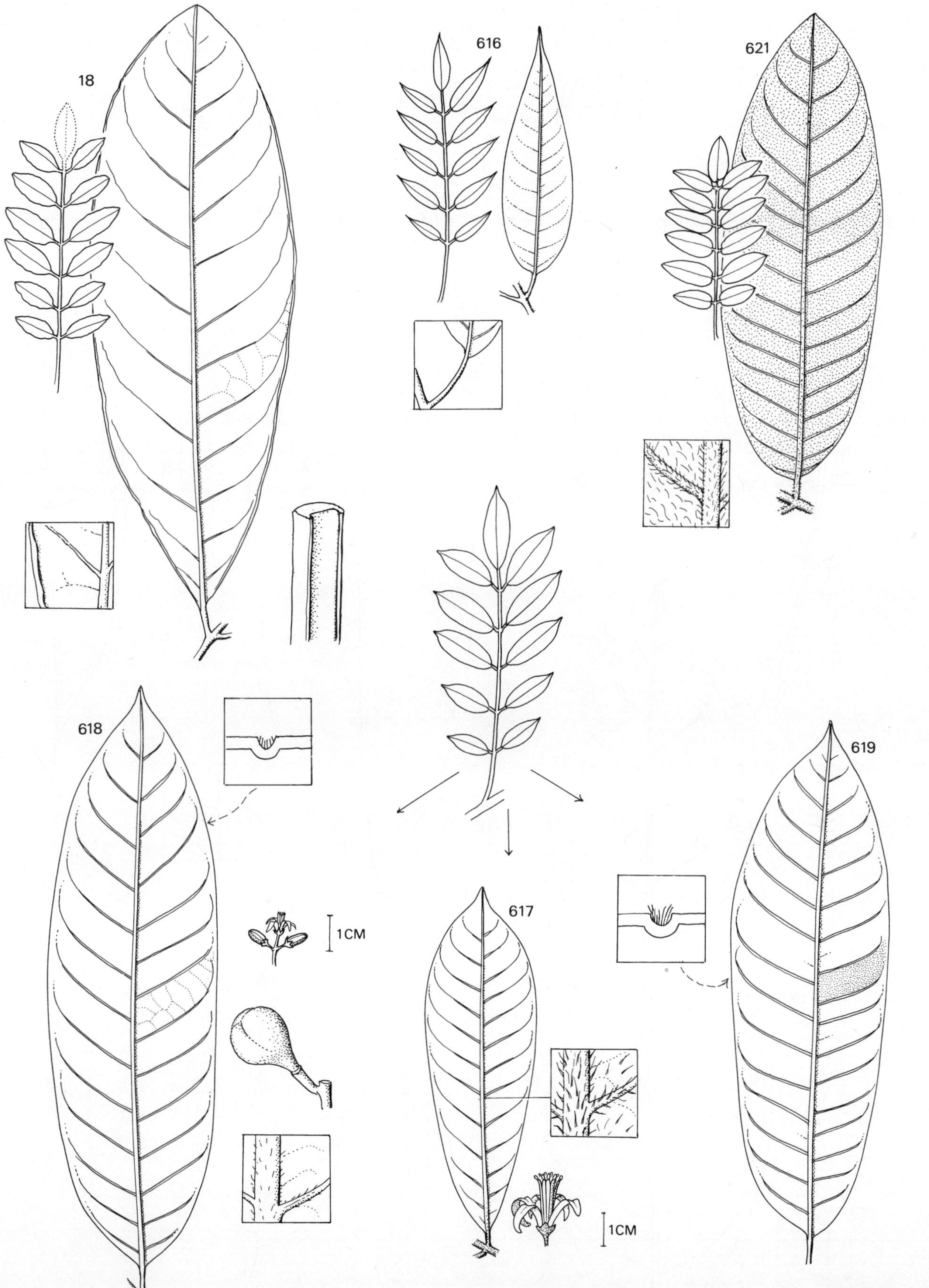
18
616
621
618
619
617
1CM
1CM
15 cm
Rosemary Wise.

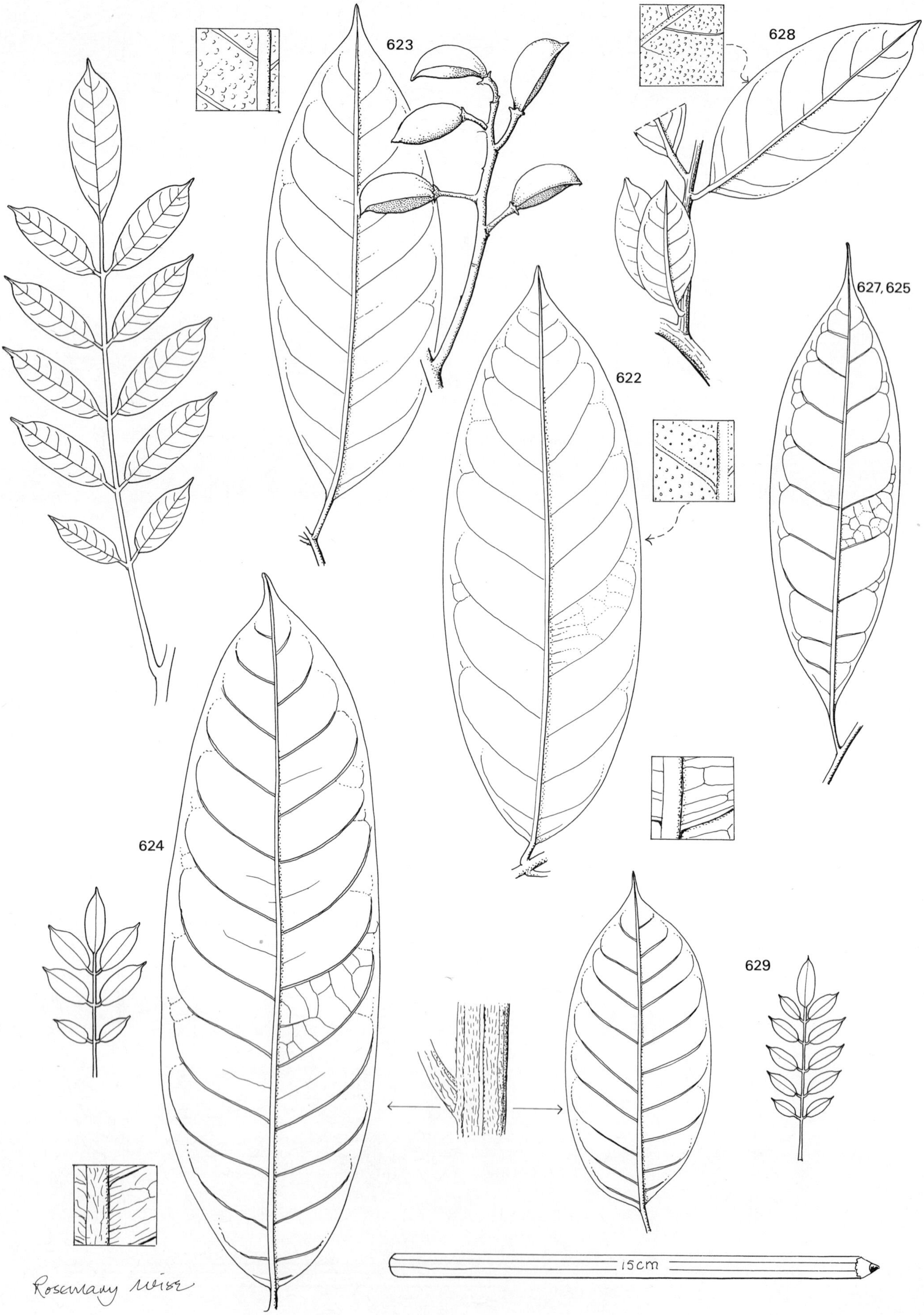
623
628
627, 625
622
624
629
15cm
Rosemary Wise

GROUP 35: ANACARDIACEAE

Trees of this family, which includes the mango and cashew, typically have resinous exudates in the bark and clustered compound lvs (the cultivated fruit trees are exceptional with their simple lvs). As in the previous few families, the flowers are small, regular, in branched inflorescences. The flowers have a swollen, ring-like structure ('disk') to which the stamens are attached. In *Trichoscypha* (4 or 5-merous flowers with 3 styles) and *Sorindeia* (5-merous with 1 style) the flowers have (1 locular ovaries with) terminal styles. The remaining species have (multilocular ovaries with) 4-5 styles around the apex. The remains of the styles can often be seen around the apex of the fruits, which are normally fleshy (drupes), often edible, with a central stone.

Genus	Fruit
Trichoscypha	Ellipsoid, *c.*2 cm but variable
Sorindeia	Ellipsoid, yellow, 1 cm long, edible
Spondias	Ovoid edible, yellow 'plum', 3 cm
Antrocaryon	5-lobed fleshy, 3 cm; woody, pitted stone rounded-conical with (5) holes
Pseudospondias	Ovoid, purple, 2 cm long, with 2-4 tiny style remains at tip
Lannea	Flattened ellipsoid, black, <1 cm, with remains of 4 styles at tip

Key to subgroups

Slash with latex (or little-branched (fertile) trees with large lvs); lflets often pustulate; **evergreen forest**	**Group 35A**
Slash without latex (or tree in **drier forests**), or with fine-transverse venation	**Group 35B**

Group 35A: *Trichoscypha*
(Slash soft (contoured) red, fibrous + latex; lvs strongly clustered)
(Only in evergreen forest)

The Ghanaian species of *Trichoscypha* are not well collected, and difficult to distinguish. The following key is therefore incomplete, pending further research. Apart from *T. arborea*, most species seem to be v. localized or of small stature. (Some species in Gp 35B produce a resinous exudate that could be described as latex.)

Rachis and young twigs with dense, very conspicuous, long, reddish hairs at first; fts (and disk) densely hairy		
— Rachis and (grooved) midrib below with v. short hairs amongst the long ones; midrib impressed above with the long hairs emerging; lflets 3-6 pairs with 7-13 prs lateral nerves; infls up to 45 cm long; flwrs white	***Trichoscypha oba***	629
— All hairs on lower surface long and spreading; lflets 2-4 prs with 12-20 prs laterals; infls up to 15 cm long; fls red	***Trichoscypha atropurpurea***	624
Rachis, leaflets, etc. with only short hairs, or a few long hairs, or glabrous; (disk sometimes glabrous)		
— Venation prominent and clearly visible below and above; lamina always papery, and **sharply acuminate**; usually with some hairs – e.g. on midrib; (disk glabrous)		
— — Leaflets around or <10 cm long, with <10 arching and joining laterals; with v. short hairs on midrib below; small understorey tree		
— — — Petals 1-2 mm long; pedicels slender; infls not v. hairy	***Trichoscypha cavalliensis***	627
— — — Petals >3 mm long; lflets pustulate; infls + dense rusty hairs	***Trichoscypha baldwinii***	625
— — Leaflets >10 cm long with >10 laterals; tufts of hairs in midrib above and some long hairs on midrib below; venation below conspicuously pale reddish and raised reticulate. Medium-sized tree; infl to 60 cm with long hairs; fls with petals >3 mm long, without pedicels	***Trichoscypha beguei***	626
— Venation sometimes not v. visible; leaves ±glabrous, often v. pustulate; often rather leathery or plasticky (but some spp. apparently produce v. varied lvs); **rachis, etc. v. corky**		
— — Basal leaflets at, or very close to base of leaf, close to twig; lflets v. pustulate above; flowers with very hairy disk, and fruits also hairy; lflets normally narrowly oblong-lanceolate; small tree	***Trichoscypha chevalieri***	628
— — Basal leaflets well separated from twigs		
— — — Veins minutely raised-reticulate below; **lvs strongly pustulate on both surfaces**; lf base ±symmetrical, with edge of lamina folded up into petiole channel. Medium-sized tree in 'coastal' **evergreen forest**, as at Cape Three Points; fls with petals <2 mm long, white, with hairy 'disk'	***Trichoscypha albiflora***	622
— — — Veins obscure, or v. laxly reticulate and raised below; papery or leathery; sometimes **with raised pustules below, but not many above**; rachis and petiolules ±corky and ridged or cracked; medium tree, widespread in **evergreen forest** and common; fls red, on infls upto 80 cm long, without hairs; fts glabrous; bark →rough; slash red, contoured, soft fibrous + spots of white latex	***Trichoscypha arborea*** [*ANAKU*]	623

Lateral nerves many and parallel, OR meeting in sub-marginal nerve v. close to margin; trees with leaves[2] strongly clustered at twig ends, like *Entandrophragma* or Neem; leaflets ±papery and sometimes serrated; slash red with white streaks with red to yellow, clear to slightly opaque exudate; trees without well-developed buttresses, with vertically fissured bark.

- Laterals many and parallel; finer venation obscure; lvs often v. hairy; slash slightly scented, red with white streaks (±orange specks), fibrous; foliage appearing delicate and lacy in (heavy) crown; distinctive conical, pitted stones of old fruits normally abundant below mature trees — ***Antrocaryon micraster*** [*APROKUMA*] 94
- Laterals not particularly many and parallel, but meeting in a nerve running close and parallel to margin; finer venation visible and lvs with many fine raised spots (pustulate) between veins; margin minutely undulate, thereby resembling slight, small serrations; nerves ±guttered above; trees sometimes with spines; slash hard, fibrous, pink with white streaks and clear, sticky exudate; **in dry forest** or savanna, but often planted for plum-like fruit (edible drupe) — ***Spondias mombin*** [*ATOA*] 575

Lateral nerves not dense nor joining in sub-marginal nerve; lvs slightly, but not strongly clustered at twig ends

- Leaflets with hairy domatia; ±nerves sidling close to each side of v. unequal base; crushed petiolules scented of unripe mangoes; young apical buds with pink-brown hairs; crooked tree, often with high, narrow buttresses or flutes; bark with fine vertical fissures, thick; slash soft, pinkish orange to yellow, gritty, with small spots of red exudate; younger trees often stilt-rooted; **tree normally gregarious in swamps or by rivers** — ***Pseudospondias microcarpa***[2] AKATAWANI 525
- Leaflets without hairy domatia; veins sunken or impressed, looking like fine cracks in leather; youngest parts of lvs, twigs, etc. with v. small 'spots' (lens) of brown stellate hairs, ±amongst longer straight hairs; slash red with white wavy lines; **bole often with conspicuous pits as if shot marks**; exudate clear and sticky; lvs of young tree become constricted near petiolules and stem when dried — ***Lannea welwitschii***[1] KUMANINI 378

NOTES: 1) ***Lannea nigritana*** is a small tree like *L. welwitschii* with v. clustered lvs, 3-5 pairs with one terminal leaflet, each generally <3 cm wide, ovate, v. asymmetric, yellow-green. It is found only in the driest types of forest (e.g. Shai hills), and in thickets. The bole of *L. nigritana* is often covered with small pits like *L. welwitshii.*

2) ***Sorindeia juglandifolia*** is a rare small tree with lateral nerves meeting in a (looping) submarginal nerve, which is often found **nr rivers in savanna**. The venation of many lflets is characteristic, with uneven veins/nerves arising nr the junction of the laterals and the midrib, and traversing to meet the next lateral nearer the margin (as in *Ficus tessellata/F. ardisioides*, Gp 19). Other lflets resemble those of *Pseudospondias*, but without domatia. ***Sorindeia zenkeri* (570)** has the same venation, is found on river banks in dry areas, and has long brown hairs on the rachis. The slash of *Sorindeia* spp. often has a characteristic scent like a mango fruit.

94
575
525
570
378
15cm

95
381
115
117
116
15 cm
Rosemary Wise

GROUP 36: SAPINDACEAE

The species of this family are normally middle-storey trees. Only *Blighia* spp. and (occasionally) *Majidea* become large trees. The wood is hard, and there are no important timber trees in the family. The slash of the larger trees is typically yellow to brown, granular and often with orange gritty streaks or contoured. In *Majidea* it has a strong, pleasant scent (like camphor) and in some of the small trees it is reddish. Leaves are usually paripinnate, although sometimes with one end-leaflet falling off, or with the leaflets not quite opposite. There is often an apical process at the end of the rachis. Several species have deeply grooved or striate twigs, and many also have conspicuous yellow-orange hairs on young parts. As well as the following species, there is the trifoliate *Allophylus africanus* (Gp 31) and the common climber *Paullinia pinnata*, with winged rachis.

The flowers are normally unisexual, always small and in inflorescences, and regular except in *Pancovia* and *Chytranthus*. The inner face of the petals is usually hairy and has at its base a small appendage, but some species lack petals. The fruits vary from dehiscent, usually 3-sided capsules with arillate fruits (as found in some Meliaceae) to drupes, as follows:

Genera with dehiscent fruits (usually with red inner surfaces)

Blighia	Leathery 3-lobed capsule; black seeds + yellow-orange aril at base of seed; in short axillary racemes
Eriocoelum	Hard, brittle, 3-lobed, orange capsule; black seeds + red aril at base; in racemes or panicles
Majidea	Brittle, inflated 3-lobed capsule; round seeds with velvety coat; in v. dense terminal panicles
Lychnodiscus	Brittle 3-lobed capsule; ellipsoid seeds with sticky red coating; in short terminal panicles
Aporrhiza	1-2 lobed, flat, velvety, leathery capsule; seeds with aril; in terminal panicle

Genera with indehiscent fruits

Placodiscus	3-lobed and fleshy; in racemes ±on older wood (fls without petals)
Chytranthus	3-many lobed and fleshy – on bole
Pancovia	3-lobed, in short racemes
Zanha	Ellipsoid, orange, edible, sweet within, pointed apex; fls in dense heads, dioecious
Lepisanthes	Ellipsoid, orange, fleshy, edible, sweet within, in pairs or 1 undeveloped; in panicles
Lecaniodiscus	Ellipsoid, orange, fleshy inside, velvety, pointed apex + calyx at base; in axillary racemes
Deinbollia	±spherical, hard, reddish orange, sweet inside; in racemes or panicles ±on older wood
Allophylus	±spherical, red; in axillary spikes

Key to subgroups

Leaflets with fine, v. regular, raised reticulation above[1]; with few or no hairs — 36D

Leaflets without *Placodiscus* venation; young twigs sometimes with (lines of) orange hair at least; lflets sometimes v. hairy or serrated

- Small, little-branched or unbranched treelets, often cauliflorous, with large or glaucous lvs clustered at top of stems — ***Chytranthus, Deinbollia*** — 36C
- Not *Chytranthus-Deinbollia*
  - Leaflets not densely hairy, nor serrated; lflets <7 pairs — 36A
  - Leaflets rough-hairy OR serrated, OR with many ±falcate leaflets mostly <2 cm wide — 36B

NOTE: 1) The very distinctive *Placodiscus* venation is visible particularly on dry lvs. In the field, therefore, it is sometimes necessary to find fallen leaflets. Outside evergreen and coastal forest it is usually safe to assume that if the tree is not fluted or if the slash is not contoured (*P. boya*) then the tree is not a *Placodiscus*.

383
484
282
281
282
392
280
15 cm
Rosemary Wise.

Group 36A: Sapindaceae
(Leaves without v. dense hairs; margin entire)

For illustrations of the first five species, see page 184.

Leaves with one terminal leaflet often aborted OR leaflets alternate OR **with conspicuous red-orange hairs on petiolules;** lflets in >3 pairs; midrib finely channelled above; petiole ±rounded; small trees with smooth, grey bark

- Petiolules normally swollen, and conspicuously yellow-hairy; tip of rachis with short blunt stump; apex of lvs of mature tree normally with 1 lflet fallen, but paired, terminal lflets also not uncommon; young stems + v. dense, orange, soft, curly hairs, with v. fine striations almost hidden by v. fine hairs; top surface with prominent venation and impressed midrib; v. common, esp. in **secondary forest**; slash brittle and contoured, orange-brown — ***Lecaniodiscus cupanoides***[1] [*DWINDWERA*] 381
- Petiolules slender, not swollen and hairs there not soft; Lflets (6-12) often v. alternate, with decurrent base; petiole often >5 cm long (to lowest lflet) and rounded;[2] hairs v. short and fine, on rachis, petiolules and ±on veins; appressed on lower surface; young twigs deeply striate to grooved; apex of rachis + fine, sharp-tipped point; slash red-brown; understorey tree of **evergreen forest** — ***Aporrhiza urophylla*** [*AKYE-FUFUO*] 95

Leaflets usually strongly paired OR apex rounded OR <3 pairs OR lvs glabrous OR petiole winged OR flattened with ±semicircular cross-section

- Leaves often with more than 3 pairs of leaflets; without tufts of hairs in nerve axils; **petiole ±semicircular in cross-section, flat above**; nerves slightly impressed above; becoming large trees, often ±fluted at base, but without large buttresses; bark smooth, grey to yellow, with many small lenticels; slash orange and gritty
  - Lvs with up to 5 prs (normally 4) of leaflets; 8-12 prs laterals on mature tree; apex of lflets often rounded, or bluntly acute; yng twigs + rachis deeply striate, *with long, dense hairs;* in **moist or drier forests, savanna**, often planted; slash thick, yellow to brown, granular, with orange gritty streaks; fts + rounded lobes — ***Blighia sapida***[3] [*AKYE*] 115
  - Lvs with up to 4 prs (normally 3) of leaflets; lflets often with >12 prs laterals; young twigs rapidly glabrous, smooth, with few lenticels; mature lflets almost glabrous; midrib and nerves below pale orange-brown or red; apex of lflets not rounded; margin often undulate, following sub-marginal nerve; petiole constricted at base; **moister forests** only; slash as above (115); flush of lvs v.pale brown; fruits with 3, sharp-edged lobes, >3 cm long, glabrous — ***Blighia welwitschii***[3] [*AKYEKOBIRI*] 116
- Leaves with 1-3 prs of lflts; OR petiole ±rounded OR lflets + tuft domatia; not large trees
  - Lflets mostly with tuft domatia, otherwise glabrous, acuminate; petiole winged; basal lflets not exactly opposite; young flush white; lower storeys of **moist forests** or **dry forest** (canopy); slash thin, crumbly, mottled brown and white; fruits + 3 sharp-edged lobes, <3 cm long + fine hairs — ***Blighia unijugata*** [*AKYE-BERE*] 117
  - Lflets never with tuft domatia
    - **Leaflets not narrowly elliptic, OR hairy**; 1-3 prs lflets; trees in or near **wet places**
      - *Lflts without marginal nerve, usually + hairs*; margin ±recurved; midrib impressed or prominent above – not flat nor channelled; slash reddish brown
        - Lvs with only short hairs, even on young parts; 1-3 prs lflets (sometimes alternate); apex of lflets often emarginate, but sometimes with a small point <1 mm long as well; twigs + reddish lenticels; **±stilt-rooted tree in swamps** in western region; racemes <8 cm long — ***Eriocoelum pungens*** [*AKYE-NAN*] 281
        - Lvs with long (>1mm), spreading hairs, especially on young parts and midrib above; lflets often >3 pairs; tip of leaf usually with thread-like point >1 mm long; racemes >8 cm long, pendulous — ***Eriocoelum racemosum*** (see 36B). 282
      - *Lflts glossy, glabrous, with nerve running around margin*; midrib prominent; petiolules 2-3 mm long; small tree in **riverine forest** — ***Pancovia bijuga***[4] 484
    - **Leaflets in 1(-2) pairs, narrowly elliptic and acute-acuminate, glossy** and glabrous; margin thickened; if in 2 pairs then the lower pair close to twigs; petiolules swollen, but petiolule and rachis flattened; rachis without sharp apical point; laterals and finer veins sometimes prominent above; bark finely scaly, thin and hard — ***Lepisanthes senegalensis***[5] [*AKYE-BUNO*] 383

NOTES:
1) ***Lecaniodiscus punctatus*** is a rare tree v. similar to ***L. cupanoides***, but with gland-dotted lvs; it is known only from the Atewa range forests; it has narrower, more oblong lflets without the orange-hairy petiolules of *L. cupanoides*.
2) Tree with stilt roots or lflets with midrib prominent or impressed but not channelled above – see ***Eriocoelum pungens*** which occasionally produces lvs with alternate lflets.
3) These two *Blighia* spp. are usually confused by tree spotters, and are hard to distinguish when not fruiting. Their seedlings have an opposite pair of trifoliate (or bifoliate) lflets, covered with fine orange hairs. In *B. welwitschii* (at least) the apex of these seedling lflets is acuminate and mucronate. (If tree in swamp, check also *Eriocoelum* spp.)
4) *Pancovia turbinata* is a similar, cauliflorous shrub from **evergreen forest**, with a raised midrib and v. smooth glossy lflets with a thickened margin.
5) *Lepisanthes* old name = *Aphania*.

Group 36B: Sapindaceae
(Hairy or serrated lflets or many narrow lflets)

Leaflets with hairs long, coarse and dense, at least on nerves		
Leaflets never toothed; lvs smaller, with fewer nerves (than *Lychnodiscus dananensis*); lower pair of lfets often rounded and clasping the stem; small trees **by rivers or in swamps**		
Leaflet tip acuminate and mucronate; 3-5 pairs, often with basal pair clasping the stem; rachis and stems striate, with flattened petiole, *c.*5 cm long **evergreen forest**; fls in pendulous racemes	***Eriocoelum racemosum***	282
Leaflet tip not mucronate; 2-3 (rarely 4) pairs; lower leaflets often rounded and clasping stems; twisted tree in waterlogged soils esp. (but not entirely) of dry areas, e.g. Togo plateau; young twigs v. coarsely hairy; flowers in panicles	***Eriocoelum kerstingii***	280
Leaflets large and hairy (often >15 cm long), sometimes toothed; venation rather scalariform; midrib impressed or guttered above with a conspicuous line of hairs; often >18 prs laterals; lflets with rough hairs below; **understorey tree** with petiole scars on bole	***Lychnodiscus dananensis*** [*KWATA-DUA*]	392
Leaflets glabrous or with short hairs		
Lflets serrated		
Serrations and apex of leaf rather blunt; tree of **savanna riversides** and **dry forests**; rachis glabrous; petiolules <2 mm long; bark flaky; slash with clear, sticky exudate	***Zanha golungensis***	665
Serrations and apex of leaflets sharply acuminate; tree of moist forests; rachis with fine hairs when young; petiolules stout, often >2 mm long; midrib finely channelled; lflets in *c.*5 prs; veins often with slightly 'glandular' puckered points along their length; bark rough, sometimes slightly spiny; slash hard, brittle, orange, granular, darkening	***Lychnodiscus reticulatus*** [*AKYE-SE*]	393
Lflets not serrated; lflets in *c.*6 prs. falcate-asymmetric, obtuse on one side of base, cuneate on the other, with hairs sometimes on midrib above; rachis slightly broadened below each pair of lflets; slash pale with orange grit and strong scent of camphor 'rob' or 'omega oil'; twigs with slightly flattened sides and small lenticels[1]	***Majidea fosteri*** [*ANKYEWA*]	405

NOTE: 1) *Majidea* seedlings are normally abundant in the vicinity of large trees; they have characteristic dark-green leaflets with a winged rachis, but less so than the similar seedlings of *Lovoa* (see Gp 34) more typical of wetter forests.

Group 36C: *Chytranthus, Deinbollia*
(Small, little-branched treelets with large lvs: often cauliflorous)

Most of these species barely qualify as trees, so their details are brief.

Leaflets glaucous or silvery reflective below; not particularly obviously hairy; midrib ±prominent above		
Larger lflets *c.*10 cm wide, *c.*14-28, alternate; bark grey, scaly	***Deinbollia grandifoliola*** (see below) [*MMATA*]	
Larger lflets >10 cm wide, and >20 cm long in *c.*4 pairs; lflets bluish or silvery below; veins a duller, more orange colour; lflets ±glabrous, with v. swollen petiolule; small arching treelet in **swampy places**	***Chytranthus macrobotrys***	163
Leaflets not glaucous below (often hairy) OR laterals in >20 pairs		
Rachis with unusually long, very dense and conspicuous hairs; up to 1m long; lflets papery, acuminate; midrib channelled above; laterals meeting in sub-marginal nerve; **evergreen forest** only	***Chytranthus cauliflorus***[1]	161
Rachis not so unusually hairy		
Lflets v. long and slender: typically 5 cm wide and 20-30 cm long: with *c.*15 laterals meeting in well-defined sub-marginal nerve; midrib channelled above with long hairs emerging from midrib channel; rachis + *c.*½ cm projection beyond terminal lflets	***Chytranthus carneus*** [*oNIBONANUA*]	159
Lflets not so unusually long and slender		
Lflets broadly elliptic (>10 cm wide) obtuse at base and abruptly acuminate; midrib impressed above; lvs up to 1 m long	***Chytranthus atroviolaceus*** [*AKYE-KoKoo*]	160
Lflets not broadly elliptic or at least cuneate at base; large lvs with papery lflets with prominent midrib above		
Petiolules stout – up to several mm wide; mature lflets + v. fine hairs or glabrous, often pustulate between finest veins; base of petiole often 1cm or more wide; calyx ±glabrous on outer surface	***Deinbollia grandifoliola*** [*MMATA*]	218
Petiolules slender, 1 mm or less wide; mature lflets with scattered, fairly stiff hairs (lens); slender lvs, plant fertile as small shrub; calyx densely hairy on outside	***Deinbollia pinnata*** [*WOAGYE-AKOA*]	220

NOTE: 1) *D. molliuscula* is a less common small tree ,with long orange hairs on the rachis and shorter hairs on the lflets.

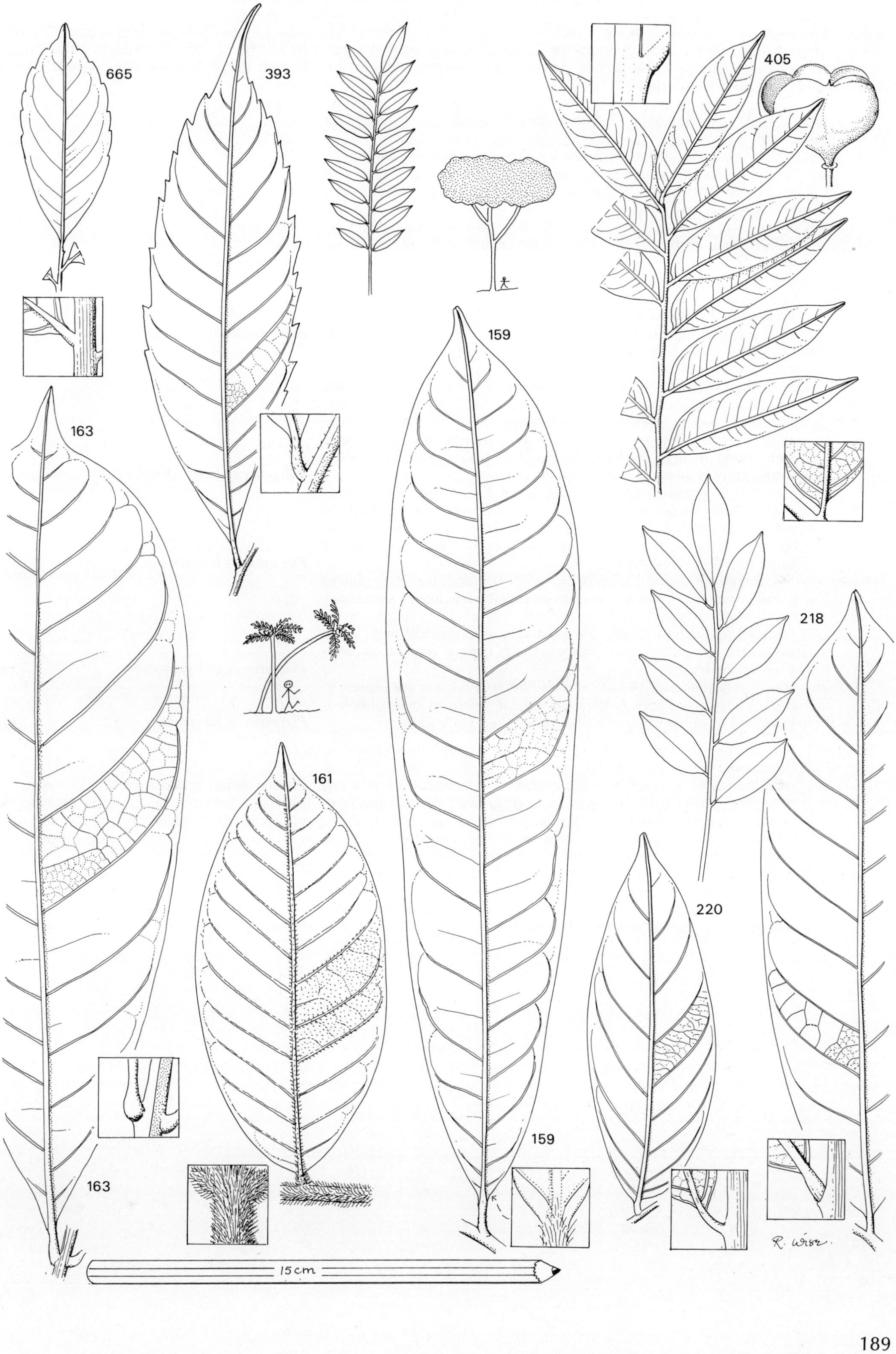
665
393
405
159
163
161
218
220
159
163
15cm
R. Wise.

Group 36D: *Placodiscus* spp.
(Venation very finely reticulate and conspicuously prominent above)

Of the following species, only ***P. boya*** is common outside the evergreen and coastal forests of the western region. The other species are not well-known. In fact, three of the Ghanaian species were described by J.B. Hall for the first time in 1980. Readers with flowering specimens can make a more reliable determination than is possible with sterile specimens by using the key in Hall's paper[1]. The flowers lack petals.

Leaves with 1-4 (-5) pairs of leaflets; if 5 then lflets broadest above middle, or slash contoured

  Leaves with 1-2 pairs of leaflets, with the basal pair sometimes clasping the stem; shoots, especially adventitious ones, often with buds in lf axils with 2 sharp points; lflets ovate, elliptic or lanceolate with a slender rachis; margin thickened (lens)

    Lowest pair of leaflets almost always clasping stem, like stipules, if they are present at all; these lower lflets cordate-ovate; rachis often without an apical point — ***Placodiscus pseudostipularis*** 515

    Lowest pair of leaflets often not clasping stem,[2] and petiole (to lowest lflets) often several cm long; lf base decurrent up to swollen petiolule; rachis usually with a sharp apical point; twigs rounded; axil buds small and glabrous; smooth-barked, cylindrical tree in **forest nr coast**[3] — ***Placodiscus attenuatus*** 510

  Leaves usually with 2-5 pairs of leaflets; commonly found *outside* zones of evergreen and coastal forests; (flowers in panicles or racemes)

    Leaflets often broadest below middle, and usually <20 cm long; young axil buds v. hairy, with small sharp tips, but midrib below, etc. completely lacking glandular hairs; yng twigs often sl. flattened; (flwr stalks 5-10 mm long); small, sinewy, **twisted tree widespread and even found in dry, fire-zone forests**; bark rough and flaky; slash brittle, gritty, contoured, rapidly darker — ***Placodiscus boya***[3] [*KAFUOSO*] 512

    Leaflets mostly broadest above middle, up to 25 cm long; young twigs v. thick, with deeply striate apical bud; lflet margin often irregularly wavy; (flwr stalks <5 mm long, and flwr buds + dense short hairs); erect shrub to small tree often in **wet places** (e.g. streamsides) in **moist and dry**[3] **forest zones** — ***Placodiscus bracteosus*** 513

Leaves with more than 4 pairs of leaflets or lvs >90 cm long; leaflets ±oblong or falcate; small trees only known in **evergreen forest**; new leaves sometimes bright red

  Rachis rounded and rather smooth; lflets not very asymmetrical, with acute to acuminate apex; midrib with glandular hairs (strong lens) below with bright red young leaves — ***Placodiscus oblongifolius*** 514

  Rachis remarkably triangular in outline with fine grooves; mature leaflets v. asymmetric with acute apex (without glandular hairs); cylindrical-boled, little-branched or unbranched tree with hot-tasting, gritty slash[4] — ***Placodiscus bancoensis*** 511

NOTES: 1) For further information: Hall, J.B., 1980. 'New and little-known species of *Placodiscus* (Sapindaceae) in West Africa'. *Adansonia* ser. 2, 20: 286-295.

2) Hall (see note 1) notes that the crown of *P. attenuatus* has a characteristic tufted appearance, associated with a rhythmic growth cycle. In each cycle of growth, the leaf produced first has a long petiole (from the stem to first leaflets), the second leaf has a shorter petiole, up to the sixth or seventh leaves which have only one pair of leaflets or even a single leaflet, with successive leaves separated by very short internodes. These later, reduced leaves thus form a pseudowhorl which covers the terminal bud.

3) The fruits of *P. boya* are pear-shaped, up to 5 cm long, yellow and are said to smell of rotting apples. *P. attenuatus* fruits are *c.*2.5 cm long, and not wrinkled when dry. *P. bracteosus* fruits are trilobed and 2-5 cm in diameter.

4) The sapling habit of this, and possibly other *Placodiscus* spp., recalls the 'litter-bin' habit of *Pycnocoma* and certain *Ouratea* spp. (see notes for Gp 16).

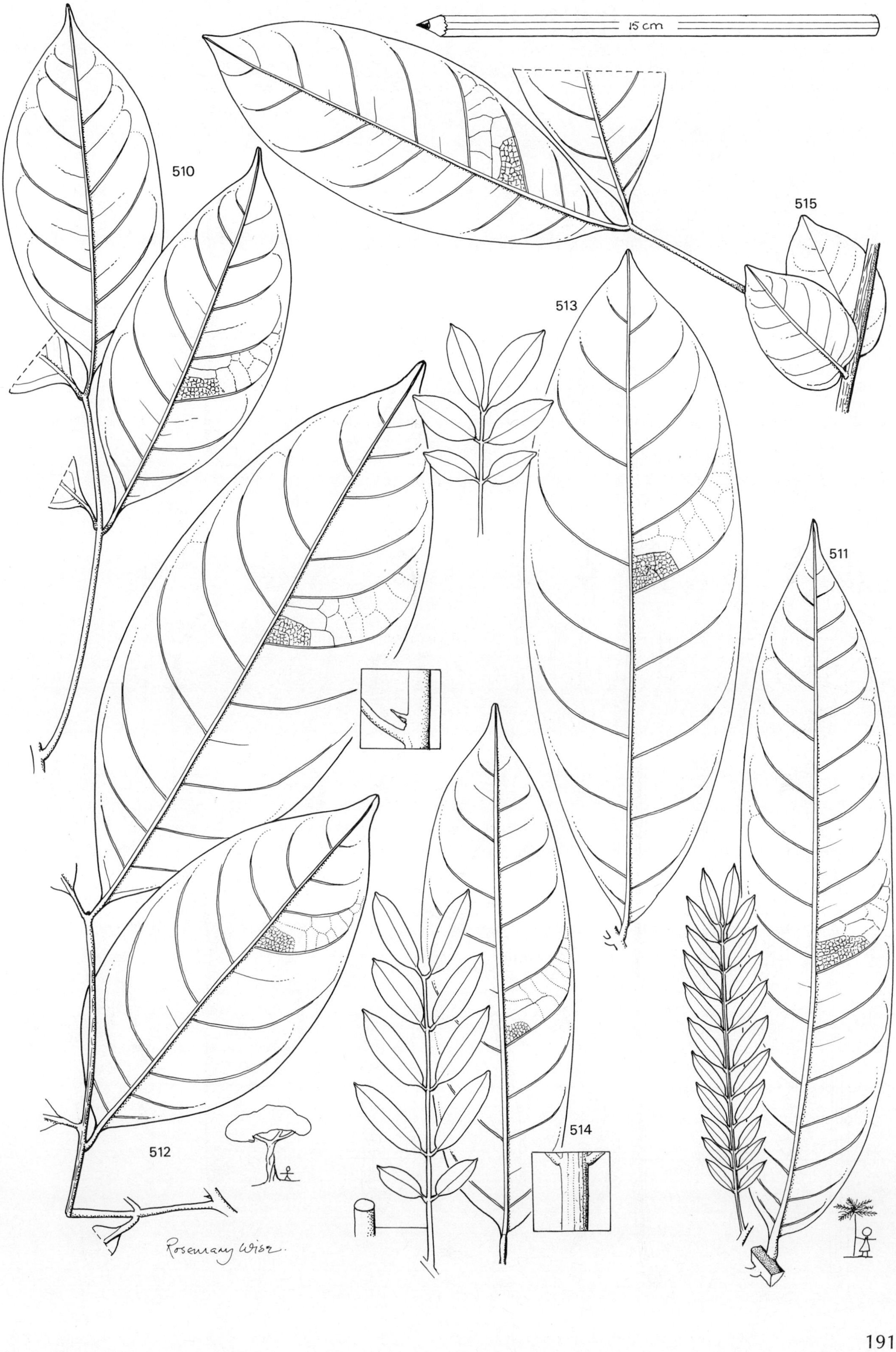
15 cm
510
515
513
511
514
512
Rosemary Wise

FRUITS OF THE LEGUMES (i)

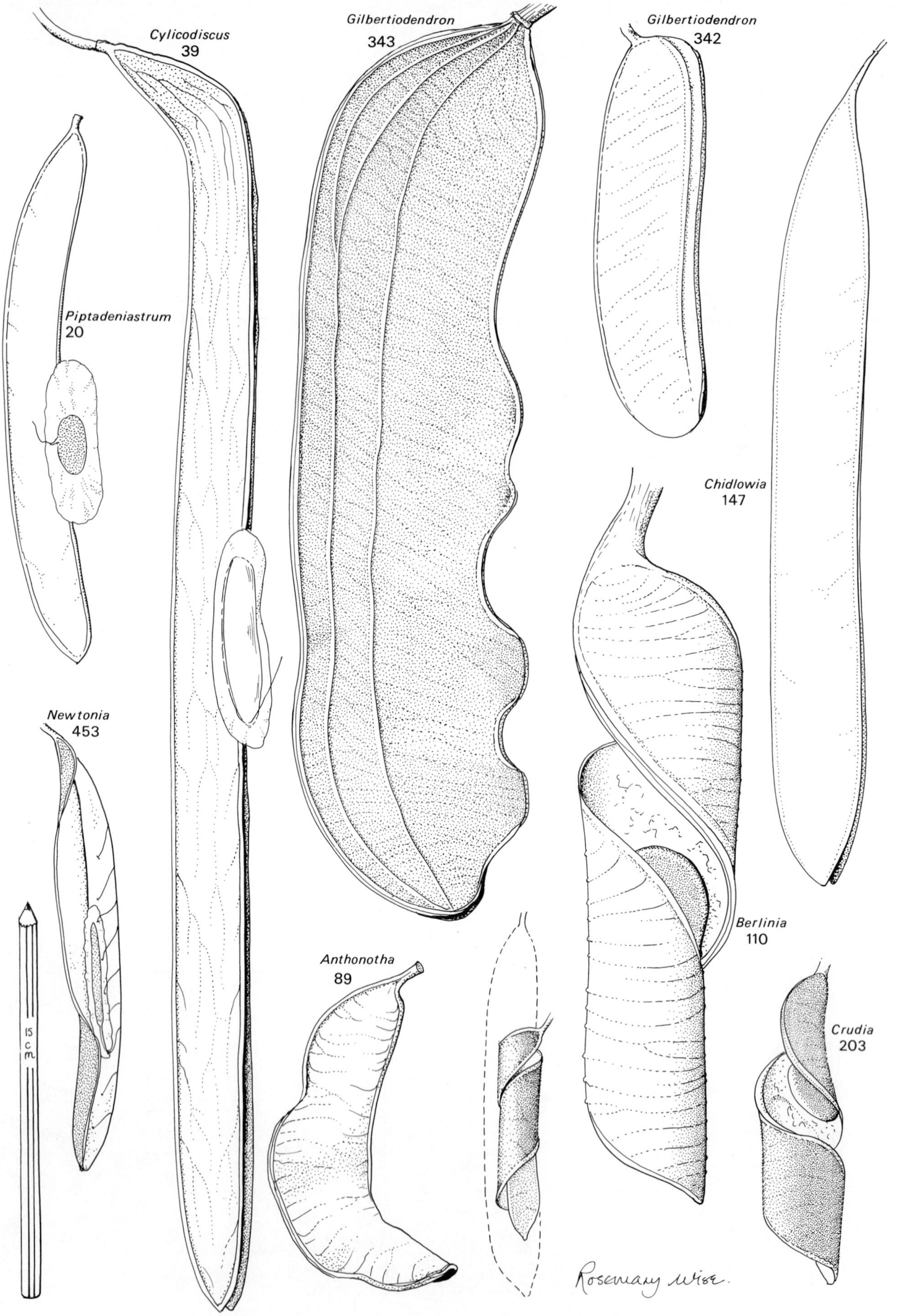

GROUP 37: LEGUMES (part 1)
(Leaves once-pinnate)

Key to subgroups and Group 38 (Legumes, part 2)

- Lvs simple (from Gp 13) — Group 37A
- Lvs compound
  - Lvs bipinnate (check carefully) — **Group 38** (Legumes, part 2)
  - Lvs once-pinnate
    - **Leaflets all in precisely opposite pairs; leaves never with stipels**
      - Leaflets in one pair — Group 37B
      - Leaflets in more than one pair
        - Margin with small glandular teeth or notches; lvs often large or with persistent stipules tree in **wet places** — Group 37F (*Gilbertiodendron* spp.)
        - Margin entire
          - Leaflets often >12 per leaf; <2 cm wide
            - Leaflets glabrous; without stipels (dry forest) — Group 37C
            - Leaflets hairy or glaucous, often thin or with stipels — Group 37I
          - Leaflets 12 or fewer per leaf or >2 cm wide
            - Leaflets glaucous, or silvery or with many long hairs below — Group 37E
            - Leaflets not discolorous, though sometimes with some fine hairs below
              - Lflets <15 cm long
                - –Lflets in 2-3 (-4) pairs OR rhombic OR all lflets 1-2 cm wide; small or riverside trees; buds usually with several overlap bud scales; (try this key first if in doubt) — Group 37C
                - –Lflets (on most lvs on tree) in more than 3 pairs
                  - Lflets usually 10-12 per lf; glabrous, thin, with reddish translucent veins and margin; *c.*3 cm wide; twisted medium-sized tree — *Chidlowia* (Gp 37E)
                  - Lflets usually in 3-5 pairs (not *Chidlowia*)
                    - Petiolules twisted OR margin thickened with marginal nerve OR apex emarginate or lvs with gland-spots — Group 37D
                    - Lflets without these distinctive features, falcate; pods often large and leathery like the sole of a shoe — Group 37F
              - Lflets >15 cm long; (pods like shoe soles, though often hairy) — Group 37F
    - **Leaflets odd in number or, at least, not precisely paired; sometimes with stipels**
      - Leaflets alternate or almost opposite; often with many fine brownish hairs below
        - Finer venation not conspicuously prominent; slash without spots of red exudate — Group 37G
        - Finer venation conspicuously prominent above or below; leaflets rarely opposite; OR slash with spots of reddish exudate — Group 37H
      - Leaflets, except for a terminal odd one, arranged precisely in pairs — Group 37I

The legumes are one of the most ecologically important families of trees in Ghana forests, although relatively few species are exploited for timber, partly because the wood is often very hard. Taxonomists recognize three main Groups, defined by the flower structure. These Groups are sometimes treated as separate families – the **Caesalpiniaceae, Papilionaceae and Mimosaceae** – but in different books or herbaria they are treated as 'subfamilies' of the family Leguminosae, e.g. Leguminosae subfamily Caesalpinioideae. As usage of the second system becomes a bit tiresome, the three family method is used here, with the informal term 'legumes' used to refer to them all.

The flowers of the Caesalpiniaceae are represented by the ornamental 'flamboyant tree', *Delonix regia*, although few species (Gp 37F) produce individual flowers as conspicuous as these. The yellow flowers of *Cassia* species are, perhaps, more typical. The Papilionaceae, on the other hand, have flowers represented by the pea and bean plants. Flowers of the Mimosaceae are individually inconspicuous, but clustered into spherical heads, or elongated spikes, typified by the **roadside** climbers or **savanna** trees of *Acacia* or *Mimosa*, or the introduced 'rain-tree', *Samanea saman*.

All three families produce pods (with exceptions in each family). Pods (and other types of fruits of the legumes) exist in a wide variety of shapes and sizes. As they are often persistent, either on the tree or on the ground below, they can be helpful in identifying trees in the field and are therefore described below:

FRUITS OF THE LEGUMES (ii)

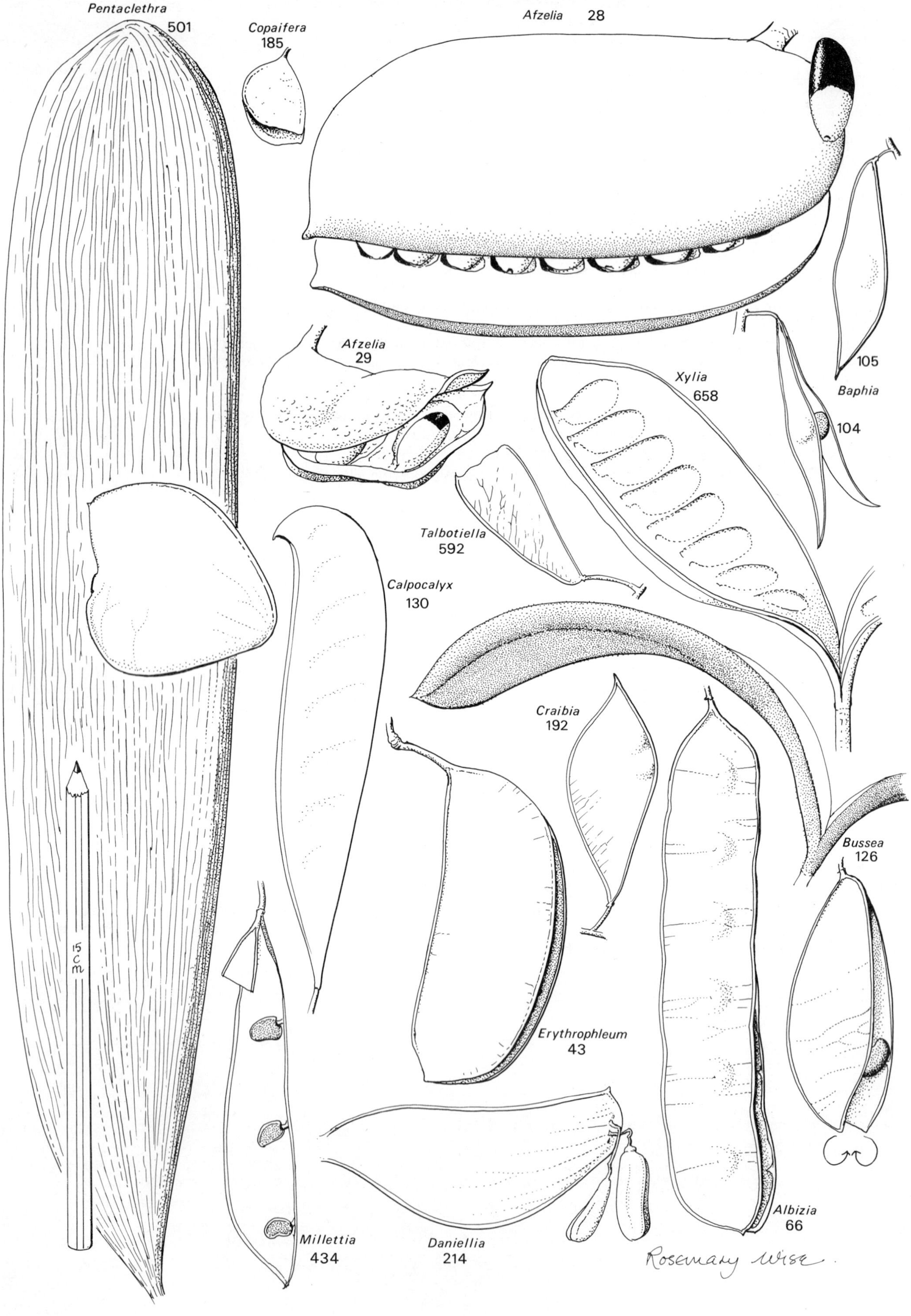

Fruits of the legumes (Groups 37-38)

A convenient (taxonomic and ecological) distinction can be made between fruits which split open at maturity to release the seeds for dispersal (dehiscent fruits) and fruits which do not split open on the tree, but which are dispersed as a whole. The first (A), summarized below, are either dispersed by the wind (especially in those with winged seeds) or by animals (especially in the genera with arils on the seeds). Some, however, may have no specialized means of long-distance dispersal, but seeds are sometimes ejected a short distance from the parent tree when the pods curl open explosively at maturity. Many of these trees grow gregariously, and are shade-tolerant as saplings.

Genera with indehiscent fruits are listed overleaf (B). 'Pods' of very similar appearance are in some cases (e.g. *Millettia, Albizia*) dehiscent and in other cases (e.g. *Lonchocarpus, Aubrevillea*) indehiscent.

(A) Dehiscent, ±flattened fruits ('pods' – splitting open when ripe)

		Group	Illustrated species
Winged, flat seeds released when thin papery to leathery pod splits usually along one edge			
Pods <0.5 m long (see p. 192)			
Attachment to pod by thread at apex of seed	*Newtonia*	(38A, B)	453
Attachment to pod by thread on (long) edge of seed	*Piptadeniastrum*	(38C)	20
Pods often >0.5 m long, v. softly orange-brown hairy; seed attached at its apex	*Cylicodiscus*	(38A)	39
Unwinged, ±flat seeds in ±flat papery or leathery pods			
Pods long and slender (>25 cm long, <10 cm wide), broadest towards 4 cm wide, pointed apex, glossy	*Chidlowia*	(37E)	147
Pods not so long or >10 cm wide for most of length			
Pods oblong, brittle-papery with fine lines (whole pod potentially wind-blown); thickened margin, with seeds attached to margin by fine 'thread' (see p. 194)	*Albizia*	(38B)	66
Pods not so thinly papery, or thickened only along upper edge, or asymmetric			
1) Pod papery, short; seeds remaining attached to one side of pod on *c.*2 cm threads (see p. 194)	*Daniellia*	(37D)	214
2) Pod ±smooth + rounded ends or obovate; not densely hairy (see p. 194)			
–(8x4 cm in *E. ivor.*, 15x6 cm in *E. suav.*); 4-10 seeds	*Erythrophleum*	(38A)	43
–1(-2) seeded, very elastic	*Cryptosepalum*	(37C)	
–With a longitudinal raised line; valves coiling inwards at dehiscence	*Didelotia*	(37A)	235
3) Pods small, slender, pointed at apex, few cm wide, 1-3 seeded often coiling inwards from apex >4 times as long as wide, oblanceolate or linear	*Baphia*	(37A)	104, 105 (see p. 194)
4) Pods not so slender, with an abrupt 'beaked' apex nearer 'backbone' than 'belly' of pod (see p. 194)			
>8 cm long but <3 cm wide	*Millettia*	(37I)	434
<8 cm long or >3 cm wide			
Cynometra ananta & Pellegriniodendron		(37B)	
Talbotiella, Hymenostegia		(37C)	592
Craibia		(37G)	192
5) Thick pods often about the size of adult's shoes soles, OR velvety OR v. wrinkled; halves curling spirally (edges inwards) when ripe (see p. 192)			
–With few raised lines on the side running end to end	*Gilbertiodendron*	(37F)	343, 342
–Smooth or with many fine diagonal bands	*Berlinia*	(37F)	110
–Irregular, pronounced diagonal wrinkles, or contorted	*Anthonotha*	(37E)	89
–Pods with seeds visible as bulges; softly hairy	*Crudia*	(37G)	203
Hard, thick and woody dehiscent pods – not easily snapped or bent (see p. 194)			
Seeds flat and without arils; pods normally with fine longitudinal lines or cracks			
Pods <25 cm long, cracking open explosively and peeling outwards (as banana-skin); comma (') or S-shaped			
Velvety, held above crown; 1-2 seeds	*Bussea*	(38A)	126
Hanging within and below crown; up to 10 seeds			
Often without seeds in bottom ⅓rd	*Calpocalyx*	(38A)	130
Usually with one or more seeds in bottom ⅓rd	*Xylia*	(38A)	658
Pods up to 40 cm long, oblong with seeds >5 cm across	*Pentaclethra*	(38B)	501
Seeds not v. flat, with red to yellow arils; pods not S or comma-shaped, splitting open on tree to reveal hard seeds			
Several seeds with arils like hats on end	*Afzelia*	(37D)	28, 29
Single seed (2-3cms) covered completely in aril	*Copaifera*	(37D)	185

Fruits cylindrical or resembling string of beads (dehiscent or indehiscent) – see next page...

(B) Fruits indehiscent or not flattened (not pod-like)

Indehiscent fruits may either be wind-blown as a whole, sometimes with the whole fruit resembling winged seeds inside some of the dehiscent pods mentioned above, or they may be adapted for dispersal by other means. *Detarium* drupes, with their coarsely fibrous inner layer resembling that of *Balanites* and *Panda*, are clearly adapted for elephant dispersal, whereas the pulpy, fruity-tasting material inside *Dialium* and *Parkia* (and the savanna *Tamarindus*) are known to be attractive to other animals, including primates. The dry, globose fruits of *Cynometra megalophylla* and *Pterocarpus santalinoides* are probably adapted to dispersal along watercourses.

		Group	Illustrated species
Fruits as a whole flattened, or papery-winged, surface often with raised veins			
–1(-2) seeds in middle; ±oblong elliptic, without very thickened margin			
With marginal nerve; <10 cm long, rounded to broadly elliptic	*Guibourtia*	(37B)	22
With marginal nerve; >10 cm long, oblong			
Usually 1-seeded with lax, v. raised reticulations	*Stemonocoleus*	(37D)	576
Usually >1-seeded with fine, not v. obvious reticulations	*Aubrevillea k.*	(38C)	99
Without m.n.; central veins on ft ±reticulate but ±parallel nearer margin	*Amphimas*	(37H)	76
–Usually 2+ seeded, with very thickened margin (typical pod shape); brittle-papery			
Pods usually dehiscent (although seeds not always released from wind-blown pods)		see *Albizia* (above)	
Pods indehiscent			
Usually <5 seeds; hairy when young, apiculate, with v. thickened margin			
With very dense, persistent soft hairs, margin undulate between seeds	*Lonchocarpus*	(37I)	391
Mature pods not densely hairy			
Stalk of ft >5 mm long; spine and young pod ±hairy	*Distemonanthus*	(37G)	42
Stalk of ft <5 mm long to joint with inflorescence; young pod hairy	*Pericopsis*	(37G)	11
Ft ±oblong with rounded ends and v. raised venation	*Aubrevillea p.*	(38B)	100
Usually >5 seeds – with thickened margin, leathery and persistent on tree			
Long, slender, many-seeded, in heads; seeds *c.*1 cm	*Parkia*	(38C)	489
Flattened with very thickened margin; seeds ½ cm or less	*Samanea*	(38C)	558
Cylindrical or string-of-beads or long and indehiscent fruits			
Pods long and slender, leathery, with fruity pulp between seeds	*Parkia*	(38C)	489
Constricted between fat seeds, coiled and splitting up into segments	*Cathormion*	(38C)	142
Contricted, like string of beads, but pod splitting lengthwise to reveal red seeds	*Erythrina*	(31C)	
Hard, brittle; cylindrical like long (to 50 cm) sausage; numerous globose seeds	*Swartzia*	(37G)	583
Hard, brittle; v. strongly winged; not splitting open	*Tetrapleura*	(38B)	600
Globose indehiscent fruits			
Not winged			
Small, velvety, button-like disks, or ellipsoid, + tasty pulp	*Dialium*	(37H(I))	228
Hard, not velvety			
V. coarsely fibrous inside, not v. bumpy; 1-seeded	*Detarium*	(37D)	226
Not coarsely fibrous inside			
Fleshy, with *c.*1 cm stalk	*Mildbraediodendron*	(37I)	432
Rounded, v. bumpy	*Cynometra megalophylla*	(37B)	211
With short 'prickles' and v. slight wing, wrinkled	*Pterocarpus santalinoides*	(37H)	532
With broad rounded papery wing around edge	*Pterocarpus*	(37H)	

FRUITS OF THE LEGUMES (iii)

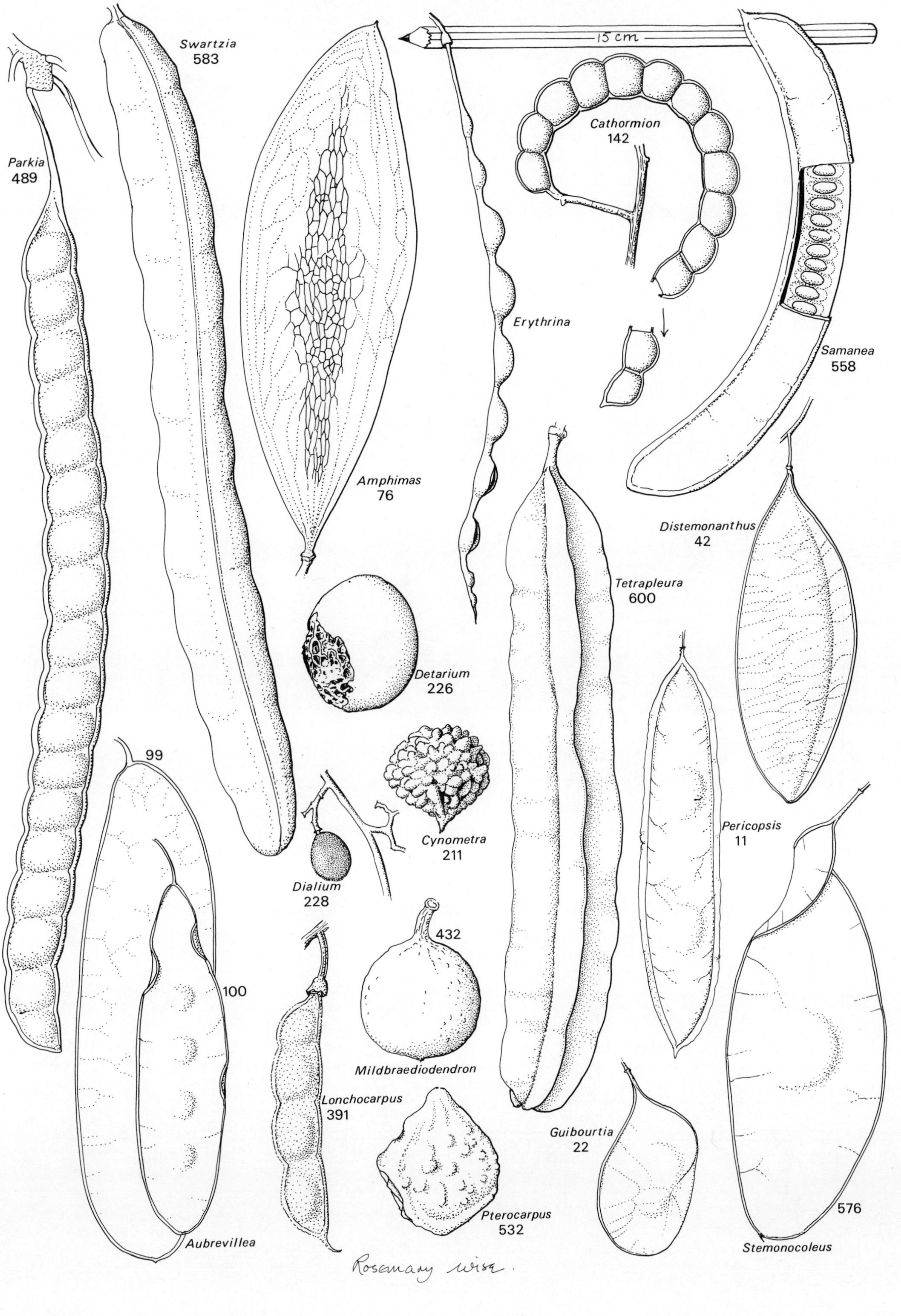

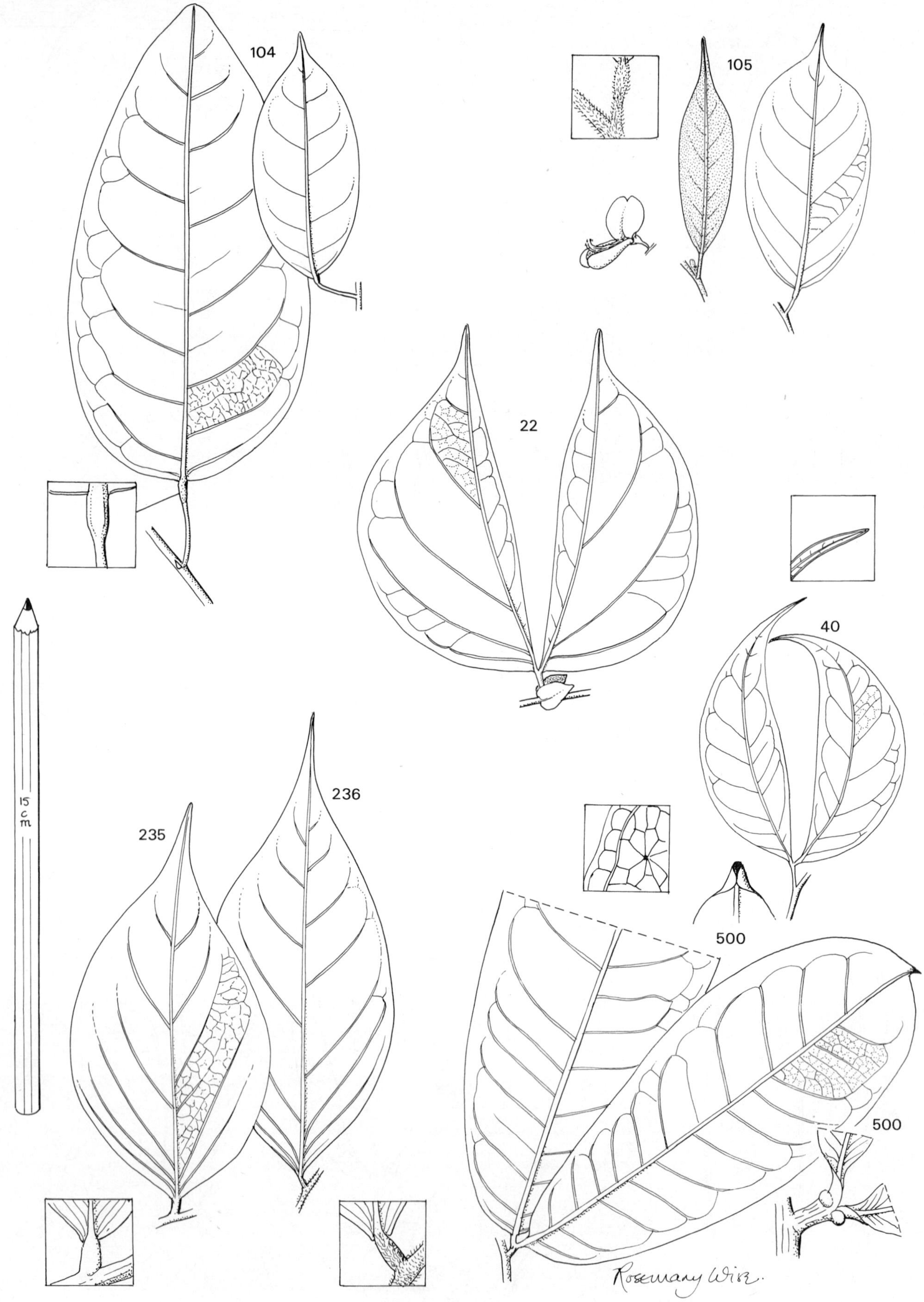
104
105
22
40
15 cm
235
236
500
500
Rosemary Wise.

Group 37A
(Simple-leaved legumes)

These two genera are unusual for legumes in having simple leaves. They do, however, produce flattened, dehiscent pods (and *Baphia* produces typical, bean-like flowers) so the family should be easily recognized when the trees are fertile. Both *Baphia* spp. are very common – *B. nitida* is Ghana's commonest forest tree. Their slash has the beany smell common amongst legumes. By sharp contrast, the last two species are extremely rare in Ghana as a whole.

Petiole slightly swollen towards tips; twigs with small crescent stipule scars at nodes; slash smelling of green beans or vegetables; very common small to medium-sized trees

- Leaves ±glabrous, except when young; petiole often >1.5 cm long; leaves with finely reticulate venation slightly prominent, especially when dried; slash yellow or creamy, fibrous, with slightly oily appearance, with hard cream or orange-ish sapwood — ***Baphia nitida*** (PAPI) [*oDWEN*] 104
- Leaves normally densely hairy, especially so when young, often sl. glaucous below; slash (pale brown/) orange-red, fibrous, + dark lines, with a pinker, more granular inner bark, and wet sapwood — ***Baphia pubescens*** (PAPI) [*oDWENKOBIRI*] 105

Petiole not swollen, but sometimes with two small, sharp, gland-like projections at top (stipellae); lf base with several, poorly-defined basal nerves; slash red, fibrous with slow, reddish, sticky exudate; new lvs red; straight, unbuttressed trees in **wet evergreen forest** (Ankasa forest reserve)

- Medium-sized trees of **swamps and riverbanks**; leaf base v. obtuse; lvs ovate with few or no hairs; twigs of flush of new lvs with many persistent stipules just above the petiole — ***Didelotia unifoliolata***[1] (CAES) 236
- Tall tree; leaf base cuneate; lvs narrowly ovate; twigs and nerves hairy — ***Didelotia idae***[1] (CAES) 235

NOTE: 1) Very few collections of the last two species have been made, so it is not at all clear how reliable will be the distinctions listed in the key.

Species in the next two Groups are often gregarious.

Group 37B (Caesalpiniaceae (part)
(Leaves with a single pair of leaflets)

Leaflets with <5 main pairs of laterals which do not meet; apex acuminate; stipules leafy and persistent; fresh lvs sometimes with translucent spots visible; bole ±cylindrical, with small neat buttresses; outer bark greyish to yellowish, often with raised 'crease-marks'; slash yellow-orange or pink-brown, brittle, granular and gritty, with a darker inner bark and shiny, damp orange-cream sapwood, slightly sweet-scented and bitter tasting; with a slow gummy, sweet exudate; very common in **northwestern semi-deciduous forest**, where the seedlings with the characteristic foliage are often extraordinarily common; pods flat, *c.*4 cm long, covered in raised glands — ***Guibourtia ehie***[1] [*ANOKYE-HYEDUA*] 22

Leaflets with >5 main pairs of laterals which meet in a sub-marginal nerve; evergreen forest (or swamps); slash reddish.

- Leaflets with (long) acuminate tips; slash reddish over yellow, hard fibrous or leathery; sapwood + ripplemarks; straight bole becoming orange-ish with age, with thin buttresses creeping along the ground (like *Dahoma*); crown large, spreading; fts flat, with sl. raised nerves oblong to 8 cm and pointed; **evergreen forest** — ***Cynometra ananta***[1] [*ANANTA*] 40
- Leaflets with short, thickened or folded acumen or margin with glandular notches; **small trees**
  - **Swamp** tree often outside evergreen forest, with large stipules; juvenile lvs — See *Gilbertiodendron (limba)*, Gp 37F
  - **Evergreen for.** tree with thickened acumen which is often contorted or rolled up; venation prominent on both sides; sometimes with stipels on petiolule; slash v. hard, thick, fibrous, dark, reddish brown; **evergreen forest** — ***Pellegriniodendron diphyllum*** [*FeLeFeLe*] 500

NOTES: 1) The seedlings of *Cynometra ananta* have no stipules, undeveloped cotyledons and opposite first lvs; *Guibourtia* seedlings resemble those of mature trees, with stipules, and they have leafy cotyledons.

2) Species in Groups 36A and 37F sometimes produce only two lflets per leaf.

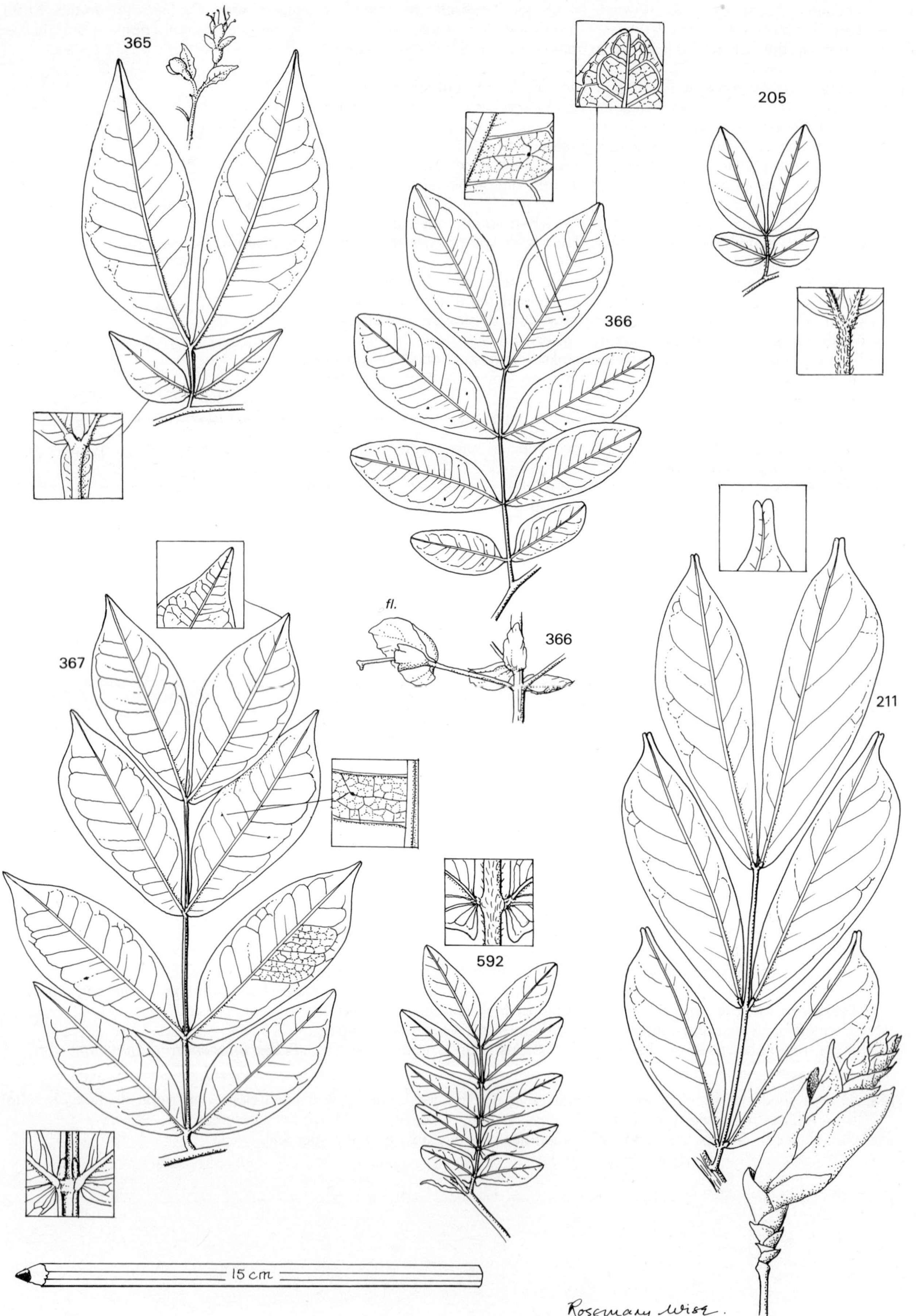
365
205
366
fl.
366
367
211
592
15 cm
Rosemary Wise.

Group 37C: Caesalpinaceae (part)
(2-4 prs of leaflets or lflets narrow)

Most of the following species have overlapping, shiny bud scales which are often still visible at the base of new shoots. The bark is smooth, and the slash often has reddish patches. The trees are often locally abundant.

Lvs with up to 4 pairs of leaflets; lflets normally >3 cm long; trees often in **moist forests**		
Rachis strongly winged between the 2 pairs of leaflets; the end pair easily the largest; rachis very winged between leaflet pairs; slash gritty yellow with darker inner bark and red patches, with sweet taste; widespread and common even in dry forests; pods flat, dehiscent	***Hymenostegia afzelii*** [*TAKROWA*]	365
Rachis not strongly winged, or lflets >2 pairs		
Leaves usually with two pairs of leaflets with conspicuous hairs on nerves, etc.; bark smooth, greenish with brown lenticels lengthened sideways; slash pink-red, hard-fibrous, with brownish exudate; tall, straight tree in **evergreen forest**; pod flat, smooth, rounded 6×2 cm, 1-(2) seeded	***Cryptosepalum tetraphyllum***	205
Leaves often with more than two pairs of leaflets, with few hairs; often with glands; often by rivers or on rocky soils		
Leaflet apex not emarginate, often acute; lvs typically with 4 pairs of leaflets; rachis deeply grooved, and young twigs sometimes grooved as well; rachis often with a small pair of tiny, waxy, gland-like structures attached to rachis at base of middle pair(s) of leaflets; sometimes gland-dotted at base of lamina as well; **riversides in evergreen forest**	***Hymenostegia gracilipes***	367
Leaflet apex with a fine notch (emarginate); usually 3 pairs		
Leaflet tip not sharply acuminate around the apical notch; lamina, especially on young flush, with glandular spots in several places between the laterals; sometimes slightly hairy; tree **typically by rivers** in western region, with many adventitious shoots and not v. straight; in Krokosua hills; slash reddish	***Hymenostegia aubrevillei*** [*ABABIMA-KoKoo*]	366
Leaflet tip markedly acuminate, with a sharp notch in the acumen (apex therefore with a minute fork); buds with many overlapping stipules (bud scales); new lvs pinkish, pendulous, with many long stipules; **usually by rivers**; low-branched; fruits knobbly and fat, *c.*4 cm diameter	***Cynometra megalophylla*** [*ANANTA-AKOA*]	211
Lvs often with more than 4 pairs of small, ±oblong or rhombic lflets, (<3 cm long)		
Leaflets 4-7 pairs; rhomboidal, with deep lobe like small butterfly wing at base; small, *c.*1 cm wide, ±glabrous; youngest twigs with leafy stipules at nodes when flushing, and with brown hairs, but soon glabrous; gregarious tree **between dry forest and savanna**, along the Afram or Volta basins; slash rather fibrous-peelable, reddish with orange-brown inner bark; pods almost triangular, flat-ended, *c.*5 cm long	***Talbotiella gentii***[1] [*TAKROWA-NUA*]	592
Leaflets different: consider bipinnate spp. Group 38B		

NOTE: 1) **Tamarindus indica** is often planted in drier areas; it has 10-15 pairs of lflets otherwise similar to those of this sp.

Group 37D
(Leaflets < 15 cm long; with marginal nerves or gland spots; 3 or more pairs)

Leaf apex with a small notch (emarginate – sometimes acute as well); unbuttressed trees

- **Leaves with leaflets always, evenly alternate**; often with obvious translucent spots
  - **Moist to evergreen forest** tree; laterals rather irregular, often in < 12 pairs joining in loopy sub-marginal nerve; translucent gland-dots not particularly obvious on mature lvs; lvs ±ovate, and acute or rounded apex with notch, pale green, thin-papery both surfaces; bole straight, ±cylindrical; outer bark fairly smooth; slash pale brown (yellower nr wood) fibrous, not hard with faint vertical bands; inner bark scented like green beans; fts yellowish, thin and papery to 15 cm, indehiscent, winged with raised veins with marginal nerve — ***Stemonocoleus micranthus*** 576
  - **Dry forest**, often low-branched tree; laterals from arm's length closely parallel, often in > 12 pairs, **with v. obviously gland-dotted lvs**; lf apex rounded, not drawn out towards apical notch; many laterals joining marginal nerve; crown dark-green, spreading; bark with large, prominent, round lenticels (3 mm wide); slash thick-fibrous, crumbly and gritty, red-brown, sometimes + paler contours; fts thick, rounded, sl. flattened, *c.*3 cm diam. v. fibrous around seeds — ***Detarium senegalense***[1] [*TAKYIKYIRIWA*] 226
- **Leaflets almost, or exactly opposite**, oblong-elliptic, sometimes v.[2] asymmetric at base; apex of leaf obtuse; veins very conspicuous; margin thickened; petiolule twisted; rachis with yellowish hairs; bole v. cylindrical above root spurs, except for old branch scars; bark rough; slash orange to pinkish brown, very hard fibrous, sometimes contoured, with distinctive sweet scent like almonds or pipe tobacco; pods elliptic, rather 'fat', sticky *c.*3 cm long; single black seed with red aril — ***Copaifera salikounda*** [*ENTEDUA*] 185

Leaflet apex not emarginate; acute to acuminate

- *Petiolules twisted*; lf without hundreds of fine glands, but some glands normally visible beside base of midrib, and sapling shoots sometimes with *c.*10 others along lamina (visible without lens); lf margin v. thickened; fts woody, with hard black seeds with reddish, waxy aril on one end
  - Lflets falcate – narrowly elliptic or lanceolate; margin thickened, but not fusing with many adjacent veins; venation regular; shade lvs f. acuminate; bark rather smooth, with hoops or scars; slash yellow to pink, granular, crumbly with gritty orange streaks, scented like tobacco or almonds; pods curved round; **not in driest forests** — ***Afzelia bella***[3] [*PAPAO-NUA*] 29
  - Lflets barely falcate – broadly elliptic or ovate; marginal nerve fusing with many adjacent veins; venation conspicuous and irregular (especially sun lvs), pale and prominent on both surfaces; **dry forests**, rocky hills and steep slopes etc.; bark rough with scales falling to leave rounded, paler areas; slash pink-brown, fibrous and granular or gritty; boughs large and low; pods oblong with rounded ends, *c.*10-15 cm long, not curved round — ***Afzelia africana*** [*PAPAO*] 28
- *Petiolules not twisted*
  - Leaflets glandless; lflets elliptic-oblanceolate, lower lflets smaller than upper — See Group 37E
  - Leaflets with hundreds of regular dark spots at the end of finest veins, within the vein network; lflets ±oblong or lanceolate, apex acute, base v. asymmetric NOT falcate; young twigs with **long stipules at apex, falling v. rapidly leaving scars**; *tree with v. hard,* **v. cylindrical, unbuttressed bole**, with large lenticels in rings; slash granular, reddish brown, with slight sweet scent; ripple marks in sapwood v. broad and easily visible; tree producing gum often to be found in soil at base of bole; pod flat and elliptic, 1-seeded, *c.*6 cm long
    - **Evergreen forest** tree with glabrous inflorescences; slash dampening soon — ***Daniellia thurifera*** [*SOPI*][4] 215
    - **Semideciduous or evergreen forest** tree with hairy inflorescence; slash remaining dry (?for several hours) — ***Daniellia ogea*** [*eHYEDUA*] 214

NOTES:
1) Two types of *D. senegalense* fruits can be recognized: according to Aubréville (1959) some trees produce sweet edible fruits, whilst others produce toxic ones. This is an interesting parallel to *Irvingia* (Gp 13C).
2) *Copaifera* is much commoner than the previous two, rare species. Sapling lflets, often gregarious in the shade around parent trees, have more obvious, fine translucent spots.
3) The two *Afzelia* spp. are confused under the names *PAPAO* (or *OKRO-SANTE*) by tree spotters and the timber trade. Saplings are similar, but often have more obvious glands along the lamina than the adults.
4) *Daniellia* spp. are probably indistinguishable under most circumstances. Savill and Fox (*Trees of Sierra Leone*) note that *Daniellia ogea* produces a longer utilizable bole than *D. thurifera*. Saplings, found mostly in exposed sites, have glaucous lvs and the very conspicuous sheath of stipules around the twig tips.

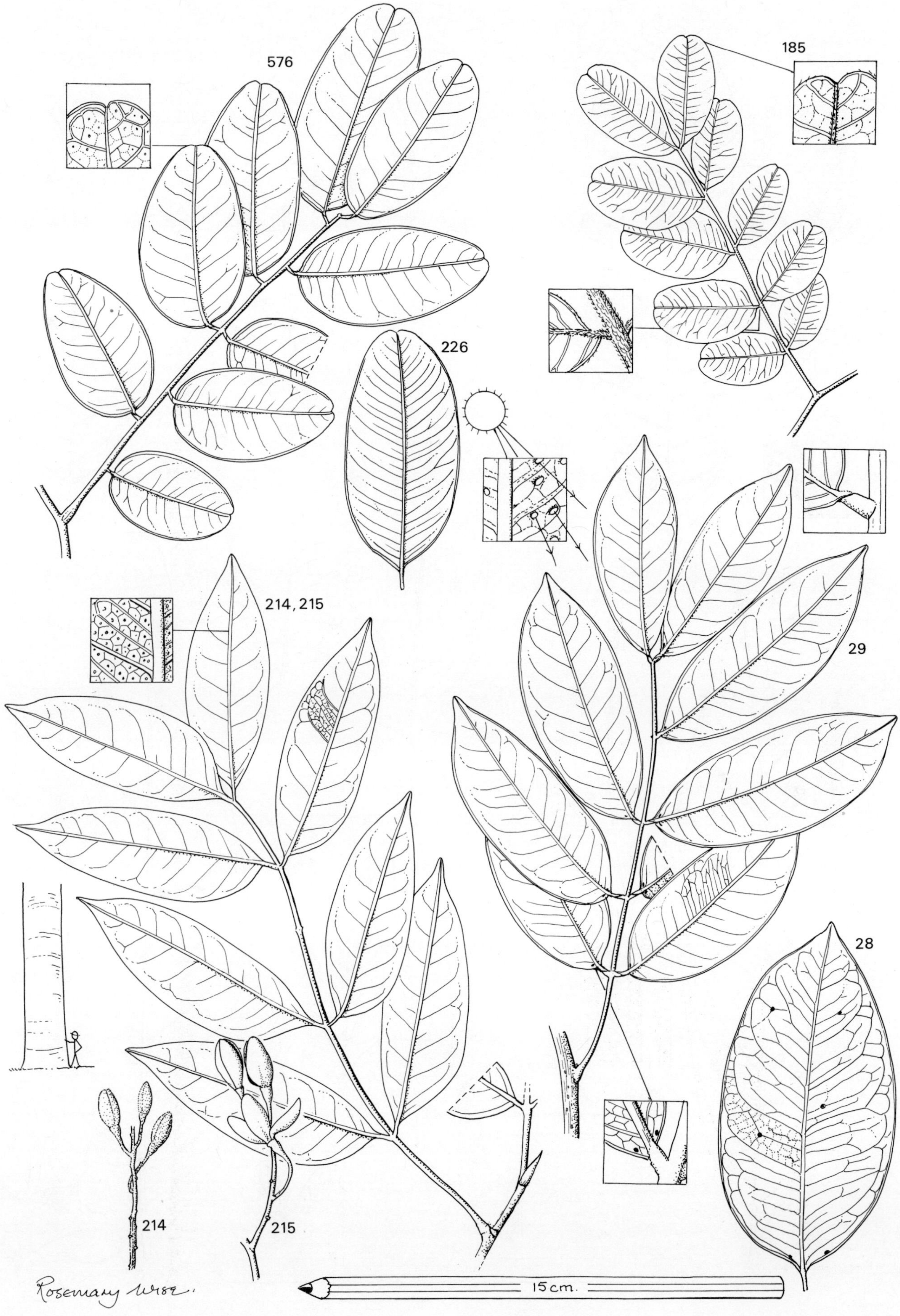
576
185
226
214, 215
29
28
214
215
15cm.
Rosemary Wise.

88
89
90
147
112
110
111
15cm
Rosemary Wise

Group 37E: Caesalpiniaceae (part)
(Lflts discolorous or hairy or glaucous below; usually <15 cm, symmetrical)
(Pods velvety, flattened, *c.*1 cm thick, with diagonal ridges or wrinkles)

Leaflets (and crown) conspicuously red-brown, discolorous; laterals many (>15), prominent, and parallel, meeting in sub-marginal nerve; rachis brown-hairy; margin recurved; apex acute; **moist forest**[2] canopy tree with brown crown due to dense hairs on lvs; bark brown, ±smooth with scales and fissures; slash fibrous with orange-gritty outer layer, slightly peelable, darkening, with whitish, **sometimes latex-like exudate nr sapwood**; pods >8 cm long, *c.*1-seeded, with conspicuous raised wavy lines — ***Anthonotha fragrans*** [*TOTORONINI*] 88

Leaflets not densely brown hairy OR with fewer, less regular laterals; small trees sometimes in drier forests

- LFLETS normally with hairs, without marginal nerve
  - Lflets (2-3 pairs) thin, long acuminate (drip-tipped, often apiculate) with a few glands as knots at end of venation towards apex of lf; with rather coarse hairs at first, becoming less hairy; not discolorous; small, uncommon tree often by **riversides**, in **evergreen forest** or in **coastal forests on sand**; slash red-brown, brittle; pods <8 cm long, rough — ***Isomacrolobium vignei***[1] [*TUTUABO*] 90
  - Lflets (3-4 prs), silvery or whitish below with v. small hairs; often with many **irregular** 'knotted veins' (but this is possibly insect attack); bole often crooked, with many adventitious shoots and low boughs; slash orange-brown or red contoured with brown, fibrous and gritty, darkening, ±brown exudate; pods >15 cm long, *c.*4 seeded; common and widespread understorey tree — ***Anthonotha macrophylla*** [*TOTORO*] 89
- LFLETS thin, not hairy, merely glaucous-discolorous; often in more than 4 pairs; lflets with orange (esp. when young) translucent veins and marginal nerve; tree twisted or fluted, with many adventitious stems; bark v. lenticellate; slash thin, red-brown to cream near sapwood, darkening, fibrous, with fruity acidic taste; pods strap-like, *c.*½ m long, slender, shiny and glabrous; lvs dry reddish brown — ***Chidlowia sanguinea*** [*ABABIMA*] 147

NOTES: 1) *I. vignei* was previously known as *Anthonotha vignei*. Unlike *Anthonotha* species, which have one posterior petal larger than the rest, *I. vignei* has four ±equal-sized petals. *Chidlowia* has slender, conspicuously red inflorescences of small flowers.

2) *A. fragrans* is most typically abundant in evergreen forest, but is sometimes found in low-lying parts of the landscape elsewhere.

Group 37F
(Pods usually approximating to shoe soles in shape and size; each side curling inwards when mature)

Lflets with one or more (glandular) 'notches', holes or teeth on the margin or at the apex OR stipules large and leafy; often v. long; often broadest around the middle; usually with large, leaf-like stipules; pods with longitudinal ridges; bole cylindrical

Leaflets with dense, fine hairs below (like spots under lens); discolorous, and up to ½ m long; in 2-4 pairs; stipules up to 5 cm long; lflet base very asymmetric, one side often cordate; medium-sized tree in **swamps** in **evergreen forest**; pod v. large and hairy, with 3 longitudinal ridges — ***Gilbertiodendron splendidum*** [*AGYAMERA*] 343

Leaflets without dense, fine hairs below, (OR outside evergreen forest swamps)

Rachis without dense, soft, chocolate brown hairs; large trees

Lflets in 3-5 pairs, with a few, inconspicuous appressed hairs; **stipules falling** soon; lflets acuminate, with a ±cuneate, slightly asymmetric base; venation fine-reticulate, and prominent; pods with 1 or 2 longitudinal ridges; large tree gregarious in **evergreen forest** canopy, or **riversides**; bark with yellowish scales; slash hard and pale brown to reddish — ***Gilbertiodendron preussii*** [*TETEKON-GYAMERA*] 342

Lflets up to 7 prs; velvety pods with 2-3 ridges; **stipules persistent often *c.*2-5 cm long**; tree in **swamps** often outside evergreen forest; bark v. flaky; slash hard and pink — ***Gilbertiodendron bilineatum*** [*TETEKON-NUA*] 340

Rachis (at first) with dense, chocolate brown hairs; lf with (1-)2-4 prs leaflets and short petiole; stipules kidney-shaped, *c.*2 cm long; pod with only 1 ridge (and many fainter diagonal lines); small tree **in swamps**, widespread — ***Gilbertiodendron limba***[1] [*TETEKON*] 341

Lflets with no marginal glands, nor persistent stipules; often broadest above middle; base of lamina on one side of petiole overlapping on top of petiole base of lamina on the other side; lflets typically slightly falcate. Pods ±many and diagonal ridges

Lflets >15 cm long; slash without smell of curry powder, but usually stringy fibrous with beany smell — (illustrations on p. 204)

Lflets with fine hairs flat against the lower surface; tall tree in **evergreen forest** with dense, dome-shaped, dark crown on cylindrical bole; slash thick fibrous, orange-brown, paler inwards, darkening, with a strong 'green', vegetable-like smell, a little like green beans; **evergreen forest**; pods glabrous; flwrs + 5 equal petals — ***Berlinia occidentalis***[1] [*KWATAFOMPABOA*] 111

Lflets glabrous; venation finely reticulate and prominent on both surfaces; lflets paler below; shaded lflets with v. long drip tips; medium-sized, very spreading tree in **swamps, riverbanks**, etc. **or evergreen forest**; pods with soft, dense, orange-brown hairs; one petal much larger than the others, and deeply 2-lobed — ***Berlinia tomentella***[2,3,4] [*KWATAFOMPABOA-BERE*] 112

Lflets <15 cm long; rachis normally <5 cm long; **with ginger hairs especially at lflet junctions**; lflet ±v. fine hairs; bole not normally cylindrical, but often slanted or otherwise irregular; bark with large scales; slash pinkish brown over paler brown inner bark, hard-fibrous, darkening, slightly sticky, with strong, distinctive sweet and slightly bitter smell of curry powder or sawmills; young lvs bright red; pod glabrous; one petal larger than the rest, and slightly notched; widespread — ***Berlinia confusa***[4] [*KWATAFOMPABOA-NINI*] 110

NOTES: 1) There is considerable confusion arising also from the great regional variation in the application of the names *TETEKON* and *KWATAFOMPABOA* (the latter referring to the pods like 'leper's sandals').

2) If the leaflets are NOT FALCATE, nor strongly oblanceolate at this point, then check *Isomacrolobium*, etc. in Group 37E.

3) *Paramacrolobium coeruleum* has not been discovered yet in Ghana, but it may occur here. It is a medium-sized tree with 3-5 pairs of glabrous, glossy lflets which will probably key to this point, but which has *half-cylindrical* stipules immediately above the petioles on the twigs. It may well be found along rivers in the drier forest zones.

4) Leaflets shorter than 15 cm are sometimes produced by *B. tomentella*, so check the rest of the description carefully. These two species can be very hard to tell apart when infertile, but the slash scent of *B. confusa* seems distinctive. *B. confusa* is the commoner tree in **semideciduous forest**, particularly away from rivers and swamps.

343
340
341
340
342
341
15 cm
R. Wise

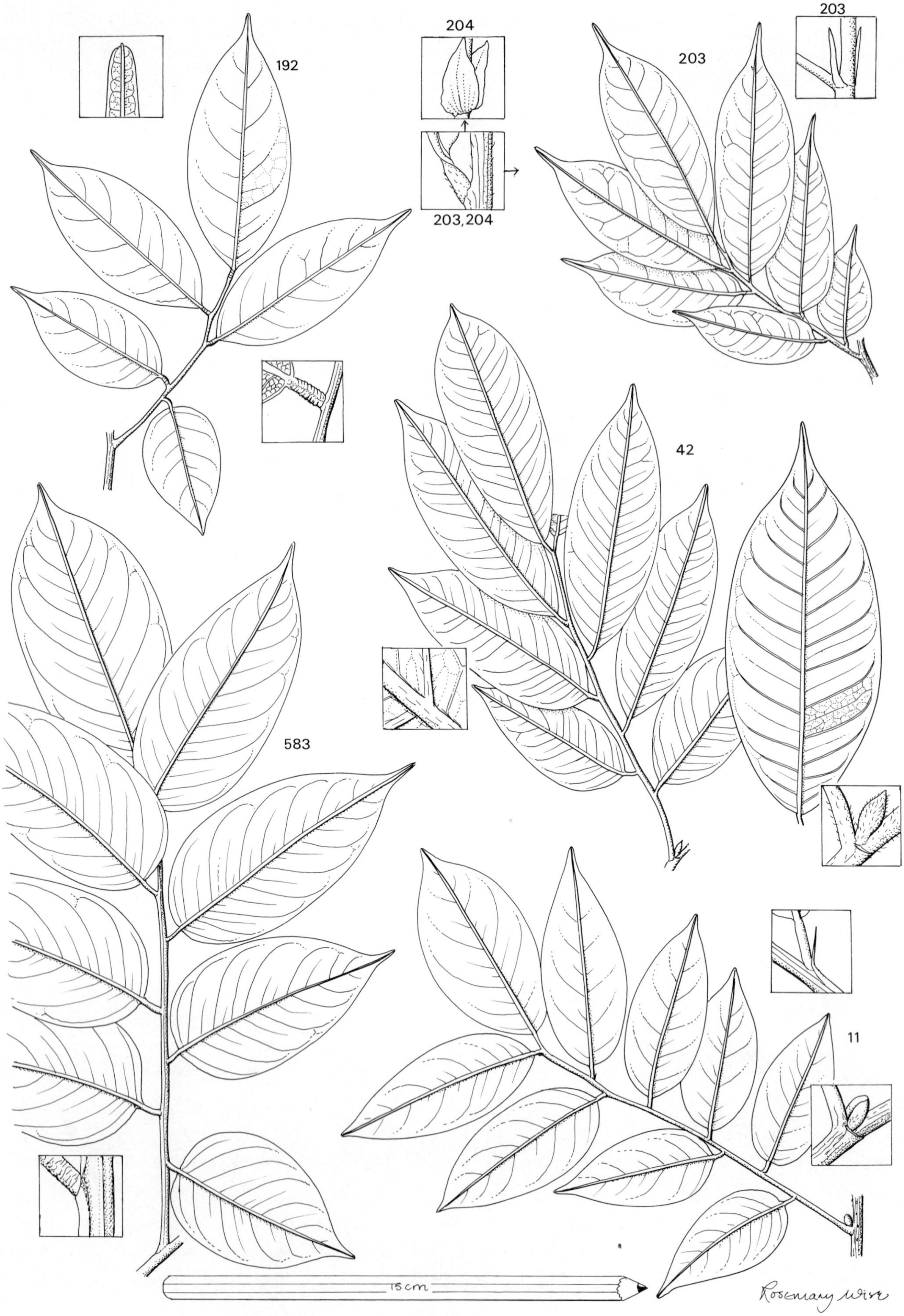
192
204
203
203,204
203
42
583
11
15 cm
Rosemary Wise

Group 37G: Caesalpiniaceae-Papilionaceae (part)
(Leaflets not paired; usually with many fine (brown) hairs below)

It is very easy to misinterpret the bipinnate nature of species such as *Bussea* in Group 38A; several of the larger leafleted species in that group have pinnae which resemble the whole leaves of this group. Outside evergreen forest, trees keyed correctly to this Group will usually be one of the last two species, which are important timber trees. Both have graceful crowns, with outward-sweeping plumes of foliage. Stipels distinguish the leaves of *Pericopsis*. The red colouration of the younger bark of *Distemonanthus* is generally in larger patches, and more flaky in texture than the smaller, redder patches of *Pericopsis*.

Leaves without stipels; venation normally clearly visible below
- Petiolules twisted; lflets usually <3 cm wide, acuminate; laterals slightly erratic, barely raised above, often joining; rachis twig-like, not grooved; pods densely hairy at first, flattened, wrinkled and bulging with 2x4 cm seeds; with sharp beak
  - **Evergreen forest** tree with straight bole, compact crown; stipules *NOT* v. persistent, but sometimes linear, 1 cm long in pairs on youngest twigs; lvs with hairs lying flat below; bole straight; bark rough with age; slash orange-brown, contoured, hard-fibrous, brittle, scented (unripe chilli pepper); pods to 20 cm long; 2-5 seeds — ***Crudia gabonensis*** (CAES) [*SAMANTAANINI*] — 203
  - **Riverside or thicket** tree with v. spreading crown, stout twigs with thick buds; lvs glabrous; stipules leafy and persistent, up to 2 cm long; bark scaly, rough and brown; slash red-brown, stringy-fibrous; pods + 1-2 seeds — ***Crudia senegalensis*** (CAES) — 204
- Petiolules not twisted but sometimes swollen
  - Small, often many-stemmed tree in southern type of **dry forest** (Shai Hills), with swollen, v. rough petiolules, with (c.5) ovate-acuminate, GLABROUS, lflets; often + a tiny mucronate point, drying yellowish below; pods dehiscent, 1-seeded — ***Craibia atlantica*** (PAPI) — 192
  - NOT *Craibia* – taller trees of **moister forests**; lflets + hairs: pods indehiscent; lflets usually >3 cm wide, acute or rounded at apex; laterals regular and parallel
    - Lflet base obtuse to cordate; ±ovate; rachis grooved; bole not v. straight, unbuttressed; slash cream with yellow-brown streaks and copious watery exudate; fruits long and cylindrical — ***Swartzia fistuloides***[1] (PAPI) [*ASOMANINI*] — 583
    - Lflet base obtuse to cuneate; laterals v. prominent above; crown blue-green; **bark vivid red-brown on younger parts becoming pale green or yellowish, flaking**, and lenticellate below the flakes; larger trees with high, narrow buttresses; slash (with green outer layer) orange to pink-brown with pale contours, with brittle and fibrous layers, sticky, with savoury smell; sapwood cream with obvious ripple marks; fts thin, flat, hairy at first, with slightly raised reticulation — ***Distemonanthus benthamianus*** (CAES) [*BONSAMDUA*] — 42

Leaves with stipels; venation not v. clear below with naked eye because most veins very fine; underside ±glaucous; **bole often uneven with bark scales falling to leave red patches;** slash (green/) yellow, darkening; pods flat + pointed ends + v. reticulated surface, 2-4 seeds — ***Pericopsis elata*** [*Afrormosia, KOKRODUA*] (PAPI) — 11

NOTE: 1) *Swartzia* was previously in Caesalpiniaceae but has recently transfered to Papilionaceae.

Group 37H: Caesalpiniaceae-Papilionaceae (part)
(Imparipinnate; fine veins prominent; v. few or no hairs)

The note at the top of Group 37G about confusion with Group 38A applies here as well. Trees of this Group (and Gp 37I) typically produce a red exudate in the slash, as well as having the musky vegetable scent (e.g. green beans) common in the bark of many legumes.

Leaflets not acuminate; with acute or rounded apex OR lvs with stipels
  Leaflet apex emarginate, with twisted petiolule or marginal nerve — See Group 37D
  Leaflet apex obtuse to acute, or lacking both twisted petiolules and marginal nerves
    Leaflets often more than 5/leaf; without conspicuous dots along veins OR tree in **moist forest**
      Leaf with stipels; young leaves with leafy stipules at base, v. hairy on rachis, with dense red-brown hairs on young stems; stipules leaving ½ ring scars at nodes; saplings and crown like *Entandrophragma*, but crown spreading and flattened; bole straight, with small, thick buttresses; bark rough with rectangular fissures; slash orange to brown, thick, soft, brittle-fibrous, with red grit and blood-red exudate; fts flat, indehiscent with papery wings and raised veins, *c.*20 cm long — ***Amphimas pterocarpoides***[1] (PAPI) [*YAYA*] 76
      Leaf with no stipels nor leafy stipules; oblong, with rounded or emarginate or acute apex; slash cream with red spots of exudate; fts flattened papery discs *c.*10 cm diameter — ***Pterocarpus mildbraedii***[2] (PAPI) 531
    Leaflets 5 (usually) per leaf; ±conspicuous dots along veins on lower surface; venation v. reticulate, raised above; twigs v. lenticellate; slash cream, with red spots; wood with ripplemarks; fts velvety, rounded (*c.*1 cm diam.), flattened with 1 seed and fruity taste; **v. dry forest or savanna** — ***Dialium guineense*** (CAES) [*ASENAA*] 229

Leaflets usually acuminate; lvs without stipels
  Lflets with twisted petiolule – go to Group 37G (*Crudia* spp.)
  Lflets with untwisted petiolules
    Leaflets 5 or fewer per leaf; glabrous, ±pustulate below; lft base usually cuneate; **margin recurved**; thicker than paper and v. glossy above, with prominent, regular reticulations; *fresh lvs with fruity acidic* (like oxalic acid) taste; bole often slightly fluted, or with large, high buttresses; bark with large plate scales; slash reddish, with spots of red exudate, and fine ripple marks; tall spreading tree with dark foliage in crown; fts flattened, + ridges, 2 cm diameter, glabrous — ***Dialium aubrevillei*** [*DUABANKYE*] 227
    Leaflets more than 5, with a few, scattered hairs; base usually ±rounded; lflet ±ovate; papery with prominent, regular reticulate venation; small **riverside** trees (rarely v. large or away from water); fts barely winged, rounded with spiny centre, 3 cm diameter — ***Pterocarpus santalinoides***[2] (PAPI) [*HoTE*] 532

NOTES: 1) *Aganope leucobotrya* (*YAYA-AKOA*) is a small straggly tree with foliage very similar to that of *Amphimas*. It can be distinguished by its very slender, sharp-pointed stipules and by lack of the dense red hairs on the young stem which are so conspicuous in *Amphimas*. Young *Amphimas* resemble *Entandrophragma* saplings, with the large compound leaves clustered at the top of unbranched stems. Crown leaves of *Amphimas* are much smaller, with less prominent venation.

2) If most leaflets >10 cm long, and the key has lead to *P. santalinoides* for a sapling leaf, then the tree is probably a (shaded form) of *Pterocarpus mildbraedii*. *P. mildbraedii* seems rare in Ghana, whereas *P. santalinoides* is a very common small tree of riversides. Very confusing sterile specimens which match closely those of *P. santalinoides* have been collected from large trees on dry land, however and *P. mildbraedii* may be more common than suggested here.

76
531
227
227
229
229
532
15 cm
Rosemary Wise

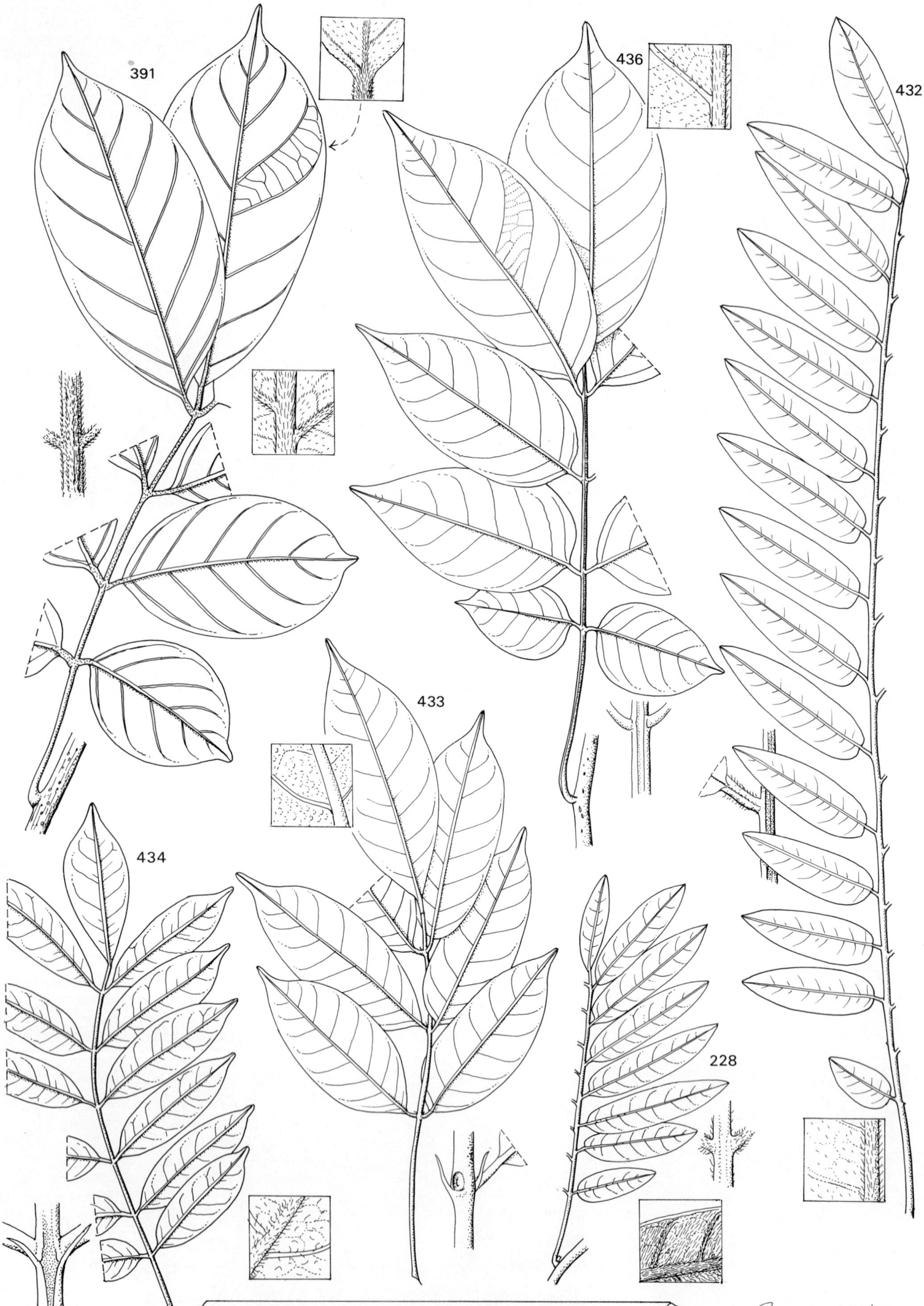
391
436
432
433
434
228
15cm
Rosemary Wise.

Group 37I: Papilionaceae-Caesalpiniaceae (part)
(Lflets (except terminal one) precisely paired, OR >12 per lf)

The slash of trees in this Group (as well as Gp 37H) typically produces a red exudate and smells of beans, peas or other vegetables, although these characters are not always found, especially in the dry season.

Leaflets 12 or fewer per leaf, OR >2 cm wide; always paired, except for terminal one; >1 cm wide

- Stipels absent or v. small; finer veins clearly visible; venation often prominent above; lflets often >3 cm wide
  - Rachis and lflets with hairs brown and dense; hairs in midrib above, and usually abundant on rest of lf as well; finer veins impressed-reticulate below, forming a matrix of ±square islands; tree often in wet places; bole ±cylindrical; slash complex, many-layered with gritty streaks, yellow to brown, with red to black spots or lines of exudate, spongy; pods to 12 cm, v. hairy — ***Lonchocarpus sericeus*** (PAPI) [*SANTE*] 391
  - Rachis almost hairless; lf with very fine, appressed hairs, slightly glaucous below; apex acuminate; rachis with two sharp ridges on upper surface; common, untidy often straggling or climbing small tree in **secondary forest**; slash v. pale; pods 14 cm long, + few ginger hairs, with wavy margin; flower buds with dense golden hairs — ***Millettia zechiana***[1] (PAPI) [*FAFRAHA*] 436
- Stipels present; finer veins sl. obscure
  - Lflets >2 cm wide; tree with coarse red-brown hairs on young parts and large stipules; tree or sapling with *Entandrophragma* habit; venation clearly visible; lflet arrangement variable — ***Amphimas pterocarpoides*** (Gp 37H) (PAPI) 76
  - Lflets <2 cm wide or hairs pale yellow below; rachis glabrous; lflets often pustulate below; small tree in **evergreen forest**; slash orange, with fine vertical lines; pods to 10 cm, indehiscent, glabrous — ***Millettia griffoniana*** (PAPI) [*TETETOAKOA*] 433

Leaflets >12 per leaf; *c.*1 cm wide, oblong, oblanceolate or sl. lanceolate

- **Lvs with stipels**; leaflets precisely opposite; glaucous or silvery below, with longer hairs on nerves; rachis grooved, petiolules not v. grooved; rachis, lflets with many long rust-coloured hairs; bright green in crown; laterals normally <10 pairs; slash yellow-orange, fibrous-peelable, slightly gritty, often with darker inner bark due to red to brown watery exudate, with v. peppery taste — ***Millettia rhodantha*** (PAPI) [*TETETOA*] 434
- **Lvs without stipels**; leaflets often not exactly opposite, often with translucent spots (on new lflets)
  - Hairs on mature leaflets and rachis not conspicuous; midrib and petiolule with a fine groove; midrib deeply striate below; rachis grooved; laterals not v. distinct from finer venation, but >10 meeting in vague sub-marginal nerve; slash '?' without red exudate; ft rounded, rather leathery and fleshy — ***Mildbraediodendron excelsum*** (PAPI)[2] 432
  - Hairs on mature leaflets and rachis brown, dense and conspicuous; petiolule not channelled; rachis not grooved; laterals clearly defined, often with >10 meeting; bole uneven, 'bumpy', with small, steep buttresses; bark thin, gritty and brittle orange-brown or white turning brown, with spots of red exudate; sapwood white with ripple marks; fts ellipsoid, velvety with acidic pulp — ***Dialium dinklagei*** (CAES) [*DWEDWEEDWE*] 228

NOTES: 1) *Millettia thonningii* is a **savanna or extreme dry forest** tree often planted in villages; it has a petiolule *c.*5 mm long; veins ±obscure; lflets ±glabrous, except for a **small fringing tuft at base**, on lower side, next to midrib; the apex is not acuminate; flowers are decorative and purple; the pods like those of *M. zechiana*.

2) *Mildbraediodendron* has recently been transferred from Caesalpiniaceae to Papilionaceae. These and similar genera represent the grey area between the families which can be used to justify the change in rank of the legume families to subfamilies.

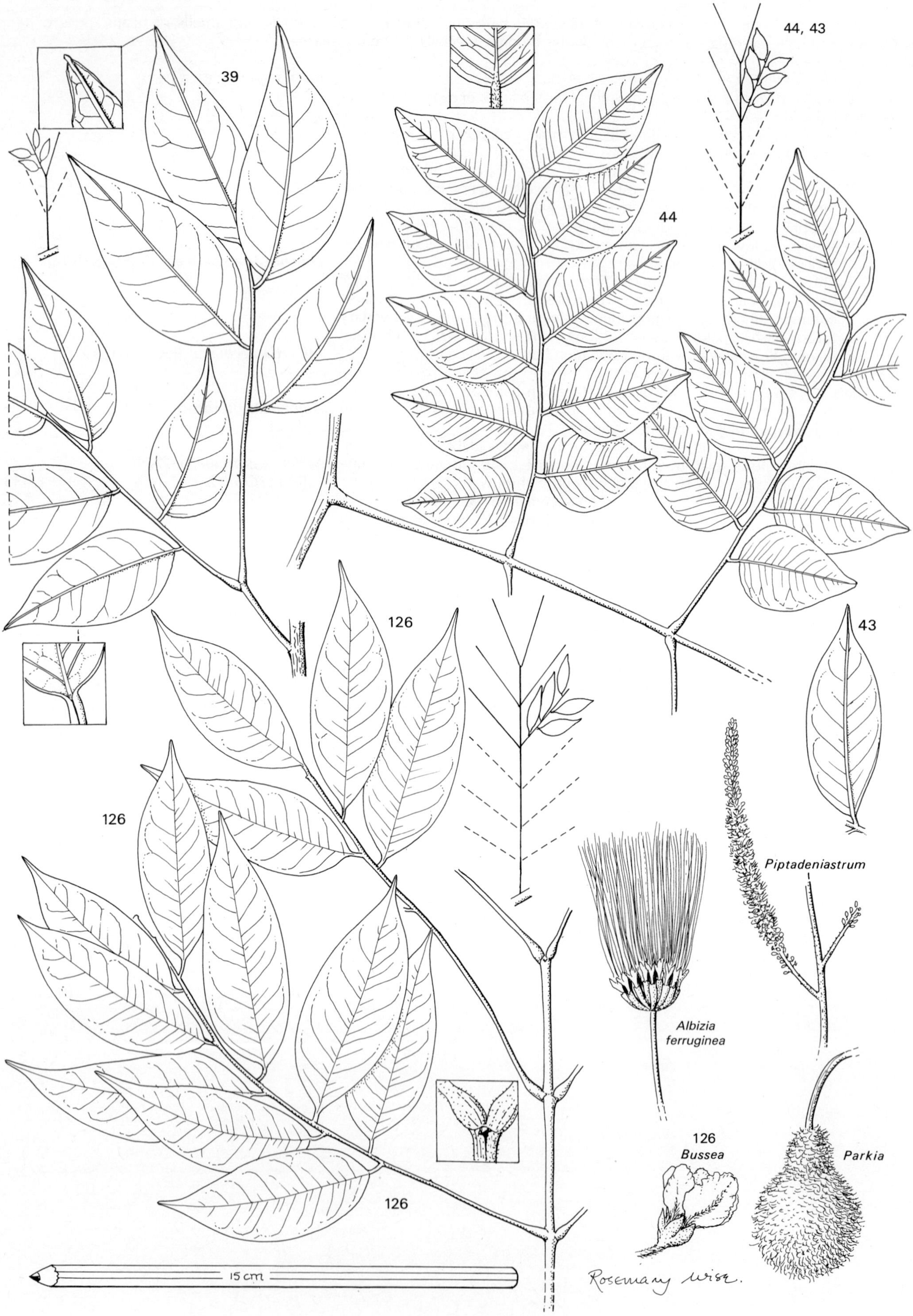

39
44, 43
44
126
43
126
Piptadeniastrum
Albizia
ferruginea
126
Bussea
Parkia
126
15 cm
Rosemary Wise.

GROUP 38: LEGUMES (part 2)
(Lvs bipinnate)

The following species are all compound-leaved legumes with 'branched-axis' (bipinnate) leaves. Most species of Groups 38 and 37 conform to Troll's architectural model of Hallé *et al.* (1978). Unlike most legumes in Group 37, however, most species in Group 38 have a very spreading, umbrella-like crown, with foliage in one or several fine-grained layers. This evidently predisposes the species to being either tall species emergent above the rest of the canopy, or lower-branched species of rivers or recently disturbed forest. The species more typical of lower storeys tend to have compact or plumed crowns. These lower storey species are all in Groups 38A (*Bussea, Calpocalyx, Xylia*) or 38B (*Newtonia* and possibly *Aubrevillea*): none of these species have feathery-fine foliage. *Xylia* and, to a lesser extent *Aubrevillea* and *Pentaclethra,* have leaves very clustered at the ends of the branches.

Two species, *Cathormion* and (young) *Cylicodiscus* are armed with spines or prickles. There are several species of v. prickly lianes in the genus *Acacia* which have very fine foliage.

The flowers of the Mimosaceae listed below are small, clustered into spikes (most species) or conspicuously large, globose heads (*Parkia* spp.) or smaller heads (*Albizia, Xylia, Cathormion* and *Samanea*). The various types of pods are discussed at the begining of Group 37.

Key to subgroups

Leaflets large; >15 mm long or >4 mm wide (>18 mm long for saplings)
- Leaflets not rhombic, and >15 mm wide; often ovate or lanceolate and acuminate; (alternate unless only one pair of pinnae or 1 pair of lflets per pinna) 38A
- Leaflets rhombic (with diagonal midrib) OR <15 mm wide; (opposite or <2 cm long) 38B

Leaflets small (<15 mm (rarely 18mm) long and <4 mm wide) and linear 38C

Species in Group 38A are often mistaken for once-pinnate leaves by the less circumspect, but otherwise the feathery or fern-like foliage is immediately recognizable as belonging to this Group, and to a 'mimosoid legume'.

NOTES for Group 38A (next text page)

1) *Erythrophleum* seedlings are bipinnate, like small versions of the parent, whereas *Cylicodiscus* seedling have the first few leaves once-pinnate, with (1-3) long-acuminate lflets. Young *Cylicodiscus* trees have spines.
2) These local names are often used interchangeably. *ATROTRE* and *PREPRE* are onomatopoeic references to the sound of the dehiscing fruits; *ATAA* is any 'bean' in a woody pod; *SAMAN*=spirit – a reference to non-edibility by mortals. *C. brevibracteatus*, in Nigeria at least, has been renamed *C. winkleri*, but the revision has not been applied to Ghanaian plants (Lock, 1989, *Legumes of Africa: a check list*)

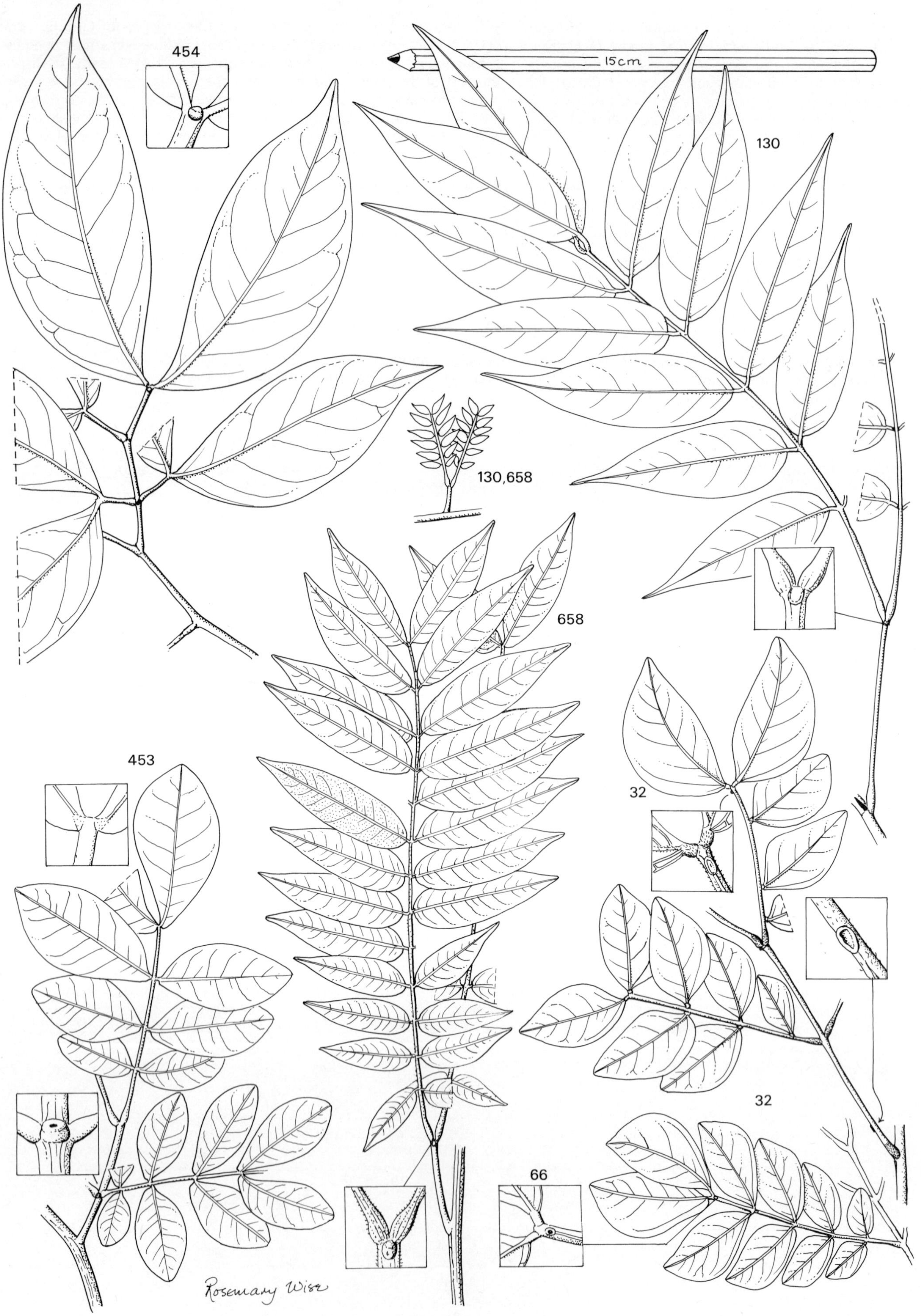
454
15cm
130
130,658
658
453
32
32
66
Rosemary Wise

Group 38A: (Caesalpiniaceae to Mimosaceae)
(Leaflets >3 cm long; ovate or acuminate)

NOTE: *Erythrophleum* and *Bussea* are in Caesalpiniaceae; all other species are Mimosaceae. *Erythrophleum* has small flowers crowded into slender spikes, like many Mimosaceae, but *Bussea* has more conspicuous yellow-petaled flowers *c.*1 cm across. *Notes 1 and 2 are on previous text page.*

Leaflets in more than one pair per pinna (>8 per leaf)

Leaflets alternate; lflets either ovate or pinnae in >1 pair

Pinnae in 1 to 4 pairs, without rusty brown rachis; leaflets ±ovate and acute; **immense emergent trees with rough, shaggy brown bark and spreading crown (DO NOT TASTE SLASH!)** (illustrations on p. 214)

–Lvs with 1(-2 pairs) of pinnae; base of lamina meeting on top of petiolule with edges of petiolule channel; twigs NOT v. lenticellate; lflet apex acuminate, and often mucronate; v. conspicuous fine venation; **gland at top of petiole**; young tree with spines; older trees with knee-like outgrowths on thick buttresses and small adventitious roots nr leaf litter around base; **slash hard-fibrous, reddish orange or yellow, with orange lines towards outer bark; with yellowish sticky exudate and foetid smell like rotten cabbage or garlic**; pods up to 1 m long, strap-like, with raised nerves; seeds flat with a thin papery wing — ***Cylicodiscus gabonensis***[1] [*DENYAo*] 39

–Lvs with 2-4 pairs of lflets; petiolule not v. channelled, with lamina meeting it at edge; without conspicuous venation; twigs lenticellate, with fine stipules at young nodes; lflet apex not v. sharp; **petiole without glands**; tree without spines, but often with large, thick buttresses; bark of older trees rough and shaggy; **slash complex (POISONOUS!), mottled pink and white, fibrous-crumbly and gritty, with reddish outer layer, darkening with bitter, red exudate** (often whitish exudate as well); sapwood cream, ripple-marked; pods flat and brittle-leathery, *c.*6-seeded

Lflets ±lanceolate-oblong, with one side ±cuneate at base; venation lax, not appearing finely transverse above; petiolules and midribs ±glabrous; lflets drying blackish; common tree in **moist forests**; pods usually <10 cm long; petals with hairs dense over surface — ***Erythrophleum ivorense***[1] [*PoTRoDOM*] 43

Lflets ±ovate, with both sides of base usually obtuse, and with venation finely transverse (above); petiolules and midribs with scattered long hairs (lens); lflets drying green; **dry or fringing forest**; pods often >10 cm long; petals with hairs in fringes on margin and midrib — ***Erythrophleum suaveolens*** [*ODOM*] 44

Pinnae in 2-6 pairs, with rusty-brown young rachis; lflets ±lanceolate but with midrib often curved, and acuminate; yng twigs with dense brown hairs; bole usually crooked and without buttresses (±flutes at base); crown dense and deep; bark smooth but with many prominent lenticels; slash thin, brittle-fibrous, very gritty, orange-brown in fine contours over smooth, wet, dark sapwood with ripple marks; with copious, watery exudate (sometimes slightly reddish), with a strong, sweet and savoury smell like ginger; pods velvety, woody, broadest at apex, held above the crown and curling open explosively — ***Bussea occidentalis*** [*KOTOPRePRe*][2] 126

Leaflets opposite; often oblong or lanceolate, and acuminate; pinnae always only 1 pair, with raised gland where they meet; **pods woody, almost S-shaped, dehiscing explosively on tree**

Petiolules 3 mm or longer; lflets not papery, never cordate, often long acuminate; <9 pairs per pinna; bole cylindrical, usually unbuttressed; with steeply ascending boughs and small **crown in plumes**; bark with dense, raised, oval lenticels and v. raised horizontal lines; slash hard-fibrous ±brittle but often peelable, crisp and slightly gritty, red or orange-brown with white streaks; sapwood hard, orange striate with sweet then bitter taste (like tonic water); often gregarious (illustrations opposite) — ***Calpocalyx brevibracteatus*** [*ATROTRe*][2] 130

Petiolules <3 mm long; lflets obtuse-cordate at base, thin-papery + fine hairs below; >12 pairs/pinna; **lvs strongly clustered at twig ends**; bole often irregular or fluted with concave buttresses; bark scaly, with scales leaving concentric-ridged pits; slash hard-fibrous, stringy, (red/) brown and contoured, darkening, with bitter taste and sweet smell — ***Xylia evansii*** [*SAMANTAWA*[2], *ABOBABEMA*] *658*

Leaflets in one pair per pinna; pinnae 2 or 4 (always <9 lflets/leaf); lflets obovate-elliptic, and ±acuminate, 6-12 cm long; crown dense, flat, not v. spreading, with narrow, tall buttresses or flutes; bark smooth, dark, with reddish lenticels, and horizontal 'creases' on buttresses; slash thin, orange-brown, with whiter, fibrous, peelable inner bark; with yellowish sl. sticky exudate, and foetid smell; **evergreen forest** only; pods thin, to 20 cm, splitting along one margin, with winged seeds — ***Newtonia duparquetiana*** [*ADADABA*] 454

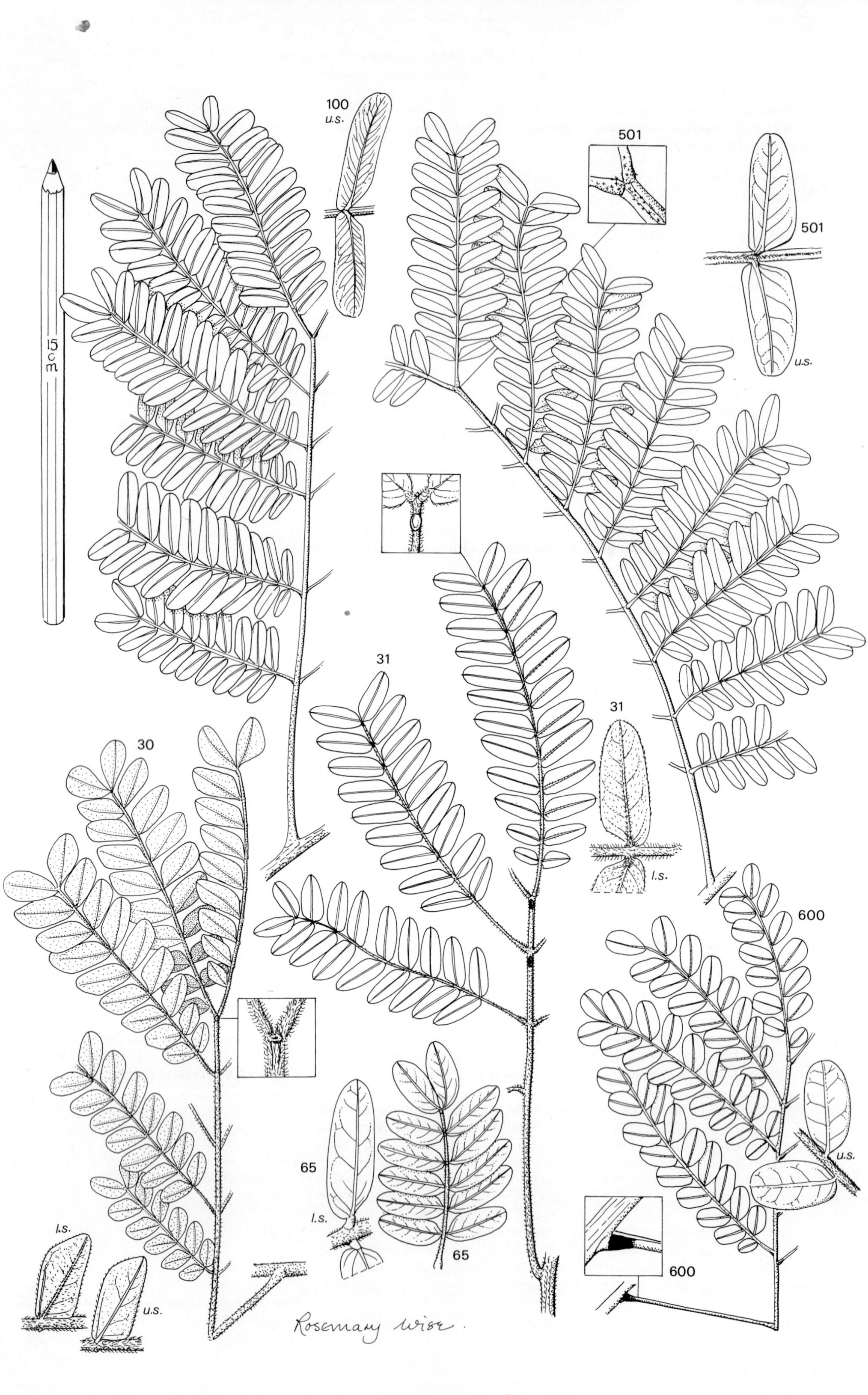
100
u.s.
501
501
u.s.
15 cm
31
31
l.s.
30
600
600
u.s.
65
l.s.
65
l.s.
u.s.
Rosemary Wise

Group 38B: Mimosaceae
(Lflets medium-sized rhombic or small-elliptic)

Leaflet (almost) rhombic, or v. asymmetric at base; with petiolule on 90° corner of lflet and midrib reaching diagonally opposite corner[2]

–NOT *Albizia adianthifolia*: lvs without glands OR **midrib** of lflets >2 cm long (from corner to corner); lflets not densely hairy

Lflets <80[1] on whole leaf; lvs usually with glands, (or tree + yellow exudate)

–Lflets without basal nerves; sometimes with glands; bole with thin, sinuous, but not v. spreading buttresses; bark grey to orange, smooth or wrinkled on buttresses; slash thin, yellowish-brown, fibrous-peelable with sticky, translucent yellow-brown exudate and rancid or foetid smell; commonest in **evergreen forest**; pods thin, to 15 cm, with thin, papery-winged seeds — ***Newtonia aubrevillei*** (see p. 216) [*ADADABA-NUA*] 453

–Lflets with 1-2 nerves arising at base; lvs with glands between (top) pinnae; bark lenticellate; slash not fibrous-peelable and without yellowish exudate; unbuttressed trees; **widespread**

Petiolules v. slender, small, cylindrical, *c.*1 mm long, yet distinct; lflets often + 2 ±perpendicular nerves at base; bark smooth, yellowish with thin peeling scales; bole straight ±small buttresses; slash with musty vegetable or tobacco scent; yellow with orange grit and bitter; pods thin, brittle, up to 25 cm long — ***Albizia glaberrima*** (see p. 216) [*oKORA-AKOA*] 66

Petiolules barely present; conical swellings, not cylinders; lflets usually with one basal nerve; bark ±smooth, with slight fissures; slash (green/) red-brown granular and fibrous, with orange lines and orange gritty streaks ±slow, brownish gum; pods <20 cm long — ***Albizia zygia*** (see p. 216) [*oKURO*] 32

Lflets >130 on whole leaf[1]; lf without glands, clustered at twig tips, red or brownish when flushing; lflet tip often emarginate or minutely mucronate

Rachis without stellate hairs, but sometimes with straight, brown hairs; leaflets with many (>10) straight, ±parallel, nerves prominent above; **often broadest towards apex**; crown dense and rather rounded, with lvs clustered; **bole straight, with narrow sinuous, sometimes wandering buttresses**; bark smooth; yellowish, thin, granular with redder-brown inner bark; pods flat and papery, to 20 cm long, broadest nr. apex, indehiscent, with raised reticulations — ***Aubrevillea platycarpa*** 100

Rachis with well-defined, grooved crest, with dense **stellate hairs**, with few hairs elsewhere on mature lvs (but + straight hairs dense on yng lvs); lflets with often <10 main, not v. straight; laterals **broadest nr base**, emarginate or mucronate; tree spreading; **bole usually twisted or fluted**; bark scaly, with yellow pits and vertical lines of lenticels; slash thick, fibrous-spongy, orange; pod woody to 40cms, broadest nr. apex, bursting on tree and peeling at apex — ***Pentaclethra macrophylla*** [*ATAA*] 501

–Lflets papery, soft with dense yellow hairs; **midrib** usually <2 cm long; white spots especially on top surface; **crater-like glands on petiole and rachis**; common tree of disturbed forest with flat top and cylindrical unbuttressed base; slash thin (green/) creamy fibrous-brittle, gritty with a yellower inner bark and clear gummy exudate — ***Albizia adianthifolia*** [*PAMPENA*] 30

Leaflet elliptic, with obtuse, or slightly asymmetric base[2]

Lflets opposite; attached to rachis by indistinct cushion-like petiolule; unbuttressed trees

Lflets softly hairy, with long ginger hairs especially around the petiole; with a tiny thickened acumen, with pinkish glands on petioles and rachis etc.; bole straight ±small fat buttresses; bark rough and scaly, thick; slash fibrous, yellow-orange, contoured, darkening, with brown gritty streaks; often with red exudate appearing in layers; obvious ripple marks in sapwood; sap + v. soapy-bitter taste; pods + round ends; flwrs + green to orange filaments long exserted — ***Albizia ferruginea*** [*AWIEMFOSAMINA*] 31

Lflets ±glabrous; slash hard and reddish brown; bark v. dark and flaky; in **dry forest** only; pods ±acute ends; fls + slightly exserted red filaments[2] — ***Albizia coriaria*** [*AWIEMFOSAMINA-AKOA*] 65

Lflets alternate; ±glabrous, yellowish, with nerves prominent above but ±obscure on lower, sl. glaucous surface; with slender distinct petiolule; bole ±straight or with sinuous flutes or buttresses on old trees; bark smooth, thin; slash (brown/) pink or orange with white streaks, brittle, with strong distinctive smell (like linseed oil or caramel) and bitter taste; yellow sapwood with ripple marks; fruits, dark brown, *c.*20 cm and **4-winged**. — ***Tetrapleura tetraptera*** [*PReKeSE*] 600

NOTES: 1) Multiplying the no. of lflets on one side of (any) pinna x 2 and x the no. of pinnae on a (any) lf gives a reasonable estimate of total no. lflets.

2) If lflets linear, 3-5 mm wide x 15-20 mm long, see next Group.

Group 38C: Mimosaceae
(Lflets small (<15 mm long))

• Of the following, very feathery-leaved trees, only *Parkia* and *Piptadeniastrum* are common. These can be distinguished when mature by the orange-ish, smooth bark, with thin brittle slash, and wandering plank buttresses in *Piptadeniastrum*, contrasted with the rougher, more irregular bole, with reddish fibrous, sticky slash in *Parkia*, which also lacks the distinctive wandering buttresses of *Piptadeniastrum*.

Leaflets >10 mm long or >1.5 mm wide

Leaflets mostly apiculate (lens), linear-oblong; **lvs without petiole glands**; petiole 6 cm or longer, v. smooth and ±glabrous (like rachis) or + fringe of long hairs on younger lvs; (rachis similar); venation prominent and clearly visible below; bark dark grey to orange ±rough, scaly; slash pink-orange, thin, hard, yellow + more fibrous inner layer with slight foetid smell; crown dark; lvs slightly clustered; high and then convex, wandering plank buttresses; pods *c.*20 cm long, broadly strap-like, broadening from base to round apex, v. thin and papery, indehiscent; **dry forests** only — ***Aubrevillea kerstingii*** [*DAHOMA-NUA*] 99

Leaflets NOT apiculate, but sometimes acute; slightly S-shaped or lvs with glands on petiole; rachis v. rusty hairy (OR TREE ARMED)

Petiole gland flat or sunken; petiole up to 5 cm long[3]; tree unarmed; 2° rachis velvety, sl. winged nr base, sharply ridged; **lflets ±S-shaped;** bole slightly irregular, often with large, high buttresses; bark irregularly scaly with many small lenticels; lf flush red; slash red-brown, thick, stringy-fibrous (yellower inner bark); with sticky orange-ish exudate; pod strap-like, *c.*20 cm with diagonal seeds visibly bulging; common flat-topped emergent — ***Parkia bicolor***[1,2] [*ASOMA*] 489

Petiole gland a raised crater; petiole to 3 cm long; **glands also on (ends of) 2° rachis between lflets**; lflets ±oblong; twigs with fine, darker ridges + many obvious lenticels; **prickles** on older stems; bark dark, with thin scales; slash creamy, soft fibrous; low-branched tree typically on **riverbanks**; pods constricted and splitting between seeds, and curled up — ***Cathormion altissimum***[3] [*ABOBONKAYERE*] 142

Lflets smaller than 1cm long and <1.5 mm wide

Lflets >3 mm long

Younger lvs with tufts of ginger hairs visible below (lens) at base of lflets; margin ±ciliate; veins ±obscure; twigs dark, with orange-brown lenticels or lines or spots; petiole 1-2 cm long, **sometimes (not always) with gland**; bark smooth, orange-ish; slash pale yellow, thin, brittle end, fibrous, striate over wet, white sapwood, with slight musty smell; wandering plank buttresses spreading, snake-like along ground for many metres, and forming small walls; crown dark in graceful, spreading layers; pod *c.*20 cm flat, strap-like, brown, splitting along one margin; flat, winged seeds attached by thread to pod; v. common emergent except in drier forests — ***Piptadeniastrum africanum***[2] [*DAHOMA*] 20

Most leaflets <3 mm long (<1 mm wide)

Curly ginger hairs on rachis; petiole c.1 cm long, with raised gland at base; **raised gland between upper rachises also**; bole fluted and twisted, or with high plank buttresses; bark rough, with pale reddish patches below dark scales and vertical lines of raised red-brown lenticels; slash yellow, v. fibrous with sticky exudate and hot, sweet taste; on **riverbanks** or pale sandy soils; pods flat, up to 20 cm, rounded ends, with many small seeds — ***Samanea dinklagei***[3] [*'SAMANEA'*] 558

NOTES: 1) ***Parkia filicoidea*** [*'ASOMA-NUA'*- **708**] has a red, fibrous, gummy slash like that of *P. bicolor*; it has recently been discovered in Krokosua hills f.r. and some dry forests, especially near rivers. It has double or bilobed petiole glands and longer lflets than *P. bicolor*, and has leaflets with clearly visible venation with **two well-defined sub-marginal nerves parallel to the midrib and margins**. The leaflet apices are sometimes notched.

2) Sapling *Piptadeniastrum* has raised glands on the petioles and rachises which are less commonly seen on older leaves, and paired slender stipules. The seedling branches early, and is slender and 'drooping', whereas *Parkia* seedlings are stout, straight and unbranched.

3) *Samanea* and *Cathormion* have recently been included in the genus *Albizia*, in spite of the considerable differences in the pods of the Ghanaian representatives.

99
99
u.s.
l.s.
u.s.
l.s.
489
489
142
142
u.s.
708
20
20
558
u.s.
+
−
15cm
Rosemary Wise

103
811
257
258
103
811
72
15cm
Rosemary Wise

GROUP 39: MONOCOTYLEDONS (part 1 and key)

Monocotyledeonous trees can be separated from all other trees in this guide on the basis of their leaves which, as in grasses, have strongly longitudinal venation – i.e. many nerves running parallel from the bottom to the top of the lamina. The 'Tree fern' (which is not a 'monocot.', nor even a flowering plant), is immediately recognizable as both a fern and as a tree.

Key to Groups 39, 40

Leaves 'compound'; unbranched trees (or spiny climbers) with large leaves clustered at top		
–'Tree fern' i.e. unbranched, slender tree with very finely divided (tripinnate) foliage; young parts with rusty scales. In Atewa range (+ possibly other **upland evergreen**) forests	***Cyathea (=Alsophila) manniana*** [*DUA-AYAA*]	72
–Palm trees or climbing palms ('cane'), with deeply divided but unbranched leaves resembling pinnate leaves	***Palmae*** (Gp 40)	
Leaves simple, often narrowly elliptic, and not appearing compound; not palms nor ferns		
Lvs (±glaucous below) with spiny margins; little-branched trees often with stilt roots and prickles and usually by the sea or in **swamps or rocky places** (trees dioecious; fleshy fruits clustered and slightly fused in a cone-like head; inflorescences at the end of branches)	***Pandanus***[1] [*NTON*]	811
Lvs without spiny margins		
Plants with many, hollow, unbranched main stems arising in a clump; lvs with sheath at base sheathing stem over many cm; tree often in **riverine forest**	***Bambusa vulgaris***[2] [*BAMBOO*]	103
Plants not 'bamboos'; usually branched and base of lf only clasping stem for a short distance – if at all	***Dracaena*** spp. (below)	

***Dracaena* spp.** (AGAVACEAE)
Only two species reach the size of trees. These might be confused with *Anthocleista* (Gp 4) or *Elaeophorbia* (see Gp 22) from a distance, because of their long, slender leaves strongly clustered at branch ends, but the venation and other details are unmistakeable. The slash is soft and whitish, ±fibrous, but brittle. The trees normally have cylindrical boles with stilt roots at the base.

The flowers are tubular, in panicles. The fruits are fleshy, 1-3 seeded, reddish and 1-2 cm long.

Lvs 6-7 cm broad, sometimes 50 cm long; small tree	***Dracaena arborea*** [*NToNME*]	257
Lvs 2-4 cm wide, *c.*15-20 cm long; twigs with v. conspicuous crescent scars where lvs fall; becoming a medium-sized tree	***Dracaena perrottetii*** [*KESENE*] =*D. mannii*	258

In addition to these, there are several species of shrubby *Dracaena*, including *D. camerooniana* and *D. surculosa* with elliptic leaves (not narrowly so): *D. surculosa* has a more abrupt lf base, which is *c.*1 cm to a point from the stem where the lf is ½ cm wide; *D. camerooniana* has more a gradually cuneate base, still <½ cm wide 2 cm from the stem. Apart from these, the species are almost herbaceous; *D. cylindrica* might be mistaken for sapling *D. arborea*. *D. elliotii* has a decurrent lf base leading into a broadly ovate lamina, with a winged petiole up to 15 cm long.

NOTES: 1) There is at least one recently named species in this genus – *P. abbiwii* Huynh – previously referred to under *P. candelabrum*. The genus is under revision.
2) The **savanna** bamboo, *Oxytenanthera abyssinica*, has larger leaves 25 cm or more long, but does not occur in forest.

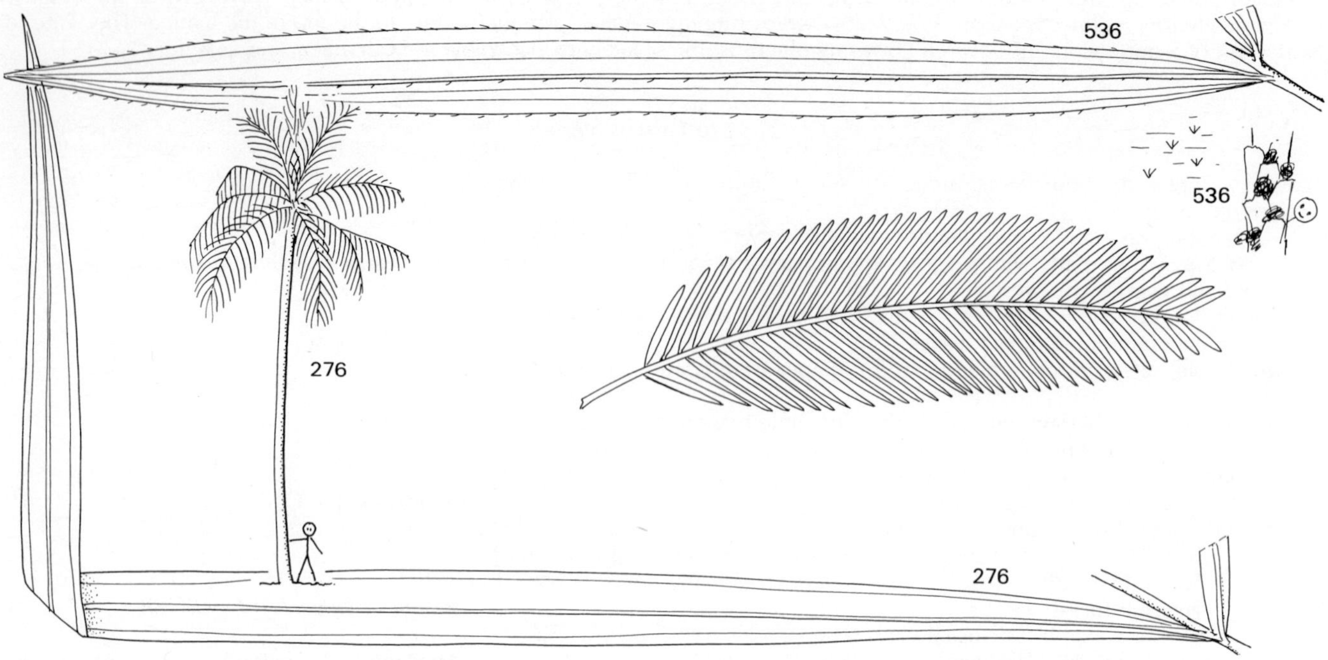
536
536
276
276

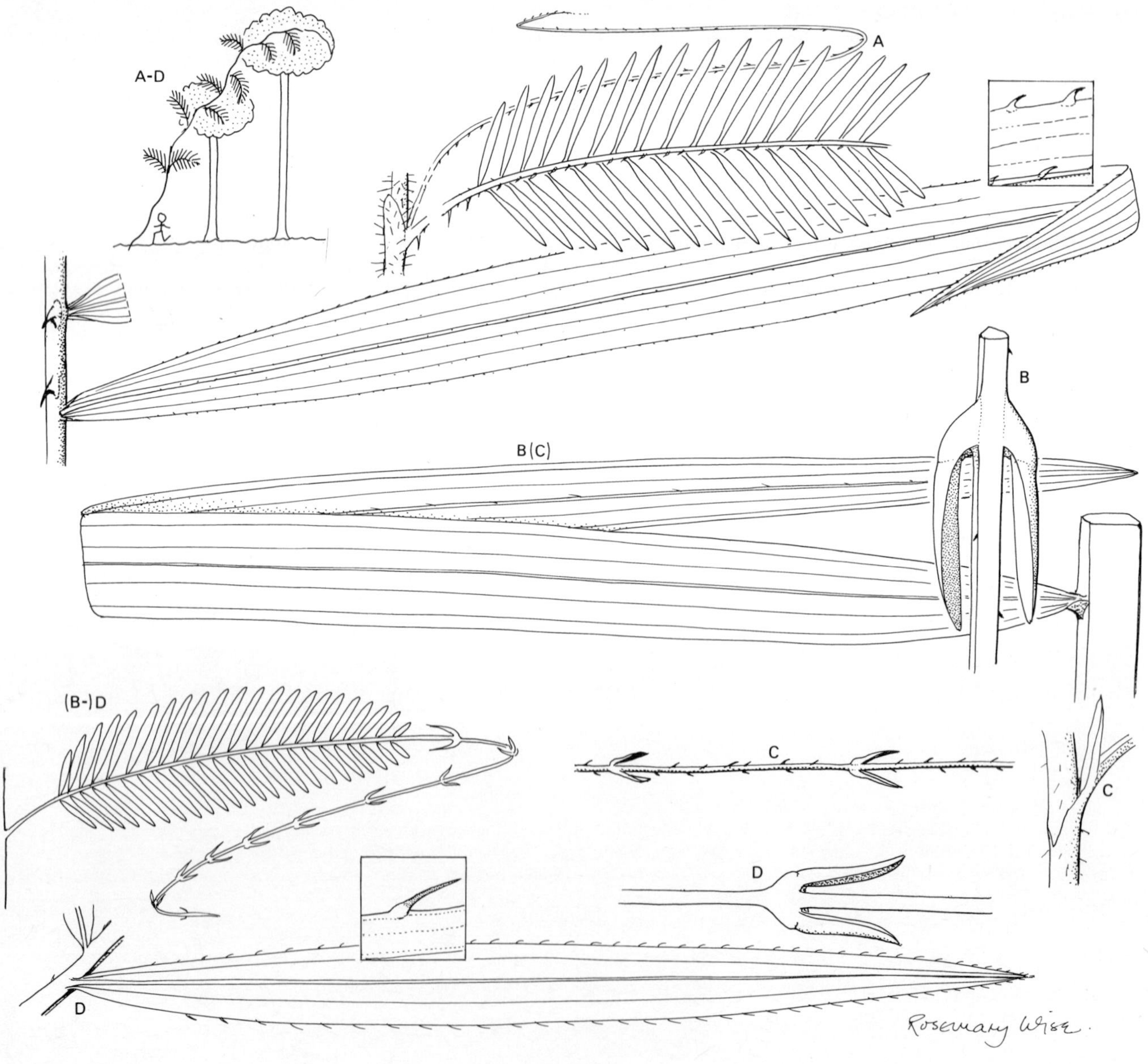
A-D
A
B
B(C)
(B-)D
C
C
D
D
Rosemary Wise.

Group 40: MONOCOTYLEDONS (part 2 – palms)

This group is of exceptional value in terms of 'minor forest products', from palm wine and oil seeds to constructional materials. Climbing palms ('rattans') have been included because of their value in the production of baskets, 'cane' furniture and light construction as much as for their biological interest. **Savanna** palms include *Borassus*, with palmate leaves and swollen upper stem, and *Hyphaene*, which branches dichotomously.

The flowers have parts in threes, often in very large inflorescences. The fruits of *Raphia* and the rattans have scales pointing towards the base of the fruit. The fruits of the other species lack scales. Single stems of *Laccosperma* (previously *Ancistrophyllum*) and *Raphia* spp. flower only once, as the apex of the tree develops into the inflorescence (Holttum's model of Halle' *et al.* (1978)). Growth of these species can sometimes continue after flowering by growth of suckers from the base of the flowering trees/rattans (as in the banana – 'Tomlinson's model'). In other species, exemplified by the oil and coconut (*Cocos nucifera*) palms, the inflorescences are lateral and growth of the tree continues after flowering (Corner's model).

Erect trees, with main stem and leaves clustered at top		
— Leaf blade folded at base so that the creased edge faces normally downwards; tufted slender palm of **coastal savanna** (Tomlinson's architectural model)	***Phoenix reclinata***	
— Lamina base with crease as upper edge		
— — Leaves occasionally with prickles, but not on the leaflets; extremely common and very variable tree widely cultivated, but common in forests from **fire zone** to **swamps** in **evergreen forest**; without black fibre-tangle at leaf bases	***Elaeis guineensis*** Oil Palm[1] [*ABe*]	276
— — Leaflets with small prickles along midrib; species gregarious in **swamps**		
— — — Dark-green leaved tree with enormous leaves, typically dominant in **swamps**[2]; base of leaves prickly, and with dense tangle of long, black, springy fibres around the stems there	***Raphia hookeri*** [*ADOBe*]	536
— — — Smaller palm[2] with pale foliage recorded only from swamps in **evergreen forest** zone, not especially in forest	***Raphia palma-pinus***	537
Climbing palms, usually with leaves dispersed along the stem, and often with recurved thorns at the tip of the 'rachis'		
— Leaves normally with 'leaflets' at tip, and not with extended spiny process; fine hair-like spines on margins and nerves; lamina long, v. narrowly elliptic with spines on older rachis rather blunt curves; leaf sheath spiny; whole rachises at apex of plant with short prickles; mature leaves almost 2 m long; fts *c.*2 cm long, pointed with *c.*15 spiralling rings of yellowish scales	***Calamus deeratus*** [*DEMMERE*]	A
— Leaf apex with many sharp spiny projections instead of leaflets **(i.e. a 'cirrhus')**; without teeth on veins		
— — **Leaf cirrhus with two sizes of prickles: short sharp ones and large, yellowish, blunt, anchor-shaped ones**; Leaf sheaths (around stem), and therefore stem, with sharp spines; margin often with only hair-like prickles, or entire; fts reddish, almost spherical, with *c.*12 spiralling rings of shiny scales		
— — — Rachis very large and thick, often 2 cm diameter; lflets long and slender, leaflets strongly reduced towards base of lf; rachis 'anchors' up to 4 cm long, with sharp-edged rachis; **in drier areas**	***Laccosperma (=Ancistrophyllum) secundiflora***[3] [*AYIKe-AKOA*]	B
— — — Rachis smaller, with many v. sharp black-tipped thorns; lflets often broadly elliptic, especially at base of lf; slender rattan rarely more than 20 m high, but lvs still up to 2 m; widespread	***Laccosperma (=Ancistrophyllum) opacum***[4] [*eYEe*]	C
— — **Leaf cirrhus with only anchor-like recurved prickles, often regularly arranged and of same size**; leaf sheath and stem not spiny; margin with sharp, regular spines >1 mm long; lflets slender and narrowly lanceolate, except on young lvs which have 2 forward-pointing broad lobes and conspicuous cross-venation; fts >2 cm long, pale brown, ellipsoid with 15-20 spiralling rings of scales	***Eremospatha macrocarpa*** [*MFIA*]	D

NOTES:
1) Many 'unusual' palms seen in villages around the forest zone are no more than cultivated varieties of the oil palm.
2) A small palm *Sclerosperma mannii* occurs only in the swamps of the **evergreen forest** zone. It has long (*c.*2 m) leaves almost undivided except for being bilobed at the apex.
3) The only Ghanaian specimen of *L. secundiflorum* seen is from the Mampong area. There is a need for more basic research into the biology of rattans in Ghana.
4) ***Laccosperma laeve***, known only from Neung f.r. in the **evergreen forest** zone is v. similar to *L. opacum* but has smaller (1 cm long) fruits, spines, etc.
5) *Eremospatha hookeri* has recently been discovered in Ghana. It resembles *E. macrocarpa* by having smooth, not spiny stems (leaf sheaths), but differs in that the cirrhus projecting from the end of the leaf has both prickles and anchors (like *Laccosperma* spp.). The leaflets of *E. hookeri* are distinctively broad and short, typically rhombic to obovate. It occurs in evergreen forest, but may be more widespread. The late addition of this species emphasizes the poor state of our knowledge of these important plants.

SECTION 3

200 MAIN SPECIES KEY

SECTION 3
'200 MAIN SPECIES KEY'

SUBSIDIARY KEY TO 200 OF THE COMMONER, LARGER TREES USING CHARACTERISTICS OF THE WHOLE TREE

Before using the '200 Main Species Key' it may be worth checking the following few, very common trees which have rather unique features, but which are otherwise more or less 'hidden' in the keys. These notes, and the 200 Main Species Key itself, should not be seen as a final means of identification, but more as a pointer to some of the commoner species. The 200 Main Species Key should be treated as a system to generate short-lists of species, and final identification should be made where possible with (fallen) leaves or leaflets. This is particularly true where the user feels any doubt about the meaning of the statements.

Brief notes about some very common, distinctive trees for beginners to consider before using key

Crown of very untidy, drooping branches with foliage tattered with holes: slash red + white	***Pycnanthus angolensis*** (*OTIE*)
Crown deeply domed, dark green, umbrella-like, with regular 'shell' of leaflets arranged like spokes in umbrella; small stilt-rooted tree common by waysides	***Musanga cecropioides*** (*oDWUMA*)
Crown and bole immense, with buttresses often wider than a (large) car is long; without latex	
Bole (from distance) smooth, grey, not perfectly cylindrical above buttresses; often with small prickles on buttresses; bark somewhat like metal from a distance; larger almost horizontal branches meeting bole not with an abrupt corner on lower side, but in a curve; largest tree in Ghana	***Ceiba pentandra*** (*ONYINA*)
Bole slightly rough and scaly , grey to yellow-grey; slash cream to yellowish	
Leaves in crown distinctly star-shaped; upper bole rarely perfectly cylindrical	***Triplochiton scleroxylon*** (*WAWA*)
Leaves clustered; bark with large silvery scales; branches ± in layers	***Terminalia superba*** (*oFRAM*)
Buttresses thin and 'wandering' along ground, + branches, forming low walls in forest understorey; bark somehow orange-reddish; leaves feathery	***Piptadeniastrum africanum*** (*DAHOMA*)
Outer bark on upper stems distinctly red-brown; branches ± ascending; graceful tree	***Distemonanthus benthamianus*** (*BONSAMDUA*)
Outer bark with large scales falling to leave bright red patches, otherwise smooth; foliage very graceful, with delicate branchlets and drooping, flowing foliage	***Pericopsis elata*** (*KOKRODUA*)
Outer bark yellowish-grey, but extremely smooth; slash with latex	***Ficus*** spp. (*DOMINI* etc.)
Slash with very conspicuous brown and cream or yellow bands (like abdomen of some kinds of wasp)	***Celtis*** spp.
Slash extraordinarily bright golden-yellow, scented, with blackish, smooth outer bark	***Enantia polycarpa*** (*DUASIKA*)
Slash bright yellow, often with black outer bark but not sweet-scented	
Large tree with broad crown; outer bark not brittle, fissured	***Terminalia ivorensis*** (*EMIRE*)
Smaller trees with narrow crowns; outer bark sometimes like black stone	***Diospyros*** spp. (*SANZA-MULIKA*)
Slash very leathery, peelable, damp, bitter, darkening rapidly; bole rather sinuous, usually with adventitious shoots revealing opposite leaves	***Corynanthe pachyceras*** (*PAMPRAMA*)

Characteristics used in the key

The '200 Main Species Key' starts with more obvious, less variable characteristics. Having dealt with the species with spines, prickles (on bole), or stilt roots, the key then asks about the most obvious feature of the slash – the presence or absence of latex (see Introduction p. 1). Other features of the slash are left until the species with distinctive crowns have been excluded. 'Crown' characteristics are very useful in some cases, but often very difficult to ascertain, so only the most obvious features are keyed out.

Tree crown types

Hallé, Oldeman and Tomlinson (1978) have described in detail a range of tree architectural models, based on patterns of trunk, branch and flower formation. I have not followed this classification of tree 'types' here, because too often trees of apparently similar form (to the eye of a person hoping to name a tall tree in the forest) have different models, and trees with a superficially very different overall form have the same architectural model. Nevertheless, some of the principles of tree construction have been recognized, and referenced in the crown-type keys below. Larger- and smaller-than average leaves (and leaflets) stand out in the crown, so keys to trees with these features have been added as well.

Some models described by Hallé *et al.* are easily recognized, but are characteristic mainly of smaller trees. It seems worth mentioning a few such tree types here.

The simplest type of tree has a single unbranched stem, with (usually large) leaves clustered at the top. Usually in Ghana's forests such trees produce flowers in the leaf axils or on the stem (Corner's model). This tree type is typical of palms, etc. (Gps

BOLE AND BASE TYPES

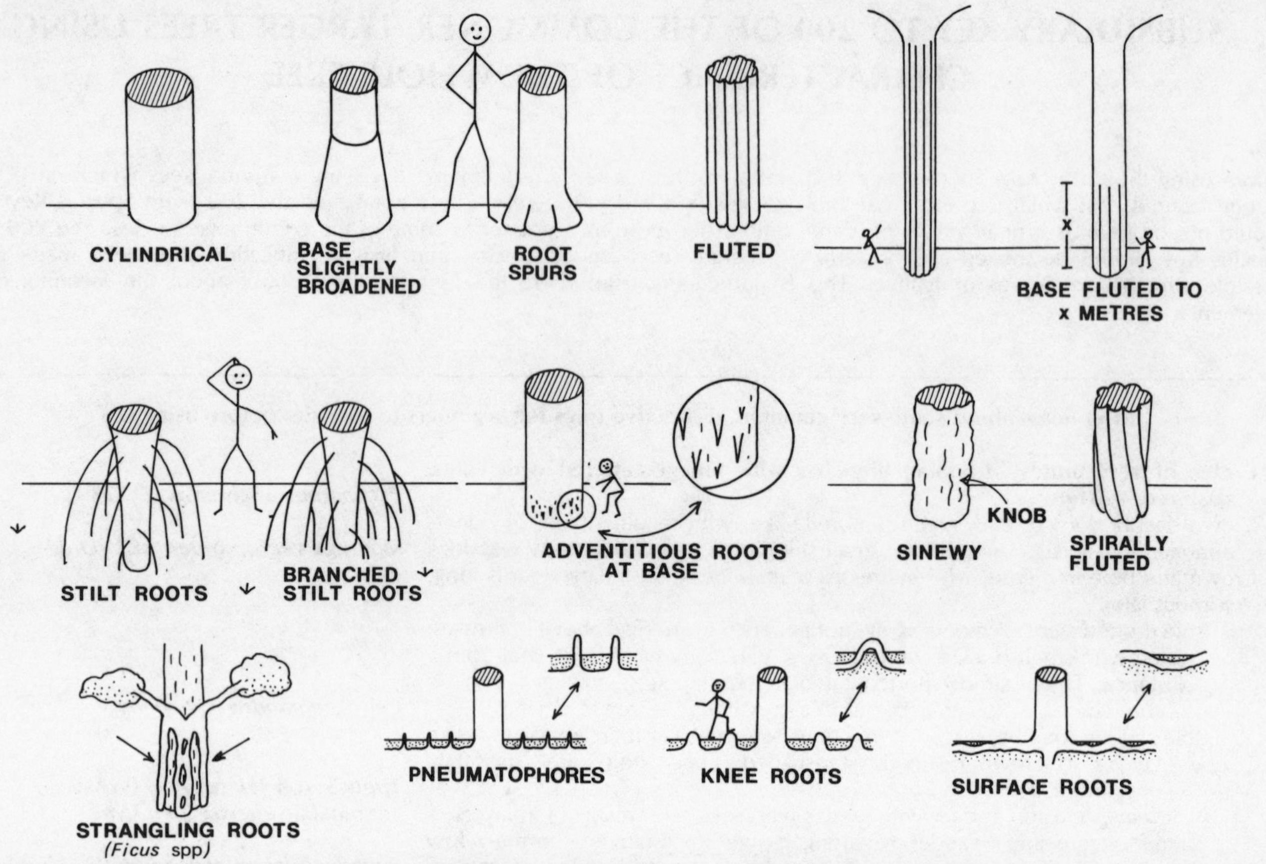

BUTTRESS TYPES

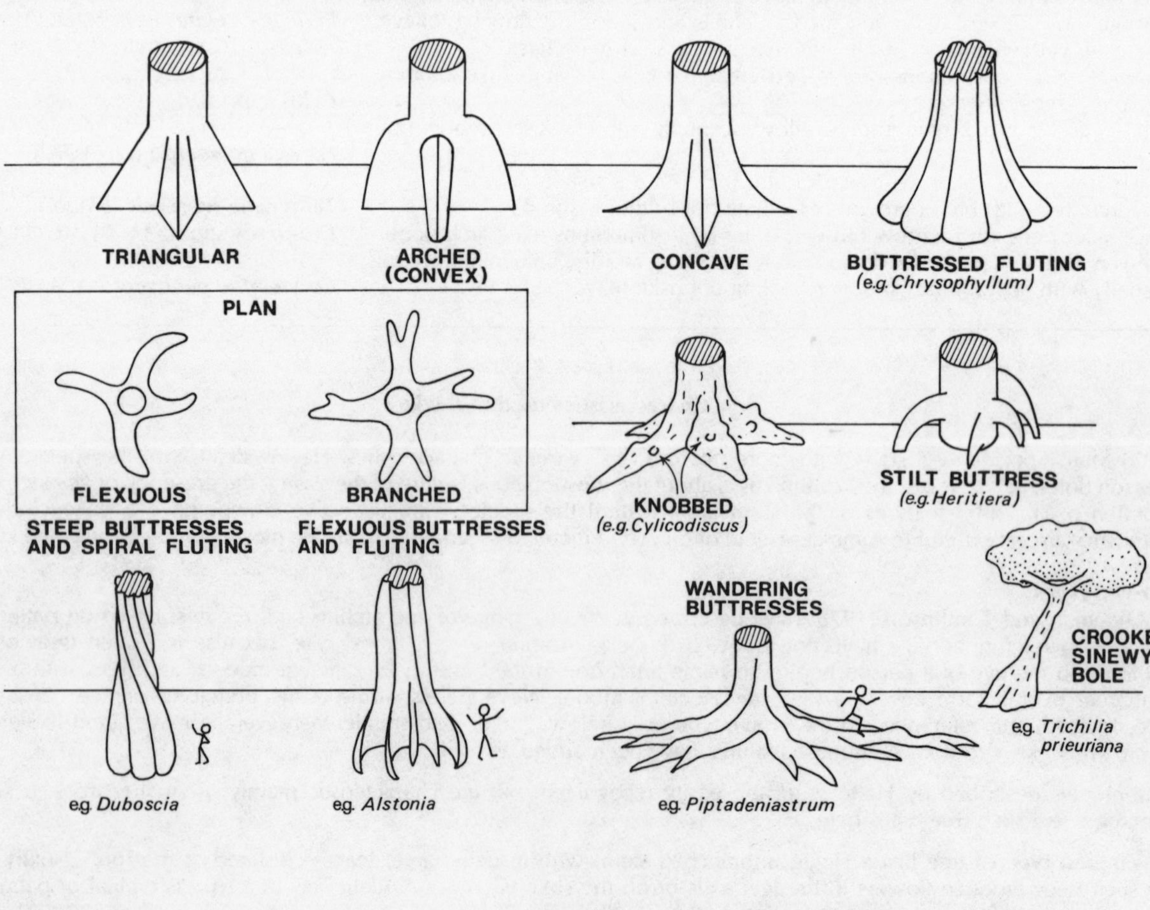

39, 40) and found also in the *Cola* spp. of Group 28 and certain Euphorbiaceae (e.g. *Pycnocoma* spp.) with large simple leaves. Compound-leaved species of this type include *Chytranthus* spp. etc. in Group 36D and sometimes *Placodiscus* spp.. Most such species are, however rather small, and saplings of many trees with '*Utile*-like' crowns (see key below) resemble strongly trees of Corner's model, but they do not flower when unbranched.

A next step up in complexity is Leeuwenberg's model, exemplified by the frangipani tree and cassava. There is no main stem above the first branches, and the branching pattern occurs in a Y-like, dichotomous fashion. Several of the smaller trees of Apocynaceae (Gp 10) grow like this, including *Rauvolfia*, *Voacanga* spp. and *Tabernaemontana* spp. In other Groups the model is shown by *Anthocleista* spp. (Gp 3), *Solanum erianthum* (Gp 26), *Cussonia* (Gp 28), and *Dracaena arborea* (Gp 39).

Some species with simple leaves branch in such a repeatable way that the branches resemble large compound-leaves. The best example, common by roadsides and unmistakably almost like a palm or tree fern, is *Psydrax subcordata* (see Gp 1A). The pattern is less obvious in, but still typical of *Panda* and the three species of Passifloraceae in Group 17. Other species form regular layers of branches, but of less predetermined size. Most have been mentioned in the Tiered Crown Key below. In *Rothmannia* and *Aulacocalyx* the (opposite) leaves are arranged in regular patterns of three (see Gp 1). Trees without whorled boughs (and quite probably with less rhythmic growth patterns), yet with a narrow crown of horizontal layers, include many Annonaceae (Gp 12), *Coula* (Gp 27B), and many opposite-leaved trees (Gps 1,2,6-9).

Most species which have simple leaves with long petioles (the majority of Gps 20-28) grow rhythmically. Many have clustered leaves and whorled branches and grow in various patterns (models) listed in the Tiered Crown Key. Many more of the long-petioled species grow according to Rauh's model, but with less obviously tiered crowns. By contrast, short-petioled simple leaves less often have clustered leaves, and tend to grow according to other architectural models, especially Troll's, Roux's and Petit's models. Many of the trees with obviously clustered, compound leaves (Gps 29-36) are listed below in the *Utile* Crown Key. Most legumes (Gp 37-38) have a different architectural strategy (Troll's model). Some of the finer-leaved legumes (Gp 38) are keyed in a Feathery-foliage Key.

The above notes, together with the main headings of the following key, cover most of the most obvious crown characteristics. Some other crown types are mentioned in more secluded parts of the keys.

There are species with highly discoloured (usually reddish) crowns mentioned in Latex Key C and the Red Slash Key. Note, however, that this refers to permanently discoloured, whole crowns, and not to the species with red flushes of new leaves (only part of the crown, or less commonly the whole crown of a deciduous tree at the start of a growing season). A few species, like *Petersianthus* have leaves which turn bright reddish-brown before falling, but this is not as common as in temperate forests.

Tree bole and base types

Some aspects of the general form of the bole or base of a tree are shown in the illustration opposite. These features are mentioned throughout this guide, particularly in the 200 Main Species Key. It may be useful to provide references for some of these characters. Slight fluting, sinews or small buttresses are produced by many species, and so are not listed.

Stilt roots	See STILT ROOT KEY in 200 Main Species Key (and Gp 3 for mangroves)
Strangling roots	See *Ficus* spp. (Gp 19C)
Pneumatophores	See Groups 1B and 3
Fluting or v. sinewy	Common in trees which produce latex – see Latex Key if slash produces latex; otherwise see: Gp 12A (*Hexalobus*); Gp 13D (*Dichapetalum*); Gp 20 (*Duboscia*); Gps 29 and 30 (slight fluting); Gp 31 (*Balanites*); Gp 34C (*Trichilia prieuriana*); Gp 38B (*Pentaclethra*).
Knobbed buttresses	*Cylicodiscus* (Gp 38A)
Wandering buttresses	Group 38B, C (*Aubrevillea* spp., Piptadeniastrum)
Large buttresses	(at least to chest height) are indicated next to species names in the following keys. Species with high narrow buttresses are indicated as †; species with more 'triangular', convex or concave buttresses (see opposite) are indicated as *. Note that buttresses may only occur on larger individuals of a species.

(K1)

THE '200 MAIN SPECIES' STARTING KEY

Note that, although there is some cross-referencing in the following keys, it is important to try and rule out the options in the order given below. The 'K' numbers are to assist rapid tracing of the 16 keys.

Tree with spines or prickles	See *ARMED SPECIES KEY* (K2)
Tree unarmed	
Tree with stilt roots	See *STILT ROOT KEY* (K3)
Tree without stilt roots	
Slash with white to yellow latex	
Slash not, or barely red	
Lvs compound	See *LATEX KEY C* (K5)
Lvs simple, or not visible	
Slash extremely gritty, or completely granular and crumbly	See *LATEX KEY B* (K4ii)
Slash gritless, or with only scattered streaks, with some fibres	
Crown dark brown or silvery	See *LATEX KEY B* (K4ii)
Crown not discoloured	
Bole with highly irregular, curvaceous, deep fluting	See *LATEX KEY B* (K4ii)
Bole with regular fluting, or not fluted	
Lenticels large and conspicuous; latex usually copious and watery, often browning later	See *LATEX KEY B* (K4ii)
Lenticels not v. conspicuous; slash often banded or fleshy, or outer bark with rectangular scales	See *LATEX KEY C* (K5)
Slash reddish	
Latex immediately thick, *bright yellow* like paint, in spots	See *LATEX KEY A* (K4i)
Latex white, dirty yellow brown, or copious, often darkening	See *LATEX KEY C* (K5)
Slash without latex	
Crown in very distinct horizontal, shallow layers (tiers)	See *TIERED CROWN KEY* (K7)
Crown not distinctly tiered	
Foliage feathery fine OR crown broadly spreading and shallow	See *SPREADING CROWN KEY* (K8)
Foliage not feathery fine AND crown not very broad-and-shallow	
Crown with leaves (or lflets of compound lvs) notably large, unusually long, or obviously lobed	See *LARGE LEAF KEY* (K9)
Crown with leaves not unusually large nor lobed	
Slash with red to brown watery exudate	See *RED EXUDATE KEY* (K6)
Slash without reddish exudate	
Crown of large, compound leaves clustered at ends of loopy twigs	See *UTILE* CROWN KEY (K10)
Crown not '*Entandrophragma*'-like	
Slash sweet or hot scented like perfume or black pepper (or both) OR 'cedar' scented' or like camphor, 'ROB' or 'VIC'	See *PERFUMED SLASH KEY* (K11)
Slash not perfume-like, although sometimes with unpleasant or vegetable-like scent	
Slash with 'musky vegetable' scent, e.g. like pipe tobacco, green beans, peas, garlic, curry powder, sweet carrots, almonds, tuna oil, or mild fish	See *MUSKY VEGETABLE SLASH* (K12)
Slash sometimes mildly sweet or other scented, but scent not of anything like these 'musky-vegetable' types	
Slash red, or with conspicuous reddish vertical bands	See *RED SLASH KEY* (K13)
Slash yellow to orange-brown, although sometimes with thin red outer layer	
Slash with very conspicuous orange gritty streaks, easily visible from 3 m	See *GRITTY SLASH KEY* (K15)
Slash not gritty, or with gritty streaks not visible at 3 m	
Slash very obviously fibrous, and spongy or peelable in long ribbons	See *FIBROUS SLASH KEY* (K14)
Slash not spongy nor ribbon-peelable; granular or brittle-fibrous	See *REMAINDER KEY* (K16, K17) (The remaining species)

ARMED SPECIES KEY (tree with spines or prickles)

		Group No.
TREE WITHOUT STILT ROOTS		
TREE WITH PRICKLES (prickles often conical or pyramid shaped)		
Prickles on bole large (>2 cm wide at base)		
Bark v. gritty fibrous with a very strong fruity-acid taste		
Leaves and leaflets large (>½ m and >10 cm long)	***Zanthoxylum gilletii*** [*oKUO*]	(31D)
Leaves smaller; crown dense and compact like Neem tree	***Zanthoxylum*** spp. [*oKUO-NINI*]	(31D)
Bark not gritty – with conspicuous broad pores in the inner bark smelling of beans	***Erythrina*** spp. [*OSOROWA*, etc.]	(31C)
Prickles small (<2 cm wide) and rounded at base		
Slash with unpleasant smell like garlic, onions or rotten cabbage	***Cylicodiscus*** (SPREADING CROWN KEY)*	(38A)
Slash without such a smell		
Bark smooth from a distance, grey. Slash hard, predominantly yellowish, sometimes with pink vertical bands, darkening	***Ceiba pentandra*** [*oNYINA*]*	(28C)
Bark brownish, and not metallic-grey; slash soft, pink-red, with white-ish bands, darkening	***Bombax buonopozense*** [*AKATA*]*	(28C)
Bark smooth, grey, on fluted or ±cylindrical bole; lvs large or tree sparsely branched	***Anthocleista*** spp.	
TREE WITH SPINES (spines branched or needle-shaped)		
Spines branched or forked; bole irregularly fluted; slash granular, pale yellowish	***Balanites wilsoniana*** [*KROBODUA*]†	(31)
Bole not fluted or spines straight		
Tree v. large – see *Klainedoxa, Cylicodiscus*, otherwise obtain leaves and **see Groups 15 and 1G**		
Tree not v. large, but with large leaves and little branched	***Anthocleista*** spp.	(4)
TREE WITH (small) STILT ROOTS or many adventitious roots at base		
Prickles small and paired; trees with leaves clustered at twig ends – slash thick, granular, fibrous and gritty, darkening rapidly	***Anthocleista*** spp. [*BONTODEE*]	(4)
Spines sharp and thin, merging in appearance with the adventitious roots		
Bark rough, dark, fissured, red-pink, brittle-fibrous, with sweet, camphor-like smell (petiole short)	***Bridelia grandis*** [*oPAMKOTOKRODUKESE*]	(15)
Bark not like this – trees never very large (petiole long); slash barely scented	***Macaranga*** spp. [*oPAM*]	(23)

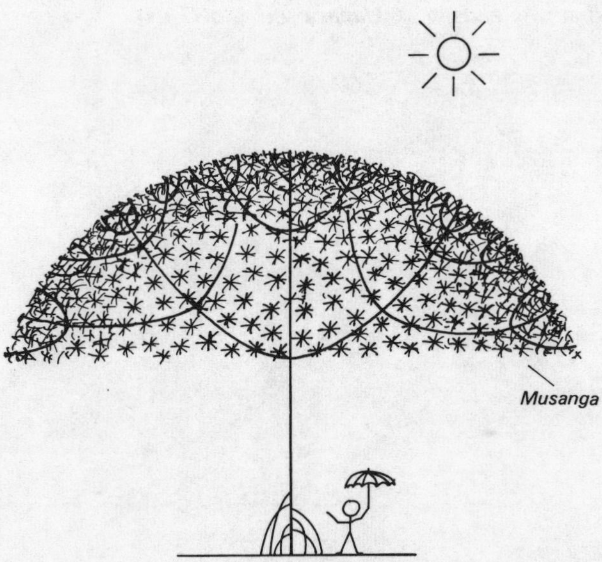

Musanga

Anthocleista

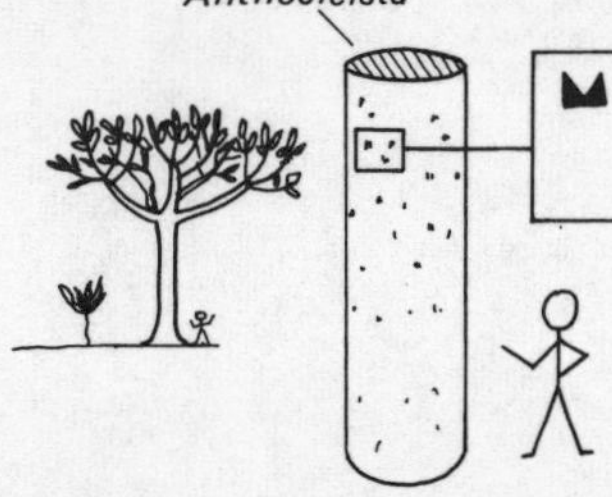

Bridelia, Macaranga

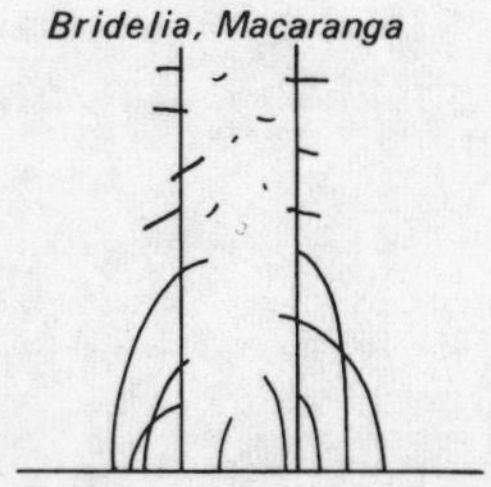

* = potentially buttressed trees. See p. 231

(K3)

STILT ROOT KEY (trees unarmed, with stilt roots)

N.B. Many species, even *Celtis mildbraedii*, produce stilt-root-like outgrowths under exceptional circumstances.

		Group No.
TREE with latex; roots arising usually high up tree and strangling a host tree	***Ficus*** spp.	(19C)
TREE without latex		
Crown a very distinctive umbrella of dark green digitate leaves; very common and widespread small tree of secondary forest	***Musanga cecropioides*** [*oDWUMA*]	(28A)
Not ***Musanga***		
Tree with stilt buttresses or only small stilt roots		
Tree with stilt buttresses; slash red and v. fibrous; crown discoloured, slightly brown-golden; **evergreen forest**	***Heritiera utilis*** [*NYANKOM*]*	(28B)
NOT **NYANKOM**; crown not golden or slash not so or tree in **drier forests**		
Slash without strong peppery scent		
Slash yellow to orange, without red exudate		
Lvs large palmate, serrated; common small tree	***Myrianthus arboreus*** [*NYANKUMA-BERE*]	(28A)
Lvs neither palmate nor serrated (or too high to see)		
Slash gritty, darkening rapidly		
Little-branched, 'cabbage tree'	***Anthocleista*** spp. [*BONTODEE*]	(4)
Small **evergreen forest** tree + sweet exudate	***Eriocoelum pungens***	(28I)
Slash not gritty-darkening		
Slash 'hissing' when cut; wetter forests only	***Protomegabaria stapfiana*** [*AGYAHERE*]	(24)
Slash not hissing, damp; long lvs	***Dracaena*** spp. [*NToNME,KESENE*]	
Slash red, with reddish exudate or yellow latex		
A) Lvs compound; high-buttressed; slash soft	***Pseudospondias*** [*AKATAWANI*]†	(35B)
B) Lvs simple, v. clustered; slash granular		
Lf apex rounded	***Spondianthus*** [*AWORA-TWEANKA*]	(24)
Lf apex pointed (**evergreen for.** only)	***Coelocaryon*** [*ABRUMA*]	(3A)
C) Lvs simple, bk. dark, cylindrical, YELLOW LATEX	See LATEX KEY A (K4i)	
Slash with strong, hot (mustard) scent and taste		
Slash reddish, extremely pungent	***Drypetes pellegrinii*** [*oPAHA-KoKoo*]	(17A)
Slash yellow to brown-orange		
Scent more similar to paint than perfume; crown and bole generally rather irregular	***Drypetes*** spp.	(17A)
Scent more of perfume; crown narrow, regular at top of straight bole, slash v. fibrous	***Xylopia*** spp.	(12C)
Tree with large, ±cylindrical, arching stilt roots		
Slash peelable in long strips, sweet-scented, darkening rapidly; lvs rather small, and spread evenly along twigs	***Xylopia staudtii*** [*oBAA KoKoo*]	(12C)
Slash not v. peelable nor scented; lvs v. clustered at twig ends	***Uapaca*** spp. [*KoNTAN*]	(24)

* = potentially buttressed trees. See p. 231

(K4i)

LATEX KEY A (Guttiferae: latex yellow and slash often reddish; crown typically regular with horizontal branches)

	All species are in Group 8
Tree with stilt roots; **evergreen forest** or **swamps** elsewhere	
A) Outer bark blackish, with small flakes or fissures; slash red-brown →darker	***Pentadesma*** [*ABOTOASEBIE*]
B) Outer bark dark, often with red patches; slash thin, yellow; crown small	***Symphonia*** [*EHUREKE*]
C) Small tree with small lvs	***Garcinia afzelii/epunctata*** [*NSOKO*]
Tree without stilt roots	
Tree in swamps, esp. in **evergreen for.**; crown of short horizontal branches	***Symphonia globulifera*** [*EHUREKE*]
Not *Symphonia*; crown often immense and dense	
1) EXUDATE AT FIRST CLEAR, becoming yellow; tasting fruity-acidic; slash fleshy with pinkish fibres in outer layers; outer bark flaky	***Allanblackia floribunda*** [*SONKYI*]
2) LATEX YELLOW from first appearance	
Bark rough with many lenticels and scales which fall to leave swirling ridge patterns and pits; slash reddish and gritty	***Mammea africana*** [*BOMPAGYA*]
Bark smooth + fine-flakes; slash contoured brown to white ±grit, brittle; slightly savoury scent	***Garcina kola*** [*TWEAPIA*] (+*G.* spp.)
3) LATEX thick and ORANGE like paint; bark v. flaky, peelable	
Slash pale yellow at first	***Harungana*** [*KOSOWA*]
Slash starting orange-brown even before latex appears	***Vismia guineensis*** [*KOSOWA-NINI*]

(K4ii)

LATEX KEY B (Moraceae-Apocynaceae)

		Group No.
BOLE NOT v. CYLINDRICAL, fluted with v. sinuous buttresses or irregular OR lvs whorled		
Many aerial roots reaching to ground—often strangling other trees	***Ficus*** spp.	(19C)
Not ***Ficus***		
Slash v. gritty; base usually with curvaceous buttresses; bark v. lenticellate; crown v. strongly tiered	***Alstonia boonei*** [*SINDURO*]†	(9B)
Slash not gritty, but fibrous or fleshy; darkening rapidly with orange-brown latex		
Slash with broad vertical bands; **lvs large**, asymmetric at base	***Treculia africana*** [*BReBReTIM*]†	(19A)
Slash without vertical bands; **lvs small**, symmetric at base	***Trilepisium madagascariense*** [*oKURE*]*	(19A)
BOLE CYLINDRICAL; not fluted, but sometimes with small root spurs or straight buttresses		
Fallen leaves silvery below: **see next group, or younger trees of species above**		
Fallen leaves not silvery		
Slash granular, but NOT GRITTY, or at least thin and v. crumbly; bole usually appearing smooth from 5m		
A) Latex forming rubbery balls when rubbed between fingers	***Funtumia elastica*** [*FRUNTUM*]	(9B)
B) Latex forming sticky mess when rubbed between fingers	***Funtumia africana*** [*OKAE*]	(9B)
C) Latex remaining pure white; small tree gregarious in **swamps**	***Anthostema aubryanum*** [*KYRIKUSA*]	(22)
Slash partially fibrous or highly gritty, OR TREE >70 CM DBH OR WITH ROUGH BARK		
Large or numerous lenticels conspicuous on pale greyish or RED-BROWN outer bark; bark generally rough and fine-scaly; slash with v. dense HARD ORANGE GRIT predominant; latex watery and staying whitish for 30+ seconds	***Milicia*** spp. [*ODUM*]	(19B)
Not *ODUM*: (BARK not flaky, or slash not gritty but usually peelable)		
1) Large lenticels in short vertical lines; slash v. fibrous, + PALER BANDS BELOW lenticel lines; off-white latex; spreading dense crown	***Morus mesozygia*** [*WONTON*]	(19A)
2) Slash with v. fibrous inner layer, but granular outer layer; lenticels often arranged horizontally; raised hoops around smooth bole; outer slash ±thin orange lines; LATEX v. rapidly CREAM-COFFEE coloured; boughs regular ±horizontal; crown of many rather rounded-elliptic dark lvs	***Antiaris toxicaria*** [*KYEN-KYEN*]*	(19A)
3) Slash rather fleshy; base of large trees usually slightly fluted; darkening very rapidly with copious latex; crown dense + fine, dark-green foliage	***Trilepisium*** [*OKURE*]*	(19A)

* = potentially buttressed trees. See p. 231

LATEX KEY C (SAPOTACEAE, etc.)

		Group No.
CROWN DISCOLOURED, brownish – fallen leaves or leaflets discoloured below, or v. hairy. SEE NOTE‡		
Simple leaves: slash v. fibrous, pale cream or darker, with vertical bands; smell like fresh paint or condensed milk		
Bole regularly fluted, especially towards base; slash usually banded, not v. red; bark often furrowed		
A) Fallen leaves small, cordate and acuminate; outer bark flaky, pink-brown	***Breviea leptosperma*** [*KANKABIM*]*	(10C)
B) Fallen leaves large, elliptic and not cordate at base; with dense soft red-brown hairs below	***Chrysophyllum perpulchrum*** [*ATABENE*]*	(10C)
C) Fallen leaves reddish on some, whiter or more silver on others	***Chrysophyllum giganteum*** [*KUMFENA*]*	(10C)
D) Fallen leaves without coarse or silky hairs, but still metallic silvery below (usually a small tree)	***Chrysophyllum subnudum*** [*ADASEMA*]*	(10C)
Bole straight; slash with dark outer bark, and v. pale yellow to white fibrous inner layer		
E) Fallen leaves large, coarse red-hairy below, ASYMMETRIC at base	***Chrysophyllum beguei***	(10C)
F) Fallen leaves with many parallel laterals, glabrous	***Chrysophyllum pruniforme***† [*DUATADWE*]	(10A)
G) Not (E) or (F), then consider especially (D) above		
Compound leaves: slash reddish brown or orange, unscented or with slight cedar scent; NOT FLUTED		
Latex v. sparse white, arising nr. sapwood; slash red-brown, fibrous; bk very scaly; in (mostly) **evergreen forest** zone	***Anthonotha fragrans*** [*TOTORO-NINI*]	(37E)
Latex brownish, in fleshy bark; slash streaky and fleshy; outer bark with rectangular scales	***Trichilia tessmannii*** [*TANURO-NINI*]	(34D)
CROWN AND LEAVES NOT DISCOLOURED		
Outer bark unusually smooth, pale yellowish to greenish (lvs trinerved or serrated)	See *Ficus* spp.	(19B)
Outer bark not like this – Not *Ficus*		
Leaves very large, and clustered at branch ends; tree usually in **swampy places**; slash thick-fleshy, banded; base ±cylindrical; bole often with little tufts of cauliflorous flowers or large round fruits	***Omphalocarpum ahia*** [*DUAPOMPO*]	(10D)
Leaves (or leaflets) not unusually large		
Bole straight AND slash pinkish or red, at least in bands		
Outer bark dark and rough with deep vertical fissures; slash v. fibrous; foliage (at least in younger trees) in tiers	***Tieghemella heckelii***	(10A)
Outer bark not deeply fissured, often scaly (or a young tree without tiered crown)		
Slash deep red, contoured, sometimes with browner layers; outer bark flaky; compound lvs v. clustered at twig ends; in **evergreen forest**	***Trichoscypha arborea*** [*ANAKU*]	(35A)
Slash not contoured fibrous, but ±fleshy OR ELSE DARKENING RAPIDLY		
Slash v. rapidly darkening to brown, pinkish or streaked with white; outer bark rough and fissured or with rectangular scales	***Trichilia spp*** [*TANDRO* etc.]	(34D)
Slash red-brown; bole with scales and sometimes with cauliflorous flowers/fruits; lvs simple, clustered; (fallen lvs oblanceolate, drying leathery brown)	***Omphalocarpum procerum/ elatum*** [*ESONODOKONO*]	(10D)
Bole fluted or buttressed or slash without significant red component		
Slash with some grit, often yellow and orange contoured, paler towards sapwood with spots of latex nr mid-bark; stringy-fibrous	***Guarea thompsonii****	(34D)
Slash without grit		
1) Bole not fluted, and usually without bark scales and adventitious shoots		
Outer bark pale, smooth except for vertical fine short fissures (bark quilted); slash with strong vertical bands, cream to brown darkening; often with straight-edged, low TRIANGULAR BUTTRESSES		
Fallen leaves with obvious orange hairs below	***Aningeria robusta*** [*ASAMFENA-NINI*]*	(10B)
Fallen leaves almost or entirely hairless (**drier for.**)	***Aningeria altissima*** [*ASAMFENA-BERE*]*	(10B)
(Rare tree in hilly areas; fallen lvs without marginal nerve, but with raised wavy venation above	***Aubregrinia taiensis***	(10A)
Outer slash with thin black line – bole straight	See (F) above	
2) Outer bark with scales falling to leave reddish-orange pits; base usually fluted; slash sometimes slightly pinkish in lines, darkening to rust colour	***Afrosersalisia afzelii*** [*BAKUNINI*]*	(10A)
3) Bole slightly fluted, with many adventitious shoots; slash fibrous spongy, darkening	***Ituridendron bequaertii***	(10D)

‡ NOTE: These species can cause considerable confusion, especially when young and when (fallen) leaves are not obtained.
* = potentially buttressed trees. See p. 231

(K6)

RED EXUDATE KEY (slash with red to red-brown watery exudate)

Smaller trees with red-brown exudate are common in the Rubiaceae, which are strongly grit-speckled in the slash (***Rothmannia***, ***Psydrax***) or fibrous-spongy (***Tarenna***); ***Psydrax*** spp. have distinctive regular, shallow crowns. Many legumes (Gps 37,38) produce red-brown gums which take a long time (>5 minutes) to appear, and which therefore are not on the whole mentioned below. However, in some of those that are mentioned below the rate with which the red exudate appears is variable, and can be easily missed in the dry season.

Trees v. large with rough bark and v. spreading umbrella crown and slash red, granular; exudate often watery and translucent at first	See *Maranthes* spp., *Erythrophleum* (SPREADING CROWN KEY, K8)	
Not ***Erythrophleum*** nor ***Maranthes***		**Group No.**
–Bole cylindrical with whorled branches; slash orange-brown or reddish with orange grit with watery red exudate, large sections easily removed without fragmenting, unpleasantly scented; lvs digitate	***Ricinodendron*** [*WAMA*]*	(28B)
–Not *WAMA*		
EXUDATE WATERY and usually copious; slash without ripple marks		
Bole cylindrical, straight, with scaly outer bark; crown of whorled branches with extremely drooping branch tips, v. untidy with most leaves tattered with holes; slash soft gritty and granular	***Pycnanthus*** [*OTIE*]	(13A)
Not *OTIE*: (bole buttressed or bark smooth or slash fibrous)		
Evergreen forest (or swamp); tree with soft fibrous orange slash with darker vertical bands; branches ±horizontal; bole straight to base or with small stilt roots	***Coelocaryon*** [*ABRUMA*]	(13A)
Tree **outside evergreen forest** (or swamp), or slash not like this		
Slash thick, fibrous-spongy, usually with musty smell, exudate darkening the fibres; bole not v. buttressed		
Leaves medium to large	See *Cordia* spp., or *Mansonia*	
Leaves small, and bright green esp. when flushing; slash often ±pinkish	***Pteleopsis hylodendron***	(25)
Slash not thus (not v.'TWENEBOA-like')		
Bark smooth; slash thin and pale yellow granular	***Tetrorchidium*** [*ANENEDUA*]	(13A)
Bark with crescent scars; slash fibrous-leathery but brittle, slightly peelable; slash contoured, yellow orange to brown, peppery	***Millettia rhodantha*** [*TETETOA*]	(37I)
EXUDATE thick or sticky, appearing slowly, in spots; slash sometime ripple-marked		
Slash brittle, gritty, with large pores, pinkish brown turning black; outer bark rough with rounded scales falling to leave discoloured patches; **evergreen forest**	***Strephonema*** [*AWURUKU*]	(21)
Slash not of this type OR tree outside evergreen forest		
Bole straight and cylindrical to base; bark fairly smooth, but with v. numerous large lenticels; slash thick and complex, fibrous spongy, but partially brittle, with black spots below lenticels, gritty, contoured, scent of sweet vegetable like carrots	***Lonchocarpus sericeus*** [*SANTE*]	(37I)
Bole buttressed or irregular, OR bark thin OR rough OR without large lenticels		
Slash hard fibrous to brittle, not fleshy		
Leaves not clustered at twig ends (or not visible because tree v. large); OR slash with ripple marks		
Bark dark and rough scaly; bole slightly irregular, often with high buttresses		
Slash with red grit, yellow to orange brown, darkening with exudate, smell of beans	***Amphimas*** [*YAYA*]*	(37H)
Slash with orange grit + inner layer with soft, soapy exudate and tobacco scent	***Albizia ferruginea****	(38B)
Bark rather smooth, except for small scales; lenticels abundant		
EXUDATE YELLOW-BROWN; bole smooth ±horizontal lines; **evergreen forest**		
Slash foetid, fibrous-peelable, yellowish	***Newtonia*** spp. [*ADADABA*]*	(38A)
Slash hard, brittle, pink-red	***Cryptosepalum tetraphyllum***	(37C)
EXUDATE in REDDISH SPOTS; slash thin, brittle-fibrous,		
Slash contoured; usually a low-branched tree of **wet places**; outer bark with small flakes	***Pterocarpus santalinoides*** [*HOTE*]	(37H)
Slash not strongly contoured; ripple marks v. fine		
Crown of many small pale (hairy) leaflets; bole uneven with small sharp buttresses	***Dialium dinklagei*** [*DWEDWEEDEWE*]†	(37I)
Crown with many glabrous dark green leaflets; bole with low thick buttresses; many small dark lenticels; fallen fresh lflets acidic	***Dialium aubrevillei*** [*DUABANKYE*]†	(37H)
Leaves simple, v. clustered at twig ends; slash hard red-brown, brittle, gritty, without ripple marks; small, low-branching tree of **swamps**; sometimes stilt-rooted or with low buttresses	***Spondianthus*** [*AWORA-TWEANKA*]	(24)
Slash soft and fleshy-fibrous, pinkish orange to pale brown, ±gritty with creamier inner bark, without ripple marks; crooked, twisted tree typically in **swamps** with high buttresses (stilt roots when young) and rough outer bark	***Pseudospondias*** [*AKATAWANI*]†	(35B)

* = potentially buttressed trees. See p. 231

TIERED CROWN KEY (trees with whorled and ±horizontal boughs)

Many trees have whorled branches, but in many cases this feature is lost to sight in older trees. It remains more noticeable in those trees which have horizontal boughs. This pattern arises as a consequence of several, quite different architectural models. In Hallé *et al.*'s (1978) scheme, it is particularly notable in the models of Aubréville (*Terminalia* and many Sapotaceae); Nozeran (e.g. *Anthostema*); Massart (*Bombax*); and Fagerlind (*Aulacocalyx*). The pattern is most visible on smaller trees, most of which are not eligible for this '200 Main Species Key'. Small trees with strongly tiered crowns, apart from young trees of the above, include: *Napoleonaea*, *Desplatsia* spp. and *Diospyros* spp. (Massart's model); *Grossera* and *Mareya* (Koriba's model); *Rothmannia* spp. (Fagerlind). Some trees with whorled boughs in Rauh's model (e.g. *Musanga*), have very ascending, non-layered boughs (this is common in Gps 27-36), but other cases where the layering is more marked, as in *Ricinodendron*, have been added to the following key. Cultivated trees include one of the best examples of trees with this type of crown: the large-leaved *Terminalia catappa*.

Trees armed	See *Bombax* and *Ceiba* in ARMED SPECIES KEY (K2)	
Trees unarmed		**Group No.**
Crown of drooping branches and tattered lvs; slash gritty normally with red exudate	***Pycnanthus angolensis*** [*OTIE*]	(13A)
Not *OTIE*		
Tree without latex		
Slash at first red to pink to red brown, fibrous, or with red exudate		
Slash sweetly, strongly scented; medium-sized trees without red exudate; common in swamps		
Crown dense and lvs glossy, ±glabrous; in **evergreen for. or swamps**	***Beilschmeidia mannii***	(12)
Crown rather open, in obvious, flat layers; lvs hairy	***Sterculia tragacantha***	(27C)
Slash not strongly, sweetly scented		
Slash with red exudate		
–Crown rather dense; lvs simple, ±rubbery to touch; **evergreen for.**	***Coelocaryon oxycarpum***	(13A)
–Crown ascending and spreading, of clustered compound lvs	***Amphimas pterocarpoides****	(37H)
–Crown in v. distinct layers; lvs digitate	See next species	
Slash without red exudate; with red to pink streaks at first, often rapidly darkening; OR lvs digitate		
Slash scented not v. pleasantly, like unfresh meat; slash with white, pithy areas, often ±grit, sometimes with red exudate	***Ricinodendron*** (see also *Petersianthus* below)*	(28B)
Not *Wama*: Red colour often very bright; bark often + deep fissures		
Ripple marks very conspicuous under x10 lens, especially on sapwood; (lvs digitate)	***Rhodognaphalon brevicuspe****	(28C)
Ripple marks not v. conspicuous; lvs simple and ±oblong	***Sterculia rhinopetala****	(27A)
Slash at first yellowish, fibrous		
Outer bark with many regular ±rectangular scales, pale and often silvery	***Terminalia superba*** [*oFRAM*]*	(25)
Outer bark with vertical fissures, without regular scales		
Outer bark ±blackish, or slash v. bright yellow		
Crown broadly spreading; tree often ±fluted	***Terminalia ivorensis*** [*EMIRE*]†	(25)
Crown narrow; trees usually cylindrical	***Diospyros*** spp.	(12)
Outer bark greyish; slash thick spongy, dull yellow-orange		
Slash foetid; bark deeply furrowed; red lvs in crown	***Petersianthus*** [*ESIA*]	(25)
Slash mushroom or earthy scented; no red lvs in crown	***Cordia*** spp.	(26)
Trees with latex		
Latex not bright yellow		
Slash yellow-orange, granular or gritty	*Funtumia, Alstonia, Anthostema, Ficus* in LATEX KEY B (K4ii)	
Slash fibrous, banded or pink to red	See *Tieghemella* and *Afrosersalisia* in LATEX KEY C (K5)	
Latex bright yellow	See LATEX KEY A (K4i)	

* = potentially buttressed trees. See p. 231

SPREADING CROWN KEY (with widely spreading (umbrella) crown or with feathery fine foliage

Species (particularly in Gp 38) with fine, feathery (bipinnate) foliage typically have a broadly spreading crown as well. This represents a distinctive extreme of what Hallé *et al*. have called 'Troll's model' of tree architecture. Large boughs typically arch over towards the outer, thinner rim of the crown, and branches from these boughs arch away similarly. Some trees with similar crowns have been added to the following key, even where the leaves are not so finely divided. (Shade) trees planted in towns with this type of spreading crown include the large, dark-crowned 'rain tree' (*Samanea saman*) with larger rhombic leaflets like those of *A. zygia*, and the smaller flamboyant tree (*Delonix regia*) with large, conspicuous bright red flowers and pods up to ½ m long. The feathery-leaved, yellow-flowered 'Copper pod' (*Peltophorum pterocarpum*) has a deeper, narrower, dark crown.

BARK ROUGH OR slash strongly distinctively scented	**All species in Gp 38 unless stated**
Slash smelling of rotten cabbage, garlic, onions or tobacco	
Young trees with spines, older trees with thick, arching buttresses with broad, knee-like knobs on, and often with small adventitious roots nr leaf-litter; slash v. fibrous; yellow to red, smell unpleasant, like rotten cabbage (lvs not very feathery)	***Cylicodiscus*** [*DENYA*]*
NOT *DENYA*	
Bark flaky rough, scented of tobacco; exudate from inner bark very soapy on fingers; slash yellowish with brown gritty streaks	***Albizia ferruginea*** [*AWIEMFOSAMINA*]*
NOT AWIEMFOSAMINA, e.g. because bark smooth	
A) Slash fibrous, peelable, with slight yellowish exudate; **evergreen forest**	***Newtonia*** spp. [*ADADABA*]*
B) Leaflets extremely finely divided; slash brittle-fibrous	***Samanea dinklagei****
C) Slash scented like linseed oil, burnt sugar or 'ROB'; slash brittle	***Tetrapleura*** [*PREKESE*]
Slash not foetid; bark rough, often scaly-pitted and very lenticellate	
–Crown brownish or otherwise slightly discoloured (fallen leaves with basal glands);	
Slash (red) fibrous, contoured; crown slightly discoloured; fallen leaves all around very hairy with many parallel laterals	***Parinari excelsa*** [*AFAM*]* (Gp 14B)
Slash (red) granular and rather brittle over orange sapwood; leaves often hairy or discoloured, but without many parallel laterals	***Maranthes*** spp. [*AFAM-NINI*, etc.]* (Gp 14B)
–Crown and fallen leaves not of this 'AFAM' type	
Slash very red, often with white wavy lines or like corned beef, brittle, gritty, often with reddish exudate; lvs not v. feathery	***Erythrophleum*** spp. [*PoTRODOM*]
Slash red-brown, or orange-ish hard fibrous, not brittle; outer bark + large scattered scales	See next spp.
BARK NOT strongly scented in slash, and smooth (but often with scattered scales, many lenticels or raised hoops)	
Slash thin, creamy, with green outer layer; foliage pale green, and fallen leaflets with diagonal midrib; common small-medium tree in **secondary forest**	***Albizia adianthifolia*** [*PAMPENA*)
Not *PAMPENA*; (outer slash not usually green, bark ± > 5 mm thick or granular)	
A) Slash red-brown or orange, hard fibrous but barely brittle; outer bark smooth except for scales falling to leave reddish or different coloured patches; + plank buttresses	
1) leaflets v. feathery and fine	***Parkia bicolor*** [*ASOMA*]*
2) Leaflets in pairs; slash slightly leathery; **evergreen forest**	***Cynometra ananta*** [*ANANTA*]* (Gp 37B)
3) Leaflets neither fine nor paired; crown dark	***Pentaclethra*** or ***Aubrevillea****
B) Slash thick fibrous, yellow – often slightly peelable; (bole often + adventitious shoots and crooked)	
Lvs extremely finely feathery, and fallen leaflets < 1 cm long and barely 1 mm wide; uncommon, and often on sandy soils or by rivers	***Samanea dinklagei****
Lvs not so finely divided; slash yellow to red with much white cambium, and orange sapwood; bole twisted, ± thick buttresses and large lenticels	***Pentaclethra*** [*ATAA*]†
C) Slash mostly granular, often gritty, brittle or fibrous with thick gritty streaks; NOT peelable	
BUTTRESSES V. conspicuous and prominent, or bark pinkish or orange-ish; crown often immense	
1) **Evergreen forest** tree with paired leaflets several cm long	***Cynometra ananta*** [*ANANTA*]* (Gp 37B)
2) Sapwood pale creamish or white; slash v. thin, pale yellow; outer bark with a slightly orange or pinkish colouration; **foliage extremely fine** and dark green in feathery layers; buttresses generally concave, and running along the ground for many metres	***Piptadeniastrum*** [*DAHOMA*]*
3) Sapwood deep orange with vertical striations; buttresses becoming huge, unslashable greyish walls; slash medium thick, + fine gritty brown streaks, pale orange; (young trees spiny); fallen leaves a few cm long, and foliage not v. feathery	***Klainedoxa***, ***Irvingia**** (Gp 13C)
BUTTRESSES inconspicuous, rarely reaching far up, or from tree; slash v. gritty, a bit fibrous – v. LENTICELLATE	
Outer bark greyish; slash dark reddish brown to orange-ish brown, with yellow lines and orange gritty streaks; crown and fallen, fresh leaflets dark green, rhombic	***Albizia zygia*** [*OKRO*]*
Outer bark slightly yellowish with scattered scales or 'swirly' pits; ± twisted or fluted; slash pale brown + orange gritty streaks; large yellow lenticels; **slash ± tobacco-scented**; fallen leaflets yellowish green	***Albizia glaberrima*** [*OKRO-AKOA*]†

* = potentially buttressed trees. See p. 231

LARGE LEAF KEY (crown with large or lobed leaves)

The opposite extreme to the last group; large leaves (or leaflets), usually on a narrow crown. Planted trees with large leaves include Teak (*Tectona grandis*, with pale, fibrous bark, often slightly fluted towards base, and opposite, rather untidy and irregular leaves) and *Terminalia catappa*, which has more glossy leaves clustered at twig ends and in tiers, with a vegetable-scented slash. The following key includes also those species of which the leaves are visibly lobed, even if high in the crown.

		Group No.
Leaflets of compound leaves large; usually falcate and not clustered at twig ends; in wet places of **evergreen forest**	***Gilbertiodendron splendidum***	(37F)
Leaves simple, usually clustered at twig ends and rarely falcate		
Slash hissing when cut; tree usually in **evergreen forest**, and often with small stilt roots or adventitious roots	***Protomegabaria*** spp.	(24)
Slash not hissing when cut		
Leaves not long and thin		
CROWN a v. unusual deeply domed umbrella of palmate leaves	See *Musanga*	(28A)
CROWN or other features not of *Musanga*		
–Leaves not particularly large, but 5-7 lobed or digitate; the star-like pattern dominating the appearance of the crown; slash fibrous or slightly fleshy, yellow, darkening		
Slash brittle, but removable in large plates; boughs whorled; ±reddish exudate	See *Ricinodendron* (K6)*	(28B)
Not *Wama*		
A) Outer bark rough and scaly; bole usually not quite cylindrical above the high buttresses, but sinewy; bands in slash usually *c.*0.5-1 cm broad	***Triplochiton*** [*WAWA*]*	(27D)
B) Outer bark fairly smooth, lenticellate; tree medium, without high buttresses and usually cylindrical; slash v. pale and smooth, not very fibrous in appearance except in narrow vertical, ladder-like strands	***Cola millenii***	(27D)
C) See other (smaller) trees in Gp 28		
–Leaves only slightly or not at all lobed; if lobed then slash reddish		
Slash reddish, banded, darkening; leaves sometimes reddish and often with easily seen undulations on the surface where the lateral nerves are; lvs often with slight lobes or undulate margin		
(Fallen) leaves trinerved, ±lobed, dark green; slash sticky sometimes scented ('green walnuts')		
Tall, straight emergent tree	***Cola gigantea*** [*WATAPUO*]*	(27D)
Small often crooked tree with smoother leaves	***Cola lateritia****	(27D)
(Fallen) lvs not trinerved, but with many parallel laterals; tree only found in **swamps** where there are almost always several together; slash pale pink, soft	See *Hallea* (below)	
Slash pale yellow to orange		
Slash with bands, darkening to dark brown v. rapidly, soft fibrous, thick and outer bark slightly fissured		
Tree gregarious in **swamps** with bole straight to ground; slash (sometimes pinkish) bitter like quinine; lvs not trinerved	***Hallea*** spp. [*SUBAHA*]	(1B)
Tree not gregarious in swamps; slash with pale and darker yellow bands, thick fibrous; leaves rather pale green; bk. often + pits	***Mansonia*** [*oPRONO*] (but see *Christiana* if small)	(27C)
Slash not darkening rapidly, OR gritty OR hard		
Outer bark v. smooth lenticellate and pale greenish; slash almost white, not gritty, with slight foetid scent; tree of **dry forest or rocky places**	***Hildegardia*** [*KYEKYEKYEWERE*]*	(27D)
Outer bark rough, or tree not *Hildegardia* for other reasons		
Base of large trees usually with buttresses		
Slash gradually darkening over *c.*30 seconds through dirty green-brown shades; v. stringy fibrous and peelable; outer bark with scattered rectangular scales and lenticellate; smelling slightly of mushrooms or earth	***Cordia millenii*** [*TWENEBOA-NINI*]	(27C)
Slash v. pale yellow, + yellow lines and orange gritty flecks, easily peeled; outer bark smooth grey, with horizontal raised lines, especially where bole meets buttresses	***Pterygota macrocarpa*** [*KYERERE*]*	(27C)
Base of even v. large trees cylindrical, or with root spurs only		
Slash thick, yellow-orange, stringy, hard; crown deep with many ±horizontal short boughs; outer bark rough and brown	***Nauclea diderichii*** [*KUSIA*]	(1B)
Slash dominated by very dense ORANGE grit; (fallen) leaves cordate and serrated	***Homalium letestui*** [*ESONONANKROMA*]	(17E)
Leaves long and thin (length/max. width >3:1)		
Outer bark very scaly, with bright yellow layer underneath; inner slash red-brown, gritty and brittle; new leaves bright red; **evergreen forest**	***Lophira alata*** [*KAKU*]	(15)
Outer bark not scaly, and without yellow layer (trees often with stilt roots)		
Slash gritty, darkening; lvs of young trees sometimes reaching 1m in length, but adult trees with smaller rounder lvs	***Anthocleista*** spp.	(4)
Slash fibrous, pale, not v. gritty	***Dracaena*** spp.	(39)
See also Uapaca spp., which have unmistakable stilt roots and darkening slash		

* = potentially buttressed trees. See page 231

UTILE CROWN KEY (compound leaves clustered at end of looping twigs)

The majority of the following species grow, according to *Rauh's* model, in a rhythmic manner. The branches are therefore often whorled, and the leaves are typically clustered, not dispersed along the twigs. In those species with compound leaves, this can produce a very distinctive crown typified by *E. utile*, of clustered compound leaves on loopy twigs. The foliage pattern could be described as a crowd of green bats bursting from a cave, or many independent, small, upcurving palm trees each struggling to reach the light. Simple leaved trees with Rauh's growth model, including *Triplochiton* and the avocado tree (and most of Gp 27 and other Gps with long petioles), whilst having the whorled boughs, do not create the same '*Utile*' crown pattern. Interestingly, amongst the compound-leaved trees the crown pattern is typical of non-legumes and rare amongst the legumes (other than those listed below). Cultivated trees with this crown-type include the ubiquitous Neem, which is especially common on the Accra plains, with its dense dark green crown of serrated leaflets, and *Cedrela*. Many Sapindaceae (Gp 36) have crowns which do not show this pattern very convincingly and have been left out of the following key. Also, Bignoniaceae (Gp 29) have clustered compound lvs, but otherwise do not conform.

		Group No.
1) Slash foetid like onions; lvs v. long; planted tree sometimes escaping	***Cedrela***	(34)
2) Tree with large, pyramid-like prickles and hot fruity-acid gritty slash	***Zanthoxylum*** spp. (See ARMED SPECIES KEY K2)	
3) Slash with red exudate	***Amphimas*** (See RED EXUDATE KEY, K6)	
4) Tree without large prickles, without red exudate and without unpleasant oniony scent		
Slash reddish, sometimes speckled or lined with white or orange		
Tree ±small spines; dry zone tree often planted for fruit; bk v. rough	***Spondias mombin***	(35B)
Not *Spondias* (or tree in **moist semi-deciduous** or **wetter natural forest**)		
Slash v. soft, chunky and thick, red with white vertical streaks, without grit, darkening to brown; outer bark smooth or scaly with corky knobs with age; bole often with heavy buttresses	***Entandrophragma angolense*** [*EDINAM*]*	(34B)
Slash not like chunky corned beef; bark often rough or fissured; bole usually v. cylindrical		
Slash with **strong** sweet 'cedar' scent; bark v. rough, with dense large lenticels in vertical, brownish stripes or in small pits; slash yellow-pink with white lines, darkening but hard, with well-defined ripple marks	***E. cylindricum*** [*PENKWA*] (SAPELE)	(34B)
Slash without strong cedar scent		
a) Outer bark greyish, not v. scaly, but with vertical fissures	***E. utile*** [*EFOOBRODEDWO*]*	(34B)
b) Outer bark becoming rough and scaly; slash with much orange grit, in a hard-corned-beef-like backround		
–Inner slash with thick gummy exudate; slash ±contoured, not darkening v. rapidly, slightly aromatic; old seeds usually around tree, like hard nuts 3 cm wide but with several holes	***Antrocaryon micraster*** [*APROKUMA*]	(35B)
–Slash without obvious gum, often with much orange grit, darkening soon to uniform brown (especially outer slash)	***E. candollei*** [*CEDAR-KOKOTE*]*	(34B)
c) Outer bark either (when young) smooth, with raised, horizontal crescent scars or (when older) with deep concentric pits or spirals where large scales have fallen; slash pinkish orange, v. fibrous, bitter after *c.*2 seconds chewing between front teeth	***Xylia evansii****	(38A)
Slash predominantly yellow-white or pale yellow to orange, (sometimes with small areas of pink), NOT darkening rapidly		
Bark with concentric pits or large crescent scars	See *Xylia* (above)	
Bark without deep concentric pits and without crescent scars, never with large buttresses		
Slash v. sweet, incense scented, with plentiful resinous yellow to orange, hard gummy, (gritty) encrustations; outer bark with scattered large lenticels, with raised partial hoops. Foliage pale green	***Canarium*** [*BEDIWONUA*]	(33)
Slash with distinctive slightly earthy sweet scent, but not like incense; outer bark smooth with fine vertical lines, often appearing slightly rotten at (straight) base; slash slightly fibrous with darker gritty patches, soft and thick	***Hannoa*** [*FOTIE*]	(32)

* = potentially buttressed trees. See p. 231

PERFUMED SLASH KEY (slash sweetly perfume-scented, sometimes peppery as well)

N.B. Many species in the '200 Main Species Key' have sweetly perfumed slashes.

		Group No.
Small twisted, fluted tree with cedar scent, flaky bark and fibrous, darkening slash	***Trichilia prieuriana*** [*KAKADIKURO*]†	(34C)
Not *KAKADIKURO*		
SLASH orange to yellow, (or orange-red borderline) usually darkening on exposure		
Slash clearly on the yellow-side of orange at first, darkening later (lvs often simple, regular on branches)		
Crown regular, of rather horizontal boughs; scent of perfume and pepper, not medicinal like camphor, etc.		
Bark peelable in long stringy or ribbon-like pieces		
Outer bark smooth but quilted or slightly furrowed, with vertical lenticel lines; slash rather soft, pithy (chunky) with stringy streaks		
Slash with large pores; lvs drooping at twig ends	***Pachypodanthium staudtii*** [*KUMDWIE*]	(12A)
Slash slightly gritty; foliage v. shiny green	***Cleistopholis patens*** [*NGONONKYENE*]	(12A)
Outer bark red-brown, slightly flaky; slash ribbon-like, not chunky		
Base of large trees deeply fluted; crown rather deep	***Hexalobus crispiflorus*** [*DUABAHA*]†	(12A)
Base of trees not deeply fluted; crown narrow, shallow	***Xylopia quintasii*/**spp. [*OBAA*]	(12C)
Bark rather brittle, sometimes contoured, with granular encrustations between vertical fibrous bands, thick	***Xylopia villosa*/**spp.†	(12C)
Crown of compound leaves, boughs not horizontal; smell of camphor or 'ROB' or 'VIC' or turpentine		
Slash soft, pale + orange gritty bits; no ripple marks; bark smooth $\pm$hoops; with abundant seedling growth nearby (winged rachis)	***Majidea fosteri*** [*ANKYEWA*]	(36B)
Slash rather hard with broad ripple marks; bole cylindrical; bark + large lenticels; seedlings only in sunlight	***Daniellia*** spp.	(37D)
Slash borderline red-orange, darkening; with orange gritty speckles (lvs compound), v. fibrous		
Crown low-branched, v.dense and heavy; bark smooth with vertical lenticel lines; slash rather granular, red orange and brown, soft and chunky	***Turraeanthus africanus***	(34C)
Crown not low-branched; bole long and straight; outer bark rough, with lenticels in vertical lines, but with scales as well, falling to leaves conspicuous pits; slash soft, fibrous and chunky	***Guarea cedrata****	(34C)
SLASH with a predominant RED or PINK component (check also K13)		
Slash partially fibrous, thick, sometimes brittle, usually bitter (lvs compound, bole $\pm$straight and large)		
Ripple marks in slash broad and very conspicuous	See *Daniellia* spp.	(37D)
Ripple marks not conspicuous		
–Crown of many tufted, long compound lvs; slash cedar-scented, often with clear ripple marks	See *SAPELE,*[1] etc. *UTILE* CROWN KEY (K10)	
–Not *SAPELE*; slash without ripple marks		
Bark smooth, or slightly scaly, with prominent lenticels; slash thick, chunky, corned beef (like *EDINAM*) but strongly scented; fallen lflets rather rounded, leathery, with many laterals and rachis rather broad, esp. between lflets; often with abundant seedling growth (seedlings + winged rachis but see *Majidea* above)	***Lovoa*** [*DUBINIBIRI*]*	(34A)
Bark with lenticels not v. prominent; leaflets usually acuminate or drip-tipped, without close laterals		
1) Slash extremely bitter, + sweet rose-water scent[2]; leaflets small; bark usually rough, and not very pale	***Khaya ivorensis*** [*DUBINI*]*	(34A)
2) Slash bitter, extremely rose-scented; bark rather smooth, but with pits and scales	***Khaya anthotheca*** [*KRUMBEN*]*	(34A)
3) Slash slightly scented and bitter; bark rough; **dry or rocky forest**	***Khaya grandifoliola*** [*KRUBA*]*	(34A)
Slash granular or chewy-brittle; or tree uneven or twisted		
Bole rather uneven and sinewy; slash red-brittle; fallen fresh leaves usually rather glutinous when crushed between hands; **evergreen forest or swamps**	***Beilschmiedia mannii*** [*TWEANKA*]	(12)
Bole straight; lenticellate; slash red-brown, $\pm$contoured, paler towards inner bark, with resinous chewy to gritty streaks; red new leaves and scaly outer bark	***Dacryodes klaineana*** [*ADWEA*]	(33)

NOTES: 1) = *Entandrophragma cylindricum*

2) 'Rose-water-scented' may be of limited descriptive value in Ghana, but is slightly more useful than writing '*Khaya* scented'.

* = potentially buttressed trees. See p. 231

(K12)

MUSKY VEGETABLE SLASH (scent like pipe tobacco, almonds, tinned tuna fish, or curry powder or beans etc.)

		Group No.
–SMELL of RAW BACON (smoked or slightly foetid meat – see *WAMA* (K6)		
–Not *WAMA*		
Bole deeply fluted; slash fibrous, smelling of boiled fresh sweet corn or roast plantain	***Duboscia viridiflora***†	(20)
Bole not deeply fluted		
Bole straight and cylindrical almost to ground; outer bark rough, scaly + vertical fissures or with hoops, slash with a thin pinkish outer layer, orange, fine-fibrous, very hard, smelling of almonds or pipe tobacco; often with many seedlings around (with notched leaflets)	***Copaifera*** [*ENTEDUA*]	(37D)
Not *ENTEDUA* (bole buttressed, or bark smooth or at least smelling more like beans)		
CROWN brown discolorous, spreading; bark v. lenticellate; slash red + smell of rotten sugar cane	***Parinari excelsa*** [*AFAM*]*	(14B)
CROWN not brownish, or crown narrow or slash yellowish		
Slash starting v. pale yellow with orange vertical streaks, peelable, smelling of green beans; small-medium buttresses	***Sterculia oblonga, Pterygota bequaertii****	(27)
Slash not pale cream and gritty, or not easily peeled		
Slash gritty and soapy; bark rough or with large scales or pits, often with thick, heavy buttresses; slash scent of almonds or tobacco; bark v. lenticellate;		
Crown v. spreading; bark dark brown, very rough; foliage fine; lflets hairy	***Albizia ferruginea****	(38B)
Crown often heavy, not v. umbrella-like; bark yellow-orange ±plate-scales falling to leave ridged scars or pits;		
Leaflets small, almost feathery foliage	***Albizia glaberrima***†	(38B)
Leaflets not small; foliage not at all feathery	***Afzelia*** spp. [*PAPAO*], ***Blighia*** spp.*	(37D,36A)
Bark generally smooth or slash not gritty		
SLASH soft fibrous OR darkening v. rapidly; (large old leathery pods usually around)		
1) Slash smelling strongly of curry powder or saw mills; new lvs reddish, not v. large; outer bark slightly scaly	***Berlinia confusa*** [*KWATAFOMPABOA*]	(37F)
2) Slash smelling of vegetables; especially green beans; v. fibrous and thick; bark rough; bole cylindrical crown heavy and dense with large leaflets; esp. **evergreen forest**	***Berlinia occidentalis/tomentella***	(37F)
3) Slash smell not strong, pinkish; **swamps or evergreen forest**	***Gilbertiodendron*** spp. [*TETEKON*]	(37F)
4) Slash gritty, with bitter-sweet exudate, almond-scented	***Scottellia*** [*TIABUTUO*]*	(17D)
SLASH hard OR brittle fibrous, AND not darkening rapidly		
–Slash with v. conspicous green outer layer AND bole with obvious large ORANGE LENTICELS; smell of 'tincture of iodine'	***Holoptelea*** [*NAKWA*]*	(18A)
–Not *ANAKWA*; bark smooth		
Slash rather brittle soft-granular; outer bark corky-flaky; crown narrow, of horizontal boughs; scent bitter sweet, like almonds or vanilla	***Ophiobotrys*** [*AKWANA*]†	(18A)
Slash peelable or hard (or tree with pods or bean-scented)		
Foliage pale and fine; bark pinkish in small squares; slash scented of caramel (lightly burnt sugar); 4-sided brittle pods	See *Tetrapleura* (K8)	
Not *Tetrapleura*; foliage normally dark in outward and upward sweeping plumes		
Slash flexible, peelable, smelling of green beans or vegetables		
a) Slash moderately hard, fibrous, often reddish, at least in patches; very often with adventitious shoots or abundant seedling growth (4 leaflets + winged rachis)	***Hymenostegia afzelii*** [*TAKROWA*]	(37C)
b) Slash brittle-leathery, yellow turning orange, usually with blackish outer layer	***Baphia pubescens*** [*oDWENKOBIRI*]	(37A)
c) Slash not darkening, and without blackish outer layer, orange-ish, with a slightly waxy appearance	***Baphia nitida*** [*oDWEN*]	(37A)
Slash brittle, hard, contoured, with smooth bark		
–Slight gingery smell, slash very gritty and hard; contoured; sapwood orange; copious thick-watery, sweet exudate; bark lenticellate; tree often bearing its yellow flowers or up-pointed woody pods	***Bussea occidentalis*** [*KOTOPREPRE*]	(38A)
–Slight fishy smell; outer bark ±irregular scales; hard, thin, many-contoured; **evergreen forest**	***Crudia gabonensis****	(37G)

* = Potentially buttressed trees. See p. 231

RED SLASH KEY (slash red or pink, at least in bands – ignoring thin layer near outer bark)

Group No.

- Tree with large leaves — See *Cola, Mitragyna*, etc. (K9)
- Tree with *Entandrophragma* branching — See *Spondias*, etc. (K10)

Slash with broad RED-PINK and WHITE or yellowish vertical bands, soft-fibrous, the vertical bands generally corresponding to (fissures) patterns on outer bark; not v. contoured

1) Outer bark with many CONSPICUOUS PITS all over bole, easily visible from 5m; slash including orange grit in white pith patches; base with very small or no buttresses — ***Lannea welwitschii*** [*KUMANINI*] (35B)
2) Outer bark with shallow pits; slash soft corned beef — See *EDINAM* (*UTILE* CROWN KEY) (K10)
3) Outer bark without pits; slash soon darkening to unform brown; branches whorled
   - Outer bark v. rough and dark brown to black; slash turning to brown very rapidly; sapwood and bark with conspicuous broad ripple marks; base with small, low blunt buttresses — ***Rhodognaphalon*** [*oNYINAKOBEN*]* (28C)
   - Outer bark not v. rough-dark; slash v. bitter
     - Slash without obvious grit; tree often large — ***Sterculia rhinopetala*** [*WAWABIMA*]* (27A)
     - Slash v. orange-gritty, with soapy effect on tongue; small tree — ***Cola nitida*** [*BESE*] (27A)
     - Slash with usually brown bands from moment of slashing, with sweet sawmill aroma; whorls of boughs and clustered lvs v. conspicuous — ***Sterculia tragacantha*** [*SoFo*] (27C)

Slash without red or white vertical bands, or with other colours much more conspicuous, or v. contoured

A) Slash hard, with distinctive (bean, tobacco, curry etc.) scent — See MUSKY VEGETABLE SLASH KEY (K12)
B) Slash spongy, thick and pale pink — See FIBROUS SLASH KEY (K14)
C) Slash not thick-spongy, nor tobacco, etc. scented

Crown golden brown or similarly discoloured

- Slash clearly fibrous, sometimes brittle (but ±thin granular outer layer only)
  - Bark pale, and scaly, with white lenticels; large tree, slash starting dark liver red, contoured; scent of rotten sugar cane — ***Parinari excelsa*** [*AFAM*]* (14B)
  - Bark not v. pale nor lenticellate; medium-sized tree with slash starting pink, darkening later, not contoured — ***Diospyros kamerunensis*** [*oMENEWA*] (11)
- Slash predominantly granular and brittle (±thin fibrous inner layer); sapwood bright yellow-orange — ***Maranthes*** spp.* (14)

Crown not reddish nor brownish discoloured

- **Outer bark rough and flaky**; Slash (soft or crumbly fibrous), sometimes contoured
  - Slash fleshy, bole cylindrical
    - Moist, fruity acid taste; with clear gummy exudate turning yellow; branches regular and horizontal — ***Allanblackia*** [*SONKYI*] (8)
    - Slash scented of cocoa pods; tree in **swamps** or **evergreen forest** with compound lvs — ***Gilbertiodendron*** spp. [*TETEKON* etc.] (37F)
  - Slash not fleshy, or tree buttressed or not as above
    1) Bole ±straight to base, cylindrical, or slightly sinewy; no or small buttresses; slash crumbly-fibrous; outer bk rough (with age)
       - Slash with red or orange vertical streaks
         - Branches whorled with ascending tips; large palmate lvs; slash ±red exudate, with savoury smell — ***Ricinodendron*** [*WAMA*]* (28B)
         - Branches not whorled, with drooping tips, small leaves — ***Margaritaria discoidea*** [*PEPEA*] (13C)
       - Slash grit-less, darkening, contoured; tree usually with red new leaves — ***Erythroxylum emarginatum*** [*BENKYI*] (13C)
    2) Bole with thin, often high buttresses; slash pinkish, darkening to brown, with acidic taste; sapwood with fine ripple marks (lens) — ***Nesogordonia papaverifera*** [*DANTA*]† (27A)
    3) Bole fluted, with v. high buttresses, with easily peeled bark and lenticels in vertical lines; hissing when slashed, with sickly-sweet scented slash — ***Sacoglottis gabonensis***† (17C)
- **Outer bark not flaky** (usually smooth, sometimes with raised hoops or fissured)
  - Bark smooth with raised hoops;
    - Slash granular, gritty; bole with many adventitious shoots. . .
      - . . .a) bearing lvs with unusual apical teeth — ***Phyllocosmus africanus*** (17C)
      - . . .b) with opp., entire, gland-dotted lvs — ***Syzygium*** spp. (7)
    - Slash ±fibrous nr sapwood, with sweet exudate, pink brown, brittle — ***Guibortia ehie**** (37B)
  - Bark with fine fissures; slash crumbly and gritty — ***Diospyros heudelotii*** (+spp.) (11)

* = potentially buttressed trees. See p. 231

FIBROUS SLASH KEY (slash fibrous, spongy or stringy, often contoured, without obvious gritty streaks)

Group No.

Slash white to pale yellow, fibrous, ***darkening through dirty greenish shades****;* not v. buttressed, but sometimes fluted

- Outer bark pale, slightly corky, flaky and peeling; slash rather thin; trees often slightly fluted; leaves digitate and opposite (normally visible as such from ground) — ***Vitex*** spp. [*oTWENTOROWA*] (29)
- Not *Vitex* spp. (or tree too tall to see foliage)
  - Slash thick and spongy; lvs simple; bole not fluted
    - Outer bark + small scales or v. lenticellate; with peelable, spongy slash and with distinctive rather musty or mushroom-like scent; fallen simple lvs ±ovate, rough or hairy — ***Cordia/Ehretia*** spp. [*TWENEBOA*] (26)
    - Slash darkening rapidly; lvs small, in graceful, spreading crown — ***Pteleopsis hylodendron*** [*KWAEKANE*] (25)
  - Not *TWENEBOA*, etc. (e.g. lvs compound or slash thin); bole usually short and contorted or fluted; lvs compound and opposite
    1) Outer bark smooth, with hoops or scars; bole often contorted — ***Kigelia africana*** [*NUFUTEN*] (30)
    2) Outer bark v. fissured or irregularly scaly; with reddish lenticels in vertical rows; bole often v. fluted; slash with quinine bitter taste and sweet scent — ***Spathodea*** [*AKUAKO-NINSUO*]† (30)
    3) Check other trees in Gp 29 – esp. if in **dry forest**

Slash ***not darkening through dirty greenish shades****,* but sometimes darkening nevertheless

- –Bole very deeply fluted; crown bright or pale green; slash peelable in thin strips
  - Flutes tending to curve round to one side as they rise; slash only peelable in rather short strips (*c.*30 cm), with sweet smell like roast plantain or boiled sweet corn; outer bark flaky; crown pale green — ***Duboscia*** [*AKOKORAGYEHINI*]† (20)
  - Flutes not curving greatly to one side; slash peeling in very long strips, slightly sweet-scented, but not like roasted plantain, etc. — ***Hexalobus crispiflorus*** (***Xylopia***) (12A,C)
- –Bole not deeply fluted, but sometimes + fine flutes or crooked; slash not peelable in ribbon-like strips, sometimes pink, usually spongy
  1) Crown with reddish tone due to red-brown new or old lvs; trees often found outside **evergreen forest**
     - Tall tree with old lvs (on tree and ground) a rich red-brown; outer bark rough, deeply fissured, slash dirty yellow, thick and spongy, slightly foetid with bitter taste; crown broad often tiered — ***Petersianthus macrocarpus*** [*ESIA*]† (25)
     - Medium-sized, often crooked tree with whole, dense crown with subdued reddish hue due to hairs on young leaves; slash with sweet, fruity taste — ***Octoknema borealis*** [*WISUBONI*] (27B)
  2) Crown without reddish colour; outer bark scaly; trees common in **evergreen forest** and rare outside it
     - Crown compact, of many small leaves; medium sized **evergreen forest** tree; bark flaky or with pits; slash slightly pinkish foetid, darkening — ***Scytopetalum*** [*oPRIM*] (13B)
     - Crown not of small leaves; outer bark shaggy, with many scales; slash darkening slightly due to watery exudate; medium-sized tree in **evergreen forest**; sometimes + spots of latex — ***Coula edulis*** [*BoDWUE*] (27B)
  3) Crown without reddish colour; outer bark extremely smooth, yellow or greenish; slash with clear yellow exudate; (fallen) lvs extremely rough — ***Ficus*** spp.*
  4) Consider also *Terminalia* spp. (K7); *Hannoa* (K10) and *Baphia* spp. (and other legumes in K12) which have yellowish, very fibrous slashes

* = potentially buttressed trees. See p. 231

GRITTY SLASH KEY (slash with conspicuous gritty speckles or streaks)

`Small trees with very gritty slashes. **Group No.**

Slash darkening rapidly (<15 seconds); slash v. thick, (±fibrous towards inner bark); bole straight to base or low buttressed
- –Slash taste (and scent) v. strongly acidic and hot; **evergreen forest** — ***Araliopsis*** [*MEAWERE*] (28B, 31A)
- –Slash without strong taste; crown dark and regular; outer bark fine fissured or scaly; gritty portions very rusty and conspicuous — ***Anopyxis*** [*KoKOTE*]† (3)
- –(**evergreen forest** tree, like *KoKOTE*, with sweet gummy exudate = *Cassipourea gummiflua*)

–Slash not darkening rapidly, although sometimes darkening within about 30 seconds
- –Slash not very fibrous (except sometimes in thin layer near sapwood) – not strongly contoured
  - Outer bark with many rounded scales, falling to leave pale yellow patches; bole unbuttressed, cylindrical or slightly sinewy — ***Strombosia*** [*AFENA*] (13B)
  - Outer bark without many rounded scales
    - **Bole not deeply fluted**
      - Bark rough, v. lenticellate; bole ±cylindrical to base; slash hard, with many orange gritty fine lines; sapwood v. hard and deep orange, with fine vertical lines — ***Homalium letestui*** [*ESONANKROMA*] (+ *Homalium* spp., especially on hills) (17E)
      - Bark usually smooth except for raised hoop-ridges, or bole strongly buttressed
        - Slash rather soft, sometimes slightly acidic; sapwood v. pale orange or yellowish
          - –Lvs compound; bark yellowish (±scented slash) — ***Blighia*** spp. [*AKYE*], ***Albizia*** spp.* (36A, 38B)
          - –Lvs simple, v. trinerved; **evergreen forest** — ***Anisophyllea*** [*KoKOTE-AKOA*] (18A)
          - –Slash with large pores; lvs simple, >15 cm long — ***Strephonema*** [*AWURUKU*] (21)
    - **Bole v. irregular with spiral flutes**; slash with scented gummy exudate (±spines) — ***Balanites wilsoniana***† (31)
- –Slash with clear fibrous contours
  - **Dry forest** tree with compound leaves — ***Stereospermum*** [*ESONOTOKWAKOFUO*] (30)
  - Heavy crowned tree with surrounding forest often stunted; outer bark fairly smooth, but with small flakes; with a clear sweet exudate — ***Okoubaka aubrevillei*** [*oDII*] (18A)

* = potentially buttressed trees. See p. 231

REMAINDER KEY (the remaining species)

Tree unarmed, without stilt roots; leaves/leaflets of moderate size, not lobed, in untiered crown; slash not predominantly red, nor clearly perfume or strong-vegetable scented; slash not stringy nor spongy-fibrous and without dense gritty streaks visible at 3 m; the slash of the following species vary from very granular to hard, compact-fibrous.

These first three species are extremely common in semi-deciduous forest; if you are unfamiliar with the slash, and standing in semi-deciduous forest, you will almost certainly be able to find an example within five minutes.

Key	Species	Group No.
Slash with distinctive chocolate brown and yellow bands (contoured), or with broken rings of brown blotches on a yellow background		
Slash with brown blotches, NOT continuous rings; outer bark slightly rough, with many small flaky patches; crown dark green	***Celtis adolfi-friderici*** [*ESA-KOSUA*]*	(18A)
Slash with continuous or almost continuous thin brown bands; outer bark not generally v. flaky, often silvery or yellowish		
Crown v. pale or yellow to reddish; fallen lvs v. hairy with ladder-like venation; outer bark usually v. yellowish; slash usually darkening soon	***Celtis zenkeri*** [*ESA-KoKoo*]*	(18B)
Crown rather dark, with fine graceful, drooping branches; fallen lvs without scalariform venation and not hairy	***Celtis mildbraedii*** [*ESA*]*	(18B)

Not these *Celtis* spp.: slash without yellow and brown contours; not yellow with broken rings of brown blotches

(The next species have a rather soft slash, easily cut through with front teeth or fingernails. If you are confident your tree has a hard slash, then move to ***. Otherwise run through the next few questions to make sure).

Key	Species	Group No.
1) Bark smooth, greenish except for conspicuous red papery flakes, which make higher parts of the stem and main branches unsually red, ± bark scales nr base, leaving pale scars; outer slash granular, inner bark fibrous, pinkish-brown to cream in layers, with ripple marks, usually scented slightly of cooked meat	***Distemonanthus benthamianus*** [*BONSAMDUA*]*	(37G)
2) Bark smooth, except for large irregularly-shaped flakes which fall to leave reddish scars. Foliage graceful and flowing; slightly drooping foliage; slash with a greenish outer layer, yellow-orange, soft chunky-fibrous, → red brown, with ripple marks in sapwood	***Pericopsis elata*** [*KOKRODUA*]*	(37G)
3) Not *BONSAMDUA* nor *KOKRODUA*; tree without red stems /red patches below scales		
Bole (large tree) irregularly fluted or with sinuous buttresses, bk smooth; slash with fine (not broad) gritty streaks		
1) With greenish underbark , fruity-acid taste, + gingerish thin, gritty bands	***Discoglypremna*** [*FETEFRE*]*	(22A)
2) (With spines on branches); v. fluted indeed, + silvery-translucent gummy exudate; slash pale	***Balanites*** [*KROBODUA*]†	(31)
3) Only slightly fluted, with dark spots appearing nr inner slash – check..	See *Irvingia* (overleaf)	
Bole not irregularly fluted etc.		
Bole ± cylindrical, with many small lenticels; slash thick, soft, orange, sweet, ± homogeneous, darkening slowly with watery exudate, with dark outer line	***Ongokea gore*** [*BoDWE*]	(13A)
Bole with small, neat, triangular buttresses, smooth except for small creases, often with the distinctive seedlings (paired, cow-horn-like leaflets) in vicinity; with gummy translucent, slightly sweet, earthy-smelling exudate nr inner bark	***Guibourtia ehie****	(37B)

*** The following species (NEXT PAGE) have a harder, brittle, bark – it may be safer to discount the soft-barked species above if you are in doubt.

* = potentially buttressed trees. See p. 231

REMAINDER KEY (contd.)

***The following species have a harder, brittle, bark – it may be safer to discount the soft-barked species (PREVIOUS PAGE) if you are in any doubt.

		Group No.
Bole strikingly and unusually cylindrical completely to base; outer bark usually greyish with large lenticels and raised hoops; slash with v. broad and obvious ripple marks, like straight fingerprints, with a v. slight sweet resinous smell and (sometimes) exudate		
In **evergreen forests**	***Daniellia*** spp. [*eHYEDUA, SOPI*]	(37D)
In **drier forests**	***Daniellia ogea*** [*eHYEDUA*]	
Bole not of this distinctive 'concrete pillar' type (or slash without ripple marks)		
Outer slash greenish; outer bark with orange-brown lenticels; slash iodine scented; usually with high, v. narrow buttresses or flutes; sapwood + v. fine ripple marks	***Holoptelea*** [*NAKWA*]*	(18A)
Outer slash not greenish, or bark without orange-brown lenticels		
1) Tree with large, greyish, straight-edged buttresses, smooth except for sinewy lines; slash hard orange; sapwood striate, dark orange; (yng trees + spines)	***Klainedoxa*** [*KROMA*]*	(13C)
2) Slash with silvery specks of copious, thick watery, sweet exudate appearing nr sapwood; slash darkening particularly around gritty specks; aftertaste bitter	***Irvingia*** [*ABESEBUO*]*	(13C)
3) Not *Irvingia* nor *Klainedoxa*		
(At this stage, a large tree is very likely to be a legume, even if not one of the following species.) Many legumes have at least slight vegetably scents; there are many species in **evergreen forests**		
(Slash usually hard, yellow to orange, often contoured, rather brittle, partially fibrous, often with raised hoops or deep concentric rings.) *Foliage becomes increasingly important at this stage*		
Bark with concentric pattern of pits, where scales have fallen (Slash tobacco, etc. scented – see *Afzelia* spp.)		
Foliage fine and heavily clustered at twig ends; slash hard, fibrous, reddish over orange-brown layers, darkening	***Xylia evansii*** [*ABOBABEMA*]*	(38A)
Foliage in dense, dark, plumes along branches; slash brittle-fibrous and gritty, with hard, orange, striate sapwood, tasting sweet, then bitter, like quinine tonic water	***Calpocalyx brevibracteatus*** [*ATROTRE*]*	(38A)
Bark without concentric pattern of pits		
–Slash with watery exudate; often with yellow flowers or upward-pointing woody pods	***Bussea occidentalis*** [*KOTOPREPRE*]	(38A)
–Foliage (often on abundant sapling growth nearby or on adventitious shoots) of 4 leaflets and winged rachis; slash mainly orange but with reddish patches under bumps on (otherwise smooth) bark	***Hymenostegia afzelii*** [*TAKROWA*]	(37C)

* = potentially buttressed trees. See p. 231

SECTION 4

GLOSSARY AND INDEX

SECTION 4
GLOSSARY AND INDEX

GLOSSARY

The following are very brief definitions, to help with the interpretation of the illustrations in this book. Because they are so short, and because many alternative, technical words have been avoided, the reader is warned that some of the supplied meanings may be misleading outside the context of this book.

G1 or G2 etc. refer to the two illustrations in the Glossary.

* – References to illustrations marked with an asterisk are to the Groups in the Main Key, where particular examples can be found. e.g. *23 refers to the illustrations for Group 23. The word is in many cases explained in more detail opposite the illustration indicated. Numbers without an asterisk refer to Groups where further textual information can be found.

*S – Illustrations marked thus are to those opposite the Starting Key, where notes on the arrangement of leaves can also be found.

'200' – Refers to the explanatory text in Section 3, particularly the 200 Main Species Key.

'INTR' – Refers to the Introduction (Section 1).

Word	Illustration	Meaning
ACID		Taste sharp, bitter, like unripe fruit
ACRID		Smell sharp, hot, alarming, unpleasant, like dense smoke or acid
ACUMINATE	G1	Apex of leaf tapering v. gradually at tip (long acuminate = drip tip)
ACUTE	G1	Apex of leaf abruptly pointed
ADVENTITIOUS	*24	(Shoot, root) arising on bole or boughs, and so bursting through bark
ALTERNATE	*S	(Leaves) arranged one per node
ANASTOMOSING		Splitting and rejoining, like strings in a net, or like strangling fig roots
ANTHER		Section of the male part of flower (STAMEN) which contains pollen
APEX	G1	End part; part furthest away from the main axis of plant
APPRESSED		(Hairs) lying flat against surface
ARIL	*36	Thin covering of (whole or part of) seed (inside fruit) often of a different colour from seed
AROMATIC	INTR	Pleasant, rather heavy or oily scent like incense or certain spices
ARTICULATED	*33	(Petiole) swollen or deformed at a point where bent or bendable (like elbow)
ASCENDING		Arching upwards, towards top (branches in tree) or apex (nerves in leaf)
ASYMMETRIC	G1	Not symmetric
ASTRINGENT		Bitter or acidic taste which leaves mouth peculiarly dry
AURICULATE	16	With ear-like lobe, usually at base of leaf
AXIS	*S	Main stem or channel from which branches or other parts arise (e.g. twig or bole)
AXIL		The corner where a structure meets the axis from which it branches (e.g. leaf axil – between petiole and stem; nerve axil – between lateral nerve and midrib)
BARK	INTR	Partially dead covering of all except youngest parts of tree, outside sapwood
BASAL	G2	Arising at base; basal nerve = lateral nerve arising at base of lamina (near top of petiole)
BASE	200	Part of tree including any irregularities above ground level not reaching crown
BERRY		Fruit with (1-) several seeds immersed in soft flesh, but seeds without stony layer (c.f. DRUPE)
BIPINNATE	*S	Leaves with two orders of branching, usually with small leaflets (e.g. *DAHOMA*, *Acacia*)
BITTER		Taste like 'bitters' or quinine
BOLE	200	Part of tree between lowest boughs of CROWN and highest part of BUTTRESSES, etc. at base
BOUGH	200	Main, massive branches of CROWN
br		Abbreviation for branch
BRACT	*22	(Usually v.small) leaf-like or stipule-like structure at nodes of inflorescence
BUTTRESS	200	Outgrowths at base of tree extending from lateral surface roots to trunk
CADUCOUS		(Stipule) falling soon after formation
CALYX		Sepals as a whole, particularly when they are united into a tube (=calyx tube)
CAPSULE	*34	Type of DEHISCENT ft which splits open along several lines (or pores) to release seeds
CARPEL		Female (often central) part of flower including ovary, attached to style, developing into fruit
CATKIN	*22	Unbranched inflorescence (spike or raceme) of dense brush of small flowers
CAULIFLOROUS	*11	Flowers and fruits arising from bole (±boughs as well)
CHAFFY		Like 'chaff' (the residue when harvested rice, millet, etc. is cleaned)
CHANNELLED	G2	(Nerves, etc.) sunk below surface, leaving a rounded channel
CILIATE		With a fringe of hairs
CLUSTERED	*S	All arising close together from an axis (e.g. leaves, branches)
COMPOUND	*S	Leaves with more than one axis; with several midribs; with 2 or more leaflets
CONTOURED		(Slash) pattern of curved parallel lines (e.g. when layered bark is slashed)
CORDATE	G1	Leaf base which curves backwards from petiole attachment; 'heart-shaped'
CORIACEOUS		Tough, like leather or firm card
CORK		Soft material used in wine-bottle tops, or similar in texture; like pale brown soft, chewy wood
COROLLA	*9	Petals as a whole, particularly when they are united into a tube (=corolla tube)
COTYLEDON		The first pair of lvs (or lf in Gp 40) sometimes hidden in seed; usually unlike later lvs
CUNEATE	G1	Leaf base wedge-shaped; becoming gradually and regularly thinner towards petiole

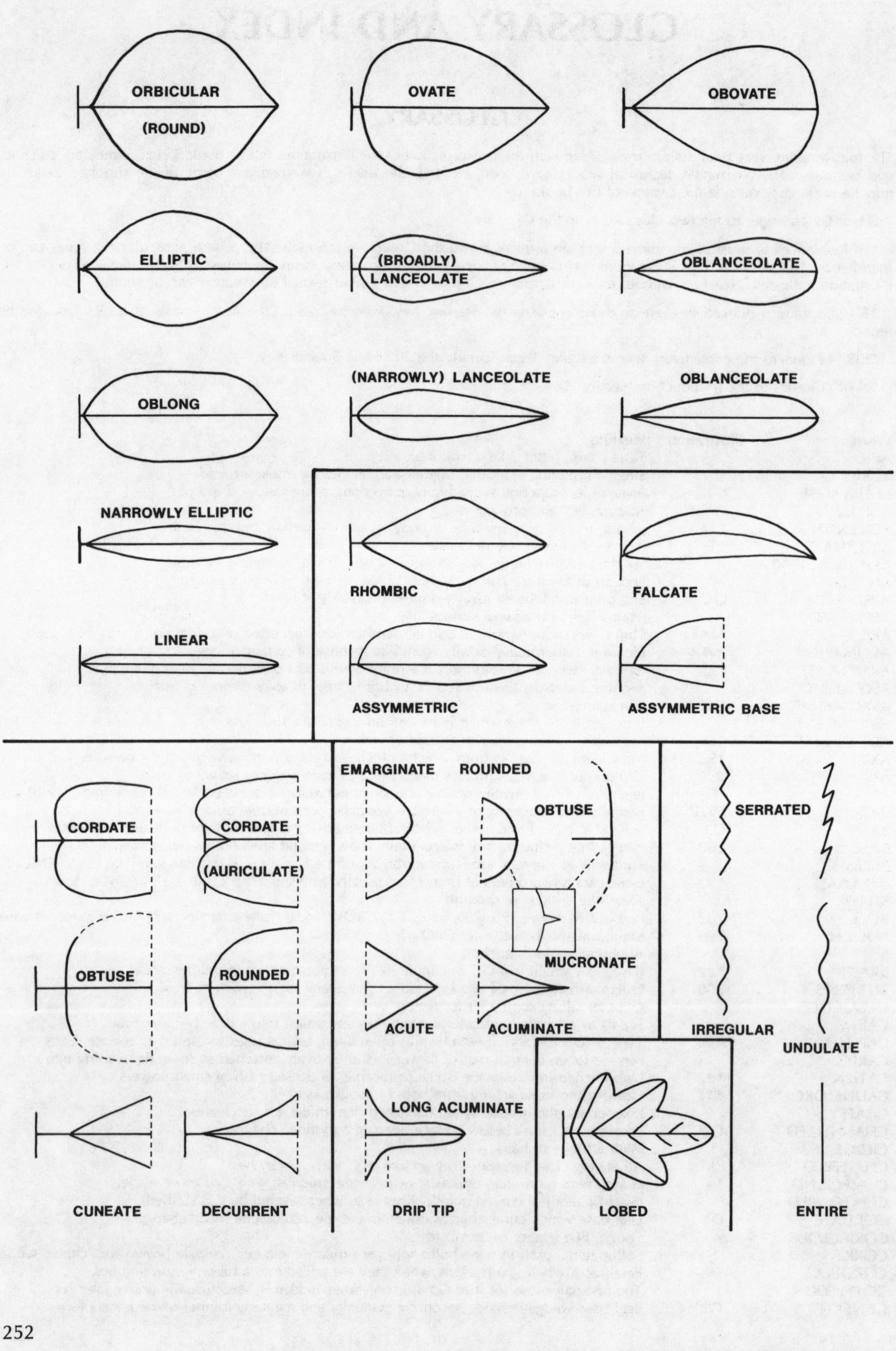
ORBICULAR
(ROUND)
OVATE
OBOVATE
ELLIPTIC
(BROADLY)
LANCEOLATE
OBLANCEOLATE
OBLONG
(NARROWLY) LANCEOLATE
OBLANCEOLATE
NARROWLY ELLIPTIC
RHOMBIC
FALCATE
LINEAR
ASSYMMETRIC
ASSYMMETRIC BASE
CORDATE
CORDATE
(AURICULATE)
EMARGINATE
ROUNDED
OBTUSE
SERRATED
MUCRONATE
OBTUSE
ROUNDED
ACUTE
ACUMINATE
IRREGULAR
UNDULATE
LONG ACUMINATE
CUNEATE
DECURRENT
DRIP TIP
LOBED
ENTIRE

Word	Illustration	Meaning
CROWN	200	System of branches at top of tree supporting leaves
CYME		Branched inflorescence without a single main axis, but with many axes ending in flowers
DECIDUOUS		Leaves all falling at one time (season)
DECURRENT	G1	Leaf base v. sharp wedge-shaped, with base of lamina running along edge of petiole
DEHISCENT	*27	(Dry fruits) splitting when ripe to release seeds (see INDEHISCENT, CAPSULE)
DENTATE	G1	With teeth (like serrate); strictly the teeth should be equal-sided
DICHOTOMOUS	G2	Splitting; branching successively into smaller halves
DICOTYLEDON		(Plant with two seedling lvs); all trees apart from the MONOCOTYLEDON (/fern) trees of Gps 39,40
DIGITATE	*S,28	Leaf with more than three leaflets, all arising from one point at end of petiole
DIGITATELY-LOBED	*27	(Simple leaf) with lobes diverging like fingers on a hand
DILATATION		(Tissue) in bark marked by broad vertical bands associated with outgrowth of bark
DIOECIOUS	19,22	Species with male and female flwrs on different trees; i.e. with male and female trees
DISK	34	(Usually fleshy) part of flowers (receptacle) of some spp. surrounding ovary; often secreting nectar
DISCOLOROUS		Differently-coloured (especially lower) side of leaf (compared to upper side)
DOMATIA	*1A	Tufts of hairs, small holes or other structures in axils of nerves
DRIP TIP	G1	Tip of leaf long drawn out – long acuminate
DRUPE	*14	Fruit with fleshy part surrounding hard, central, 1-seeded 'stone(s)' (e.g. mango, plum)
DRUPE-LIKE	*17D	Like a drupe, but with stone(s) each containing more than 1 seed
ELLIPTIC	G1	Broadest around middle with smoothly curving edges
ELLIPSOID		A solid object with elliptic outline
EMARGINATE	G1	Tip of leaf with a slight notch
ENDOCARP		Inner layer of wall of fruit (pericarp)
ENTIRE	G1	(Margin) smooth; evenly curved or straight, without teeth or undulations etc.
EVERGREEN		i) Tree which always bears foliage; i.e. never deciduous
	INTR	ii) Forest with canopy dominated by evergreen trees, in S.W. corner of Ghana
EXSERTED		Protruding, longer than surrounding object
EXUDATE	INTR	Any liquid or gelatinous substance flowing from tree (esp. when slashed)
f		Abbreviation for female
FALCATE	G1	Leaflet or Leaf with midrib and edges curved round to one side
FASCICLE	*22	Inflorescence without (obvious) stalk; dense cluster of flowers arising in one place
FERRUGINOUS		Rust-coloured
FIBRE		Long, very thin, flexible and strong like cotton thread
FIBROUS		With many fibres
FLANGE		Broad wing
FLESHY		Thick, firm yet soft, easily sliced
FLINT-BARK	(*11)	Very hard, glass or stone-like bark of certain *Diospyros* spp.
FLOWER		Reproductive structure with single axis; usually RECEPTACLE with STAMENS, OVARY(ies) and PERIANTH
FLUSH		FOLIAGE developing at one time on trees of rhythmic growth; colour $\pm$contrasting with older lvs
FLUTED	200	(Bole, Base) with many rounded, regular ascending channels
flwr		Abbreviation for flower
FOETID		Disgusting smell, like rotting meat or faeces
FOLIACEOUS		Leaf-like (e.g. stipules, bracts)
FOLIAGE	200	The leaves or leaflets on tree or branch considered as a whole
FOLLICLE	*27	Type of DEHISCENT fruit splitting along one line (like some pods)
FRAGRANT	INTR	Pleasantly and delicately scented (like many flowers)
FRUIT		Structure derived from carpel(s) containing seeds and generally encouraging their dispersal
ft		Abbreviation for fruit
GLABROUS		Hairless
GLAND		Permanent structure on leaf (or other parts) distinct from surrounding tissue and of following forms:
	*17C	i) Basal gland; close to midrib just above base of leaf
	*17C	ii) Marginal; like teeth, or notched but more specialized – conspicuously discoloured, thick etc.
	*17C	iii) Knotted vein; (raised or sunken) area in mid-lamina associated with congestion of veins
	*17D	iv) Gland dots; dark or translucent spots in a regular pattern on (lower) surface
		v) Glandular hair; hair with enlarged tip of different colour
		NB–although glands are strictly secretory organs, in this guide the term is used very loosely to cover many small structures of varied, usually unknown function
GLAUCOUS	12,11	Surface (of leaf) with unshiny (matt) texture like fine dust and pale blue or pinkish
GLUTINOUS		Slightly sticky to touch like (or coated with) thick, viscous liquid
GRANULE		Small, rounded hard particle
GREGARIOUS		Occurring together in groups
GUTTERED	G2	(Nerves, etc.) partly prominent, but with channels running beside each edge
HELIOPHILOUS		'Sun-loving'; typical of forest gaps, road-sides, savanna, etc.
HERMAPHRODITE		(Flower) of both sexes i.e. with both stamens and ovary
HOOP		(On bark) narrow circle of distinct texture or colour running round tree, twig, etc.
HYBRID		The offspring of two different species, intermediate in character between the parent species
IMPARIPINNATE	*S	(Pinnate leaf) with leaflets either alternate, or opposite with an odd terminal leaflet
IMPRESSED	G2	Raised line or ridge (of nerves, etc.) which lies as if pressed into the surface which bears it
INDEHISCENT	*36	(Fleshy fruit) not splitting open at maturity, e.g. DRUPE or BERRY
INFERIOR OVARY	4	Ovary developing below the point of insertion of petals, etc.; becoming fruit with calyx at top

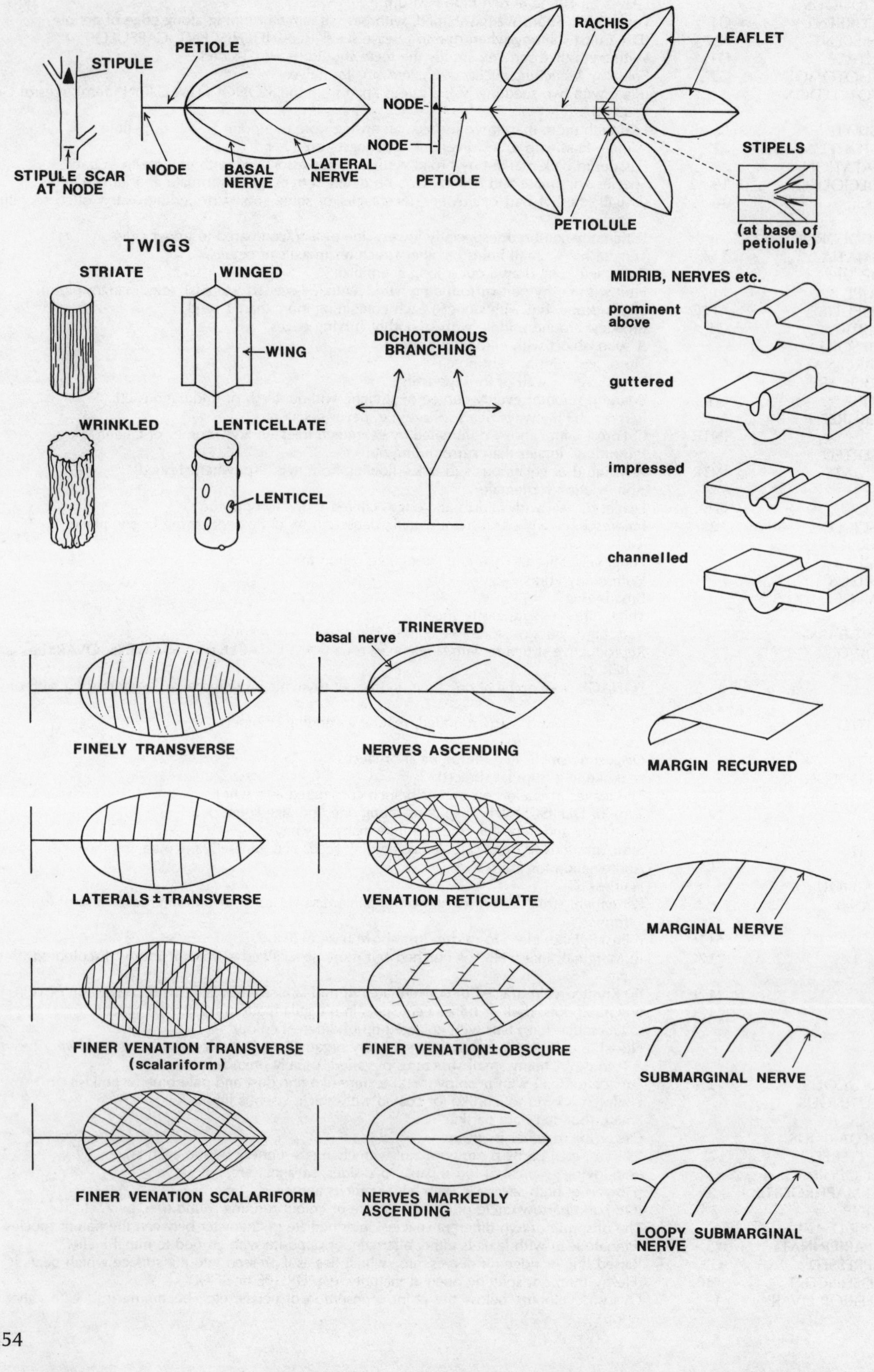
STIPULE
PETIOLE
STIPULE SCAR AT NODE
NODE
BASAL NERVE
LATERAL NERVE
NODE
NODE
PETIOLE
RACHIS
LEAFLET
PETIOLULE
STIPELS
(at base of petiolule)
TWIGS
STRIATE
WINGED
WING
WRINKLED
LENTICELLATE
LENTICEL
DICHOTOMOUS BRANCHING
MIDRIB, NERVES etc.
prominent above
guttered
impressed
channelled
basal nerve
TRINERVED
FINELY TRANSVERSE
NERVES ASCENDING
MARGIN RECURVED
LATERALS ± TRANSVERSE
VENATION RETICULATE
MARGINAL NERVE
FINER VENATION TRANSVERSE (scalariform)
FINER VENATION ± OBSCURE
SUBMARGINAL NERVE
FINER VENATION SCALARIFORM
NERVES MARKEDLY ASCENDING
LOOPY SUBMARGINAL NERVE

Word	Illustration	Meaning
infl		Abbreviation for inflorescence
INFLORESCENCE	*22	Several/many flowers arranged on a single leafless axis
INTERNODE		Part of stem between nodes
INTERPETIOLAR	*1	(STIPULES) between the petioles
JUVENILE		Younger, infertile stages in life cycle
KNEE-ROOTS	*3	Parts of root which rise out and bend back into soil/water of swamps, like knees of sunken leg
KNOBS	*22B	c. rounded, e.g. hemispherical protrusions from surface (e.g. of bole)
LAMINA	*S	The flat, green, thin part of a leaf or leaflet; i.e. exluding the petiole or petiolule
LANCEOLATE	G1	Shaped like a lance head; more than 3 times as long as broad and broadest below middle
LATERAL	G2	i) sideways ii) abbreviation of Lateral Nerve – i.e. nerve branching along midrib like ribs
LATEX	INTR	Opaque, milk-like white or yellow, sticky or rubbery exudate
LATICIFEROUS	200	Producing latex
LEAF	*S	Flattened green structure repeated many times on tree which normally develops and falls as a unit; SIMPLE or COMPOUND
LEAFLET	*S	Structure like simple leaf forming part of COMPOUND LEAF
LENTICEL	G2	Small, usually corky structure with enclosed gap on bark or young twigs
lf		Abbreviation for leaf
lflet		Abbreviation for leaflet
LOBED	G1	(Simple leaf or leaflet) with deep irregularities in margin
m		Abbreviation for male
MANGROVE	*3	Sea-side, ± salt-water swamp, or trees growing there
MARGIN	G2	Edge (of leaf)
MARGINAL		(Nerve, gland) situated on the margin
MIDRIB	G2	Central and largest channel in lamina
MONOCOTYLEDON	*39	(Plant with one seedling leaf) e.g. palms and grasses with many nerves running parallel to margin (see DICOTS.)
MONOECIOUS		With separate male and female flowers, but both sexes on one tree (see DIOECIOUS)
MOSAIC		Pattern made up of many closely-fitting pieces
MUCRONATE	G1	Tip of leaf with a thread-like process on end of midrib
NERVE	G2	(Major) channel in lamina after midrib (see MARGINAL, BASAL, LATERAL, and VEIN)
NODE	G2	Place on stem where leaves arise or have fallen
OBLANCEOLATE	G1	As LANCEOLATE, but broadest beyond middle
OBLONG	G1	(Leaf) *c.* broadest in middle with almost parallel sides
OBOVATE	G1	As OVATE, but broadest beyond middle
OBSCURE	G2	Not visible, or not at all clear
OBTUSE	G1	Flat ended (apex or base)
OPPOSITE	*S	(Leaves) arranged two per node
OVATE	G1	(Leaf) with outline (roughly) of egg and broadest below middle (2-2.5 times as long as broad)
OVOID		Egg-shaped (three-dimensional) structure
OPAQUE		'Murky'; not possible to see through
PANICLE	*22	Branched inflorescence
PARIPINNATE	*S	(Pinnate leaf) with leaflets *all* arranged in pairs, opposite each other
PEDICEL		Stalk of a single flower
PEDUNCLE		Stalk of inflorescence bearing more than one flower
PELLUCID		Letting light through (esp. gland-spots in leaf) = TRANSLUCENT
PELTATE	*28A	Leaf with petiole attached to under surface, not to edge
PERIANTH		Flattened structures of outer part of most flwrs; usually with (coloured) petals + different sepals
PETAL		One of members of inner part of PERIANTH, especially if brightly coloured; see COROLLA
PETIOLE	G2	The stalk by which the midrib or rachis of a leaf is attached to the stem
PETIOLULE	G2	The stalk by which the midrib of a leaflet is attached to the axis of a compound leaf
PINNATE	*S	Compound lf with leaflets attached to only one axis (the rachis); PARIPINNATE or IMPARIPINNATE
PITH		Very soft and airy tissue at the centre of stems
PNEUMATOPHORE	*3	Root attachment arising above water/soil surface in swamps
POD	*37	Dry, usually flattened fruit with seeds in 1 line; ± opening along 1-2 edges
pr		Abbreviation for pair
PRICKLE	*31D	Sharp (± conical or pyramid-like) broad-based outgrowth outside wood; ± removable (see SPINE)
PROCESS	*34C	Small structure arising from end of another; outgrowth
PROMINENT	G2	Raised above surrounding surface; (nerves, etc.) easily felt when stroked with finger
PROP ROOT	*3	Root growing ± vertically downwards from brs (& lower stem); c.f. STILT ROOT (synonym in other books)
PSEUDOSTIPULES	*30	Lower leaflets on PINNATE leaf which, surrounding twig, resemble stipules
PUNCTATE	*37D	Spotted with glands (or small, undefined, flat, translucent dots)
PUNGENT		Strong, slightly alarming smell like black pepper
PUSTULE	*13B	Raised bump or wart-like spot
PUTRID		Disgusting smell = FOETID
RACEME	*22	Unbranched inflorescence; individual flowers with stalks (pedicels)
RACHIS	*S	Central axis of pinnate leaf after first leaflet(s). In bipinnate lvs = pinna axis (secondary rachis), or main leaf axis after first pinnae (primary rachis)
RECEPTACLE		± fleshy, usually basal part of flower supporting PERIANTH, OVARY, STAMENS; or surrounding OVARY
RECESSED	G2	(Nerves etc.) sunk below surface (see IMPRESSED) BUT still raised above lowest part of channel

Word	Illustration	Meaning
RECURVED	G2	Margin of leaf rolled over slightly
REFLEXED		Abruptly curved or bent downward or backward
RESIN		Resinous exudate
RESINOUS		Thick, oily or sticky translucent or transparent, often aromatic (exudates)
RETICULATE	G2	Arranged like strings in a net
RETUSE	G1	Emarginate, but with a deeper notch
RHACHIS		=RACHIS
RHOMBIC	G1	Four-sided, with one pair of sides different from the adjacent pair
RIPPLE MARK		Very fine, finger-print-like transverse lines in sapwood or bark
RUMINATE	12	(Inside of seed) very contorted, or convoluted (as if chewed or wrapped up) in X-section
SCABRID	*28A	Rough, like sand-paper
SCALARIFORM	G2	Arranged like rungs in a ladder
SCALE	*21	1) On leaf – tiny (<1mm) article lying ±flat on surface, disappearing with age 2) On bark – sections of bark of varied size which break away from bole
SCURFY	*27B	With fine flaky covering like rust or dandruff
SCENT		General term for not unpleasant smell
SEED		(Part of fruit) derived from single fertilized ovule (ovule contains the female 'egg')
SEPALS		Outer part of perianth, especially if leaf-like; see CALYX
SERRATE	G1	Saw-toothed; usually with teeth arching forwards (but in this guide, any 'toothed' lf)
SESSILE		Without a stalk
SHRUB		Small woody plant (rarely or not exceeding 5 cm d.b.h., or without bole)
SIMPLE	*S	Leaves with one axis; i.e. unbranched, with one midrib
SINEWY	200	(Bole) with many rounded, but IRREGULAR channels, like straining muscles
SINUOUS		Deeply wavy (margin); highly curvaceous
sl.		Abbreviation for slightly
sp.		Abbreviation for species (plural =spp.)
SPATHULATE		(Drip tip) with round, broadened apex like a spoon
SPIKE	*22	Unbranched inflorescence; individual flowers without stalks (see raceme)
SPINE	*15	Sharp, (±branched) needle-like projection from below bark ± regularly arranged; not v. detachable
STELLATE HAIR	*27	Star-shaped hair or cluster of hairs radiating from a point
STIGMA		Top part of STYLE, to which pollen becomes attached
STILT ROOTS	*24	Woody, ±supportive roots growing from trunk (not branches as well) of tree at base to ground
STIPEL	G2	Hair-like structure at base of leaflet on some compound lvs (Gp 37)
STIPULE	G2	Hair, leaf-like or other small structure close to petiole at leaf node
STRIATE	G2	With many fine parallel lines
STYLE		Narrow neck section of female part of flower (gynoecium) connecting OVARY to STIGMA
SUB-..		Not quite; almost e.g.....
SUB-MARGINAL	G2	(Nerve) running close to margin
SUB-OPPOSITE	*25	Almost, but not always or not exactly opposite
SUCKER		New stem arising from ground at base of tree
SUPERIOR OVARY		Ovary developing above the point of insertion of PERIANTH; becoming fruit with calyx at base
SWAMP		Flat or low-lying ground permanently or regularly soaked with (standing) water
SYMMETRIC	G1	Regular, with one half equal (but usually mirrored) to the other
THORN		Used elsewhere for spine or prickle; used in this book for PRICKLE
TIERS	200	Layers; storeys, like floors in a many-floored house
TOMENTOSE		With very dense and long hairs and soft texture
TRANSLUCENT		Letting light through, but murky (esp. of exudates etc.)
TRANSPARENT		Possible to see through
TRANSVERSE	G2	(Nerves, lines on bark etc.) running perpendicular to axis (of leaf, bole etc.)
TREELET		Small tree, usually <7 m tall
TRINERVED	*18	With 2 or more strong basal nerves besides the midrib and arching forwards
TRIFOLIATE	*34A	Compound leaves with three leaflets attached to end of petiole
TRUNK		±vertical part of tree supporting crown (i.e. BASE+BOLE ±axis of CROWN)
TUFT		Brush-like cluster of hairs
TWIG		Youngest axis, often still bearing leaves or stipules
UNDULATE	G1	Margin of leaf wavy when leaf viewed in plan view (N.B. in other books it may mean when lf seen from side)
VALVE		One of the flaps/parts/sides opening on a dehiscent fruit
VEIN	G2	Channels in leaf (in this guide used for finer branches – as opposed to NERVES)
VENATION	G2	The system or pattern of veins and nerves, or ('finer venation') just veins
WAVY	*12A	With smooth, shallow curves – (In this guide, particularly when leaf viewed from side – UNDULATE)
WHORLED	*S	(Leaves, branches) arranged more than two per node along axis
WING	*31B	Thin layer, shelf or sheet along side (of petiole, fruit, etc.)

INDEX TO SCIENTIFIC NAMES

FAMILY names are abbreviated as follows:

Abbrev.	Full name (+ alternatives)	Reference Group	Abbrev.	Full name (+ alternatives)	Reference Group
AGAVA	AGAVACEAE	39	LOGAN	LOGANIACEAE	4
ANACA	ANACARDIACEAE	35	MEDUS	MEDUSANDRACEAE	13
ANISO	ANISOPHYLLEACEAE	18	MELAS	MELASTOMATACEAE	6
ANNON	ANNONACEAE	12	MELIA	MELIACEAE	34
APOCY	APOCYNACEAE	9	MELIN	MELIANTHACEAE	31
ARALI	ARALIACEAE	28	MIMOS	MIMOSACEAE (or LEGUMINOSAE subfamily Mimosoideae)	38
BALAN	BALANITACEAE	31	MORAC	MORACEAE (inc. MYRIANTHACEAE)	19
BIGNO	BIGNONIACEAE	30	MYRIS	MYRISTICACEAE	13
BOMBA	BOMBACEAE	28	MYRSI	MYRSINACEAE	22A
BORAG	BORAGINACEAE	26	MYRTA	MYRTACEAE	7
BURSE	BURSERACEAE	33	OCHNA	OCHNACEAE	16
CAESA	CAESALPINIACEAE (or LEGUMINOSAE subfamily Caesalpinioideae)	37	OLACA	OLACACEAE	13
			OLEAC	OLEACEAE	5
CAPPA	CAPPARACEAE	13	PALMA	PALMACEAE	40
CELAS	CELASTRACEAE	3	PANDA	PANDACEAE	17D
CHRYS	CHRYSOBALANACEAE (ROSACEAE)	14	PAPIL	PAPILIONACEAE (or LEGUMINOSAE subfamily Papilionoideae)	14 37
COMBR	COMBRETACEAE	25			
COMPO	COMPOSITAE	26	PASSI	PASSIFLORACEAE	17
DICHA	DICHAPETALACEAE (CHAILLETIACEAE)	13	PTERI	CYATHEACEAE	39
			RHAMN	RHAMNACEAE	2
EBENA	EBENACEAE	11	RHIZO	RHIZOPHORACEAE	3
ERYTH	ERYTHROXYLACEAE	13	RUBIA	RUBIACEAE	1
EUPHO	EUPHORBIACEAE	22	RUTAC	RUTACEAE	31
FLACO	FLACOURTIACEAE (inc. SAMYDACEAE)	17	SANTA	SANTALACEAE	18
			SAPIN	SAPINDACEAE	36
GRAMI	GRAMINEAE	39	SAPOT	SAPOTACEAE	10
GUTTI	GUTTIFERAE (inc. HYPERICACEAE)	8	SCYTO	SCYTOPETALACEAE	13
HUACE	HUACEAE	21	SIMAR	SIMAROUBACEAE	32
HUMIR	HUMIRIACEAE	17	SOLAN	SOLANACEAE	26
HYMEN	HYMENOCARDIACEAE	22	STERC	STERCULIACEAE	27
ICACI	ICACINACEAE	13	TILIA	TILIACEAE	20
IRVIN	IRVINGIACEAE	13	ULMAC	ULMACEAE	18
IXONA	IXONANTHACEAE	13	VERBE	VERBENACEAE	29
LAURA	LAURACEAE	12	VIOLA	VIOLACEAE	17B
LECYT	LECYTHIDACEAE	25			

CODE
Each species is given a unique code. Originally, important timber trees (Classes 1-3) had been allocated codes 1-53, and all other (Class 4) species allocated numbers in alphabetical order. Unfortunately, this pattern has been disrupted slightly by name changes and additions to the list. Species without codes are not (native) trees capable of reaching 5 cm dbh.

LATIN NAMES
These follow Hall and Swaine (1981) except for a few changes. Old names have also been included in the index, followed by 'SEE..(new name) '. References to old names are also made in the last column, after an '=' sign. In some cases, potential new names have not been adopted because of widespread familiarity with the old names combined with doubt as to the significance of the name change.

KEY GROUP
These are the references to the Groups where the species is mentioned in the Main Key. Bracketed references are to species which are not illustrated and which are mentioned only in the notes. Where more than one group is mentioned, the first reference is to the group associated with an illustration.

OTHER NAMES
In this column other Latin names (see above), trade names, and (very few) alternative local names are mentioned.

FAMILY	CODE	LATIN NAME	KEY GROUP	STANDARD LOCAL NAME	OTHER NAMES
EUPHO	54	Acalypha neptunica	(22A)		
CHRYS	55	Acioa barteri	14A		SEE Dactyladenia barteri
CHRYS	56	Acioa dinklagei	14A	ATWERE	SEE Dactyladenia dinklagei
CHRYS	57	Acioa hirsuta	(14A)		SEE Dactyladenia hirsuta
ACANT	58	Adhatoda robusta	-		
RUTAC	59	Aeglopsis chevalieri	31A,15	KWAE-AKENKA	
RUTAC	60	Afraegle paniculata	31A,15	OBUOBI	
CHRYS	386	Afrolicania elaeosperma	14A		SEE Licania elaeosperma
PAPIL	11	Afrormosia elata	37G	KOKRODUA	SEE Pericopsis elata, AFRORMOSIA
SAPOT	61	Afrosersalisia afzelii	10A,D	BAKUNINI	
HUACE	62	Afrostyrax lepidophyllus	21	DUAGYENNE	
CAESA	28	Afzelia africana	37D	PAPAO	AFZELIA
CAESA	29	Afzelia bella	37D	PAPAO-NUA	(AFZELIA)
PAPIL	63	Aganope leucobotrya	(37H)	YAYA-AKOA	(=Ostryoderris leucobotrya)
RUBIA	64	Aidia genipiflora	1E	OTWENSONO	
MIMOS	30	Albizia adianthifolia	38B	PAMPENA	ALBIZIA
MIMOS	65	Albizia coriaria	38B	AWIEMFOSAMINA-AKOA	
MIMOS	31	Albizia ferruginea	38B	AWIEMFOSAMINA	ALBIZIA
MIMOS	66	Albizia glaberrima	38B	oKORA-AKOA	
MIMOS	32	Albizia zygia	38B	oKORO	OKURO
EUPHO	67	Alchornea cordifolia	22A	GYAMA	
EUPHO	68	Alchornea floribunda	(17C)	GYAMANINI	
GUTTI	33	Allanblackia floribunda	8B	SONKYI	(=A. parviflora)
VIOLA	69	Allexis cauliflora	(17B)		
SAPIN	70	Allophylus africanus	31	DUA-AHABANUM	
SAPIN	71	Allophylus spicatus	-	TETEDUA	
PTERI	72	Alsophila manniana	39	DUA-AYAA	SEE Cyathea manniana
APOCY	73	Alstonia boonei	9A	SINURO	
EUPHO	74	Amanoa bracteosa	13C		
EUPHO	75	Amanoa strobilacea	13C		
CAESA	76	Amphimas pterocarpoides	37H,I	YAYA	ASANFRAN
PALMA		Ancistrophyllum spp.	40		SEE Laccosperma spp.
PASSI	77	Androsiphonia adenostegia	17C		
SAPOT	78	Aningeria altissima	10B	SAMFENA(-BERE)	
SAPOT	79	Aningeria robusta	10B	SAMFENANINI	ASANFONA
ANISO	80	Anisophyllea meniaudii	18A	KoKoTE-AKOA	
COMBR	81	Anogeissus leiocarpus	25,13D	KANe	
ANNON	82	Anonidium mannii	12E	ASUMPA	
RHIZO	34	Anopyxis klaineana	3	KoKoTE	KOKOTI
LOGAN	83	Anthocleista djalonensis	4	BoNToDEeBERE	
LOGAN	84	Anthocleista liebrechtsiana	4		
LOGAN	85	Anthocleista microphylla	(4)		
LOGAN	86	Anthocleista nobilis	4	BoNToDEe	
LOGAN	87	Anthocleista vogelii	4	AWORABoNToDEe	
CAESA	88	Anthonotha fragrans	37E	TOTORONINI	
CAESA	89	Anthonotha macrophylla	37E	TOTORO	
CAESA	90	Anthonotha vignei	37E	TUTUABO	SEE Isomacrolobium vignei
EUPHO	91	Anthostema aubryanum	22,10	KYIRIKUSA	
MORAC	21	Antiaris toxicaria	19A	KYEN-KYEN	ANTIARIS
EUPHO	92	Antidesma laciniatum	13D	FoTo-NINI	
EUPHO	93	Antidesma membranaceum	13D	NUMANUMAGYAMA	
ANACA	94	Antrocaryon micraster	35B	APROKUMA	
SAPIN	383	Aphania senegalensis	36A		SEE Lepisanthes senegalensis
SAPIN	95	Aporrhiza urophylla	36A		
OLACA	96	Aptandra zenkeri	13A	AYEMTUDUA	
RUTAC	97	Araliopsis soyauxii	28B	MEAWERE	(=A. tabouensis)
SAPOT	98	Aubegrinia taiensis	10A,D	DUATADWE-KESE	
MIMOS	99	Aubrevillea kerstingii	38C	DAHOMANUA	
MIMOS	100	Aubrevillea platycarpa	38B		
RUBIA	101	Aulacocalyx jasminiflora	1E	NTWESON	
VERBE	901	Avicennia africana	3		Olive mangrove
BALAN	102	Balanites wilsoniana	31	KROBODUA	
GRAMI	103	Bambusa vulgaris	39	MPAMPRO	
PAPIL	104	Baphia nitida	37A,13	oDWEN	
PAPIL	105	Baphia pubescens	37A,13	oDWENKOBIRI	(DWENDWERA)
LAURA	106	Beilschmiedia mannii	12	TWEANKA	Spicy cedar
RUBIA	107	Belonophora hypoglauca	1D		
SAPOT	108	Bequaertiodendron magalismontanum	10A		
SAPOT	109	Bequaertiodendron oblanceolatum	10A	NNANFURO	

FAMILY	CODE	LATIN NAME	KEY GROUP	STANDARD LOCAL NAME	OTHER NAMES
CAESA	110	Berlinia confusa	37F	KWATAFOMPABOANINI	
CAESA	111	Berlinia occidentalis	37F	KWATAFOMPABOAKoKoo	
CAESA	112	Berlinia tomentella	37F	KWATAFOMPABOABERE	
MELIN	113	Bersama abyssinica	31B	ESONODUA	
RUBIA	114	Bertiera racemosa	1C	KAKADUA	
SAPIN	115	Blighia sapida	36A	AKYE	
SAPIN	117	Blighia unijugata	36A	AKYEBIRI	
SAPIN	116	Blighia welwitschii	36A	AKYEKOBIRI	
BOMBA	118	Bombax brevicuspe	28C	ONYINAKoBEN	SEE Rhodognaphalon brevicuspe
BOMBA	119	Bombax buonopozense	28C	AKONKODIE	AKATA
MORAC	630	Bosqueia angolensis	19A		SEE Trilepisium
SAPOT	120	Breviea leptosperma	10C	KANKABIM	
EUPHO	121	Bridelia atroviridis	15	oPAMKOTOKRODU	
EUPHO	122	Bridelia grandis	15	oPAMKOTOKRODUKeSe	
EUPHO	123	Bridelia micrantha	15	BADIE	
SIMAR	124	Brucea guineensis	(32)	eSERESOKRODUA	
ANNON		Brieya			See Pipostigma
CAPPA	125	Buchholzia coriacea	27B	eSONOBESE	KONINI
CAESA	126	Bussea occidentalis	38A	KOTOPRePre	
PALMA		Calamus deeratus	40		
FLACO	127	Caloncoba echinata	27B,17E		
FLACO	128	Caloncoba gilgiana	23,17E	AWIEWU	
FLACO	129	Caloncoba glauca	(27B)		
MIMOS	130	Calpocalyx brevibracteatus	38A	ATROTRe	
		Calpocalyx winkleri	38A		
RUBIA	171	Calycosiphonia spathicalyx	1E		(=Coffea spathicalyx)
BURSE	35	Canarium schweinfurthii	33	BEDIWONUA	CANARIUM
RUBIA	131	Canthium arnoldianum	1A		SEE Pydrax arnoldiana
RUBIA	132	Canthium subcordatum	1A	TETIAUPoN	SEE Psydrax (=C. glabriflorum)
RUBIA	134	Canthium vulgare	1A	OGYAPAMNINI	SEE Psydrax parviflora
MELIA	135	Carapa procera	34C	KWAKUOBESE	
FLACO	136	Casearia barteri	13	PANUM	
FLACO	137	Casearia calodendron	17D	ATWIAWA	(=C. inaequalis)
RHIZO	138	Cassipourea afzelii	3		
RHIZO	139	Cassipourea congoensis	3	KoKoTENUA	
RHIZO	140	Cassipourea gummiflua	3	KoKoTENUA	(=C. glabra)
RHIZO	141	Cassipourea hiotou	2,3		
RHIZO	707	Cassipourea lescotiana	3		
MIMOS	142	Cathormion altissimum	38C	ABOBONKAYERe	
MORAC	803	Cecropia peltata	28A	FRENCH oDWUMA	
MELIA	801	Cedrela odorata	(34)	CEDRELA	
BOMBA	143	Ceiba pentandra	28C	ONYINA	
ULMAC	36	Celtis adolfi-friderici	18A	ESAKOSUA	
ULMAC	144	Celtis africana	18B		
ULMAC	37	Celtis mildbraedii	18B,A	ESA	
ULMAC	145	Celtis wightii	18B	ESAFUFUO, PREMPRESA	(=Celtis philippensis)
ULMAC	38	Celtis zenkeri	18A,B	ESAKoKo	
ULMAC	146	Chaetachme aristata	15	ESONO-ANKAA	(=Chaetachme)
CAESA	147	Chidlowia sanguinea	37E	ABABIMA	
OLEAC	388	Chionanthus africanus	5		(=Linociera africana)
OLEAC	389	Chionanthus mannii *var.* congestus	5		(=Linociera congesta)
OLEAC	390	Chionanthus mannii	5		(=Linociera mannii)
MORAC	1	Chlorophora excelsa	19B	ODUM	SEE Milicia excelsa IROKO,EP
MORAC	2	Chlorophora regia	19B	ODUM-NUA	SEE Milicia regia IROKO,EP
TILIA	148	Christiana africana	27C	SESEDUA	SUPRONO
CHRYS	149	Chrysobalanus icaco	14A		(subspecies atacorensis)
SAPOT	150	Chrysophyllum albidum	10C		
SAPOT	151	Chrysophyllum azaguieanum	10D		
SAPOT	152	Chrysophyllum beguei	10C	DUATADWE-NINI	(=Gambeya beguei)
SAPOT	153	Chrysophyllum delevoyi	10C		
SAPOT	154	Chrysophyllum giganteum	10C	KUMFANA	
SAPOT	155	Chrysophyllum pentagonocarpum	10A		
SAPOT	156	Chrysophyllum perpulchrum	10C	ATABENE	
SAPOT	157	Chrysophyllum pruniforme	10A	DUATADWE	
SAPOT	158	Chrysophyllum subnudum	10C	ADASEMA	
SAPIN	160	Chytranthus atroviolaceus	36C	AKYEKoKoo	
SAPIN	159	Chytranthus carneus	36C	ONIBONANUA	
SAPIN	161	Chytranthus cauliflorus	36C		(=Laccodiscus cauliflorus)

FAMILY	CODE	LATIN NAME	KEY GROUP	STANDARD LOCAL NAME	OTHER NAMES
SAPIN	162	Chytranthus ellipticus	(36C)		
SAPIN	163	Chytranthus macrobotrys	36C	TRUMWIE	SUINIA,NTWESEMA
RUTAC	164	Citropsis articulata	31B	KWAA-ANKAA	African cherry orange
RUTAC	165	Citropsis gabunensis	31B		
RUTAC	166	Clausena anisata	31	SAMANOBI	SAMANKAA
EUPHO	167	Cleidion gabonicum	17C	MPAWU	
EUPHO	168	Cleistanthus polystachyus	13B		
ANNON	169	Cleistopholis patens	12A	NGONENKYENE	
MYRIS	170	Coelocaryon oxycarpum	13A	ABRUMA	
RUBIA	804	Coffea canephora	(1D)		Coffea robusta, Coffee
RUBIA	171	Coffea spathicalyx	1E		SEE Calycosiphonia spathicalyx
RUBIA	172	Coffea togoensis	1E		
STERC	173	Cola boxiana	27A		
STERC	174	Cola caricifolia	27D	ANANSEAYA	SONKOROBIA
STERC	175	Cola chlamydantha	28B	TANA-NFRe	(=Chlamydocola chlamydantha)
STERC	176	Cola digitata	28B		
STERC	177	Cola flavo-velutina	27A	AFRAMDASA	
STERC	178	Cola gigantea	27D	WATAPUO	
STERC	179	Cola lateritia	27D	WATAPUOBERE	
STERC	180	Cola millenii	27D	ANANSEDODOWA	DODOWA
STERC	181	Cola nitida	27A	BESE	
STERC	182	Cola reticulata	27A		
STERC	183	Cola umbratilis	28B	TANANFRoBERE	
STERC	184	Cola verticillata	27A	BESETORO	
LECYT	48	Combretodendron macrocarpum			SEE Petersianthus
CAESA	185	Copaifera salikounda	37D	ENTEDUA	BUBINGA
BORAG	186	Cordia millenii	27C,26	TWENEBOA-NINI	
BORAG	187	Cordia platythyrsa	26	TWENEBOABERE	
BORAG	188	Cordia senegalensis	26	OKOSU	
BORAG	189	Cordia vignei	13D	TWENEBOA-AKOA	
RUBIA	190	Corynanthe pachyceras	1D	PAMPRANA, PAMPENAMA (NOT PAMPENA)	
OLACA	191	Coula edulis	27B	BoDWUE	Gaboon nut AFR. WALNUT
PAPIL	192	Craibia atlantica	37G		
RUBIA	193	Craterispermum caudatum	1E	DUADe	
RUBIA	194	Craterispermum cerinanthum	1E	AFRA-NI-AFEI	
RUBIA	195	Craterispermum laurinum	(1E)		
MORAC	255	Craterogyne kameruniana	(19A)		SEE Dorstenia kameruniana
CAPPA	196	Crateva adansonii	31	CHELUM PUNGA	(=Crateva religiosa)
EUPHO	197	Croton aubrevillei	(21)		
EUPHO	198	Croton penduliflorus	22A	NYAMREM	
EUPHO	199	Croton sylvaticus	(22A)		
EUPHO	200	Croton zambesicus	21	DODWATU	
EUPHO	201	Crotonogyne chevalieri	(22B)		
EUPHO	202	Crotonogyne manniana	21		
CAESA	203	Crudia gabonensis	37G	SAMANTAANINI	
CAESA	204	Crudia senegalensis	37G		
CAESA	205	Cryptosepalum tetraphyllum	37C		
ARALI	206	Cussonia bancoensis	28B	KWAEBRoFRE	
RUBIA	207	Cuviera acutiflora	1G	KWAKUO-ASRA	
RUBIA	208	Cuviera macroura	1G	TETEANKASE-KOFI	
RUBIA	209	Cuviera nigrescens	1G	KoTo-BOWERE	
RUBIA	210	Cuviera subuliflora	(1G)		
PTERI	72	Cyathea manniana	39		(=Alsophila manniana)
MIMOS	39	Cylicodiscus gabunensis	38A	DENYAo	OKAN
CAESA	40	Cynometra ananta	37B	ANANTA	
CAESA	211	Cynometra megalophylla	37C	ANANTA-AKOA	
BURSE	212	Dacryodes klaineana	33	ADWEA	
CHRYS	55	Dactyladenia barteri	14A		(=Acioa barteri)
CHRYS	56	Dactyladenia dinklagei	14A		(=Acioa dinklagei)
CHRYS	57	Dactyladenia hirsuta	(14A)		(=Acioa hirsuta)
PAPIL	213	Dalbergia setifera	-		
CAESA	214	Daniellia ogea	37D	eHYeDUA	OGEA, SHEDUA GUM COPAL
CAESA	215	Daniellia thurifera	37D	SOPI	S. LEONE FRANKINCENSE NIGER COPAL
FLACO	216	Dasylepis brevipedicellata	17D	ASRATOADUANINI	
VIOLA	217	Decorsella paradoxa	(17B)		(=Gymnorinorea abidjanensis)
SAPIN	218	Deinbollia grandifolia	36C	MMATA	
SAPIN	219	Deinbollia molliuscula	(36C)		

FAMILY	CODE	LATIN NAME	KEY GROUP	STANDARD LOCAL NAME	OTHER NAMES
SAPIN	220	Deinbollia pinnata	36C	WOAGYE-AKOA	
ANNON	221	Dennettia tripetala	12F		
ICACI	222	Desmostachys vogelii	-		
TILIA	223	Desplatsia chrysochlamys	20	eSONOWISAMFIE	
TILIA	224	Desplatsia dewevrei	20	eSONOWISAMFIENINI	
TILIA	225	Desplatsia subericarpa	20	eSONOWISAMFIEBERE	
CAESA	226	Detarium senegalense	37D	TAKYIKYIRIWA	TALLOW
CAESA	227	Dialium aubrevillei	37H	DUABANKYE	
CAESA	228	Dialium dinklagei	37I	DWEDWEEDWE	
CAESA	229	Dialium guineense	37H	ASENAA	
MELAS	230	Dichaetanthera africana	6		
DICHA	231	Dichapetalum barteri	13D	AKUSAKUSA	
DICHA	232	Dichapetalum guineense	13D		SEE D. madagascariense
DICHA	232	Dichapetalum madagascariense	13D	eSONOWEDIE	(=D. guineense)
RUBIA	233	Dictyandra arborescens	1G	KWAKUO-ASENABA	
RUBIA	234	Dictyandra involucrata	(1D)		(=Leptactina involucrata)
CAESA	235	Didelotia idae	37A		
CAESA	236	Didelotia unifoliolata	37A		
EBENA	237	Diospyros abyssinica	11	GBLITSO	
EBENA	238	Diospyros barteri	11	AHENEBANSATEA	
EBENA	239	Diospyros canaliculata	11	OTWABERE	
EBENA	240	Diospyros cooperi	11	FRENCH ATWEA-BERE	
EBENA	241	Diospyros ferrea	11	oMENEWA-NINI	
EBENA	242	Diospyros gabunensis	11	KUSIBIRI	
EBENA	243	Diospyros heudelotii	11	oMENEWABERE	
EBENA	244	Diospyros kamerunensis	11	oMENEWA	
EBENA	245	Diospyros mannii	11	ATWEAFUFU	
EBENA	246	Diospyros mespiliformis	(11)	KEKE	
EBENA	247	Diospyros monbuttensis	11,15	ATWERE-NANTIN	
EBENA	248	Diospyros piscatoria	11	oTWETO-KESE	
EBENA	41	Diospyros sanza-minika	11	SANZA-MULIKA	
EBENA	249	Diospyros soubreana	11	oTWETO	
EBENA	706	Diospyros vignei	11	oMENEWA-HOAKOA	
EBENA	250	Diospyros viridicans	11	ATWEA	
RUTAC	251	Diphasia angolensis	31A	AMUDURO	(=D. klaineana)
EUPHO	252	Discoclaoxylon hexandrum	22B	DUBRAFONINI	(=Claoxylon hexandrum)
EUPHO	253	Discoglypremna caloneura	22A	FeTeFRe	
CAESA	42	Distemonanthus benthamianus	37G	BONSAMDUA	AVAN
STERC	254	Dombeya buettneri	(27)	MFo	
MORAC	255	Dorstenia kameruniana	(19A)		(=Craterogyne kameruniana)
FLACO	256	Dovyalis zenkeri	15		(=D. afzelii)
AGAVA	257	Dracaena arborea	39A	NToNMe	
AGAVA	258	Dracaena mannii	39A	KESENE	(=D. perrotettii)
EUPHO	259	Drypetes aframensis	17A		
EUPHO	260	Drypetes afzelii	(17A)		
EUPHO	261	Drypetes aubrevillei	17A	oPAHANINI, DUAMOKO	
EUPHO	262	Drypetes aylmeri	17A	oPAHAFUFUO	
EUPHO	263	Drypetes chevalieri	17A	KATRIKA-AKOA	
EUPHO	264	Drypetes floribunda	17A	BEDIBeSA	TETSO (Ga)
EUPHO	265	Drypetes gilgiana	17A	KATRIKA(-NINI)	
EUPHO	266	Drypetes ivorensis	17A	oPAHA-BERE	
EUPHO	267	Drypetes leonensis	17A	oPAHA-NUA	
EUPHO	268	Drypetes parvifolia	17A	KATRIKABERE	
EUPHO	269	Drypetes pellegrinii	17A	oPAHAKoKoo	
EUPHO	270	Drypetes principum	17A	oPAHA	
EUPHO	271	Drypetes singroboensis	17A	oPAHA-AKOA	
TILIA	272	Duboscia viridiflora	20	AKOKORAGYEHINI	
BORAG	273	Ehretia cymosa	13D,26		
BORAG	274	Ehretia trachyphylla	13D,26	OKYINI	
MELIA	275	Ekebergia senegalensis	34C		
PALMA	276	Elaeis guineensis	40	ABe	
CELAS	277	Elaeodendron buchananii	(3)		(=Cassine buchananii)
EUPHO	278	Elaeophorbia grandifolia	22,10	AKANI	(=E. drupifera)
ANNON	279	Enantia polycarpa	12D	DUASIKA	
MELIA	3	Entandrophragma angolense	34B	EDINAM	GEDUNOHOR
MELIA	16	Entandrophragma candollei	34B	PENKWA-AKOA, CEDAR-KOKOTE, CANDOLLEI, OMU	
MELIA	4	Entandrophragma cylindricum	34B	PENKWA	SAPELE
MELIA	5	Entandrophragma utile	34B	EFOoBRODEDWO	UTILE
PALMA		Eremospatha macrocarpa	40		
STERC	577	Eribroma oblonga	27A		SEE Sterculia oblonga

FAMILY	CODE	LATIN NAME	KEY GROUP	STANDARD LOCAL NAME	OTHER NAMES
SAPIN	280	Eriocoelum kerstingii	36B		
SAPIN	281	Eriocoelum pungens	36A	AKYE-NAN, ONIBONA	
SAPIN	282	Eriocoelum racemosum	36A,B	ONIBONAKoKoo	
PAPIL	283	Erythrina addisoniae	31C		oSOROWA
PAPIL	284	Erythrina mildbraedii	31C	oSOROWA	
PAPIL	285	Erythrina vogelii	31C	oSORE	
EUPHO	286	Erythrococca africana	-	GYIGYAM	
CAESA	43	Erythrophleum ivorense	38A	PoTRoDOM	
CAESA	44	Erythrophleum suaveolens	38A	oDOM	(= E. guineense)
ERYTH	287	Erythroxylum mannii	13A,C	PEPEANINI	BENKYI
CAPPA	288	Euadenia trifoliolata	31		
RUBIA	289	Euclinia longiflora	1F	GYANEYA	
MYRTA	290	Eugenia calophylloides	(6)		
MYRTA	291	Eugenia coronata	(6)	KRAKA	
MYRTA	292	Eugenia kalbreyeri	(6)		
MYRTA	293	Eugenia leonensis	(6)		
MYRTA	294	Eugenia obanensis	(6)		
EUPHO	295	Euphorbia deightonii	-		
RUTAC		Fagara	31D		SEE Zanthoxylum
MORAC	299	Ficus ardisioides	19C		(= F. camptoneura)
MORAC	305	Ficus artocarpoides	19C		(= F. elegans)
MORAC	297	Ficus barteri	19C		
MORAC	298	Ficus bubu	19C		
MORAC	300	Ficus capensis	19B		SEE F. sur
MORAC	302	Ficus conraui	19C		(inc. F. praticola)
MORAC	296	Ficus craterostoma	19C	ANOMANI	(= F. anomani)
MORAC	303	Ficus cyathistipula	19C		
MORAC	304	Ficus elasticoides	19C		
MORAC	307	Ficus exasperata	28A	NYANKYEReNE	
MORAC	308	Ficus kamerunensis	19C		
MORAC	309	Ficus leprieurii	19C		(now F. natalensis ssp. lep.)
MORAC	310	Ficus lingua	19C		
MORAC	311	Ficus lutea	19C	FoNTo	
MORAC	312	Ficus lyrata	19C		
MORAC	314	Ficus mucuso	19B	BUMBRA	
MORAC	315	Ficus natalensis	19C	AKABOFUNI	
MORAC	316	Ficus ottoniifolia	19C		
MORAC	317	Ficus ovata	19C		
MORAC	318	Ficus polita	19C	BLoHUNYI	
MORAC	319	Ficus populifolia	19C		
MORAC	321	Ficus recurvata	19C		(= F. goliath)
MORAC	322	Ficus sagittifolia	19C		
MORAC	313	Ficus sansibarica	19C		(ssp. macrosperma: = F. macrosperma)
MORAC	306	Ficus saussureana	19C		(= F. eriobotryoides)
MORAC	300	Ficus sur	19B	NWADUA, DOMINI	(= F. capensis)
MORAC	323	Ficus tessellata	19		
MORAC	324	Ficus thonningii	19		
MORAC	301	Ficus trichopoda	19		(= F. congensis)
MORAC	325	Ficus umbellata	19		
MORAC	326	Ficus variifolia	28A		
MORAC	327	Ficus vogeliana	19B	OPANTO	
FLACO	328	Flacourtia flavescens	15		
APOCY	329	Funtumia africana	9B	OKAE	
APOCY	330	Funtumia elastica	9B	FRUNTUM	
RUBIA	331	Gaertnera paniculata	1E		
GUTTI	332	Garcinia afzelii	8B	NSoKo	
GUTTI	333	Garcinia epunctata	8B	NSoKoNUA	
GUTTI	334	Garcinia gnetoides	8C	TWEAPIAKOA	
GUTTI	335	Garcinia kola	8C	TWEAPIA	
GUTTI	336	Garcinia smeathmannii	8C	TWEPIA-BERE	(= G. polyantha)
RUBIA	337	Gardenia imperialis	1C		
RUBIA	338	Gardenia nitida	(1E)		
RUBIA	339	Gardenia vogelii	(1E)		
CAESA	340	Gilbertiodendron bilineatum	37F	TETEKON-NUA	
CAESA	341	Gilbertiodendron limba	37F	TETEKON	
CAESA	342	Gilbertiodendron preussii	37F	TETEKON-GYAMERA	
CAESA	343	Gilbertiodendron splendidum	37F	AGYAMERA	
SAPOT	344	Gluema ivorensis	10D		
TILIA	345	Glyphaea brevis	20	FoTo	
ANNON	346	Greenwayodendron oliveri	12F	DUABIRI	(= Polyalthia oliveri)
TILIA	347	Grewia mollis	20	KYAPOTORO	(inc. G. pubescens)
EUPHO	348	Grossera vignei	22B	MPEDURO	

FAMILY	CODE	LATIN NAME	KEY GROUP	STANDARD LOCAL NAME	OTHER NAMES
MELIA	17	Guarea cedrata	34C	KWABOHORO (-BERE)	SCENTED GUAREA
MELIA	18	Guarea thompsonii	34D	KWADWUMA, KWABOHORO-NINI	BLACK GUAREA
CAESA	22	Guibourtia ehie	37B	ANOKYE-HYEDUA,	HYEDUA-NINI
RUBIA	24	Hallea ledermannii	1B	SUBAHA-AKoA	(=Mitragyna ciliata)
RUBIA	25	Hallea stipulosa	1B	SUBAHA	(=Mitragyna stipulosa)
SIMAR	349	Hannoa klaineana	32	FOTIE	(=Quassia undulata) HOTROHOTRO
GUTTI	350	Harungana madagascariensis	8A	KOSOWA	
RUBIA	351	Heinsia crinita	-		
OLACA	352	Heisteria parvifolia	13B	SIKAKYIA	
STERC	15	Heritiera utilis	28B	NYANKOM	(=Tarrietia utilis)
ANNON	353	Hexalobus crispiflorus	12A,C,D	DUABAHA	
STERC	354	Hildegardia barteri	27D	AKYEReKYEWeWA	
CHRYS		Hirtella	14		SEE Magnistipula
APOCY	355	Holarrhena floribunda	9B	SESE	(=H. wulfsbergii)
ULMAC	45	Holoptelea grandis	18A	NAKWA	
FLACO	356	Homalium africanum	17E		(=H. molle)
FLACO	357	Homalium dewevrei	17E		(=H. angustistipulatum)
FLACO	358	Homalium letestui	17E	ESONONANKOROMA	
FLACO	359	Homalium longistylum	17E		(=H. aylmeri)
FLACO	360	Homalium stipulaceum	17E		(=H. neurophyllum)
APOCY	361	Hunteria eburnea	9C	KANWENE-AKOA	
APOCY	363	Hunteria ghanensis	9C		
APOCY	362	Hunteria umbellata	9C		
HYMEN	364	Hymenocardia lyrata	21		
CAESA	365	Hymenostegia afzelii	37C	TAKOROWA	
CAESA	366	Hymenostegia aubrevillei	37C		
CAESA	367	Hymenostegia gracilipes	37C	ABABIMA-KoKoo	
IRVIN	368	Irvingia gabonensis	13C	ABESEBUO	
ANNON	369	Isolona campanulata	12F		
ANNON	370	Isolona deightonii	(12F)		
ANNON	371	Isolona hexaloba	(12F)		
CAESA	90	Isomacrolobium vignei	37E		(=Anthonotha vignei)
SAPOT	372	Ituridendron bequaertii	10D		(Omphalocarpum pachysteloides)
RUBIA	373	Ixora laxiflora	1D		
EUPHO	374	Keayodendron bridelioides	13B	AKOKoSRADEe	(=Casearia bridelioides)
MELIA	6	Khaya anthotheca	34A	KRUMBEN	ANTHOTHECA
MELIA	7	Khaya grandifoliola	34A	KRUBA	AFRIC. MAHOGANY
MELIA	8	Khaya ivorensis	34A	DUBINI, DUBINKoKoo	AFRIC. MAHOGANY
BIGNO	375	Kigelia africana	30	NUFUTEN	
IRVIN	376	Klainedoxa gabonensis	13C,15	KROMA	
PALMA		Laccosperma opacum	40		(=Ancistrophyllum opacum)
PALMA		Laccosperma secundiflorum	40		(=Ancistrophyllum secundifl.)
COMBR	902	Laguncularia racemosa	3		White buttonwood (mangrove)
ANACA	377	Lannea nigritana	35B		
ANACA	378	Lannea welwitschii	35B	KUMANINI	
RHAMN	379	Lasiodiscus fasciculiflorus	2	ADAFA-NINI	
RHAMN	380	Lasiodiscus mannii	2	ADAFA	(=L. mildbraedii)
SAPIN	381	Lecaniodiscus cupanioides	36A	DWINDWERA	
SAPIN	382	Lecaniodiscus punctatus	(36A)		
SAPIN	383	Lepisanthes senegalensis	36A	AKYE-BUNO	(=Aphania senegalensis)
		Leptactina involucrata			SEE Dictyandra involucrata
ICACI	384	Leptaulus daphnoides	13B	AFENA-AKOA	
STERC	385	Leptonychia pubescens	21	FoToNUA	
CHRYS	386	Licania elaeosperma	14A		(=Afrolicania elaeosperma)
FLACO	387	Lindackeria dentata	22A		
OLEAC	388	Linociera africana	5		SEE Chionanthus africanus
OLEAC	389	Linociera congesta	(5)		(=Chionanthus mannii *var.* congesta)
OLEAC	390	Linociera mannii	(5)		SEE Chionanthus mannii
PAPIL	391	Lonchocarpus sericeus	37I	SANTE	
OCHNA	19	Lophira alata	16	KAKU	EKKI
MELIA	12	Lovoa trichilioides	34A	DUBINIBIRI, DUBINFUFUO	AFR. WALNUT
SAPIN	392	Lychnodiscus dananensis	36B	SUKYE	
SAPIN	393	Lychnodiscus reticulatus	36B	AKYE-SE	
EUPHO	394	Macaranga barteri	23	oPAM	
EUPHO		Macaranga beillei	23		
EUPHO	395	Macaranga heterophylla	23	oPAMKoKoo	

FAMILY	CODE	LATIN NAME	KEY GROUP	STANDARD LOCAL NAME	OTHER NAMES
EUPHO	396	Macaranga heudelotii	23	AWORA-oPAM	
EUPHO	397	Macaranga hurifolia	23	oPAMFUFUO	
EUPHO	398	Macaranga spinosa	23		
CAPPA	399	Maerua duchesnei	13B	KONINI-BERE	(=Ritchiea duchesnei)
MYRSI	400	Maesa lanceolata	(22A)		
EUPHO	401	Maesobotrya barteri	22B	APOTREWA	
RHAMN	402	Maesopsis eminii	2	oNWAMDUA	
CHRYS	403	Magnistipula butayei	14B		
CHRYS	404	Magnistipula zenkeri	14B	AWORA-AFAM	(=Hirtella fleuryana)
SAPIN	405	Majidea fosteri	36B,34C	ANKYEWA	
SAPOT	406	Malacantha alnifolia	10B	ASAMFENA-AKOA	
EUPHO	407	Mallotus oppositifolius	22A	ANYANYANFOROWA	
EUPHO	408	Mallotus subulatus	(22A)		
GUTTI	46	Mammea africana	8B	BOMPAGYA	
SAPOT	409	Manilkara obovata	10A	BEREKANKUM	(=Manilkara multinervis)
STERC	23	Mansonia altissima	27C	OPRONO	MANSONIA
CHRYS	410	Maranthes aubrevillei	14B		(=Parinari aubrevillei)
CHRYS	411	Maranthes chrysophylla	14B	KAJABIRI	(=Parinari chrysophylla)
CHRYS	412	Maranthes glabra	14A	AFAMNINI	(=Parinari glabra)
CHRYS	413	Maranthes kerstingii	14A		(=Parinari kerstingii)
CHRYS	414	Maranthes robusta	14B	AFAMBERE	(=Parinari robusta)
EUPHO	415	Mareya micrantha	22B	DUBRAFO	
EUPHO	416	Margaritaria discoidea	13B,15	PEPEA	(=Phyllanthus discoideus)
BIGNO	417	Markhamia lutea	30	oBOGYANEBOoBEREE	
BIGNO	418	Markhamia tomentosa	30	oBOGYANEBOoNINI	
EUPHO	419	Martretia quadricornis	(22)		
RUBIA	420	Massularia acuminata	1F	POBE	
CELAS	421	Maytenus undata	(17B)		(=M. undatus)
MELAS	422	Memecylon afzelii	6	oTWE-ANI	
MELAS	423	Memecylon barteri	(6)		SEE Spathandra barteri
MELAS	424	Memecylon blakeoides	6	oTWESENINI	SEE Spathandra blakeoides
MELAS	425	Memecylon cinnamonoides	6		SEE Warneckea cinnamonoides
MELAS	426	Memecylon fasciculare	-		=Warneckea fasciculare
MELAS		Memecylon fleuryi	6		SEE Spathandra blakeoides var.
MELAS	427	Memecylon guineense	6		SEE Warneckea guineense
MELAS	428	Memecylon lateriflorum	6	oTWESE	
MELAS	715	Memecylon membranifolium	6		SEE Warneckea membranifolium
MELAS	429	Memecylon memecyloides	6		SEE Warneckea memecyloides
MELAS	430	Memecylon normandi	6		
PANDA	431	Microdesmis puberula	17D	oFEMA	
CAESA	432	Mildbraediodendron excelsum	37I	TWEAWODO	
MORAC	1	Milicia excelsa	19B	ODUM	(=Chlorophora excelsa) IROKO,EP
MORAC	2	Milicia regia	19B	oDUM-NUA	(=Chlorophora regia) IROKO,EP
PAPIL	433	Millettia griffoniana	37I		(=Lonchocarpus griffonianus)
PAPIL	434	Millettia rhodantha	37I	TETETOA	
PAPIL	435	Millettia thonningii	37I	TAATSO	
PAPIL	436	Millettia zechiana	37I	FAFRAHA	
ANNON	437	Mischogyne elliotiana	12D		(=Uvariastrum elliotianum)
RUBIA	24	Mitragyna ciliata	1B	SUBAHA ABURA	SEE Hallea ledermannii
RUBIA	25	Mitragyna stipulosa	1B	SUBAHA-AKOA ABURA	SEE Hallea stipulosa
ANNON	438	Monocyclanthus vignei	12D		
ANNON	439	Monodora brevipes	(12F)	MOTOKORODUA-AKOA	
ANNON	440	Monodora myristica	12E	WEDEABA	
ANNON	441	Monodora tenuifolia	12F	MOTOKURADUA	
RUBIA	442	Morinda lucida	1A	KONKROMA	
MORAC	47	Morus mesozygia	19A	WoNTon	
MORAC	443	Musanga cecropioides	28A(19)	oDWUMA	
MORAC	444	Myrianthus arboreus	28A(19)	NYANKUMA (-BERE)	
MORAC	445	Myrianthus libericus	28A(19)	NYANKUMANINI	
LECYT	446	Napoleonaea vogelii	17D,13B	oBUA	(=Napoleona vogelii)
RUBIA	10	Nauclea diderrichii	1B	KUSIA	OPEPE
RUBIA	447	Nauclea pobeguinii	1B	SUKUSIA	SEE Sarcocephalus pobeguinii
EUPHO	448	Necepsia afzelii	(22B)		
SAPOT	449	Neolemonniera clitandrifolia	10D		

FAMILY	CODE	LATIN NAME	KEY GROUP	STANDARD LOCAL NAME	OTHER NAMES
ANNON	450	Neostenanthera gabonensis	12B		
ANNON	451	Neostenanthera hamata	12B		
STERC	26	Nesogordonia papaverifera	27A, 21	DANTA	DANTA
BIGNO	452	Newbouldia laevis	30	SESEMASA	
MIMOS	453	Newtonia aubrevillei	38B	ADADABA-NUA	
MIMOS	454	Newtonia duparquetiana	38A	ADADABA	
SIMAR	455	Nothospondias staudtii	32		
LOGAN	456	Nuxia congesta	3		
OCHNA	457	Ochna afzelii	16		
OCHNA	458	Ochna membranacea	16		
OCHNA	459	Ochna ovata	16		
OCHNA	460	Ochna staudtii	16	KWAASIWA	(=Ochna kibbiensis)
IXONA		Ochthocosmus spp.	17C		SEE Phyllocosmus spp.
OLACA	461	Octoknema borealis	27B,(13)	WISUBONI	
STERC	462	Octolobus spectabilis	27A	AFINAFI	(=Octolobus angustatus)
SANTA	463	Okoubaka aubrevillei	18A	ODII	
OLACA	464	Olax subscorpioidea	13A,B	AHOoHENEDUA	
SAPOT	465	Omphalocarpum ahia	10D	DUAPOMPO	
SAPOT	466	Omphalocarpum elatum	10D	ESONODoKONO	
SAPOT	372	Omphalocarpum pachysteloides	10D		SEE Ituridendron bequaertii
SAPOT	467	Omphalocarpum procerum	10D	OGYATAFONKONWA	
FLACO	468	Oncoba brachyanthera	15		
FLACO	469	Oncoba spinosa	15	ASRATOWADUA	
OLACA	470	Ongokea gore	13A	BODWE	
FLACO	471	Ophiobotrys zenkeri	18A,(17E)	AKWANA	AKWANDA
RUTAC	472	Oricia suaveolens	31A		
PAPIL	63	Ostryoderris leucobotrya	(37H)		SEE Aganope leucobotrya
OCHNA	473	Ouratea amplectens	16		
OCHNA	474	Ouratea calantha	16		
OCHNA	475	Ouratea calophylla	16	OPUNINI	
OCHNA	476	Ouratea congesta	(16)		
OCHNA	477	Ouratea flava	16		
OCHNA	478	Ouratea reticulata	(16)	ANANSEDUA	
RUBIA		Oxyanthus formosus	1F		
RUBIA		Oxyanthus pallidus	1F		
RUBIA	479	Oxyanthus speciosus	1F	KORANTEMA	
RUBIA	480	Oxyanthus unilocularis	1C	KWAETAWA	
ANNON	481	Pachypodanthium staudtii	12A	KUMDWIE	(DUAWISA)
SAPOT	482	Pachystela brevipes	10C,D	AFRAMSUA	
SAPOT	483	Pachystela msolo	10C	ASABA	
SAPIN	484	Pancovia bijuga	36A		
SAPIN	485	Pancovia turbinata	(36A)		
PANDA	486	Panda oleosa	17D	KOKROBOBA	
CAESA		Paramacrolobium coeruleum	37F		
CHRYS	487	Parinari congensis	14B		
CHRYS	488	Parinari excelsa	14B	AFAM	KWA-EDUA
CHRYS		Parinari spp.	14		SEE ALSO Maranthes spp.
MIMOS	489	Parkia bicolor	38C	ASOMA	
MIMOS	708	Parkia filicoidea	38C	ASOMA-NUA	
PASSI	490	Paropsia guineensis	17C		
RUBIA	491	Pauridiantha hirtella	1C		
RUBIA	492	Pauridiantha sylvicola	1E		
RUBIA	493	Pausinystalia lane-poolei	1D		
RUBIA	494	Pavetta corymbosa	(1F)	KRONKOO	
RUBIA	495	Pavetta lasicolada	(1F)		
RUBIA	496	Pavetta mannioides	(1F)		
RUBIA	497	Pavetta mollis	(1F)		
RUBIA	498	Pavetta mollissima	(1F)		
RUBIA	499	Pavetta owariensis	(1F)		
CAESA	500	Pellegriniodendron diphyllum	37B	FeLeFeLe	
MIMOS	501	Pentaclethra macrophylla	38B	ATAA	
GUTTI	502	Pentadesma butyracea	8B	ABOTOASEBIE	
PAPIL	11	Pericopsis elata	37G	KOKRODUA	(=Afrormosia elata)
LECYT	48	Petersianthus macrocarpus	25	ESIA	(=Combretodendron macrocarpum)
EUPHO	416	Phyllanthus discoideus	13B,15		SEE Margaritaria discoidea
EUPHO	503	Phyllanthus profusus	-	PEPEABERE	
IXONA	504	Phyllocosmus africanus	17C	AKoKORABEDITOA	(=Ochthocosmus africanus)
IXONA	505	Phyllocosmus sessiliflorus	17C		(=Ochthocosmus chippii)
APOCY	506	Picralima nitida	9C	KANWENE	
SIMAR	507	Pierreodendron kerstingii	32	FOTIE-AKOA	KWAENUAMANGO

FAMILY	CODE	LATIN NAME	KEY GROUP	STANDARD LOCAL NAME	OTHER NAMES
MIMOS	20	Piptadeniastrum africanum	38C	DAHOMA	
ANNON	508	Piptostigma fasciculatum	12B	DUASIKA-FUFUO	(=Brieya fasciculata)
ANNON	509	Piptostigma fugax	12B		
SAPIN	510	Placodiscus attenuatus	36D		
SAPIN	511	Placodiscus bancoensis	36D	KAFUOSONINI	
SAPIN	512	Placodiscus boya	36D	KAFUOSO	
SAPIN	513	Placodiscus bracteosus	36D		
SAPIN	514	Placodiscus oblongifolius	36D		
SAPIN	515	Placodiscus pseudostipularis	36D		
APOCY	516	Pleiocarpa mutica	9C	ONWENMA	
APOCY	517	Pleiocarpa pycnantha	9A		
ANNON	346	Polyalthia oliveri	12F		SEE Greenwayodendron oliveri
ANNON	518	Polyceratocarpus parviflorus	12B		
ARALI	519	Polyscias fulva	(28B)		
VERBE	520	Premna angolensis	5		
VERBE	521	Premna hispida	5		
EUPHO	522	Protomegabaria macrophylla	24	AGYAHERENUA	
EUPHO	523	Protomegabaria stapfiana	24	AGYAHERE	
EUPHO	524	Pseudagrostistachys africana	22B	SUKROMA	
ANACA	525	Pseudospondias microcarpa	35B	KATAWANI	
RUBIA	526	Psychotria gabonica	(1E)		
RUBIA	527	Psychotria guineensis	(1E)		
RUBIA	528	Psychotria succulenta	(1E)		
RUBIA	131	Psydrax arnoldiana	1A		(=Canthium arnoldianum)
RUBIA	134	Psydrax parviflora	1A		(=Canthium vulgare)
RUBIA	132	Psydrax subcordata	1A		(=Canthium subcordatum)
COMBR	529	Pteleopsis habeensis	25,5		
COMBR	530	Pteleopsis hylodendron	25,13D,5	KWAE-KANE	
PAPIL	531	Pterocarpus mildbraedii	37H		
PAPIL	532	Pterocarpus santalinoides	37H	HOTe	
STERC	533	Pterygota bequaertii	27C	KYEREYeBERE	
STERC	534	Pterygota macrocarpa	27C	KYEREYe	
MYRIS	49	Pycnanthus angolensis	13A	OTIE	
EUPHO	535	Pycnocoma macrophylla	(17C)	KAFIEKAFIE	
PALMA	536	Raphia hookeri	40	ADOBe	
PALMA	537	Raphia palma-pinus	40		
APOCY	538	Rauvolfia vomitoria	9A	KAKAPENPEN	
SCYTO	539	Rhaptopetalum beguei	13A		(=Rhaptopetalum coriaceum)
RHIZO	903	Rhizophora racemosa	3		Red mangrove
BOMBA	118	Rhodognaphalon brevicuspe	28C		(=Bombax brevicuspe)
ANACA	540	Rhus longipes	-		
EUPHO	541	Ricinodendron heudelotii	28B	WAMA	OWAMMA
VIOLA	542	Rinorea angustifolia	17B		(=R. ardisiiflora)
VIOLA	543	Rinorea brachypetala	(22B)		
VIOLA	544	Rinorea convallariiflora	17B		
VIOLA	545	Rinorea dentata	17B		
VIOLA	546	Rinorea kibbiensis	17B	DUOBOROBAE	
VIOLA		Rinorea ilicifolia	17B	APOSE-NINI	
VIOLA	547	Rinorea oblongifolia	22B	MPAWUNTUNTUM	
VIOLA	548	Rinorea prasina	17B		
VIOLA	549	Rinorea subintegrifolia	(22B)		
VIOLA	550	Rinorea welwitschii	(22B)		(=R. elliotii, R. longicuspis)
CAPPA	399	Ritchiea duchesnei	13B		SEE Maerua duchesnei
RUBIA	551	Robynsia glabrata	1G,F		
RUBIA	552	Rothmannia hispida	1F	TUKOBO	
RUBIA	553	Rothmannia longiflora	1F	SAMANKUBE	
RUBIA	554	Rothmannia megalostigma	1C		
RUBIA	555	Rothmannia urcelliformis	1F	ATIPA-TUKOBO	
RUBIA	556	Rothmannia whitfieldii	1F	SABOBE	
RUBIA		Rytigynia umbellulata	(1F)		
HUMIR	557	Sacoglottis gabonensis	17C	FAWERE	
MIMOS	558	Samanea dinklagei	38C	ABOBoNKAHYIRE	(=Albizia dinklagei)
EUPHO	559	Sapium aubrevillei	19A, 22	KETEBONTORE-NUA	
EUPHO	560	Sapium ellipticum	19A, 22	KETEBONTORE	
RUBIA	447	Sarcocephalus pobeguinii			(=Nauclea pobeguinii)
STERC	561	Scaphopetalum amoenum	21, 13C	NSOTO	
ARALI	562	Schefflera barteri	28B		
OLEAC	563	Schrebera arborea	5	KOKOFOBENE	
RUBIA	564	Schumanniophytum problematicum	1C		
FLACO	50	Scottellia klaineana	17D	TIABUTUO	(=S. chevalieri, S. coriacea) KOROKON

FAMILY	CODE	LATIN NAME	KEY GROUP	STANDARD LOCAL NAME	OTHER NAMES
SCYTO	565	Scytopetalum tieghemii	13B	OPRIM	
RUBIA	566	Sericanthe chevalieri	1F		(=Tricalysia chevalieri)
RUBIA	567	Sericanthe toupetou	1E		(=Tricalysia toupetou)
MORAC	568	Sloetiopsis usambarensis	(19A)		(=Neosloetiopsis usambarensis)
PASSI	569	Smeathmannia pubescens	17C	TURUNNUA	
SOLAN	803	Solanum erianthum	26	PePeDIAWUO	(=S. verbascifolium)
ANACA	570	Sorindeia zenkeri	35B		
MEDUS	571	Soyauxia grandifolia	13C	ABOTESIMA	
MEDUS	572	Soyauxia velutina	13D	ABOTESIMANUA	
MELAS	423	Spathandra barteri	6		(=Memecylon barteri)
MELAS	424	Spathandra blakeoides	6		(=Memecylon blakeoides, M. fleuryi)
BIGNO	573	Spathodea campanulata	30	AKUAKUO-NINSUO	
EUPHO	574	Spondianthus preussii	24	AWORATWEANKA, AWORATEe	
ANACA	575	Spondias mombin	35B	ATOA	
CAESA	576	Stemonocoleus micranthus	37D		
STERC	577	Sterculia oblonga	27A	oHAA	(=Eribroma oblonga)
STERC	51	Sterculia rhinopetala	27A	WAWABIMA	
STERC	578	Sterculia tragacantha	27C	SOFO	
BIGNO	579	Stereospermum acuminatissimum	30	ESONOTOKWAKOFUO	
COMBR	580	Strephonema pseudocola	21	AWURUKU	
OLACA	52	Strombosia glaucescens	13B	AFENA	(=S. pustulata)
LOGAN	581	Strychnos spinosa	4		
EUPHO	582	Suregada occidentalis	(17D)		
PAPI	583	Swartzia fistuloides	37G	ASOMANINI	
GUTTI	584	Symphonia globulifera	8B	EHUREKE	
SAPOT	585	Synsepalum aubrevillei	10C	ASAA-NINI	
SAPOT	586	Synsepalum dulcificum	(10)	ASAA	MIRACLE BERRY
MYRTA	587	Syzygium guineense	7		(=S. staudtii)
MYRTA	588	Syzygium rowlandii	7	ASIBENYANYA	
APOCY	589	Tabernaemontana africana	9C	OBONAWA	(=T. chippii)
APOCY	590	Tabernaemontana crassa	9C		
APOCY	591	Tabernaemontana pachysiphon	9C		
CAESA	592	Talbotiella gentii	37C	TAKOROWANUA	
DICHA	593	Tapura fischeri	13D		
DICHA	594	Tapura ivorensis	13D		
RUBIA	595	Tarenna gracilis	(1E)		
RUBIA	596	Tarenna laurentii	(1E)		
RUBIA	597	Tarenna pavettoides	(1E)		
RUBIA	598	Tarenna vignei	(1E)		
STERC	15	Tarrietia utilis	28B,21	NYANKOM	SEE Heritiera utilis
RUTAC	599	Teclea verdoorniana	31A		
RUTAC	643	Teclea sudanica			SEE Vepris eugenifolia
COMBR	13	Terminalia ivorensis	25	EMIRE	IDIGBO
COMBR	53	Terminalia superba	25	oFRAM	
MIMOS	600	Tetrapleura tetraptera	38B	PReKeSE	
EUPHO	601	Tetrorchidium didymostemon	13A, 17C	ANENEDUA	ABOAGYEDUA
EUPHO	602	Thecacoris stenosepala	-		
SAPOT	9	Tieghemella heckelii	10A,D	BAKU	MAKORE
MORAC	603	Treculia africana	19A	BReBReTIM, oTOTIM	
ULMAC	604	Trema orientalis	18B	SESEA	(=Trema guineensis)
RUBIA		Tricalysia			SEE AlSO Sericanthe spp.
RUBIA	605	Tricalysia biafrana	1E	DAKONUA	
RUBIA	606	Tricalysia bracteata	(1E)		
RUBIA	607	Tricalysia coriacea	(1E)		
RUBIA	608	Tricalysia discolor	1D	KWAEKoFI	
RUBIA	609	Tricalysia elliotii	1D		
RUBIA	610	Tricalysia macrophylla	1D		
RUBIA	611	Tricalysia oligoneura	(1E)		
RUBIA	612	Tricalysia pallens	1E	TUROMDUA	
RUBIA	613	Tricalysia reflexa	(1E)		
RUBIA	614	Tricalysia reticulata	(1E)		
RUBIA	615	Tricalysia vignei	-		
MELIA	616	Trichilia martineaui	34D	TANURO-NUA	
MELIA	617	Trichilia megalantha	34D	TANURO-KESE	
MELIA	618	Trichilia monadelpha	34D	TANURO	(=T. heudelotii)
MELIA	619	Trichilia ornithothera	34D	TANURO-BERE	
MELIA	620	Trichilia prieuriana	34C	KAKADIKURO	(=T. prieureana)
MELIA	621	Trichilia tessmannii	34D	TANURONINI	(=T. lanata)

FAMILY	CODE	LATIN NAME	KEY GROUP	STANDARD LOCAL NAME	OTHER NAMES
ANACA	622	Trichoscypha albiflora	35A		
ANACA	623	Trichoscypha arborea	35A	ANAKU	
ANACA	624	Trichoscypha atropurpurea	35A		
ANACA	625	Trichoscypha baldwinii	35A		
ANACA	626	Trichoscypha beguei	35A		
ANACA	627	Trichoscypha cavalliensis	35A		
ANACA	628	Trichoscypha chevalieri	35A		
ANACA	629	Trichoscypha oba	35A		
MORAC	630	Trilepisium madagascariense	19A	OKURE	(=Bosqueia angolensis)
STERC	14	Triplochiton scleroxylon	27D	WAWA	OBECHE
MELIA	631	Turraea ghanensis	(24)		
MELIA	27	Turraeanthus africanus	34C	APAPAYE	AVODIRE
EUPHO	633	Uapaca corbisieri	24	KONTAMMIRI	(=U. esculenta)
EUPHO	634	Uapaca guineensis	24	KONTAN	
EUPHO	632	Uapaca heudelotii	24	KONTANAKOA	
EUPHO	635	Uapaca paludosa	24		
ANNON	636	Uvariastrum insculptum	12D		
ANNON	637	Uvariastrum pierreanum	12D	oTWE-EHI	
ANNON	638	Uvariodendron angustifolium	12D	BoMMoFOKWADU	
ANNON	639	Uvariodendron calophyllum	12E	ESONOKWADU-KoKoo	
ANNON	640	Uvariodendron occidentale	12D	eSONOKWADU	(=U. mirabile)
ANNON	641	Uvariopsis globiflora	12E	ASUMPA-NUA	
RUBIA	642	Vangueriopsis spinosa	1G		
RUBIA		Vangueriopsis vanguerioides	1G		
RUTAC	643	Vepris heterophylla	31A		(=Teclea sudanica)
COMPO	644	Vernonia amygdalina	(26)	AWoNWENE	
COMPO	645	Vernonia colorata	26		
COMPO	646	Vernonia conferta	26	OWUDIFOKeTe	
COMPO	647	Vernonia richardiana	(26)		
COMPO	648	Vernonia titanophylla	(26)		
SAPOT	649	Vincentella passargei	10D		
SAPOT		Vincentella revoluta	10D		
GUTTI	650	Vismia guineensis	8A	KOSOWANINI	
VERBE	651	Vitex ferruginea	29	oTWENTOROWA	
VERBE	652	Vitex grandifolia	29	SUPOWA	DINSINKRO
VERBE	654	Vitex micrantha	29	OTWENTOROWANINI	
VERBE	653	Vitex rivularis	29	OTWENTOROWABERE	
APOCY	655	Voacanga africana	9C	OFURUMA	
APOCY	656	Voacanga thouarsii	9C	FOBA	
MELAS	425	Warneckea cinnamonoides	6		(=Memecylon cinnamonoides)
MELAS	427	Warneckea guineense	6		(=Memecylon guineense)
MELAS	715	Warneckea membranifolium	6		(=Memecylon membranifolium)
MELAS	429	Warneckea memecyloides	6		(=Memecylon memecyloides)
OLACA	657	Ximenia americana	15		
MIMOS	658	Xylia evansii	38A	SAMANTAWA	ABOBABEMA
ANNON	659	Xylopia aethiopica	12C	HWENETIA	
ANNON	660	Xylopia elliotii	12C		
ANNON	705	Xylopia parviflora	12C	oBAA-HWA	
ANNON	661	Xylopia quintasii	12C,D	oBAA	
ANNON	662	Xylopia rubescens	12C		
ANNON	663	Xylopia staudtii	12C	DUANAN	OBAAKoKoo
ANNON	664	Xylopia villosa	12C	oBAAFUFUO	
SAPIN	665	Zanha golungensis	36B		
RUTAC	666	Zanthoxylum chevalieri	31D	OYAABERE	(=Fagara pubescens)
RUTAC	667	Zanthoxylum gilletii	31D	OKUO	(=Fagara macrophylla)
RUTAC	668	Zanthoxylum lemairei	31D	OKUONINI	(=Fagara lemaireii) (OYAA)
RUTAC	669	Zanthoxylum leprieurii	31D	OYAA	(=Fagara leprieurii)
RUTAC	670	Zanthoxylum psammophilum	-		(=Fagara psammophila)
RUTAC	671	Zanthoxylum rubescens	31D		(=Fagara rubescens)
RUTAC	672	Zanthoxylum viride	(31D)	OYAANINI	(=Fagara viridis)
RUTAC	673	Zanthoxylum xanthoxyloides	31D	KANTO	(=Fagara zanthoxyloides)
RHAMN	674	Zizyphus espinosa	15		

INDEX TO LOCAL NAMES

Local names vary greatly – from place to place and from person to person – in their interpretation. The following index is mainly to the standardized names listed in the text, and so will often be very misleading if a reader, who is armed only with a local name, hopes to identify the scientific name of a plant. Often, but by no means always, the local name will point to the correct Group, even if to the wrong species. Some alternative names, and a handful of warnings about potential for confusion, are listed in the text of the appropriate Group. A few, common trade names have also been listed.

STANDARD LOCAL NAME	CODE	SUGGESTED INTERPRETATION	KEY GROUP	NOTES/OTHER NAMES
AFRICAN WALNUT				SEE Lovoa, Coula
AFRICAN MAHOGANY				SEE Group 34
AFRORMOSIA				SEE KOKRODUA
oBAA	661	Xylopia quintasii	12C,D	
oBAA-HWA	705	Xylopia parviflora/elliotii	12C	
oBAA-KoKoo		Xylopia staudtii		SEE DUANAN
oBAAFUFUO	664	Xylopia villosa	12C	
oBAAKoKo	663	Xylopia staudtii	12C	DUANAN
ABABIMA	147	Chidlowia sanguinea	37E	(But c.f. Xylia; Gp 39 notes)
ABABIMA-KoKoo	367	Hymenostegia gracilipes	37C	
BADIE	123	Bridelia micrantha	15	
BAKU	9	Tieghemella heckelii	10A,D	MAKORE
BAKUNINI	61	Afrosersalisia afzelii	10A,D	
BARK CLOTH TREE		Antiaris toxicaria		SEE KYEN-KYEN
OBECHE		Triplochiton scleroxylon		SEE WAWA
BEDIBeSA	264	Drypetes floribunda	17A	TETSO (Ga)
BEDIWONUA	35	Canarium schweinfurthii	33	CANARIUM
BENKYI		Erthroxylum mannii		SEE PEPEANINI
BEREKANKUM	409	Manilkara obovata	10A	(=Manilkara multinervis)
BESE	181	Cola nitida	27A	
ABESEBUO	368	Irvingia gabonensis	13C	
BESETORO	184	Cola verticillata	27A	
BLoHUNYI	318	Ficus polita	19	
ABOBABEMA	658	Xylia evansii	38A	
ABOBONKAYERe	142	Cathormion altissimum	38C	
ABOBoNKAHYIRE	558	Samanea dinklagei	38C	
BODWE	470	Ongokea gore	13A	
oBOGYANEBOoBERE	417	Markhamia lutea	30	
oBOGYANEBOoNINI	418	Markhamia tomentosa	30	
BOMPAGYA	46	Mammea africana	8B	
OBONAWA	589	Tabernaemontana africana	9C	(=T. chippii)
BONSAMDUA	42	Distemonanthus benthamianus	37G	AVAN
ABOTESIMA	571	Soyauxia grandifolia	13C	
ABOTESIMANUA	572	Soyauxia velutina	13D	
ABOTOASEBIE	502	Pentadesma butyracea	8B	
BRIMSTONE TREE		Morinda lucida		SEE KONKROMA
ABRUMA	170	Coelocaryon oxycarpum	13A	
BReBReTIM	603	Treculia africana	19A	oTOTOTIM
oBUA	446	Napoleonaea vogelii	17D,13B	(=Napoleona vogelii)
BUBINGA		Copaifera salikounda		SEE ENTEDUA
BUMBRA	314	Ficus mucuso	19B	
OBUOBI	60	Afraegle paniculata	31A,15	
ABURA		Hallea (=Mitragyna) spp.		SEE SUBAHA (-AKOA)
ABe	276	Elaeis guineensis	40	
BoDWUE	191	Coula edulis	27B	Gaboon nut AFR. WALNUT
BoMMoFOKWADU	638	Uvariodendron angustifolium	12D	
BoNToDEe	86	Anthocleista nobilis	4	
BoNToDEeBERE	83	Anthocleista djalonensis	4	
CANDOLLEI		Entandrophragma candollei		SEE CEDAR KOKOTE
CEDAR KoKoTE	16	Entandrophragma candollei	34B	PENKWA-AKOA
CEDRELA	801	Cedrela odorata	(34)	
CHELUM PUNGA	196	Crateva adansonii	31	(=Crataeva religiosa)
ADADABA	454	Newtonia duparquetiana	38A	
ADADABA-NUA	453	Newtonia aubrevillei	38B	
ADAFA	380	Lasiodiscus mannii	2	(=L. mildbraedii)
ADAFA-NINI	379	Lasiodiscus fasciculiflorus	2	
DAHOMA	20	Piptadeniastrum africanum	38C	
DAHOMANUA	99	Aubrevillea kerstingii	38C	
DAKONUA	605	Tricalysia biafrana	1E	
DANTA	26	Nesogordonia papaverifera	27A, 21	DANTA
ADASEMA	158	Chrysophyllum subnudum	10C	
DEMMERE		Calamus deeratus	40	
DENYAo	39	Cylicodiscus gabunensis	38A	OKAN
IDIGBO		Terminalia ivorensis		SEE EMIRE
ODII	463	Okoubaka aubrevillei	18A	

STANDARD LOCAL NAME	CODE	SUGGESTED INTERPRETATION	KEY GROUP	NOTES/OTHER NAMES
EDINAM	3	Entandrophragma angolense	34B	GEDUNOHOR
DINSINKRO	652	Vitex grandifolia	29	SUPOWA (DINSINKRO used often for Euadenia)
ADOBe	536	Raphia hookeri	40	
DODWATU	200	Croton zambesicus	21	
oDOM	44	Erythrophleum suaveolens	38A	(=E. guineense)
DUA-AHABANUM	70	Allophylus africanus	31	
DUA-AYAA	72	Cyathea manniana	39	(=Alsophila manniana)
DUABAHA	353	Hexalobus crispiflorus	12A,C,D	
DUABANKYE	227	Dialium aubrevillei	37H	
DUABIRI	346	Greenwayodendron oliveri	12F	(=Polyalthia oliveri)
DUADe	193	Craterispermum caudatum	1E	
DUAGYENNE	62	Afrostyrax lepidophyllus	21	
DUAMOKO	261	Drypetes aubrevillei	17A	
DUANAN	663	Xylopia staudtii	12C	OBAAKoKoo
DUAPOMPO	465	Omphalocarpum ahia	10D	
DUASIKA	279	Enantia polycarpa	12D	
DUASIKA-FUFUO	508	Piptostigma fasciculatum	12B	(=Brieya fasciculata)
DUATADWE	157	Chrysophyllum pruniforme	10A	
DUATADWE-KESE	98	Aubegrinia taiensis	10A,D	
DUATADWE-NINI	152	Chrysophyllum beguei	10C	(=Gambeya beguei)
DUAWISA	369	Isolona campanulata	12F	
DUBINFUFUO	12	Lovoa trichilioides	34A	DUBINIBIRI
DUBINI	8	Khaya ivorensis	• 34A	AFRICAN MAHOGANY
DUBINIBIRI	12	Lovoa trichilioides	34A	AFRICAN WALNUT
DUBINKoKoo	8	Khaya ivorensis	34A	DUBINI
DUBRAFO	415	Mareya micrantha	22B	
DUBRAFONINI	252	Discoclaoxylon hexandrum	22B	(=Claoxylon hexandrum)
ODUM	1	Milicia excelsa	19B	(=Chlorophora excelsa), IROKO,EP
oDUM-NUA	2	Milicia regia	19B	(=Chlorophora regia) IROKO,EP
DUOBOROBAE	546	Rinorea kibbiensis	17B	
ADWEA	212	Dacryodes klaineana	33	
DWEDWEEDWE	228	Dialium dinklagei	37I	
oDWEN	104	Baphia nitida	37A,13	
oDWENKOBIRI	105	Baphia pubescens	37A,13	(DWENDWERA, but see next sp.)
DWINDWERA	381	Lecaniodiscus cupanioides	36A	
oDWUMA	443	Musanga cecropioides	28A(19)	
FAFRAHA	436	Millettia zechiana	37I	
AFAM	488	Parinari exelsa	14B	KWA-EDUA
AFAMBERE	414	Maranthes robusta	14B	(=Parinari robusta)
AFAMNINI	412	Maranthes glabra	14A	(=Parinari glabra)
FAWERE	557	Sacoglottis gabonensis	17C	
oFEMA	431	Microdesmis puberula	17D	
AFENA	52	Strombosia glaucescens	13B	
AFENA-AKOA	384	Leptaulus daphnoides	13B	
AFINAFI	462	Octolobus spectabilis	27A	(=Octolobus angustatus)
FOBA	656	Voacanga thouarsii	9C	
FOTIE	349	Hannoa klaineana	32	(=Quassia undulata) HOTROHOTRO
FOTIE-AKOA	507	Pierreodendron kerstingii	32	KWAENUAMANGO
EFOoBRODEDWO	5	Entandrophragma utile	34B	UTILE
AFRA-NI-AFEI	194	Craterispermum cerinanthum	1E	
oFRAM	53	Terminalia superba	25	
AFRAMDASA	177	Cola flavo-velutina	27A	
AFRAMSUA	482	Pachystela brevipes	10C,D	
FRANGIPANI		Plumeria rubra	(9)	
FRENCH ATWEA-BERE	240	Diospyros cooperi	11	
FRENCH oDWUMA	803	Cecropia peltata	28A	
FRUNTUM	330	Funtumia elastica	9B	
OFURUMA	655	Voacanga africana	9C	
FeLeFeLe	500	Pellegriniodendron diphyllum	37B	
FeTeFRe	253	Discoglypremna caloneura	22A	
FoNTo	311	Ficus lutea	19	
FoTo	345	Glyphaea brevis	20	
FoTo-NINI	92	Antidesma laciniatum	13D	
FoToNUA	385	Leptonychia pubescens	21	
GBLITSO	237	Diospyros abyssinica	11	
GEDUNOHOR		Entandrophragma angolense		SEE EDINAM
AGYAHERE	523	Protomegabaria stapfiana	24	
AGYAHERENUA	522	Protomegabaria macrophylla	24	
GYAMA	67	Alchornea cordifolia	22A	
GYAMANINI	68	Alchornea floribunda	(17C)	
AGYAMERA	343	Gilbertiodendron splendidum	37F	
GYANEYA	289	Euclinia longiflora	1F	

STANDARD LOCAL NAME	CODE	SUGGESTED INTERPRETATION	KEY GROUP	NOTES/OTHER NAMES
oGYAPAM(-BERE)		Psydrax subcordata		SEE TETEADUPON
oGYAPAMNINI	134	Psydrax parviflora	1A	
oGYAPAMNWI	551	Robynsia glabrata	1G,F	
oGYATAFONKONWA	467	Omphalocarpum procerum	10D	
GYEDUA		Ficus spp.		SEE AMANGYEDUA, DOMINI
GYIGYAM	286	Erythrococca africana	-	
oHAA	577	Sterculia oblonga	27A	(=Eribroma oblonga)
AHENEBANSATEA	238	Diospyros barteri	11	
HOTROHOTRO		Hannoa klaineana		SEE FOTIE
HOTe	532	Pterocarpus santalinoides	37H	
HOTEKESE	531	Pterocarpus mildbraedii	37H	
AHOoHENEDUA	464	Olax subscorpioidea	13A,B	
EHUREKE	584	Symphonia globulifera	8B	
HWENETIA	659	Xylopia aethiopica	12C	
eHYeDUA	214	Daniellia ogea	37D	OFEA, SHEDUA GUM COPAL
AKABOFUNI	315	Ficus natalensis	19	
OKAE	329	Funtumia africana	9B	
KAFIEKAFIE	535	Pycnocoma macrophylla	(17C)	
KAFUOSO	512	Placodiscus boya	36D	
KAFUOSONINI	511	Placodiscus bancoensis	36D	
KAJABIRI	411	Maranthes chrysophylla	14B	(=Parinari chrysophylla)
KAKADIKURO	620	Trichilia prieuriana	34C	(=T. prieureana)
KAKADUA	114	Bertiera racemosa	1C	
KAKAPENPEN	538	Rauvolfia vomitoria	9A	
KAKU	19	Lophira alata	16	EKKI
oKAN		Cylicodiscus gabonensis		SEE DENYAo
AKANI	278	Elaeophorbia grandifolia	22,10	(=E. drupifera)
KANKABIM	120	Breviea leptosperma	10C	
KANTO	673	Zanthoxylum xanthoxyloides	31D	(=Fagara zanthoxyloides)
KANWENE	506	Picralima/Hunteria spp.	9C	
KANWENE-AKOA	361	Hunteria eburnea	9C	
KANe	81	Anogeissus leiocarpa	25,13D	
AKASA, AKASOA etc.		Chrysophyllum spp. (esp. C. giganteum)		see notes for Gp 10C
AKATA	119	Bombax buonopozense	28C	
KATAWANI	525	Pseudospondias microcarpa	35B	
KATRIKA(-NINI)	265	Drypetes gilgiana	17A	
KATRIKA-AKOA	263	Drypetes chevalieri	17A	
KATRIKABERE	268	Drypetes parvifolia	17A	
KEKE	246	Diospyros mespiliformis	(11)	
KESENE	258	Dracaena mannii	39A	(=D. perrotettii)
KETEBONTORE	560	Sapium ellipticum	19A, 22	
KETEBONTORE-NUA	559	Sapium aubrevillei	19A, 22	
EKKI		Lophira alata		SEE KAKU
KOKOFOBENE	563	Schrebera arborea	5	
AKOKORAGYEHINI	272	Duboscia viridiflora	20	
KOKROBOBA	486	Panda oleosa	17D	
KOKRODUA	11	Pericopsis elata	37G	(=Afrormosia elata) AFRORMOSIA
AKOKoSRADEe	374	Keayodendron bridelioides	13B	(=Casearia bridelioides)
KONINI	125	Buchholzia coriacea	27B	
KONINI-BERE	399	Maerua duchesnei	13B	(=Ritchiea duchesnei)
AKONKODIE	119	Bombax buonopozense	28C	AKATA
KONKROMA	442	Morinda lucida	1A	
KONTAMMIRI	633	Uapaca corbisieri	24	(=U. esculenta)
KONTAN	634	Uapaca guineensis	24	
KONTANAKOA	632	Uapaca heudelotii	24	
oKORA-AKOA	66	Albizia glaberrima	38B	
KORANTEMA	479	Oxyanthus speciosus	1F	
oKORO	32	Albizia zygia	38B	OKURO
oKORo-SANTE		Afzelia spp./ Albizia glaberrima		
KOROKON		Scottellia klaineana		SEE TIABUTUO
KOSOWA	350	Harungana madagascariensis	8A	
KOSOWANINI	650	Vismia guineensis	8A	
OKOSU	188	Cordia senegalensis	26	
KOTOPRePre	126	Bussea occidentalis	38A	(see notes for Gp)
KRAKA	291	Eugenia coronata	(6)	
KROBODUA	102	Balanites wilsoniana	31	
KROMA	376	Klainedoxa gabonensis	13C,15	
KRONKOO	494	Pavetta corymbosa	(1F)	
KRUBA	7	Khaya grandifoliola	34A	AFRIC. MAHOGANY
KRUMBEN	6	Khaya anthotheca	34A	ANTHOTHECA
AKUAKUO-NINSUO	573	Spathodea campanulata	30	
KUMANINI	378	Lannea welwitschii	35B	
KUMDWIE	481	Pachypodanthium staudtii	12A	(DUAWISA but see Isolona)
KUMFANA	154	Chrysophyllum giganteum	10C	

STANDARD LOCAL NAME	CODE	SUGGESTED INTERPRETATION	KEY GROUP	NOTES/OTHER NAMES
OKUO	667	Zanthoxylum gilletii	31D	(=Fagara macrophylla)
OKUONINI	668	Zanthoxylum lemairei	31D	(=Fagara lemairei)
OKURE	630	Trilepisium madagascariense	19A	(=Bosqueia angolensis)
AKUSAKUSA	231	Dichapetalum barteri	13D	
KUSIA	10	Nauclea diderrichii	1B	OPEPE
KUSIBIRI	242	Diospyros gabunensis	11	
KWA-EDUA		Parinari excelsa		SEE AFAM
KWAA-ANKAA	164	Citropsis articulata	31B	ge African cherry orange
KWAASIWA	460	Ochna staudtii	16	(=Ochna kibbiensis)
KWABOHORO(-BERE)	17	Guarea cedrata	34C	SCENTED GUAREA
KWABOHORO-NINI	18	Guarea thompsonii	34D	
KWADWUMA	18	Guarea thompsonii	34D	BLACK GUAREA
KWAE-AKENKA	59	Aeglopsis chevalieri	31A,15	
KWAE-KANE	530	Pteleopsis hylodendron	25,13D,5	
KWAEBRoFRE	206	Cussonia bancoensis	28B	
KWAEKoFI	608	Tricalysia discolor/spp.	1D	(See notes for Gp)
KWAETAWA	480	Oxyanthus unilocularis	1C	
KWAKUO-ASENABA	233	Dictyandra arborescens	1G	
KWAKUO-ASRA	207	Cuviera acutiflora	1G	
KWAKUOBESE	135	Carapa procera	34C	
AKWANA	471	Ophiobotrys zenkeri	18A,(17E)	AKWANDA
AKWANA-FUFO	137	Casearia calodendron	17D	(=C. inaequalis) ATWIAWA
KWATAFOMPABOA		Berlinia spp. (sometimes used for Gilbertiodendron, etc.)		
KWATAFOMPABOABERE	112	Berlinia tomentella	37F	
KWATAFOMPABOAKoKoo	111	Berlinia occidentalis	37F	
KWATAFOMPABOANINI	110	Berlinia confusa	37F	
KYAPOTORO	347	Grewia mollis	20	(inc. G. pubescens)
AKYE	115	Blighia sapida	36A	
AKYE-BUNO	383	Lepisanthes senegalensis	36A	
AKYE-SE	393	Lychnodiscus reticulatus	36B	
AKYEBERE	117	Blighia unijugata	36A	
AKYEKOBIRI	116	Blighia welwitschii	36A	
AKYEKoKoo	160	Chytranthus atroviolaceus	36C	
KYEN-KYEN	21	Antiaris toxicaria	19A	ANTIARIS
KYEREYe	534	Pterygota macrocarpa	27C	
KYEREYeBERE	533	Pterygota bequaertii	27C	
AKYEReKYEWeWA	354	Hildegardia barteri	27D	
OKYINI	274	Ehretia trachyphylla	13D,26	
KYIRIKUSA	91	Anthostema aubryanum	22,10	
AKoKORABEDITOA	504	Phyllocosmus africanus	17C	(=Ochthocosmus africanus)
KoKoTE	34	Anopyxis klaineana	3	KOKOTI
KoKoTE-AKOA	80	Anisophyllea meniaudii	19A	
KoKoTENUA	140 etc.	Cassipourea spp.	3	
KoTo-BOWERE	209	Cuviera nigrescens	1G	
MAHOGANY				SEE GROUP 34
MAKORE		Tieghemella heckelii		SEE BAKU
MANGO-AKASA	344	Gluema ivorensis	10D	
MANGROVE				SEE GROUP 3
AMANGYEDUA	299	Ficus spp.	19C	GYEDUA etc: many spp.
MEAWERE	97	Araliopsis soyauxii	28B	(=A. tabouensis)
oMENEWA	244	Diospyros kamerunensis	11	
oMENEWA-HOA	243	Diospyros heudelotii	11	(also oMENEWA-BERE)
oMENEWA-HOAKOA	706	Diospyros vignei	11	
oMENEWA-NINI	241	Diospyros ferrea	11	
MFo	254	Dombeya buettneri	(27)	
EMIRE	13	Terminalia ivorensis	25	IDIGBO
MMATA	218	Deinbollia grandifolia	36C	
MOTOKORODUA-AKOA	439	Monodora brevipes	(12F)	
MOTOKURADUA	441	Monodora tenuifolia	12F	
MPAMPRO	103	Bambusa vulgaris	39	
MPAWU	167	Cleidion gabonicum	17C	
MPAWUNTUNTUM	547	Rinorea oblongifolia	22B	
MPEDURO	348	Grossera vignei	22B	
AMUDURO	251	Diphasia angolensis	31A	(=D. klaineana)
ANAKU	623	Trichoscypha arborea	35A	
NAKWA	45	Holoptelea grandis	18A	
ANANSEAYA	174	Cola caricifolia	27D	SONKOROBIA
ANANSEDODOWA	180	Cola millenii	27D	DODOWA
ANANSEDUA	478	Ouratea reticulata	(16)	
ANANTA	40	Cynometra ananta	37B	
ANANTA-AKOA	211	Cynometra megalophylla	37C	
ANENEDUA	601	Tetrorchidium didymostemon	13A, 17C	ABOAGYEDUA
NGONENKYENE	169	Cleistopholis patens	12A	

STANDARD LOCAL NAME	CODE	SUGGESTED INTERPRETATION	KEY GROUP	NOTES/OTHER NAMES
ONIBONA	281	Eriocoelum pungens	36A	
ONIBONAKoKoo	282	Eriocoelum racemosum	36A,B	
ONIBONANUA	159	Chytranthus carneus	36C	
ANKYEWA	405	Majidea fosteri	36B,34C	
NNANFURO	109	Bequaertiodendron oblanceolatum	10A	
ANOKYE-HYEDUA	22	Guibourtia ehie	37B	SHEDUA
ANOMANI	296	Ficus spp.	19C	
NSOTO	561	Scaphopetalum amoenum	21, 13C	
NSoKo	332	Garcinia afzelii	8B	
NSoKoNUA	333	Garcinia epunctata	8B	
ENTEDUA	185	Copaifera salikounda	37D	BUBINGA
NTWESON	101	Aulacocalyx jasminiflora	1E	(NOT OTWENSONO)
NTON		Pandanus spp.	39	
NToNMe	257	Dracaena arborea	39A	
NUFUTEN	375	Kigelia africana	30	
NUMANUMAGYAMA	93	Antidesma membranaceum	13D	(Sometimes used for Discoglypremna)
NWADUA	300	Ficus sur	19B	(=F. capensis) DOMINI
oNWAMDUA	402	Maesopsis eminii	2	(But see Alstonia)
oNWENMA	516	Pleiocarpa mutica	9C	
oNYAME-DUA		Alstonia boonei		SEE SINURO
NYAMREM	198	Croton penduliflorus	22A	
NYANKOM	15	Heritiera utilis	28B,21	(=Tarrietia utilis) NIANGOM
NYANKUMA(-BERE)	444	Myrianthus arboreus	28A(19)	
NYANKUMANINI	445	Myrianthus libericus	28A(19)	
NYANKYEReNE	307	Ficus exasperata	28A	
ANYANYANFOROWA	407	Mallotus oppositifolius	22A	
ONYINA	143	Ceiba pentandra	28C	
oNYINAKOBEN	118	Rhodognaphalon brevicuspe	28C	(=Bombax brevicuspe)
OGEA		Daniellia ogea		SEE HYEDUA
oPAHA	270	Drypetes principum	17A	
oPAHA-AKOA	271	Drypetes singroboensis	17A	
oPAHA-BERE	266	Drypetes ivorensis	17A	
oPAHA-NUA	267	Drypetes leonensis	17A	
oPAHAFUFUO	262	Drypetes aylmeri	17A	
oPAHAKoKoo	269	Drypetes pellegrinii	17A	
oPAHANINI	260	Drypetes aubrevillei	17A	
oPAHA-TENE	259	Drypetes aframensis	17A	
oPAM	394	Macaranga barteri	23	
oPAMFUFUO	397	Macaranga hurifolia	23	
oPAMKOTOKRODU	121	Bridelia atroviridis	15	
oPAMKOTOKRODUKeSe	122	Bridelia grandis	15	
oPAMKoKoo	395	Macaranga heterophylla	23	
oPAMNUA	398	Macaranga spinosa	23	
PAMPENA	30	Albizia adianthifolia	38B	ALBIZIA
PAMPENAMA	190	Corynanthe pachyceras	1D	PAMPRAMA
PAMPENAMA-NUA	493	Pausinystalia lane-poolei	1D	
PAMPRAMA	190	Corynanthe pachyceras	1D	
OPANTO	327	Ficus vogeliana	19B	
PANUM	136	Casearia barteri	13	
PAPAO		Afzelia spp. (esp. A. bella)	37D	
PAPAO-BERE	28	Afzelia africana	37D	AFZELIA
PAPAO-NUA	29	Afzelia bella	37D	(AFZELIA)
APAPAYE	27	Turraeanthus africanus	34C	AVODIRE
PENKWA	4	Entandrophragma cylindricum	34B	SAPELE
PENKWA-AKOA	16	Entandrophragma candollei	34B	CANDOLLEI, OMU
OPEPE		Nauclea diderichii		SEE KUSIA
PEPEA	416	Margaritaria discoidea	13B,15	(=Phyllanthus discoideus)
PEPEABERE	503	Phyllanthus profusus	-	
PEPEANINI	287	Erythroxylum mannii	13A,C	BENKYI
POBE	420	Massularia acuminata	1F	
POBE-KESE	554	Rothmannia megalostigma	1C	
APOSE-NINI	(524B)	Rinorea ilicifolia (/spp.)		
APOTREWA	401	Maesobotrya barteri	22B	
PREMPRESA	145	Celtis wightii	18B	
OPRIM	565	Scytopetalum tieghemii	13B	
APROKUMA	94	Antrocaryon micraster	35B	
OPRONO	23	Mansonia altissima	27C	MANSONIA
PReKeSE	600	Tetrapleura tetraptera	38B	
OPUNINI	475	Ouratea calophylla	16	
PePeDIAWUO	803	Solanum erianthum	26	(=S. verbascifolium)
PoTRoDOM	43	Erythrophleum ivorense	38A	
IROKO		Milicia spp.		SEE oDUM
RUBBER TREE		Hevea brasiliensis	(22,31)	AMAN

STANDARD LOCAL NAME	CODE	SUGGESTED INTERPRETATION	KEY GROUP	NOTES/OTHER NAMES
ESA	37	Celtis mildbraedii	18B,A	
ASAA	586	Synsepalum dulcificum	(10)	MIRACLE BERRY
ASAA-NINI	585	Synsepalum aubrevillei	10C	
ASABA	483	Pachystela msolo	10C	
SABOBE	556	Rothmannia whitfieldii	1F	
ESAKOSUA	36	Celtis adolfi-friderici	18A	
ESAKoKo	38	Celtis zenkeri	18A,B	
SAMANKAA		Clausena anista		SEE SAMANOBI
SAMANKUBE	553	Rothmannia longiflora	1F	
SAMANOBI	166	Clausena anisata	31	SAMANKAA
SAMANTAANINI	203	Crudia gabonensis	37G	
SAMANTAWA	658	Xylia evansii	38A	ABOBABEMA (but see Chidlowia)
ASAMFENA		Aningeria spp. (esp. A. robusta)	10B	(see notes for Gp)
ASAMFENA-AKOA	406	Malacantha alnifolia	10B	
ASAMFENA-BERE	78	Aningeria altissima	10B	
SAMFENANINI	79	Aningeria robusta	10B	ASANFONA
ASANFRAN		Amphimas pterocarpoides		
SANTE	391	Lonchocarpus sericeus	37I	
SANZA-MULIKA	41	Diospyros sanza-minika	11	
ASENAA	229	Dialium guineense	37H	
eSERESOKRODUA	124	Brucea guineensis	(32)	
SESE	355	Holarrhena floribunda	9B	(=H. wulfsbergii)
SESEA	604	Trema orientalis	18B	(=Trema guineensis)
SESEDUA	148	Christiana africana	27C	SUPRONO
SESEMASA	452	Newbouldia laevis	30	
SHEDUA		Daniellia ogea		SEE HYEDUA
ESIA	48	Petersianthus macrocarpus	25	(=Combretodendron macrocarpum)
ASIBENYANYA	588	Syzygium rowlandii	7	
SIKAKYIA	352	Heisteria parvifolia	13B	
SINURO	73	Alstonia boonei	9A	
SOFO	578	Sterculia tragacantha	27C	
SOFOSE	387	Lindackeria dentata	22A	
ASOMA	489	Parkia bicolor	38C	
ASOMA-NUA	708	Parkia filicoidea	38C	
ASOMANINI	583	Swartzia fistuloides	37G	
SONKOROBIA		Cola caricifolia		SEE ANANSEAYA
SONKYI	33	Allanblackia floribunda	8B	(=A. parviflora)
ESONO-ANKAA	146	Chaetachme aristata	15	
eSONOBESE	125	Buchholzia coriacea	27B	KONINI
ESONODUA	113	Bersama abyssinica	31B	
ESONODoKONO	466	Omphalocarpum elatum	10D	
eSONOKWADU	640	Uvariodendron occidentale	12D	(=U. mirabile)
ESONOKWADU-KoKoo	639	Uvariodendron calophyllum	12E	
ESONONANKOROMA	358	Homalium letestui	17E	
ESONOTOKWAKOFOo	579	Stereospermum acuminatissimum	30	
eSONOWEDIE	232	Dichapetalum madagascariense	13D	(=D. guineense)
eSONOWISAMFIE	223	Desplatsia chrysochlamys	20	
eSONOWISAMFIEBERE	225	Desplatsia subericarpa	20	
eSONOWISAMFIENINI	224	Desplatsia dewevrei	20	
SOPI	215	Daniellia thurifera	37D	S. LEONE FRANKINCENSE, BUT see Gp notes
OSORE	285	Erythrina vogelii/spp.	31C	
oSOROWA	284	Erythrina mildbraedii/spp.	31C	
ASRATOADUANINI	216	Dasylepis brevipedicellata	17D	
ASRATOWADUA	469	Oncoba spinosa	15	
SUBAHA	24	Hallea ledermannii	1B	ABURA, (=Mitragyna ciliata)
SUBAHA-AKOA	25	Hallea (Mitragyna) stipulosa	1B	ABURA
SUINIA		Chytranthus macrobotrys		SEE TRUMWIE
SUKROMA	524	Pseudagrostistachys africana	22B	
SUKUSIA	447	Nauclea pobeguinii	1B	
SUKYE	392	Lychnodiscus dananensis	36B	
ASUMPA	82	Anonidium mannii	12E	
ASUMPA-NUA	641	Uvariopsis globiflora	12E	
SUPOWA	652	Vitex grandifolia	29	DINSINKRO
SUPRONO		Christiana africana		SEE SESEDUA
SUTWESE	418	Spathandra blakeoides	6	(=Memecylon blakeoides)
ATAA	501	Pentaclethra macrophylla	38B	
TAATSO	435	Millettia thonningii	37I	
ATABENE	156	Chrysophyllum perpulchrum	10C	
TAKOROWA	365	Hymenostegia afzelii	37C	
TAKOROWANUA	592	Talbotiella gentii	37C	
TAKYIKYIRIWA	226	Detarium senegalense	37D	TALLOW
TANA-NFRe	175	Cola chlamydantha	28B	(=Chlamydocola chlamydantha)
TANANFRoBERE	183	Cola umbratilis	28B	

STANDARD LOCAL NAME	CODE	SUGGESTED INTERPRETATION	KEY GROUP	NOTES/OTHER NAMES
TANUNINI	621	Trichilia tessmannii	34C	
TANURO	618	Trichilia monadelpha	34D	(=T. heudelotii)
TANURO-BERE	619	Trichilia ornithothera	34D	
TANURO-KESE	617	Trichilia megalantha	34D	
TANURO-NUA	616	Trichilia martineaui	34D	
TANURONINI	621	Trichilia tessmannii	34C	(=T. lanata)
TETEADUPON	132	Psydrax subcordata	1A	(=Canthium subcordatum)
TETEANKASE-KOFI	208	Cuviera macroura	1G	
TETEDUA	71	Allophylus spicatus	-	
TETEKON	341	Gilbertiodendron limba	37F	
TETEKON-GYAMERA	342	Gilbertiodendron preussii	37F	
TETEKON-NUA	340	Gilbertiodendron bilineatum	37F	
TETEOAKOA	433	Millettia griffoniana	37I	(=Lonchocarpus griffonianus)
TETETOA	434	Millettia rhodantha	37I	
TIABUTUO	50	Scottellia klaineana	17D	(=S. chevalieri, S. coriacea) KOROKON
OTIE	49	Pycnanthus angolensis	13A	
ATIPA-TUKOBO	555	Rothmannia urcelliformis	1F	
ATOA	575	Spondias mombin	35B	
OTOTIM	603	Treculia africana	19A	
TOTORO	89	Anthonotha macrophylla	37E	
TOTORONINI	88	Anthonotha fragrans	37E	
ATROTRe	130	Calpocalyx brevibracteatus	38A	
TRUMWIE	163	Chytranthus macrobotrys	36C	SUINIA, NTWESEMA
TUKOBO	552	Rothmannia hispida	1F	
TUROMDUA	612	Tricalysia pallens	1E	
TURUNNUA	569	Smeathmannia pubescens	17C	
TUTUABO	90	Isomacrolobium vignei	37E	(=Anthonotha vignei)
oTWE-ANI	422	Memecylon afzelii	6	
oTWE-EHI	637	Uvariastrum pierreanum	12D	
ATWEA	250	Diospyros viridicans	11	
OTWEABERE	239	Diospyros canaliculata	11	
ATWEAFUFU	245	Diospyros mannii	11	
TWEANKA	106	Beilschmeidia mannii	12	Spicy cedar
TWEAPIA	335	Garcinia kola	8C	
TWEAPIAKOA	334	Garcinia gnetoides	8C	
TWEAWODO	432	Mildbraediodendron excelsum	37I	
TWENEBOA		Cordia spp.	26	
TWENEBOA-AKOA	189	Cordia vignei	13D	
TWENEBOA-NINI	186	Cordia millenii	27C,26	
TWENEBOABERE	187	Cordia platythyrsa	26	
OTWENSONO	64	Aidia genipiflora	1E	(*NOT* NTWESON)
oTWENTOROWA	651	Vitex ferruginea(/spp.)	29	
OTWENTOROWABERE	653	Vitex rivularis	29	
OTWENTOROWANINI	654	Vitex micrantha	29	
TWEPIA-BERE	336	Garcinia smeathmannii	8C	(=G. polyantha)
ATWERE	56	Dactyladenia dinklagei	14A	(=Acioa dinklagei)
ATWERE-NANTIN	247	Diospyros monbuttensis	11,15	
oTWESE	428	Memecylon lateriflorum	6	
oTWESE-NINI	423	Warneckea spp.	6	
oTWETO	249	Diospyros soubreana	11	
oTWETO-KESE	248	Diospyros piscatoria	11	
AVAN		Distemonanthus benthamianus		SEE BONSAMDUA
AVODIRE		Turraeanthus africanus		SEE APAPAYE
WAMA	541	Ricinodendron heudelotii	28B	OWAMMA
WATAPUO	178	Cola gigantea	27D	
WATAPUOBERE	179	Cola lateritia	27D	
WAWA	14	Triplochiton scleroxylon	27D	OBECHE
WAWABIMA	51	Sterculia rhinopetala	27A	
oWEBIRIBI	360	Homalium stipulaceum/spp.	17E	(See Gp notes)
WEDEABA	440	Monodora myristica	12E	
WEDEABA-HOA	518	Polyceratocarpus parviflorus	12B	
AWIEMFOSAMINA	31	Albizia ferruginea	38B	ALBIZIA
AWIEMFOSAMINA-AKOA	65	Albizia coriaria	38B	
AWIEWU	128	Caloncoba gilgiana	23,17E	
AWIEWU-NUA	127	Caloncoba echinata	27B	
WISUBONI	461	Octoknema borealis	27B,(13)	
WOAGYE-AKOA	220	Deinbollia pinnata	36C	
AWORA-AFAM	404	Magnistipula zenkeri	14B	(=Hirtella fleuryana)
AWORA-oPAM	396	Macaranga heudelotii	23	
AWORABoNToDEe	87	Anthocleista vogelii	4	
AWORATEe	574	Spondianthus preussii	24	
AWORATWEANKA	574	Spondianthus preussii	24	
OWUDIFOKeTe	646	Vernonia conferta	26	

STANDARD LOCAL NAME	CODE	SUGGESTED INTERPRETATION	KEY GROUP	NOTES/OTHER NAMES
AWURUKU	580	Strephonema pseudocola	21	
WoNTon	47	Morus mesozygia	19A	
AWoNWENE	644	Vernonia amygdalina	(26)	
OYAA	669	Zanthoxylum leprieurii	31D	(=Fagara leprieurii)
OYAABERE	666	Zanthoxylum chevalieri	31D	(=Fagara pubescens)
OYAANINI	672	Zanthoxylum viride	(31D)	(=Fagara viridis)
YAYA	76	Amphimas pterocarpoides	37H,I	ASANFRAN
YAYA-AKOA	63	Aganope leucobotrya	(37H)	(=Ostryoderris leucobotrya)
AYEMTUDUA	96	Aptandra zenkeri	13A	
eYEe		Ancistrophyllum opacum	40	(=Ancistrophyllum opacum)
AYIKe-AKOA		Laccosperma secundiflora	40	(=Ancistrophyllum secundifl.)

(3249/89) Printed by Hobbs the Printers of Southampton

GROUPS 1–10

A catapult is the most convenient means of obtaining leaves from large trees. The tree in the centre is ***Okoubaka aubrevillei***, a parasite which stunts nearby trees and which is famous for its magic powers. (Gp 1)

Psydrax subcordata: (photo M. Swaine) tree with slender bole and characteristic crown (compare *Xylopia quintasii*). N.B. Ants nest (half way up) and scars (all over) on bole. (Gp 1)

Euclinia longiflora: clustered leaves and white *Rothmannia*-like flowers. (Gp 1)

Oxyanthus unilocularis: large leaves and clusters of slender flowers. (Gp 1)

Morinda lucida: slash yellow with a hot taste. (Gp 1)

Anthocleista djalonensis: silhouette of young, large-leaved tree, (centre foreground) and little-branched older trees. (Gp 4)

Anthocleista sp.: large-leaved young individual. (Gp 4)

Anthocleista djalonensis: granular, orange slash darkening slightly. (Gp 4)

Mammea africana: pink fleshy slash with spots of yellow latex. (Gp 8)

Alstonia boonei: deeply fluted bole and slash with copious white latex. (Gp 9)

Alstonia boonei: crown in layers; bole developing from sub-apical branches. (Gp 9)

Tieghemella heckelii: immense, cylindrical bole. (Gp 10)

Tieghemella heckelii: old tree with rough, vertically fissured bark. (Gp 10)

Tieghemella heckelii: red slash with white latex (young tree). (Gp 10)

Chrysophyllum subnudum: fluted, mature tree with white latex (+ silvery lvs from epicormic shoot). (Gp 10)

Afrosersalisia afzelli: rough bark and orange slash. Latex is sparse during the dry season (but see upper slash in shadow). (Gp 10)

GROUPS 10–14

Omphalocarpum ahia: orange-brown slash with white latex. (Gp 10)

Chrysophyllum giganteum: orange-brown slash with white latex. (Gp 10)

Omphalocarpum elatum: cauliflorous flowers (with stalks) on the cylindrical bole. (Gp 10)

Omphalocarpum elatum: cylindrical bole with red slash and white latex. (Gp 10)

Omphalocarpum ahia: large leaves clustered at tips of thick twigs, and large cauliflorous fruits. (The tree in a swampy site with ***Gilbertiodendron limba*** (Gp 37F) (foreground). (Gp 10)

Diospyros canaliculata: slash (unscented) with black outer layer and rapidly darkening yellow inner layers. (Gp 11)

Diospyros sanza-minika: outer bark unmistakable: black and hard, like stone or coal, and difficult to slash. (Gp 11)

Uvariopsis globiflora: (photo M. Swaine): cauliflorous flowers, unusual for Annonaceae because of 4-petals (most have 3 or 6). (Gp 12)

Hexalobus crispiflorus: deeply fluted, with scented slash and slashed bark peeling in long strips. (Gp 12)

Xylopia quintasii: slender bole surmounted by shallow crown of spirally arranged, regular, slender boughs. (Gp 12)

Xylopia parviflora: old tree with irregularly fluted bole and scented, orange slash. (Gp 12)

Pycnanthus angolensis: crown characteristically untidy, with whorled boughs and pendulous thin branches. (Gp 13)

Margaritaria discoidea: reddish slash and rather flaky bark. (Gp 13)

Irvingia gabonensis: orange slash with short greyish vertical streaks which darken rapidly. (Gp 13)

Parinari excelsa: lenticellate bark with reddish-brown slash often smelling of rotting sugar cane. (Gp 14)

Dactyladenia sp.: (Fure Headwaters – see notes Gp 14) bark vertically striate, slightly flaky, with red-brown slash. (Gp 14)

GROUPS 15–19

Chaetacme aristata: small tree with spines, especially on short shoots emerging from bole. (Gp 15)

Bridelia grandis: bole with adventitious roots at base and spines higher up. Slash red and highly scented. (Gp 15)

Lophira alata: the rough, flaky outer bark covers a vivid, sulphur-yellow layer visible in the slash. (Gp 16)

Drypetes floribunda: dense clusters of yellow cauliflorous flowers, in the driest forest types. (Gp 17)

Phyllocosmus africanus: bole normally with many adventitious shoots at base and red-brown slash. (Gp 17)

Scottellia klaineana: small, tidy crown and smooth bole. (Gp 17)

Scottellia klaineana: slightly leathery slash scented like ***Drypetes***, and sometimes slightly red. (Gp 17)

Okoubaka aubrevillei: bark slightly flaky with orange-borne slash – see also Plate 1, first photo. (Gp 18)

Holoptelea grandis: yellow-orange, iodine-scented slash with greenish outer layer and lenticellate surface. (Gp 18)

Celtis adolfi-friderici: young tree with rounded crown, a dark green cloud of foliage. (Gp 18)

Celtis adolfi-friderici: slash of young tree, with brown blotches rather than bands. (Gp 18)

Celtis zenkeri: bole with yellow-green tint; slash with yellow and brown 'contours', darkening rapidly. (Gp 18)

Celtis mildbraedii: bole usually smoother, and slash usually darkening more slowly than the v. similar ***C. zenkeri***. (Gp 18)

Ficus vogeliana: very smooth yellowish bark (produced also by other ***Ficus*** of Gp 19B). (Gp 19)

Ficus vogeliana: red slash with cream coffee-coloured latex. (Gp 19)

Strangling fig (***Ficus*** sp. of Gp 19C). Fallen leaves are essential for identification. (Gp 19)

Ficus umbellata: sometimes, free-standing figs appear to become strangled by their own aerial roots. (Gp 19)

Trilepisium madagascariense: fluted, smooth bole. (Gp 19)

Trilepisium madagascariense: leathery, darkening slash with slightly brown, creamy latex. Bole smooth with raised lenticels and lines. (Gp 19)

Milicia excelsa [ODUM]: bark very rough, with conspicuous orange-brown lenticels. Very gritty slash with white latex. (Gp 19)

Treculia africana: immense fruit. (Gp 19)

Duboscia viridiflora: deeply fluted bole with darkening slash smelling slightly of roast plantain. (Gp 20)

Leptonychia pubescens: typical Sterculiaceae pale yellow fibrous slash with vertical fibrous bands, but darkening through pinkish shades. (Gp 21)

Uapaca sp.: large stilt roots (photo by M. Swaine). (Gp 24)

Protomegabaria stapfiana: a tree that hisses when slashed, with crowd of adventitious shoots at base. (Gp 24)

Terminalia ivorensis (centre picture): outward sweeping crown of dark, tiered boughs and clustered leaves. ***Morus mesozygia***: dark crown on left (Gp 19) and ***Pycnanthus*** (Gp 13A, top left) (Gp 25)

Terminalia superba: young tree. Note clustered leaves and tiered boughs. (Gp 25)

Terminalia superba: older tree. Bark with silvery scales. Tiered boughs still visible. (Gp 25) ***Elaeis guineense*** palm trees below. (Gp 40)

Pteleopsis hylodendron: cylindrical bole with fibrous slash darkening with exudate which turns from greenish brown to (at length) black. (Gp 25)

Cordia platythyrsa: young tree (centre picture) with greenish, ***Ficus***-like bark and distinctive crown (in very disturbed forest). (Gp 26)

Sterculia tragacantha: crown (broken off above lowest tier) of lvs clustered at twig tips. (Gp 27)

Octoknema borealis: uneven bole with slightly scaly bark, and hard fibrous slash with fruity taste. (Gp 27)

Christiana africana: (***Mansonia***-like) yellowish slash, with vertical bands, darkening rapidly. (Gp 27)

Hildegardia barteri: greenish (or yellowish) bole with distinctive buttresses in dry, rocky forest. (Gp 27)

Hildegardia barteri: slash whitish yellow, darkening, soft, with vertical fibrous bands matching patterns in outer bark. (Gp 27)

Hildegardia barteri: crown of large leaves – careful allowance must be made for the height of the tree in estimating leaf size. (Gp 27)

Cola gigantea: uneven bole with greyish, slightly fissured (fibrous) outer bark. (Gp 27) Digitate lvs of ***Myrianthus arboreus*** (Gp 28) visible.

Cola gigantea (uneven bole) (Gp 27) and ***Terminalia superba*** (Gp 25, scaly, more even bole)

Cecropia peltata (centre): in roadside vegetation with ***Musanga*** in background. (Gp 28A)

Musanga cecropioides (photo M. Swaine): 1-2 year-old tree with stilt roots and leaf scars on lenticellate bole. (Gp 28A)

Musanga cecropioides: distinctive umbrella crown of digitate leaves from below. (Gp 28A)

Musanga cecropioides: young trees from side, with ***Anthocleista*** sp. (Gp 4) at centre. (Gp 28A)

Ceiba pentandra: parts of the crown often develop leaves and fruits out of phase with the rest of the crown. (Gp 28C)

Ceiba pentandra: large, grey buttresses. (Gp 28C)

Ceiba pentandra: slash yellow orange, with vertical bands (dilatation tissue). (Gp 28C)

Ceiba pentandra: prickles, especially on parts of buttress. (Gp 28C)

Bombax sp.: (on forest-savanna boundary) with tiered crown and red flowers. (Gp 28C)

Vitex sp.: slightly fluted tree with flaky bark; exudate darkening through greenish shades. (Gp 29)

GROUPS 30–36

Slash with greenish, darkening exudate (probably Gp 29-30) next to a spiny treelet (Gp 31 or 15).

Stereospermum acuminatissimum: slash very fibrous, but with gritty streaks. (Gp 30)

Zanthoxylum gilletii: (nr. centre) with '***Utile***'-like crown growing up in gap. (Gp 31)

Zanthoxylum gilletii: prickles (varied in precise shape) on young tree. Slash gritty, with strong taste. (Gp 31)

Zanthoxylum leprieurii: prickles on mature tree, typical of all mature ***Zanthoxylum*** spp. (Gp 31)

Zanthoxylum gilletii: sapling (centre, bottom) (Gp 31) with similar sapling of ***Entandrophragma angolense*** (nr. centre). (Gp 34B)

Canarium schweinfurthii: whorled boughs and '***Utile***'-like crown of clustered compound leaves. (Gp 33)

Canarium schweinfurthii: fissured bark with very scented, resinous slash. (Gp 33)

Khaya anthotheca: bark fairly smooth, with red, scented slash. (Gp 34)

Khaya grandifoliola: immense crown, and rough bark, in dry forest. (Gp 34)

Khaya grandifoliola: very rough bark with red, scented slash. (Gp 34)

Entandrophragma cylindricum: young tree (slightly burnt) with diamond-marked bark and v. scented, pink slash. (Gp 34)

Antrocaryon micraster: cylindrical bole with rough, dark bark and reddish slash. (Gp 35)

Blighia sapida: crown with fruits. (Gp 36)

Blighia sapida: slash yellow with orange gritty speckles. (Gp 36)

Chytranthus macrobotrys: small tree with cauliflorous fruits. (Gp 36)

Guibourtia ehie: rather rounded, small buttresses, with many parallel raised lines. (Gp 37)

Guibourtia ehie: orange slash with slow, gummy, slightly scented exudate. (Gp 37)

Talbotiella gentii: gregarious, evergreen tree of dry rocky hills. (Gp 37)

Talbotiella gentii: bark fibrous, red to yellow. (Gp 37)

Stemonocoleus micranthus: bole cylindrical, fibrous slash pale brown and yellowish with vertical lines. (Gp 37)

Daniellia (thurifera): (slash after 20 min) very cylindrical bole; slash darkening slowly, with gummy exudate. (Gp 37)

Chidlowia sanguinea: conspicuous inflorescences of bright red flowers. (Gp 37)

Berlinia occidentalis: heavy crown of dense foliage atop cylindrical bole. (Gp 37)

Berlinia occidentalis: very rough bark with thick, fibrous slash with strong vegetable smell. (Gp 37)

Berlinia tomentella: conspicuous golden-brown hairy pods in crown. (Gp 37)

Crudia gabonensis: slash (scented) hard, fibrous, contoured with scaly outer bark. (Gp 37)

Distemonanthus benthamianus: bark with papery flakes, and conspicuous red-brown patches. (Gp 37) (On left, foliage of ***Drypetes gilgiana*** (Gp 17A).)

Distemonanthus benthamianus: slash clearly contoured (with green outer layer); bark lenticellate away from flakes. (Gp 37)

Millettia thonningii: bole, slash and foliage. Only in the driest forests. (Gp 37)

Erythrophleum ivorense: very rough bark with liver red slash. (Gp 38)

Erythrophleum ivorense: slash complex, fibrous but crumbly, with red exudate nr. sapwood. (Gp 38)

Cylicodiscus gabunensis: massive, outward sweeping boughs support an immense crown of finely-divided foliage. (Gp 38)

Cylicodiscus gabunensis: bole with rough bark. Thick buttresses with curious 'knobs' (e.g. in patch of sunlight in this picture) (Gp 38)

Cylicodiscus gabunensis: slash hard, yellow, fibrous with foetid, old cabbage smell and yellowish exudate. (Gp 38)

Crown of fine feathery foliage characteristic of Gp 38B-38C (? = young ***Piptadeniastrum***).

Albizia adianthifolia: common, feathery-leaved tree. (Gp 38)

Aubrevillea kerstingii: rough-barked tree with feathery foliage. (Gp 38)

Aubrevillea platycarpa: foliage more 'globular' and not as feathery as other species of group. (Gp 38)

Aubrevillea platycarpa: orange-yellow, fibrous slash. (Gp 38)

Albizia glaberrima: yellow-grey, rough bark. (Gp 38)

Albizia glaberrima: very lenticellate bark with fibrous and gritty, scented slash. (Gp 38)

Piptadeniastrum africanum (photo M. Swaine): distinctive 'wandering' buttresses, and the bole almost glow with a golden orange colour. (Gp 38)

Cyathea manniana: tree fern in the Atewa range. Note the leaves unfurling at the top of the tree. (Gp 39)

Bambusa vulgaris: a clump of bamboo, like many giant feathers. (Gp 39)

Bambusa vulgaris: in forests bamboo is usually restricted to rivers. (Gp 39)

Dracaena perrottetii: white to greenish slash, and small stilt roots. (Gp 39)

Dracaena perrottetii: characteristic foliage, on a young tree. (Gp 39)